Azulene-Based Materials

Scrivener Publishing
100 Cummings Center, Suite 541J
Beverly, MA 01915-6106

Publishers at Scrivener
Martin Scrivener (martin@scrivenerpublishing.com)
Phillip Carmical (pcarmical@scrivenerpublishing.com)

Azulene-Based Materials

Design, Synthesis, and Applications

Edited by

Neha Rani Kumar

Department of Chemistry, Dhemaji College, Dhemaji, Assam, India

and

Abhijeet R. Agrawal

Institute of Chemistry, The Hebrew University of Jerusalem, Israel

Scrivener Publishing

WILEY

This edition first published 2026 by John Wiley & Sons, Inc., 111 River Street, Hoboken, NJ 07030, USA and Scrivener Publishing LLC, 100 Cummings Center, Suite 541J, Beverly, MA 01915, USA

For more information about Scrivener publications please visit www.scrivenerpublishing.com.

Wiley Global Headquarters
111 River Street, Hoboken, NJ 07030, USA

For details of our global editorial offices, customer services, and more information about Wiley products visit us at www.wiley.com.

The manufacturer's authorized representative according to the EU General Product Safety Regulation is Wiley-VCH GmbH, Boschstr. 12, 69469 Weinheim, Germany, e-mail: Product_Safety@wiley.com.

Library of Congress Cataloging-in-Publication Data

ISBN 978-1-394-15915-4

Front cover images courtesy of Wikimedia Commons
Cover design by Russell Richardson

Set in size of 11pt and Minion Pro by Manila Typesetting Company, Makati, Philippines

Printed in the USA

10 9 8 7 6 5 4 3 2 1

Contents

Preface **xiii**

Part 1: Fundamental Aspects of Azulene Chemistry 1

1 Discovery and Structural Insights into Azulene **3**
Sahid S.K., Nasseb Singh, Manda Sathish, Alamgir Ahmad Dar and Neha Rani Kumar
1.1 Introduction 4
1.2 Discovery and Synthesis of Azulene 5
1.3 Structural Insights 8
1.3.1 Photophysics and Symmetry of the Azulene 8
1.3.2 Glidewell–Lloyd Rule 10
1.3.3 Azulene Moiety as Electron Reservoir 12
1.3.4 Photodissociation of Azulene Cation 13
1.4 Applications of Azulene and Its Derivatives 13
1.5 Conclusion 15
References 15

2 Azulene-Containing Natural Products **19**
Abhilash Sharma, Hemanta Hazarika and Dhrubajit Borah
2.1 Introduction 19
2.2 Occurrence and Isolation of Guaiazulene and Chamazulene 21
2.3 Synthesis of Guaiazulene 28
2.4 Synthesis of Chamazulene 38
2.5 Applications 44
2.5.1 Guaiazulene 44
2.5.2 Chamazulene 44
2.6 Conclusion 45
References 45

3 Synthesis of Azulene and Its Functionalized Derivatives **53**
G.B. Dharma Rao, Aditi Tiwari, A. Udayasri and Anjaneyulu Bendi
3.1 Introduction 54
3.2 Traditional Process for the Synthesis of Azulene 55
3.3 Synthesis of Azulene Derivatives 57
3.4 Conclusion 76
Acknowledgment 76
References 77

4 Azaazulenes: Synthesis, Reactivity, and Applications **81**
R. Balavardhana Rao Adapa
4.1 Introduction 82
4.2 Types of Azaazulene Isomers 82
4.2.1 Monoazaazulenes 82
4.2.2 Diazaazulenes 83
4.2.3 Triazaazulenes 83
4.3 Synthesis of Azaazulenes 85
4.3.1 Syntheses from Reactive Troponoids 85
4.3.1.1 Synthesis from 2-Aminotropones 86
4.3.1.2 Synthesis from 2-Aminotropones and Ketene Dimer 86
4.3.1.3 Synthesis from 2-Aminotropone and Heterocumulenes 87
4.3.1.4 Synthesis from 2-Troponylhydrazides 88
4.3.1.5 Synthesis from 2-Methoxytropones 88
4.3.1.6 Regiospecific Synthesis of 1,3-Diazaazulene 89
4.3.1.7 Synthesis from 2-Chlorotropones and Acyl Hydrazides 90
4.3.1.8 Synthesis from 2-Substituted Tropones and N-Silylenamine or the Enamine 90
4.3.1.9 Synthesis from Tropone and Pyridinium Ylide 91
4.3.2 Synthesis from the Oximes by Amino-Heck Reaction 92
4.3.3 Synthesis from Azaheptafulvenes 93
4.3.4 Synthesis from 5-Nitrosotropolone and Pyridine 94
4.3.5 Synthesis by Buchner Ring Expansion Reaction 96
4.3.6 Synthesis by Abnormal Bischler–Napieralski Reaction 96
4.3.7 Synthesis by Intramolecular Cyclization 96
4.3.7.1 Synthesis by Flash Vacuum Pyrolysis 96
4.3.7.2 Synthesis by Platinum-Catalyzed Intramolecular Cyclization of the Alkynes 97

4.3.8 Synthesis from Intramolecular Nucleophilic Attack by Organolithium Reagents 97
4.3.9 Synthesis from Acid-Catalyzed Condensation of Dehydrotryptophan Ester with Aldehydes 98
4.3.10 Synthesis by Bischler–Napieralski Reaction of the Carboxamides 99
4.3.11 Syntheses Utilizing Intermolecular Condensations 100
4.3.12 Synthesis from Dialdehyde and Ketone 100
4.3.13 Synthesis from 1,2-Diacid Chloride and n,n'-Diphenylacetamidine 100
4.3.14 Synthesis from 1,2-Diamine and Ethyl Acetoacetate 101
4.4 Reactions of Azaazulene 102
4.4.1 Coupling Reactions 102
4.4.2 Suzuki Coupling Reactions 102
4.4.3 Negishi Coupling 104
4.4.4 Sonogashira–Hagihara Reaction 105
4.4.5 Hartwig–Buchwald Reaction 106
4.4.6 Cycloaddition and Annulation 107
4.4.7 Aza–Wittig Reaction 108
4.5 Applications 109
4.5.1 Phosphodiesterase 2 Inhibitors 109
4.5.2 Antiviral Activity 110
4.5.3 Cholecystokinin-1 Receptor Antagonists 110
4.5.4 Epigenetic Inhibitors 111
References 111

Part 2: Azulene in Functional and Bioactive Materials 117

5 Azulene-Based System in Optoelectronics 119
Parth, Shivani and Amrit Kaur
5.1 Introduction 120
5.2 Applications in Nonlinear Optics 121
5.3 Applications in Molecular Devices 125
5.4 Applications in Organic Field Effect Transistors 128
5.4.1 Azulene-Based Polymers in Organic Field Effect Transistors 129
5.4.2 Azulene-Based Small Molecules in Organic Field Effect Transistors 132
5.5 Application in Organic Photovoltaics 135

5.5.1 Azulene-Based Small Molecules in Organic Photovoltaics 136
5.5.2 Azulene-Based Small Polymers in Organic Photovoltaics 137
5.6 Miscellaneous Studies 138
5.7 Conclusions 141
References 142

6 Azulene-Based Systems as Molecular Switches 151
Sariful Molla and Sheelbhadra Chatterjee
6.1 Introduction 151
6.2 Azulene-Fused Molecular Switches 153
6.2.1 Azulene-Functionalized Stilbene Systems 154
6.2.2 Azulene-Functionalized Diarylethene Systems 161
6.3 Azulene-Functionalized Azobenzene Switches 166
6.4 Conclusion and Future Prospects 173
References 173

7 Azulene-Based π-Functional Materials and Their Application as Sensors 179
Bagmita Bhattacharyya and Kangkana Deka
7.1 Introduction 179
7.2 Application of Azulene and Its Derivatives for Sensing of Heavy Metals 181
7.3 Azulene-Based Sensors for Detection of Anions 186
7.3.1 Sensors for Phosphate 186
7.3.2 Sensors for Fluoride 187
7.3.3 Sensors for Nitrite 191
7.4 Bioimaging 192
7.4.1 Use of Azulene in Bioimaging 193
7.5 Conclusion 195
References 195

8 Exploring the Synthesis and Characteristics of Nanographenes and PAHs with Azulene Embedment 199
Dileep Kumar Singh and Bijoy P. Mathew
8.1 Introduction 200
8.2 Azulene-Based Polycyclic Aromatic Hydrocarbon 202
8.3 Azulene-Embedded Nanographenes 226
8.4 Conclusion 237
Acknowledgments 238
References 238

9 Azulene-Based Systems of Biological Significance **245**
Sunita Singh, Pragati Kushwaha, Pankaj Kumar Chaurasia, Shashi Lata Bharati, Ayman Younes Fathy Allam, Dhananjay Kumar Singh and Naveen Kumar
9.1 Azulenes, Their Sources, and Therapeutic Innovations 246
9.2 Azulene and Its Derivatives with Biological Significance 248
9.2.1 Antidiabetic Action of Azulenes 248
9.2.2 Antimicrobial Action of Azulene and Related Compounds 250
9.2.3 Anticancer Effects of Azulene and Its Derivatives 252
9.2.4 Anti-Inflammatory Action of Azulene Derivatives 256
9.2.5 Anti-Ulcer Properties of Azulene Derivatives 259
9.2.6 Antioxidant Effects of Azulene and Its Derivatives 261
9.2.7 Dermatological Effects of Azulene and Its Derivatives 262
9.2.8 Cardioprotective Effects of Azulene and Its Derivatives 262
9.2.9 Hepatoprotective Effects of Azulene and Its Derivatives 263
9.2.10 Neuroprotective Effects of Azulene and Its Derivatives 263
9.3 Challenges and Future Prospects 265
References 266

Part 3: Azulene in Porous and Coordination Chemistry **275**

10 Azulene-Based Porous Materials: Synthesis and Applications **277**
Debajit Bora, Bikash Chandra Mushahary, Chayanika Goswami and Sanjeev Pran Mahanta
10.1 Porous Materials 278
10.2 Azulene and Its Electronic Properties 279
10.3 Functionalization of the Azulene Derivatives 281
10.4 Synthetic Methods to Prepare Porous Materials 282
10.4.1 Solvothermal Method 283
10.4.2 Microwave-Assisted Method 283
10.4.3 Mechanochemical Method 284
10.4.4 Liquid–Liquid Interface Polymerization Method 284
10.5 Different Types of Azulene-Based Porous Materials and Their Applications 284

10.5.1 Metal-Organic Frameworks 284
10.5.2 Covalent Organic Frameworks 287
10.5.3 Conjugated Microporous Polymer 294
10.5.4 Hydrogen-Bonded Organic Frameworks 297
10.5.5 Covalent Triazine Framework 298
10.6 Conclusion 302
References 303

11 Calixazulenes and Azuliporphyrins 311
Nitika Grover
11.1 Introduction 312
11.2 Azuliporphyrin 313
11.2.1 Synthetic Developments of Azuliporphyrins 314
11.2.2 Reactivity of Azuliporphyrins 318
11.2.3 Expanded Azuliporphyrins 321
11.3 Calixazulenes 322
11.3.1 Synthesis and Reactivity of Calixazulenes 323
11.3.2 Supramolecular Host-Guest Interaction 326
11.4 Conclusion and Outlook 329
References 330

12 Functionalization and Utilization of Azulene in Organometallic and Coordination Chemistry 333
Sulendar K. Mahato and Rakesh Kumar Rai
12.1 Introduction 333
12.2 Metallaazulenes 336
12.2.1 Complexes of Azulene with Iron 336
12.2.2 Complexes of Azulene with Rhodium and Iridium 336
12.2.3 Synthesis of Metalla-Dual-Azulenes 337
12.2.4 Organometallic Complexes of Azulene 338
12.2.5 Azuliporphyrins 338
12.2.6 Synthesis of Selenium and Tellurium Core-Modified Azuliporphyrinogens 341
12.3 Metal Carbonyl Complexes of Azulene 342
12.4 Coordination Complexes of Azulene 343
12.5 Conclusions 348
References 348

Part 4: Theoretical and Future Perspectives **351**

13 Theoretical Insights into Azulene Chemistry **353**
Manash Protim Borpuzari and Abhijit Boruah
13.1 Introduction 353
13.2 Computational Methods to Study Azulene Chemistry 356
13.2.1 Hartree–Fock Theory (HF) 356
13.2.2 Second-Order Møller–Plesset Perturbation Theory 356
13.2.3 Complete Active Space-Based Method 358
13.2.4 Density Functional Theory 358
13.2.5 Time-Dependent Density Functional Theory (TDDFT) 359
13.2.6 Equation of Motion 359
13.3 Recent Advances 360
13.3.1 Structural Properties 360
13.3.2 Noncovalent Interaction Between Azulene Dimer 361
13.3.3 Excited State Properties 363
13.3.4 Rydberg States 365
13.3.5 Electron Transport 366
13.3.6 Reaction Mechanism 369
13.3.7 Hyperpolarizability 371
13.4 Conclusion and Outlook 372
References 372

14 Future Perspectives of Azulene Chemistry **377**
Kalyan Jyoti Kalita and Neha Rani Kumar
14.1 Introduction 377
14.2 Advances in Synthesis 379
14.2.1 Green Synthetic Routes for Azulene and Its Derivatives 379
14.2.2 Challenges in Regioselective Synthesis and Recent Solutions 380
14.2.3 Functionalization Techniques 382
14.3 Applications and Future Directions 382
14.3.1 Energy Storage and Conversion 382
14.3.2 Organic Optoelectronics 383
14.3.3 Integration into Wearable and Flexible Devices 384
14.3.4 Enhancing Efficiency in Energy Harvesting Devices 385
14.3.5 Intelligent Optoelectronic Systems and Sensors 385
14.3.6 Biomedical Applications 387

14.4 Role of Computational Techniques, Including Artificial Intelligence, in Azulene Chemistry 389
14.5 Conclusion 389
References 390

Index **395**

Preface

Organic electronics is a field of material chemistry or broadly saying a field of organic chemistry dealing with the design, synthesis and characterization of novel conjugated systems which find application in organic devices. In the era of digitalization Organic electronics is believed to shape our future. Design of new conjugated systems is at the heart of a synthetic organic chemist and understanding the chemistry of existing systems is necessary to design novel pathways. Azulene which is an isomer of naphthalene is an integral part of organic electronics field. This is evident from a simple scifinder search with the word Azulene which shows more than 9000 articles published so far. Besides this there are several reports involving azulene chemistry indirectly. However, if one takes a look at the literature, there is no single source where one can find the major details of azulene and its derivatives including not only their synthesis but also their application. Several review articles dealing with individual topics have been published but finding all the important information at a single hand source will be of enormous use to researchers from all background working in Azulene chemistry. "Azulene-Based Materials: Design, Synthesis and Applications" is a single hand guide to azulene chemistry.

Beginning from the history of identification of azulene and its isolation from natural sources the book at first provides an insight into the important synthesis methods of azulene. Next a dive into the synthesis of functionalized azulene derivatives has been presented. This is followed by detailed chapters on applications of azulene and its derivatives in novel advanced materials such as molecular switches, anion receptors/sensors, electrochromic material, organic/polymeric conductors, azulene embedded nanographene's and fused systems, biological applications, azulene containing porous systems etc. This book is believed to address the challenges of organic and material chemists working in azulene chemistry.

It is divided into four parts

Part 1: Fundamental Aspects of Azulene Chemistry
This section encompasses **Chapter 1, Chapter 2, Chapter 3 and Chapter 4** and will delve into the origins, structural peculiarities, and foundational aspects of azulene chemistry. It begins with an exploration of its discovery and unique nonbenzenoid aromaticity, followed by a discussion of naturally occurring azulene derivatives. The focus then shifts to the synthesis of azulene and its functionalized derivatives, including aza-azulenes, highlighting their reactivity and applications.

Part 2: Azulene in Functional and Bioactive Materials
Azulene's integration into advanced functional materials is the focus of this section. The **Chapters 5, 6, 7, 8, and 9** explore its role in molecular switches, π-functional materials for sensor applications, and its incorporation into nanographenes and polycyclic aromatic hydrocarbons. The biological significance of azulene-based systems is also discussed, emphasizing its potential in medicinal and biochemical applications.

Part 3: Azulene in Porous and Coordination Chemistry
This section which includes **Chapter 10, 11, and 12** highlights the expanding role of azulene in materials science. It covers the synthesis and applications of azulene-based porous materials, the chemistry of calixazulenes and azuliporphyrins, and its utilization in organometallic and coordination chemistry, showcasing its adaptability in various functional frameworks.

Part 4: Theoretical and Future Perspectives
The final section i.e. **Chapter 13 and Chapter 14** offers theoretical insights into azulene chemistry, unraveling its electronic properties and reaction mechanisms through computational studies. The concluding chapter provides an outlook on future advancements, exploring the potential directions and emerging applications of azulene chemistry in various scientific domains.

This book will act as a single hand guide to Researchers in academia, Material Chemists and physicists, Industry and government laboratories as well as advanced undergraduate and graduate students working in organic synthesis, organic electronics, organic semiconducting systems. It will not only provide detailed insight into the various synthesis methods of azulene systems, but will also highlight its derivatives and all their major applications in details as individual chapters.

Part 1

FUNDAMENTAL ASPECTS OF AZULENE CHEMISTRY

1

Discovery and Structural Insights into Azulene

Sahid S.K.[1], Nasseb Singh[1], Manda Sathish[2], Alamgir Ahmad Dar[3] and Neha Rani Kumar[4]*

[1]*Department of Medicinal Chemistry, National Institute of Pharmaceutical Education and Research, Kolkata, West Bengal, India*
[2]*Centro de Investigación de Estudios Avanzados del Maule (CIEAM), Vicerrectoría de Investigación y Postgrado, Universidad Católica del Maule, Talca, Chile*
[3]*Research Centre for Residue and Quality Analysis (RCRQA), Sher-e-Kashmir University of Agricultural Sciences and Technology of Kashmir, Shalimar, Campus, Srinagar, India*
[4]*Department of Chemistry, Dhemaji College, Dhemaji, Assam, India*

Abstract
The history of azulene dates back to the 15th century, when it was first identified as the azure-blue intriguing chromophore produced by steam-distilling German chamomile. It was named by Septimus Piesse in 1864. The structure of azulene was first reported by Ruzicka, and its synthesis was reported by Pfau and Plattner in 1939, utilizing indane and ethyl diazoacetate. Azulene is a non-benzenoid aromatic hydrocarbon and is a structural isomer of naphthalene with 10π electrons. It has a planar ring system with a fused pentagon and heptagon, and its ground-state electronic structure is dominated by two Kekulé structures, where as its zwitterionic form has minor contribution to the ground state. Some experimental data indicates that azulene is a non-rigid structure and possesses antisymmetric Kekulé type coordinates, resulting in localized planar Kekulé formations. Due to its unique structural features and properties, azulene is often referred to as aromatic chameleon by organic chemists. Besides that, it also has been a topic of wide interest not only to organic chemists but material scientists and biologists, as well. This chapter highlights the discovery of azulene followed by insights into its structure and its varied applications.

Keywords: Azulene, Kasha's rule, guaiazulene, vetivazulene, organic electronics

**Corresponding author*: nehakumar0926@gmail.com

Neha Rani Kumar and Abhijeet R. Agrawal (eds.) Azulene-Based Materials: Design, Synthesis, and Applications, (3–18) © 2026 Scrivener Publishing LLC

1.1 Introduction

Azulene is a naturally occurring, blue-colored, non-benzenoid organic moiety found in various plants and mushrooms. Examples of some of these species that contain azulene include *Matricaria chamomilla*, *Artemisia absinthium*, *Achillea millefolium*, and *Lactarius indigo*. Azulene and its derivatives have been widely studied because of their potential applications in diverse fields such as medicine, cosmetics, optoelectronic devices, and essential oils. The name azulene found its origin from the Spanish word "azul," which means blue. In the 15th century, azulene was identified as an azure-blue chromophore extracted from German chamomile using steam distillation. In 1864, Septimus Piesse named azulene after discovering it in yarrow and wormwood. The research conducted by Ruzicka laid the foundation for determining azulene structures [1]. In 1937, azulene was first synthesized by Pfau and Plattner [2]. Around 30 years later, Nozoe developed a new method for synthesizing multifunctional azulene derivatives and then azulene gained momentum [3].

Azulene is now a widely studied system in organic, inorganic, pharmaceutical, and theoretical chemistry domains, owing to its unique structure and peculiar properties [4]. Due to its numerous applications, the azulene industry was estimated to be worth USD 360 million in 2023, and the global market for azulene-based industrial products is projected to reach USD 1 billion by 2036. Azulene is an isomer of naphthalene with a similar odor, but it is a blue-colored non-benzenoid aromatic molecule whereas naphthalene is a colorless benzenoid aromatic compound. Vetivazulene (IUPAC name: 4,8-dimethyl-2-isopropylazulene) and guaiazulene (IUPAC name: 1,4-dimethyl-7-isopropylazulene) are classic examples of two terpenoids containing an azulene core structure in their molecules (Figure 1.1) [5]. These compounds are perhaps the unique constituents of pigments in many marine invertebrates, wood oils, and mushrooms. Unlike naphthalene, which is a neutral molecule (μ = 0D), azulene exhibits a permanent dipole (μ = 1.08D) because azulene exists as a dipolar organic molecule, with an electron-deficient seven-membered ring carrying a unit positive charge, and an electron-rich five-membered ring carrying a unit negative charge. The dipolar character of azulene can be explained by the stability of the cyclopentadienyl anion and tropylium cation, which acquire 6π electrons and obey Hückel's $(4n+2)\ \pi$ rule of aromaticity. This is likely the primary driving factor for the loss of one electron from the seven-membered ring, while the five-membered ring acquires one electron, resulting in increased stability of the fused hepta-pentacyclic non-benzoid aromatic

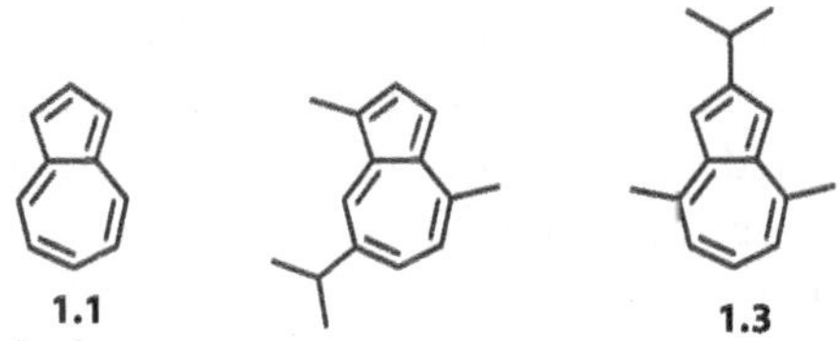

1.1 Azulene **1.2** Guaiazulene **1.3** Vetivazulene

Figure 1.1 Structure of azulene, guaiazulene and vetivazulene.

ring system, known as azulene. Reactivity experiments on azulene and its derivatives also reveal that the seven-membered ring due to its electron-deficient nature is electrophilic, while the five-membered ring due to its electron-rich nature is nucleophilic.

Azulene shares similar aromatic features, peripheral bond lengths, and Friedel-Crafts substitution processes as naphthalene. It is noteworthy that the stability benefit derived from aromaticity in azulene is around half that of naphthalene. The ground state is dipolar, resulting in a rich color that is uncommon for small unsaturated aromatic systems [6]. This behavior of the azulene ring system has driven synthetic chemists to explore various substituted azulenes with diverse applications in synthetic organic chemistry, medicinal chemistry, and materials science [7, 8].

1.2 Discovery and Synthesis of Azulene

The discovery and structural insights into azulenes have evolved over the past 500 years. Although there has been even earlier documentation of its presence in essential oils. Azulenes are characterized by their intense blue color, which is retained even in high dilution. The blue color of azulenes was initially thought to arise from the copper contamination coming from the distillation step but this was later proven to be intrinsic to the azulene molecules. The azulenes have a boiling temperature significantly higher than sesquiterpenes. However, they co-distill with them, resulting in a deep blue color of the fraction. In Semmler's experiments, around 20 oils, including yarrow, camomile, and cubeb, had blue coloration upon distillation. By 1936, it was estimated that 20% of the 260 known essential oils contained azulenes or their precursors. Finally, in 1864, the term "azulene" was first introduced by Piesse designating the blue oils, which eventually became the parent compound, azulene with the molecular formula $C_{10}H_8$ [9].

Azulenes were initially named by prefixing the name of the source oil to which they were derived. For example, the azulene derivative obtained from guaiol was named guaiazulene, and the azulene obtained from vetiver oil was named vetivazulene. These names were later consolidated into single forms when identical azulenes were found in different sources. Azulenes have been extracted from a variety of plants, including chamomile oil (*Matricaria chamomilla L.*), *Alpinia japonica*, vetiver oil, and cajeput oil (*Melaleuca leucadendron L.*). Besides that, azulenes have also been commonly found in by-products like lignite oil and acetylene pyrolysis. Azulene itself, $C_{10}H_8$, was first obtained through the dry distillation of calcium adipate by Pfau and Plattner. However, its natural occurrence remains limited to sources like coucal oil and tobacco smoke [9].

The physical properties of azulenes include a boiling point significantly higher than that of sesquiterpenes and their intense coloration. Initially, Semmler proposed that the intense blue color of azulene arose from a dimeric sesquiterpene structure, similar to indigo. This hypothesis was believed to be true until structural studies confirmed that azulene was indeed a monomer. Kremers and Augspurger confirmed azulene's bicyclic nature through catalytic hydrogenation, yielding decahydroazulene, although some proposed structures failed to explain azulene's color. Ruzicka had done significant research on the structural elucidation of azulene. Structural investigation of azulenes through oxidation of partially hydrogenated azulenes yielded inconclusive results, as azulenes themselves produced only small fragments upon ozonization. However, these results paved the path for the exact structural determination of azulene. Further research by Pfau and Plattner focused on isolating pure azulene products for spectral analysis. Ruzicka and Rudolph made the most appropriate description of azulene in 1926 when they concluded that this intriguing molecule had a bicyclic ring system, albeit not a six-membered aromatic ring, and the bicyclic system had a structural similarity to sesquiterpenes [1].

The first synthesis of azulene dates back to 1937 that was put forward by Plattner and Pfau but over the time this approach was considered disadvantageous as it involved a tedious dehydrogentation step (Scheme 1.1a) [2]. Ziegler-Hafner's method was a useful approach for azulene synthesis, in cases when substituents were to be introduced on the seven-membered ring (Scheme 1.1b) [10]. Nozoe and co-workers demonstrated the most effective azulene synthesis, which entailed treating an active troponoid

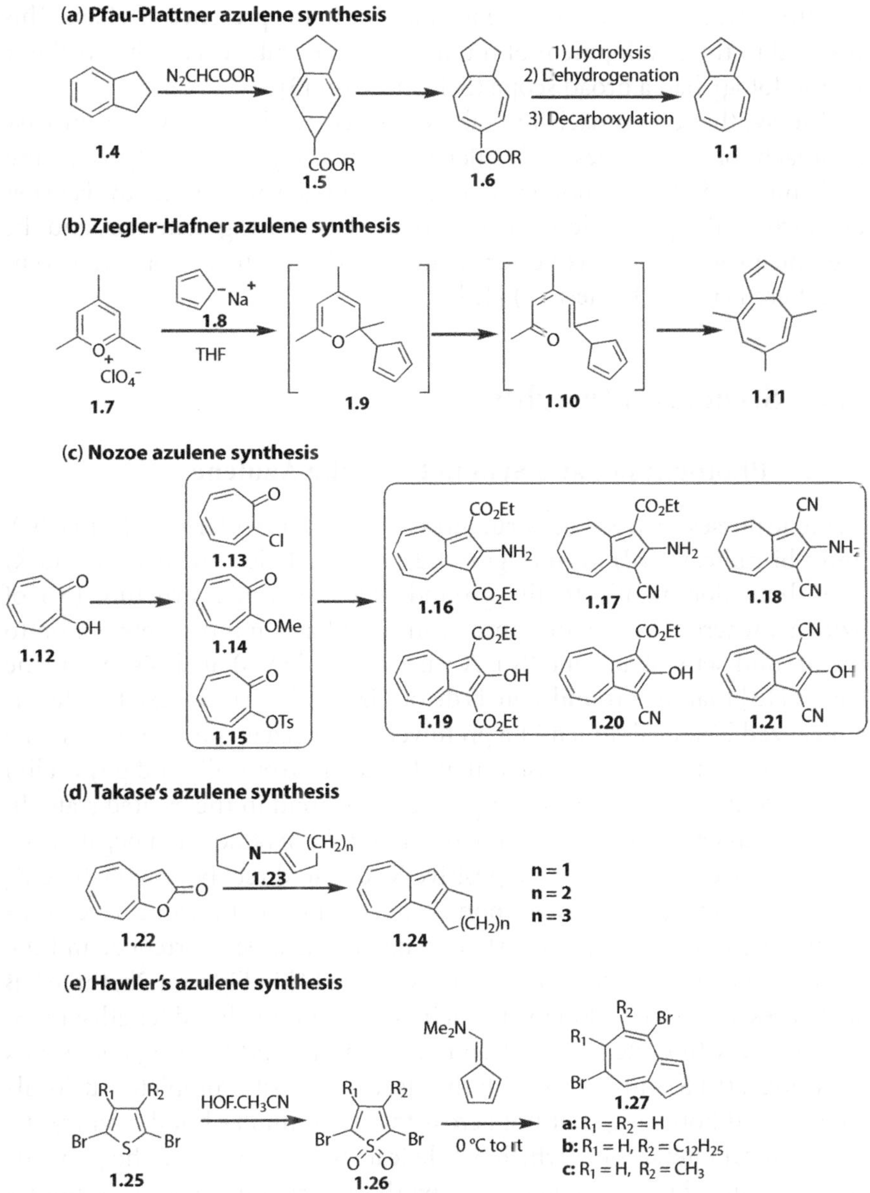

Scheme 1.1 Key synthesis methods of azulene.

precursor with an active methylene molecule in the presence of a base. This resulted in the development of multifunctional azulene, and the synthetic methodology had a broad scope (Scheme 1.1c) [3].

The synthetic approach by Takase and co-workers provided an easy approach for the synthesis of azulene system as a part of a polycyclic unit (Scheme 1.1d) [11]. Another synthesis approach was given by Hawker *et al.* where thiophene derivatives were used as starting materials, and the method provided the scope for inclusion of substituents on the seven-membered ring (Scheme 1.1e) [12].

1.3 Structural Insights

1.3.1 Photophysics and Symmetry of the Azulene

Azulene possess two Kekulé structures (L and R), a zwitterionic form (Zw), and a bi-radical (Brd) form (Figure 1.2a). The Kekulé structures (L and R) contribute dominantly to the ground-state electronic structure (S_0) of azulene, whereas the zwitterionic form (Zw) has a minor contribution to the ground state of azulene (S_0). A valence bond (VB) analysis of azulene shows [13] that the ground state is delocalized, whereas the excited state is a localized biradical, much like prefulvene in benzene photo physics. The transannular π bond is almost a single bond (electronically and physically) in the ground state and nearly a pure double bond in the excited state. In the ground-state VB structure, the trans-annular π bond has a negative K_{ij}, whereas the excited state has a positive K_{ij} due to direct bonding where K_{ij} = exchange integrals between non-orthogonal orbitals. To achieve comparable K_{ij} values, the ground-state structure must be shortened, and the excited state must be elongated. As illustrated [14] in Figure 1.2b, three VB structures are shown, along with their corresponding bond lengths: (a) S_1 minimum, with quite localized double bonds around the ring and across the bridge, (b) S_0 minimum, with the electron density completely delocalized around both rings but not across the transannular bond, and (c) the S_1/S_0 conical intersection, which is a delicate balance between the previous two structures but dominated more by the S_1 structure. The transannular bond, which has two bonds in S_1 and one in S_0, is the most significant structural modification. Smith illustrated [13] that the fulvene S_0/S_1 crossover minimum, or the lowest energy point on S_1, correlates to Klessinger's 90° twisted geometry [15]. The existence of such an accessible conical junction

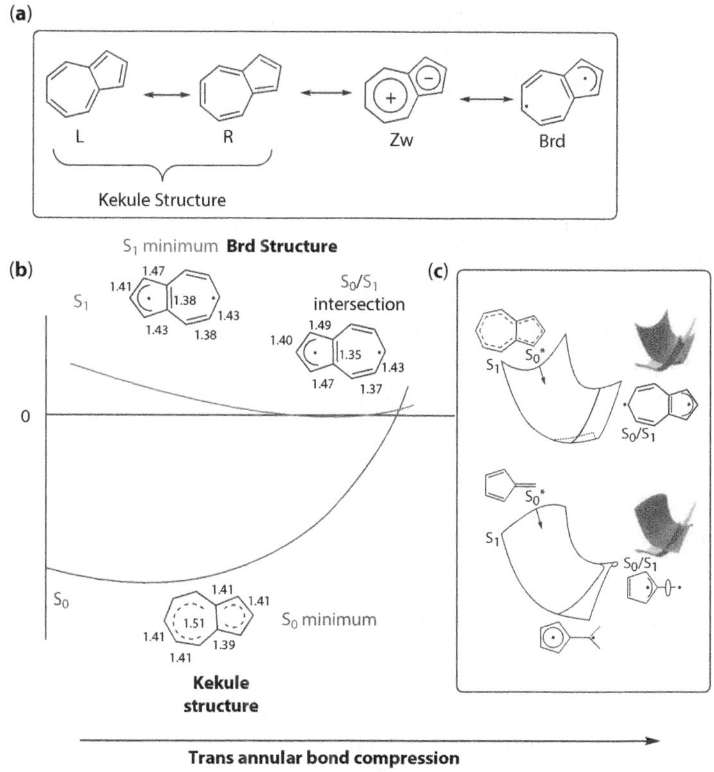

Figure 1.2 (a) Different forms of canonical structure of azulene; (b) schematic presentation of S0/S1 crossing in azulene molecule; (c) a schematic representation of the S1 potential energy surfaces of azulene and fulvene. Reproduced with permission [13]. Copyright 1996, American Chemical Society.

between fulvene's S_0 and S_1 states explains the absence of fluorescence, resulting in efficient and irreversible decay. Semiclassical dynamics calculations with molecular mechanics valence bond (and energies computed at CASMP2/6-31G*) suggest that, for vibrationally excited S_1 fulvene, S_1 to S_0 decay is actually more likely to occur in the planar azulene-like crossing region, located about 12 kcal mol^{-1} higher in energy than the twisted crossing minimum but about 30 kcal mol^{-1} lower than the vertically excited geometry S_0* (Figure 1.2c). As with azulene [15], the S_1-S_0 degradation occurs in femtoseconds before a single oscillation over the crossing line is finished. The severe twisting gradient on S_0 suggests that isomerization of exocyclic methylene may not be noticed until 0-0 excitation to S_1 [13].

Further structural insights into azulene can be obtained by its crystal structure. The crystal structure of azulene was reported by Brandenberger *et al.* The space group was predicted to be $P2_1/a\{C^5\ 2h)$. The unit cell had two azulene molecules in it, which was indicative of a centrosymmetric molecule [16]. However, the chemical findings predicted the things to be different. The following explanations were given for these disagreements: "(a) the implied center of symmetry is only statistically true and is due to rotating molecules or a disordered structure, or (b) the absent spectra leading to the deduction of the space group $P2_1/a$ are accidental and the true space group is of lower symmetry." Insights into these were given by Günthard, and a comparison of the entropy of azulene and naphthalene with identical lattice dimensions was indicative of disorder in the crystal structure of azulene [17, 18]. An independent study by Robertson and Shearer also revealed that the space group of the crystal structure was indeed $P2_1/a$, and the structure was bicyclic, comprising of a fused five- and seven-membered ring, though the disorder was present [18].

1.3.2 Glidewell–Lloyd Rule

Platt's perimeter rule has been used as an important tool to assign the global aromatic character of polycyclic systems, but this can be used only for ground-state compounds [19]. However, this rule fails to provide any insights into local aromaticity. The local aromaticity in benzenoid aromatic systems can be understood effectively by making use of Clar's sextet rule [20, 21]. The Glidewell–Lloyd rule is a more efficient alternative to understand effectively the local aromaticity in polycyclic systems. Its advantage also emerges from the fact that it applies to all types of systems regardless of ring size, and it is even applicable to odd-membered rings [22]. This rule states that "the population of π-electrons in conjugated polycyclic systems tends to form the smallest 4n+2 groups while avoiding the formation of the smallest 4n groups." This rule has been applied to comprehend the effect of local aromaticity on the stability of indole and other related isomers [23]. Azulene, which comprises of a fused five- and seven-membered ring has distinct aromatic properties compared to naphthalene, and this has been attributed to its unique structure [24–26]. Platt's perimeter model fails to explain the local aromaticity of azulene, though the same applies to naphthalene. However, the 10π electrons of azulene are consistent with Platt's perimeter model. The aromatic character of azulene is explained based on the two-charged rings that include the tropylium cation and the cyclopentadienyl anion [27], and the same is also responsible for the unusually high dipole moment of azulene ($\mu = 1.08$ D) and small singlet-triplet gap.

Further, a better understanding of the electronic properties of azulene was provided by Ricardo. Ricardo [28] employed the magnetic criterion in conjunction with magnetically induced current density maps, ring current strengths, nucleus-independent chemical shift ($NICS_{zz}$) (1), and the bifurcation value of three-dimensional $NICS_{zz}$ surfaces. The delocalization criterion was used to calculate the multicenter index (MCI) and the electron localization function (ELF_{π}). These findings revealed that the aromatic character of the five-membered ring is greater than that of the seven-membered ring and the free $C_5H_5^-$ ring. On the contrary, free $C_7H_7^+$ was found to be more aromatic than the seven-membered ring. The intramolecular transfer of electron from the seven-membered ring to the five-membered ring effectively explains the local aromaticity of azulene unit. Besides that, both the aromaticity, as well as the relative stability of azulene, can be effectively assigned by the Glidewell–Lloyd rule, as evidenced by the preference for the formation of five-membered rings in resonance structures that explain its properties (e.g., dipole moment and relative stability). Figures 1.3 (a–d) shows the magnetically induced current density vector map at 1.0 a.u. above the molecular plane. A global diatropic (clockwise) current is observed; it is consistent with Platt's model for aromatic compounds.

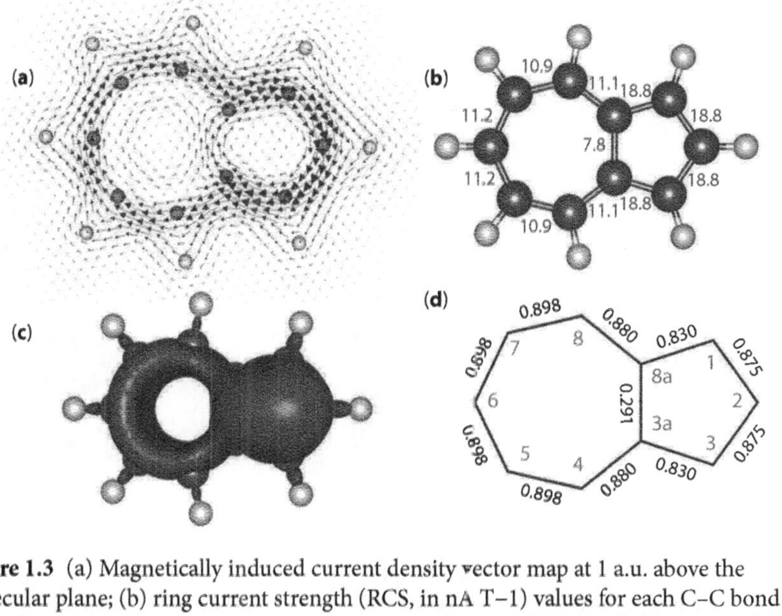

Figure 1.3 (a) Magnetically induced current density vector map at 1 a.u. above the molecular plane; (b) ring current strength (RCS, in nA T−1) values for each C–C bond for azulene; (c) NICSzz isosurface at −30 ppm isovalue; and (d) ELFπ bifurcation values. All calculations performed at the B3LYP/6-311G** level of theory, reproduced with permission [29]. Copyright 1999, Royal Society of Chemistry (Great Britain).

1.3.3 Azulene Moiety as Electron Reservoir

In the case of azulene, only the x-axis is symmetrical, whereas in naphthalene, the symmetry is observed in both x- and y-axis (Figure 1.4a). This results in a notable variation in the chemical reactivities of the two rings present in azulene [29, 30]. Theoretical insights into the frontier molecular orbital diagram of azulene show that owing to stabilization by resonance (Figures 1.4c and 1.4d), the odd positions of azulene (1-, 3-, 5-, and 7-) are prone to undergo electrophilic substitution reaction and the even positions (2-, 4-, 6-, and 8-) are prone to nucleophilic substitution (Figure 1.4b) [31]. Over time, different methods have been explored for the synthesis of functionalized azulene. Nozoe's method is considered the most remarkable for the synthesis of multifunctionalized azulene. Earlier, the synthesis of 2,6-difunctionalized azulene was considered the toughest, however, the last decade has witnessed a rapid surge in publications dealing with the 2,6-connectivity of azulene. The 2,6-connectivity of azulene is considered the most advantageous as that results in the maximum utilization of its resonance structure along the C_{2v} dipolar axis and results in systems of low HOMO-LUMO gap.

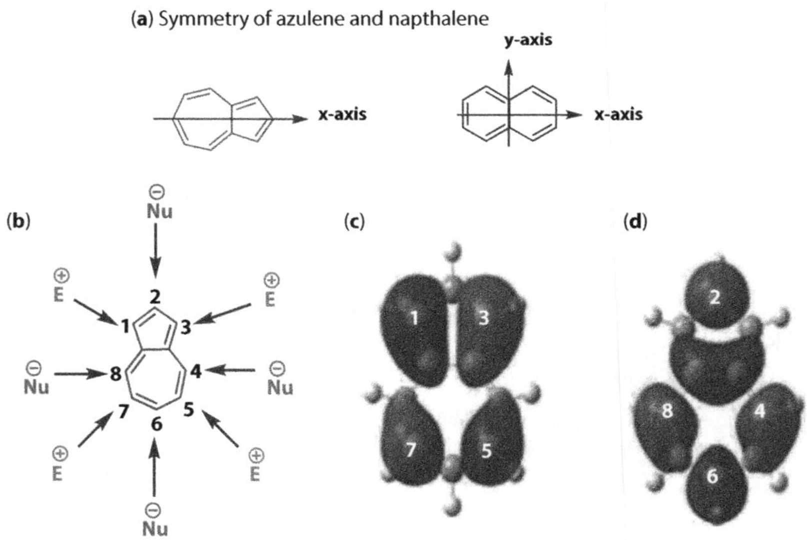

Figure 1.4 (a) Symmetry of azulene and naphthalene; (b) sites for functionalization of azulene skeleton; (c) & (d) HOMO & LUMO of azulene calculated using Gaussian 09 program with B3LYP functional and 6-31G; (d) basis set.

1.3.4 Photodissociation of Azulene Cation

Mebel and colleagues [32] employed the ab initio/Rice–Ramsperger–Kassel–Marcus method to provide mechanistic insights into the photodissociation of the azulene cation at various photon energy levels. The Rice–Ramsperger–Kassel–Marcus method and microcanonical variational transition state theories were employed to calculate the rate constants for specific reaction paths, after mapping out the reaction pathways leading to different breakdown products at the G3(MP2, CC)//B3LYP level. The relative product yields (branching rations) of the dissociation pathway were calculated by applying steady-state method. It was revealed that the azulene cation easily undergoes isomerization to a naphthalene cation upon photoexcitation. Among various isomerization mechanisms considered, the dominant reaction pathway is the norcaradiene-vinylidene mechanism that leads to formation of 50% of naphthalene. The main dissociation pathways involve the elimination of acetylene, H_2 molecules, and atomic hydrogen. As photon energy increases, the branching ratio for the H-atom elimination channel decreases, while the branching ratios for H_2 and acetylene eliminations increase. Across all photon energies considered, the primary $C_8H_6^+$ fragment is a pentalene cation. As the excitation energy increases, its yield slightly decreases, while the branching ratios for fragments like phenylacetylene cation and benzocyclobutadiene cations also increase.

1.4 Applications of Azulene and Its Derivatives

Azulenes with their distinctive chemical and physical properties are a unique class of compounds with a wide range of applications across different scientific and industrial fields. Azulene and its derivatives are becoming potential candidates in **organic electronics**, particularly in **organic light-emitting diodes** and **organic solar cells**. The strong absorption profile of azulene is beneficial for enhancing light-harvesting efficiency in solar cells. Additionally, their **electronic structure** and **photophysical properties provide the resulting materials with a small HOMO-LUMO gap, thus** making them suitable for **organic** semiconductors and providing an alternative to traditional inorganic materials.

Azulenes and its derivatives have also emerged as potential materials in medicinal chemistry and pharmacology owing to their **antibacterial, antioxidant**, and **anti-inflammatory** properties. They have also been employed in bioimaging. The unique blue color of azulenes makes them useful for applications in dyes and pigments and is also used in cosmetics to impart a blue hue to products such as shampoos, creams, and lotions. Azulenes and their derivatives are also gaining attention in **environmental applications** such as sensing owing to their ability to change color in response to different environmental stimuli. They are also now included in advanced materials such as nanoparticles, nanostructures, and **thin films** for use in **nanotechnology**, where they contribute to the creation of highly functional nanomaterials for applications in **electronics** and **biotechnology**. The electron acceptor and the nature of the electron donor of azulenes make them suitable candidates for **homogeneous and heterogeneous catalysis**. Additionally, they have been used as precursors for the **synthesis of other organic compounds**, which further broadens their potential applications in the fine chemical industry.

Thus, azulenes exhibit a diverse array of applications spanning **medicine**, **pharmacology**, **materials science**, **electronics**, **environmental monitoring**, and **chemical industry**. The ongoing research into their properties continues to uncover new uses, making them a valuable and versatile class of compounds with significant potential in both existing and emerging fields. The further chapters of the book give detailed insights into the synthesis, as well as the applications of azulene and its derivatives.

- **Chapter 2**, **Chapter 3**, and **Chapter 4** will shed light on the synthesis of azulene containing natural products, azulenes and its functionalized derivatives, as well as aza-azulenes.
- **Chapter 5**, **Chapter 6**, **Chapter 7**, and **Chapter 8** will focus on the applications of azulene-based systems in organic electronics, sensors including azulene-embedded nanographenes and polyaromatic hydrocarbons.
- **Chapter 9** will highlight the medicinal importance of azulene and its derivatives.
- **Chapter 10**, **Chapter 11**, and **Chapter 12** will give insights into the broad use of azulene in inorganic chemistry including porous systems, azulene containing organometallic systems, and azuliporphyrins.
- **Chapter 13** and **Chapter 14** will focus on the theoretical aspects of azulene chemistry and how the blend of experimental and theoretical chemistry can pave the future prospects of azulene chemistry.

1.5 Conclusion

The unique structure and exceptional properties of azulene distinguish them from regular aromatic hydrocarbons; hence, they have been a fascinating class of compounds to researchers. The fusion of five- and seven-membered rings in azulene results in a dipole moment of 1.08D in contrast to azulene, hence its blue color. The photophysical properties of azulenes further emphasize their divergence from other aromatic hydrocarbons. Azulenes are also known for their anomalous emission from S_2 to S_0, defying Kasha's rule and providing a valuable point of interest in their study. Moreover, the localized aromaticity and relative stability of azulenes, as explained by the Glidewell–Lloyd rule, further reinforce their distinctive nature within the realm of conjugated polycyclic compounds. With their potential still untapped, azulenes are chemicals with enormous potential for use in a variety of sectors, including materials science, organic electronics, and medicinal chemistry. In the end, azulenes present a multitude of prospects for further scientific investigation, rendering them a crucial field of interest for investigators aiming to discover novel facets in organic chemistry and other fields.

AI disclosure statement: The author acknowledges the use of AI-based tools for minor language editing or grammatical correction. However, all intellectual contributions, analyses, and conclusions presented in this chapter are entirely the author's own.

References

1. Ruzicka, L. and Rudolph, E.A., Höhere terpenverbindungen XXVII. Zur kenntnis der azulene. *Helv. Chim. Acta*, 9, 131, 1926.
2. Pfau, A.S. and Plattner, P.A., Zur kenntnis der flüchtigen pflanzenstoffe VIII. Synthese des vetivazulens. *Helv. Chim. Acta*, 22, 202, 1939.
3. (a) Nozoe, T., Seto, S., Matsumura, S., Asano, T., Synthesis of azulene derivatives from troponoids and cyanoacetic ester. *Proc. Jpn. Acad.*, 32, 339, 1956. (b) Nozoe, T., Seto, S., Nozoe, S., Syntheses of azulene and 1-azaazulanone derivatives by the application of cyanoacetamide to tropolone and 4-methyltropolone methyl esters. *Proc. Jpn Acad.*, 32, 472, 1956. (c) Nozoe, T., Seto, S., Matsumura, S., Murase, Y., The synthesis of azulene derivatives from troponoids. *Bull. Chem. Soc. Jpn.*, 35, 1179, 1962. (d) Nozoe, T., Takase, K., Shimazaki, N., The synthesis of diethyl 2-hydroxyazulene-1, 3-dicarboxylate from troponoids and some reactions of 2-hydroxyazulene derivatives.

Bull. Chem. Soc. Jpn., 37, 1644, 1964. (e) Nozoe, T., Takase, K., Fukuda, S., The synthesis of 1-phenylazulene and 2-phenylazulene from the troponoid compound. *Bull. Chem. Soc. Jpn.*, 44, 2210, 1971.

4. Xin, H., Hou, B., Gao, X., Azulene-based π-functional materials: Design, synthesis, and applications. *Acc. Chem. Res.*, 54, 7, 1737, 2021.
5. Akram, W., Tagde, P., Ahmed, S., Arora, S., Emran, T.B., Babalghith, A.O., Sweilam, S.H., Gandara, J.S., Guaiazulene and related compounds: A review of current perspective on biomedical applications. *Life Sci.*, 316, 121389, 2023.
6. (a). Michl, J. and Thulstrup, E.W., Why is azulene blue and anthracene white? A sinple MO picture. *Tetrahedron*, 32, 2, 205, 1976. (b). Tétreault, N., Muthyala, R.S., Liu, R.S.H., Steer, R.P., Control of the photophysical properties of polyatomic molecules by substitution and solvation: The second excited singlet state of azulene. *J. Phys. Chem. A.*, 103, 15, 2524, 1999.
7. Dong, J.-X. and Zhang, H.-L., Azulene-based organic functional molecules for optoelectronics. *Chin. Chem. Lett.*, 27, 8, 1097, 2016.
8. Zeng, H.N., Png, Z.M., Xu, J., Azulene in polymers and their properties. *Chem. Asian J.*, 15, 13, 1904, 2020.
9. Gordon, M., The azulenes. *Chem. Rev.*, 50, 1, 127, 1952.
10. (a) Ziegler, K. and Hafner, K., Eine rationelle synthese des azulens. *Angew. Chem.*, 67, 301, 1955. (b) Poronik, Y.M., Mazur, L.M., Samoć, M., Jacquemin, D., Gryko, D.T., 2,5-Bis(azulenyl)pyrrolo[3,2-b] pyrroles-the key influence of the linkage position on the linear and nonlinear optical properties. *J. Mater. Chem. C*, 5, 2620, 2017.
11. Yang, P.W., Yasunami, M., Takase, K., The formation of azulene derivatives by the reaction of 2H-cyclohepta[b]furan-2-ones with enamines. *Tetrahedron Lett.*, 12, 4275, 1971.
12. Amir, E., Amir, R.J., Campos, L.M., Hawker, C.J., Stimuli-responsive azulene-based conjugated oligomers with polyaniline-like properties. *J. Am. Chem. Soc.*, 133, 10046, 2011.
13. Robb, M.A., In this molecule there must be a conical intersection, in: *Advances in Physical Organic Chemistry*, I.H. Williams and N.H. Williams (Eds.), pp. 189–228, Academic Press, 2009.
14. Bearpark, M.J., Bernardi, F., Clifford, S., Olivucci, M., Robb, M.A., Smith, B.R., Vreven, T., The azulene S_1 state decays *via* a conical intersection: A CASSCF study with MMVB dynamics. *J. Am. Chem. Soc.*, 118, 1, 169, 1996.
15. Dreyer, J. and Klessinger, M., Excited states and photochemical reactivity of fulvene. A theoretical study. *J. Chem. Phys.*, 101, 10655, 1994.
16. Gunthard, H.H., Plattner, P.A., Brandenberger, E., Über die Kristallstruktur des Azulens *Experientia*, 4, 425, 1948.
17. Gunthard, H.H., These Doc. Sci. tech. Über die Kristallstruktur des Azulens. Zurich, *Neographic*, in 8°, 73, 1949.

18. (a) Gunthard, H.H., Dielektrische messungen an polaren gemischen. 3. Mitteilung. thermodynamische theorie der solvatation kugelförmiger ionen in binären lösungsmittelgemischen. *Helv. Chim. Acta*, 38, 1955. (b) Robertson, J.M. and Shearer, H.M.M., Structure of azulene. *Nature*, 885, 1956.
19. Platt, J.R., Classification of spectra of cata-condensed hydrocarbons. *J. Chem. Phys.*, 17, 484, 1949.
20. Clar, E., The aromatic sextet, p. 49, Wiley, New York, 1972.
21. Solà, M., Forty years of Clar's aromatic π-sextet rule. *Front. Chem.*, 1, 4, 2013.
22. Glidewell, C. and Lloyd, D., MNDO study of bond orders in some conjugated bi- and tri-cyclic hydrocarbons. *Tetrahedron*, 40, 4455, 1984.
23. Pino-Rios, R. and Sola, M., The relative stability of indole isomers is a consequence of the Glidewell Lloyd rule. *J. Phys. Chem. A*, 125, 230, 2021.
24. Mattohti, R. and Kerim, A., A study of the aromaticity and ring currents of azulene and azaazulenes. *RSC Adv.*, 6, 108538, 2016.
25. Aihara, J. and Oe, S., Graph-theoretical re-interpretation of NICS values for polycyclic aromatic hydrocarbons: Naphthalene and azulene. *Bull. Chem. Soc. Jpn.*, 76, 1363, 2003.
26. Poater, J., Heitkämper, J., Poater, A., Maraval, V., Chauvin, R., Zwitterionic aromaticity on azulene extrapolated to carbo-azulene. *Eur. J. Org. Chem.*, 46, 6450, 2021.
27. Lvov, A.G. and Bredihhin, A., Azulene as an ingredient for visible-light- and stimuli-responsive photoswitches. *Org. Biomol. Chem.*, 19, 4460, 2021.
28. Rodrigo, B.G. and Ricardo, P.R., Is azulene's local aromaticity and relative stability driven by the Glidewell-Lloyd rule? *Phys. Chem. Chem. Phys.*, 26, 12162, 2024.
29. Xin, H. and Gao, X., Application of azulene in constructing organic optoelectronic materials: New tricks for an old dog. *ChemPlusChem*, 82, 945, 2017.
30. Razus, A.C., Azulene moiety as electron reservoir in positively charged systems; A short survey. *Symmetry*, 13, 526, 2021.
31. Lash, T.D., Out of the blue! Azuliporphyrins and related carbaporphyrinoid systems. *Acc. Chem. Res.*, 49, 3, 471, 2016.
32. Dyakov, Y.A., Ni, C.-K., Lin, S.H., Lee, Y.T., Mebel, A.M., Ab initio and RRKM study of photodissociation of azulene cation. *Phys. Chem. Chem. Phys.*, 8, 1404, 2006.

2

Azulene-Containing Natural Products

Abhilash Sharma[1*], Hemanta Hazarika[2] and Dhrubajit Borah[3]

[1]Department of Chemistry, B. Borooah College, Guwahati, Assam, India
[2]Department of Chemistry, Nabajyoti College, Kalgachia, Barpeta, Assam, India
[3]Department of Chemistry, N. N. Saikia College, Titabor, Jorhat, Assam, India

Abstract
Azulene is a bicyclic aromatic hydrocarbon found in various natural products with a unique structural motif, which contributes to diverse biological properties. Azulene and its derivatives are widely found in various plants and mushrooms, possessing fascinating medicinal properties. Natural products containing azulene scaffolds, such as guaiazulene and chamazulene, are used in the fields of pharmacology, natural product chemistry, and medicinal chemistry. The distinct characteristics of azulene, including its anti-inflammatory, antimicrobial, and antioxidant properties, render these natural products highly promising for drug discovery and development. Even though several academicians and researchers have studied this topic, a more thorough investigation is still needed. Thus, this chapter highlights a comprehensive overview of the occurrence, isolation and synthesis of azulene derivatives, and the application of azulene-containing natural products.

***Keywords*:** Azulene, guaiazulene, chamazulene, natural product

2.1 Introduction

Azulene is the structural isomer of naphthalene. Both are aromatic and bicyclic, containing five conjugate double bonds. Although azulene and naphthalene are structurally similar, they render different physiochemical properties [1]. Unlike naphthalene, azulene is a polar molecule and has a large inherent dipole moment due to the polarization of π electrons towards the five-membered ring. Additionally, it is a blue compound while

**Corresponding author*: sharma.abhilash72@gmail.com

Neha Rani Kumar and Abhijeet R. Agrawal (eds.) Azulene-Based Materials: Design, Synthesis, and Applications, (19–52) © 2026 Scrivener Publishing LLC

naphthalene is colorless. Remarkably, the electronic properties of azulene derivatives can be easily tuned by incorporating various chromophores in their structures. For this unique property, azulene derivatives are used as photocatalysts, as active materials in organic electronics, and as photoswitches. Besides, azulene and its derivates gained prominence in medicinal chemistry due to their use as antiulcer, antidiabetic, and anticancer drugs. Most of these derivatives are isolated from diverse plant extracts and mushrooms [2]. Of these, guaiazulene (**2.1**) and chamazulene (**2.2**) in Figure 2.1 are the most essential azulene-containing natural products, which found applications in diverse biological applications.

Guaiazulene is a sesquiterpene with a molecular formula $C_{15}H_{18}$. Most of its derivatives, alongside guaiazulene, were isolated from natural sources such as gorgonian soft corals and marine plants [3]. These compounds usually impart colors and are responsible for the brilliant colors of soft corals [4]. Due to its excellent photophysical properties and non-toxic nature, guaiazulene is used as a color additive in cosmetic items. Additionally, guaiazulene has emerged as a preferred choice for photoactive components in optoelectronic devices due to its low cost and unique optical properties [5, 6].

Beyond these, guaiazulene and its derivatives have also received increasing attention from medicinal chemists due to their diverse biological applications. During the last 40 years, concentrated efforts have been made to decipher their pharmacological effects against several diseases. These research efforts divulged that guaiazulene and its compounds had remarkable anti-inflammatory properties. For example, sodium guaiazulene sulfonate, the active component of compound glutamine granules, shows remarkable anti-inflammatory properties and can be used to heal stomach ulcers [7]. Moreover, guaiazulene-based sulfonamide derivatives have also demonstrated strong anti-inflammatory characteristics in addition to potential anticancer effects [8]. Many guaiazulene-based heterocyclic

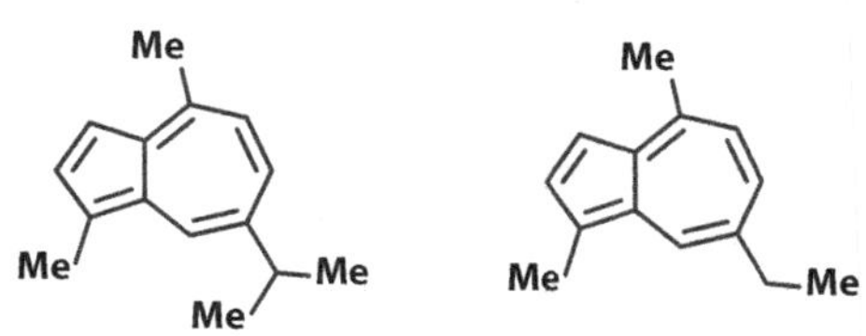

2.1 **2.2**

Figure 2.1 Structure of guaiazulene and chamazulene.

derivatives have been synthesized due to their substantial biological activity; however, evaluations of their biological activity have been relatively less explored. Furthermore, despite the known antiviral activity of azulene derivatives, no studies on guaiazulene derivatives have been published. Beyond these, it has antibacterial, antioxidant, epithelializing, antimutagenic, immunomodulatory, fungicidal, expectorant [9], diuretic, diaphoretic, demulcent, and bitter stimulant effects [10].

Regarding the derivatives of azulene, another much-discussed structural motif is chamazulene. It is a bicyclic sesquiterpenoid polyalkene that is commonly used as an anti-inflammatory agent in herbal products and cosmetics [11]. The chemical formula for this bicyclic polyalkene is $C_{14}H_{16}$. When wormwood oil was distilled under reduced pressure, some fraction rich in chamazulene was produced, with a boiling point ranging from 92–100°C. The breakdown of matricin in heat-exposed plants produces chamazulene, which cannot be obtained directly from natural sources. Chamazulene, depending on its concentration, turns the essential oil blue. At low concentrations, it produces a bright blue color; at high concentrations, it makes a dark blue color. Chamazulene is isolated from essential oil extracted from a limited number of plants around the globe [12–17]. Like other azulene derivatives, chamazulene has also been investigated for its potential use in biological applications. Thus, this chapter aims to provide an overview of natural sources of guaiazulene and chamazulene and their isolations. Given the importance of these compounds in various aspects of life, concerted efforts have been made to develop synthetic methods to access these bioactive compounds. Therefore, we also discuss methodologies adopted for synthesizing these molecules.

2.2 Occurrence and Isolation of Guaiazulene and Chamazulene

Guaiazulene and its derivatives are a class of natural products, which gained considerable interest due to their various biological properties. This interesting class of compounds were obtained from natural sources like gorgonians and soft corals (Figure 2.2). The most abundant sources of guaiazulene derivatives include gorgonian genera, such as *Calicogorgia*, *Acalycigorgia*, *Euplexaura*, *Alcyonium*, and *Paramuricea*. Besides, these compounds were also isolated from terrestrial plants, including the fungal genus *Lactarius*.

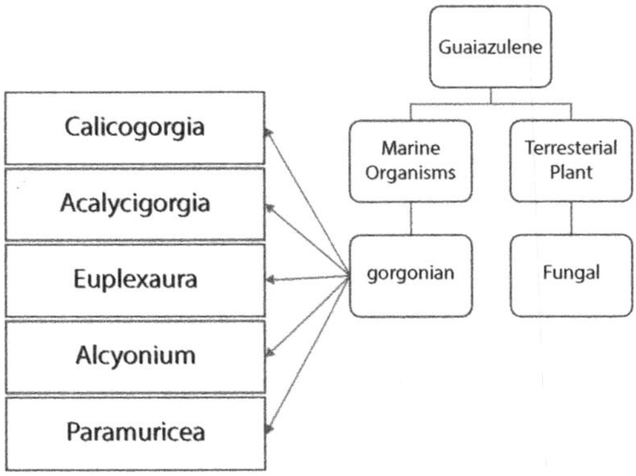

Figure 2.2 Natural sources of guaiazulene.

Fusetani *et al.* first isolated the guaiazulene (**2.1**, Figure 2.3) from Japanese marine invertebrates, gorgonian *Euplexaura erecta* [18]. To isolate the desired compound, initially, *E. erecta* (20g) were ground and extracted with diethyl ether. Thereafter, the solvent was evaporated at 30°C, yielding a brownish-blue oil. The brownish-blue oil was loaded with a silica column and eluted with diethyl ether followed by hexane. Finally, a blue band was collected that contained guaiazulene (40 mg) as a bluish violet oil.

Linderazulene (**2.3**) is another interesting purple pigment, structurally similar to guaiazulene (Figure 2.3). This bioactive compound was isolated by Imre *et al.* from the Marmara Sea gorgonian *Paramuricea chamaeleon* [19]. After collecting the marine invertebrate species, it was preserved in acetone and extracted with the same solvent. The combined extracts were concentrated under reduced pressure, and the remaining aqueous portion was extracted with ether, yielding an oily residue. The oily residue was loaded with a silica column and eluted with petrol. The petrol elutes, which contain purple pigment, were further purified by preparative layer chromatography on silica gel using the same eluent. Finally, pure pigment was obtained after recrystallizing it from aqueous alcohol. The identity of the chemical structure was confirmed by mass-spectrometry and nuclear magnetic resonance analysis.

Linderazulene (**2.3**) was also isolated from an olive gorgonian, *placogorgia* by the research group led by Scheuer [20]. Moreover, they also isolated guaianolide (**2.4**), a yellow color pigment from the same specimen (Figure 2.3). The typical isolation procedure involves the extraction of *placogorgia* specimens with hexane followed by the separation of the mixture into constituent compounds by chromatographic techniques. After extraction, hexane was evaporated to yield a dried residue. The residue was

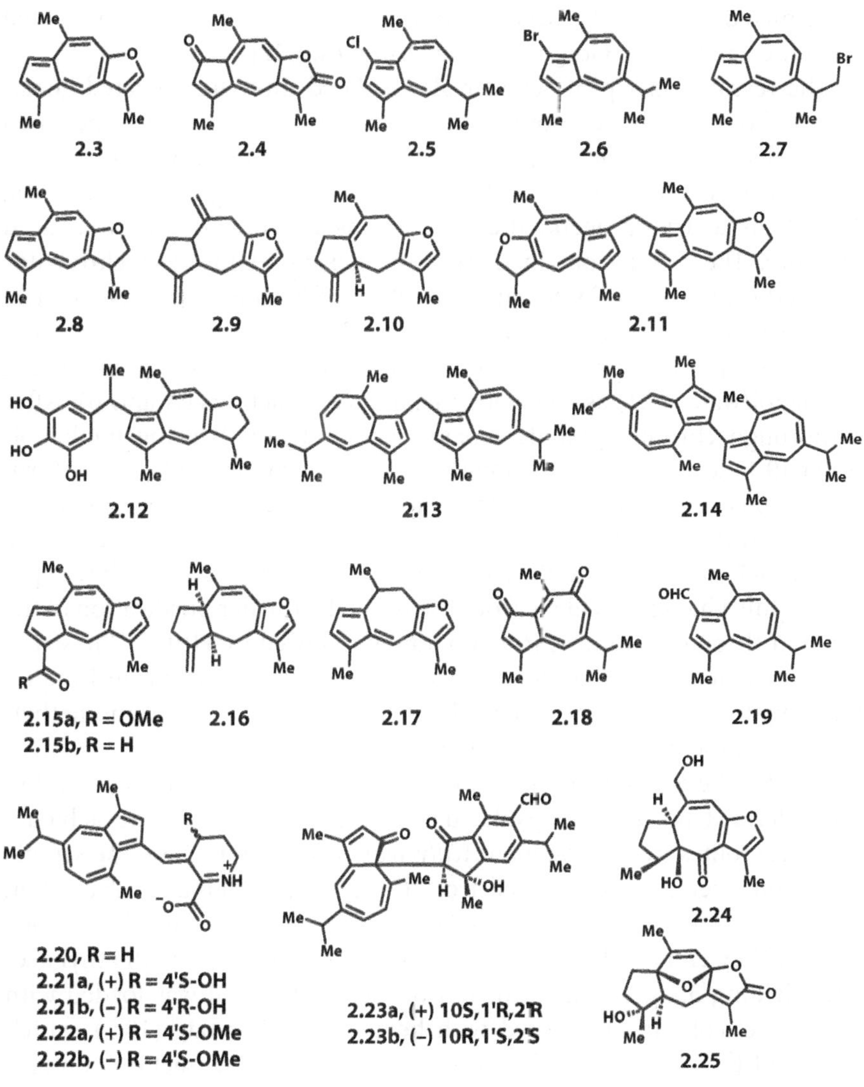

Figure 2.3 Examples of guaiazulene derivatives isolated from natural sources.

partitioned into two fractions using biobeads SX-8, employing a solvent mixture of benzene and hexane (1:1) as an eluent. The purple fractions furnished linderazulene. Further purification of the remaining yellow fraction using a neutral alumina column, employing hexane/EtOAc (3:1) eluent, yields a yellow residue. A yellow precipitate was obtained when the yellow residue was mixed with hexane/EtOAc (4:1). Finally, pure guaianolide was obtained through the recrystallization of the yellow precipitate using the same mixed solvent [21]. Notably, compounds (**2.3, 2.4**) were also isolated from the Far-Eastern Gorgonaria *Stenogorgia* sp [22].

The research group of Scheuer also isolated three halogenated guaiazulene from a deep-sea gorgonian, *paramuriceidae* [23]. The specimen was collected from the deep sea and immediately preserved in dry ice. Afterwards, it was extracted with hexane. After drying the extract using Na_2SO_4 as a drying agent, the solvent was evaporated under reduced pressure. The residue obtained from the hexane extract was chromatographed on biobeads SX-8 using a mixture of hexane and benzene (1:1) as an eluent, resulting in seven fractions. Fraction 4 was further separated through high-pressure liquid chromatography (HPLC) employing methanol/water (9:1) as an eluent, yielding two fractions (A and B). After a thorough chromatographic separation of fraction B using methanol/water (9:1) as an eluent, researchers obtained two pigments in the form of blue oils. With various spectroscopic techniques, these two pigments were identified as 3-chloro-7-isopropyl-1,4-dimethylazulene (**2.5**) and 3-bromo-7-isopropyl-1,4-dimethylazulene (**2.6**) (Figure 2.3). They separated a third haloguaiazulene from fraction B by further purification of the fraction through column chromatography followed by HPLC. The structural elucidation of the compound using various spectroscopic techniques unambiguously confirmed that the compound is 7-(1′-bromomethyl) ethyl-1,4-dimethylazulene (**2.7**) (Figure 2.3).

In 1987, the research group led by Higa isolated a new pigment, 2,3-dihydrolinderazulene (**2.8**) alongside guaiazulene and linderazulene as bioactive metabolites from gorgonian *Acalycigorgia sp* (Figure 2.3). The sample of marine species was collected from Cape Zanpa, Okinawa. Thereafter, the species was extracted with acetone. After concentrating the acetone extract, they extracted the remaining aqueous suspension with ethyl acetate, yielding 3.4 g of oil. A portion of the resulting oil was loaded with a silica gel column and eluted with heptane-ethyl acetate (10:1), which separated the oil into two fractions. Of these, the first fraction contains blue pigments; hence, a further chromatographic separation technique was used to isolate the compounds present in the mixture. Initially, the portion was filtered in a sintered glass filter filled with a bed of reverse phase adsorbent (RP-8), employing methanol and acetone eluates. Afterward, the filtration process was repeated using methanol eluate to furnish 320 mg of a mixture comprising three pigments. Subsequently, the resulting mixture was separated into five fractions by eluting with heptane-ethyl acetate (10:1) in a Lobar S-60 column. Further separations of fractions containing three pigments through preparative thin layer chromatography gave three pigments. Subsequently, HPLC was used to separate three pigments into its constituent compounds. Finally, after several tedious purification steps, they obtained 93 mg of guaiazulene, 7.8 mg of linderazulene, and 47 mg of 2,3-dihydrolinderazulene (**2.8**) [24]. The same group also isolated echinofuran (**2.10**) from the gorgonian *Echinogorgia praelonga* and explored its biological applications. For isolating the product, the fresh specimen was extracted with methanol. Subsequently, the methanol extract was concentrated to furnish an aqueous suspension. The aqueous suspension was further extracted with ethyl acetate to give a dark oil. Afterwards, the dark oil was loaded with a silica column to yield pure echinofuran. Notably, the compound decomposed upon standing and needs to be characterized immediately after purification [25].

Following this work, Ochi *et al.* isolated two new pigments along with guaiazulene, linderazulene, and 2,3-dihydrolinderazulene from a gorgonian *Acalycigorgia sp.* Structural elucidation with various spectroscopic techniques revealed that both compounds belong to the condensed guaiazulenoid pigment family, namely gorgiabisazulene (**2.11**) and gorgiagallylazulene (**2.12**) (Figure 2.3). After collecting specimens from Sukumo Bay, the material was mixed homogeneously with methanol and allowed to stand at room temperature for a few hours. The resulting mixture was filtered, and methanol was evaporated to furnish an aqueous suspension. The aqueous suspension was further extracted with dichloromethane. Finally, compounds were separated by sequential application of three chromatographic techniques: Sephadex LH-20 (MeOH/CH_2Cl_2), silica gel (Hexane/EtOAc), and reverse phase liquid chromatography (MeOH/H_2O) [26].

Another intriguing sesquiterpene that has a guaiane structure is bebryazulene (**2.9**), which was extracted from the gorgonian coral *Bebryce grandicalyx* by Gaydou *et al.* (Figure 2.3). The species was collected from the lagoon of Mayotte, located in the Indian Ocean. The specimen was extracted with chloroform-methanol (2:1), which contains this previously unknown sesquiterpene [27].

Shin *et al.* isolated three pigments belonging to the guaiazulene family from the gorgonian *Calicogorgia*. After isolation and purification, using combined chemical and various spectroscopic techniques, these pigments were identified as guaiazulene, 2,2′-diguaiazulenylmethane (**2.13**) and 2,2′-diguaiazulenyl (**2.14**) (Figure 2.3). To isolate these compounds, specimens were extracted with dichloromethane, followed by silica vacuum flash chromatography, and HPLC was used to yield three blue and bluish-green pigments [28].

In 2005, Wright *et al.* reported the isolation of two new linderazulene derivatives (**2.15a, 2.15b**) along with linderazulene from a deep-sea gorgonian, *Paramuricea* (Figure 2.3). After collecting the specimen, it was immediately frozen and stored in the frozen state until workup. Before workup, the frozen specimen was transformed into small slices by chopping and thereafter, the specimen was extracted with ethanol. After concentrating the combined ethanol extract by distillation under reduced pressure, they obtained an aqueous residue. Subsequently, aqueous residue was extracted with ethyl acetate. The removal of solvent from ethyl acetate extract furnished a reddish-brown solid. Finally, compounds (**2.15a**, **2.15b**, **2.3**) were obtained after purification of the solid mixture by applying sequential vacuum column chromatography and HPLC separation techniques [29].

Interestingly, besides linderazulene, gorgonian coral is also a rich source of bioactive hydrogenated linderazulene. For instance, tricyclic furanosesquiterpene (**2.16**, **2.17**) were isolated from gorgonian *Echinogorgia complexa* collected from the West Indian coast (Figure 2.3) [30].

In 2012, Lin *et al.* demonstrated that Chinese gorgonian *Anthogorgia sp.* is a rich source of various value-added products. They isolated 15 new structurally diverse guaiazulene-based terpenoids from this marine species. Remarkably, they reported, for the first time, the isolation and identification of guaiazulenyl trimers from marine organisms. Besides guaiazulene, they isolated guaiazulene-3,5-dione (**2.18**), ketolactone (**2.4**), and formylguaiazulene (**2.19**) using various separation techniques (Figure 2.3) [31]. Moreover, gorgonian *Muriceides collaris*, collected from the South China Sea, proved to be a rich source of guaiazulene alkaloids. The research group led by Liu isolated three novel compounds with guaiazulene core, muriceidines A-C (**2.20–2.22**) from this species. Notably, they also isolated muriceidone A (**2.23**) (Figure 2.3) alongside guaiazulene and formylguaiazulene (**2.19**) [32].

Apart from gorgonian, guaiazulene and its derivatives are also isolated from terrestrial plants. For instance, Ahmad *et al.* isolated taraxacin and a known compound, ketolactone (**2.4**), from *Taraxacum wallichii*. They gathered the plant from Islamabad to extract these chemicals. The plant material was crushed, submerged in methanol for seven days, and then extracted using methanol. Afterwards, the combined methanolic extract was concentrated under reduced pressure and further extracted with hexane and ethyl acetate. The ethyl acetate soluble-portion was concentrated under reduced pressure, and the resulting residue was loaded with a silica column to purify the compounds. Finally, pure compounds were obtained for spectroscopic characterization after further purification of the residue obtained from column chromatography by preparative thin-layer chromatography using a mixture solvent of chloroform and methanol as eluent [33].

Meanwhile, Kawazoe *et al.* isolated ketolactone (**2.4**) along with secoguaianolide sesquiterpene from the Artemisia *gilvescens* plant [34]. Moreover, ketolactone (**2.4**) was also isolated alongside Zedoardiol (**2.24**), a new furanoguaiane sesquiterpenoid obtained from rhizomes of *Curcuma xanthorrhiza* (Figure 2.3) [35]. Meanwhile, the research group led by Phan isolated a new guaianolide (**2.25**) from *Curcuma kwangsiensis* (Figure 2.3) [36].

On the other hand, chamazulene, a bicyclic sesquiterpene, is another important bioactive molecule belonging to the azulene family. Naturally, this compound is isolated from medicinal plants such as chamomile, yarrow, and wormwood. Chamazulene is one major component of essential oils extracted from chamomile. It is formed at a high temperature from matricin when essential oil is extracted by steam distillation.

Although several isolation methods were adopted for accessing chamazulene, we summarize only the recent works that deal with the subject matter. Various conventional extraction techniques such as steam distillation and soxhlet techniques were used to extract essential compounds from plant materials. Although these conventional techniques effectively separate organic components that are non-volatile, they are not suitable for extracting volatile chemicals as they require elevated temperatures to isolate the extract, leading to the degradation of non-volatile components in the extract. To overcome this issue, researchers adopted a supercritical fluid extraction method to isolate essential components from plant materials [37]. For instance, Knez *et al.* isolated chamazulene, matricin, and bisabolol from chamomile flower heads using a supercritical fluid extraction method. They demonstrated that extraction yields depend upon the solvent density, temperature, and pressure of carbon dioxide [38]. Indeed, conventional methods like the SECD (Solvent Extraction Clevenger Distillation) were also used to extract blue essential oil from chamomile flowers. This work disclosed that the choice of solvents for extraction determines the yield and the quality of the extract. Among the solvents used for the extraction, dichloromethane was found to be the most suitable while the extraction was quite inefficient in hexane. The constituents present in the extracted essential oil were identified by using gas chromatography-mass spectrometry (GC-MS) analysis [39].

Chamazulene alongside other bioactive compounds was also isolated from the essential oil and volatile components extracted from chamomile tea prepared from flower heads of *Matricaria recutita L.* For isolating the volatile constituent of the extract, hydro distillation, and SPE solid phase extraction methods were employed. Finally, constituent compounds were identified by GC-MS analysis [40]. Using similar separation methods,

El-Assri extracted essential oil from wild chamomile (*Matricaria recutita* L), found in Morocco. The chemical analysis of the essential oil extract showed the presence of chamazulene alongside numerous other bioactive compounds [41]. The essential oil extract produced from chamomile (*Matricaria recutita* L), found in central Albania also contains chamazulene as one of the constituents [42]. Moreover, the research group led by Maffei isolated chamazulene from essential oil extracted from Chamomile *Matricaria spp.* using a column chromatography technique [43].

Recently, liquid-solid extraction using a metal-organic framework emerged as an effective extraction strategy to isolate bioactive compounds from medicinal plants due to its low cost, and high efficiency in isolating the compounds. Moreover, using this technique, pure bioactive compounds can be isolated in one step without using chromatographic separation methods. For instance, Abdelhameed *et al.* demonstrated the efficacy of copper benzene-1,3,5-tricarboxylate metal-organic framework (Cu-BTC MOF) in extracting chamazulene from chamomile extract [44].

Apart from chamomile, *Artemisia arborescens*, a brushy shrub, found in the Mediterranean area, is also a rich source of chamazulene. The essential oil extracted from this plant contains about 52% chamazulene. The essential oils from aerial parts of this plant were extracted employing a hydro distillation technique using a Clevenger-type apparatus. Finally, constituent compounds contained in the essential oil were identified using GC-MS analysis [45].

2.3 Synthesis of Guaiazulene

Guaiazulene is one of the natural compounds containing azulenes and has applications in a variety of biological functions. Recognizing the importance of guaiazulene in various chemical industries, chemists have been developing efficient chemical methods to access this bioactive compound. In 1950, in a pioneering work, Plattner reported, for the first time, the synthesis of guaiazulene. Since then, several methods have been developed to synthesize this molecule [46]. Nonetheless, the following processes are used to synthesize guaiazulene, which includes dehydration, dehydrogenation, 6+4 cycloaddition, thermolysis, and rearrangement reactions.

According to studies established in the 19th century, Adams *et al.* reported the synthesis of guaiazulene (**2.1**) *via* dehydrogenation of hexahydrohelenalin (**2.26**) using 10% Pd-C (Scheme 2.1) [47].

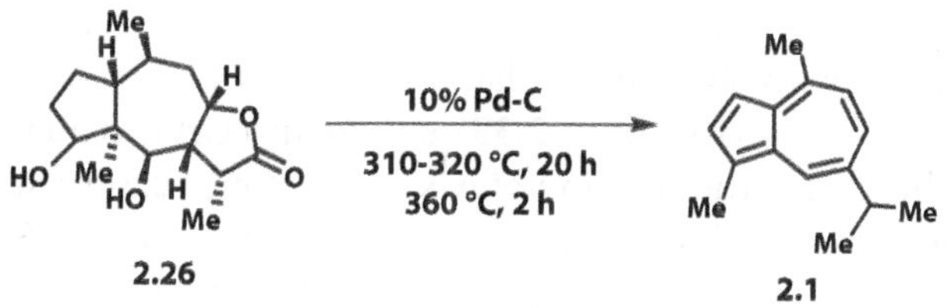

Scheme 2.1 Synthesis of guaiazulene from hexahydrohelenalin.

On the other hand, Sorm *et al.* published a sequential reaction strategy for guaiazulene total synthesis for the first time in 1951 [48]. The reported synthesis involves seven steps to access the target compound. They initiated the synthesis by using 1,4-dimethyl-6,7-dihydroxyindane (**2.27**) as the starting material where compound **2.27** was transformed into 1,4-dimethylhydrindane-6,7-diol (**2.28**) by performing a catalytic hydrogenation reaction using Raney-Nickel as a catalyst. Upon oxidation of dio (**2.28**) by periodic acid gave *β*-(2-carboxy-3-methylcyclopentyl)-butyric acid (**2.29**). Subsequently, the introduction of carbon atoms on both side chains produced *γ*-(2-carboxymethyl-3-methylcyclopentyl)-valeric acid (**2.30**). Thereafter, Fe powder and $Ba(OH)_2$ are used to perform ketonic ring closure on the valeric acid, resulting in 2,8-dimethyl bicyclo-(5.3.0)-decan-5-one (**2.31**). Alkylation of **2.31** using isopropyl lithium furnished tertiary alcohol (**2.32**). Subsequently, **2.32** undergoes dehydration in the presence of $KHSO_4$, yielding 2,8-dimethyl-5-isopropylbicyclo-(5.3.0)-decene (**2.33**). Finally, guaiazulene as a trinitrobenzene adduct was achieved by dehydrogenating **2.33** with sulfur at 230°C (Scheme 2.2).

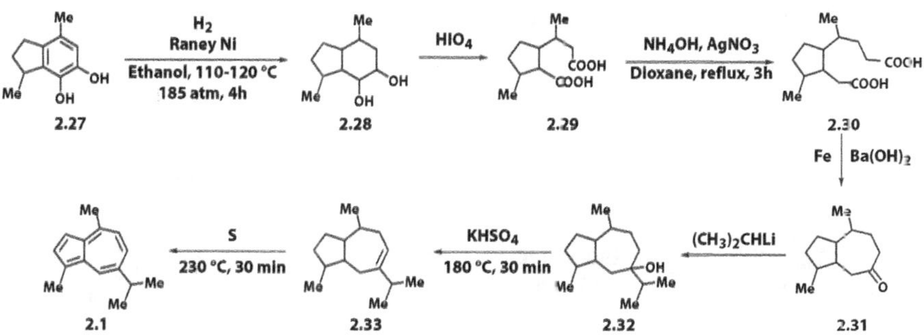

Scheme 2.2 Total synthesis of guaiazulene.

Later, Ukita *et al.* employed α-kessyl alcohol to synthesize guaiazulene. The methodology involves the oxidation of α-kessyl alcohol (**2.34**) into its corresponding α-kessyl ketone (**2.35**) employing CrO_3 oxidant. As illustrated in Scheme 2.3, the reaction of α-kessyl ketone (**2.35**) with isopropyl magnesium iodide and subsequent dehydrogenation of **2.36** over Pd-C catalyst resulted in the formation of guaiazulene (**2.1**) and 2-isopropylguaiazulene (**2.37**) as trinitrobenzene complex [49].

Based on the dehydrogenation of guaianolides, Sorm *et al.* described the synthesis of guaiazulene and chamazulene from Helenalin, which was isolated from *Helenium autumnale* [50]. Helenalin (**2.38**) readily underwent dehydrogenation in the presence of Se at 290 – 300°C (Scheme 2.4), yielding a mixture of guaiazulene (**2.1**) and chamazulene (**2.2**) as trinitrobenzene adducts.

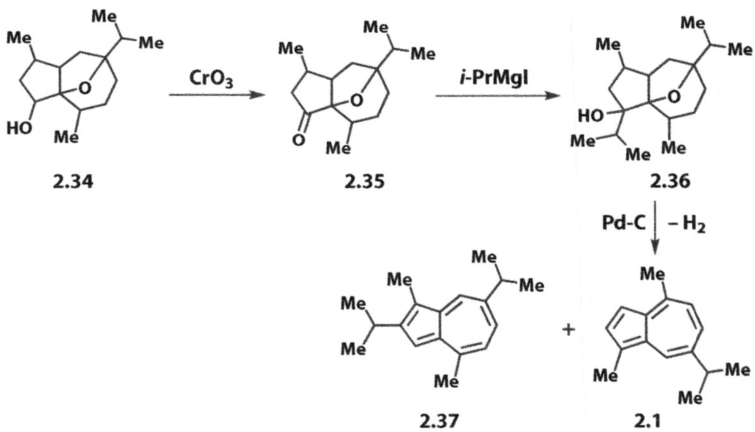

Scheme 2.3 Synthesis of guaiazulene from hexahydrohelenalin.

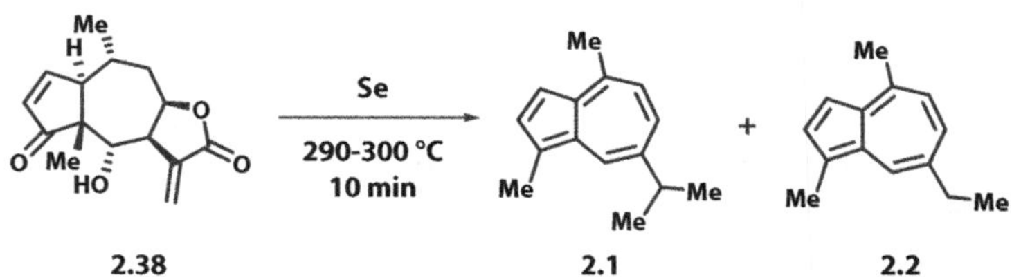

Scheme 2.4 Guaiazulene and chamazulene synthesis from helenalin.

Another route for the synthesis of guaiazulene was explicated by Jacob *et al.* It involved the reaction of *β*-keto ester, 2-carbethoxy-4-isopropyl-7-methylsuberone (**2.39**) with ethyl crotonate, which resulted in the formation of diester, ethyl(2-carbethoxy-4-isopropyl-7-methylsuberone)-2-*β*-butyrate (**2.40**). Hydrolysis of diester with CH_3COOH and HCl followed by the treatment with ethanol and H_2SO_4 led to the formation of ketoester, ethyl(4-isopropyl-7-methylsuberone)-2-*β*-butyrate (**2.41**). An equivalent *δ*-lactone (**2.42**) is obtained by cyclizing the ketoester with $NaBH_4$. Heating it in polyphosphoric acid (PPA) at 65°C produces the unsaturated ketone (**2.43**). The unsaturated ketone is reduced by $LiAlH_4$ and combined with $KHSO_4$ to form hexahydroguaiazulene. The required guaiazulene as trinitrobenzene complex is formed by dehydrogenating hexahydroguaiazulene with sulfur at 220°C (Scheme 2.5) [51].

A succinct method for guaiazulene based on the Huang-Minlon reduction of lactol (**2.44**) at 170 – 175°C was revealed by Mathur *et al.* in 1965. Similarly, guaiazulene (**2.1**) was formed as a trinitrobenzene adduct in 8% yield upon sulfur dehydrogenation of monol (**2.45**) (Scheme 2.6) [52].

Additionally, Herz *et al.* reported the synthesis of guaiazulene (**2.1**) from virginolide (**2.46**), a guaianolide isolated from *Helenium virginicum* Blake. This transformation involves the reduction of **2.46** by $LiAlH_4$ followed by dehydrogenation of the resulting compound using a 10% Pd-C catalyst (Scheme 2.7) [53].

Hoffmann *et al.* also produced guaiazulene by hydroformylating ketone derivative (**2.47**) with rhodium catalyst and then converting it into hexahydroazulene derivative (**2.49**) through intramolecular aldol condensation of

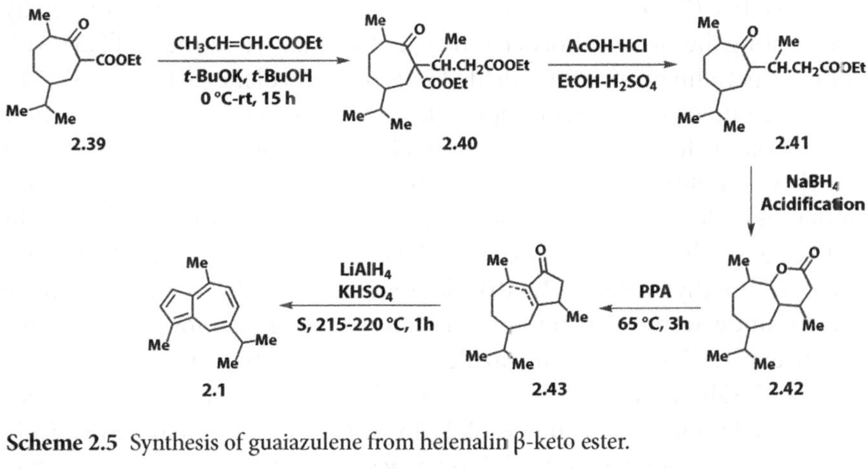

Scheme 2.5 Synthesis of guaiazulene from helenalin β-keto ester.

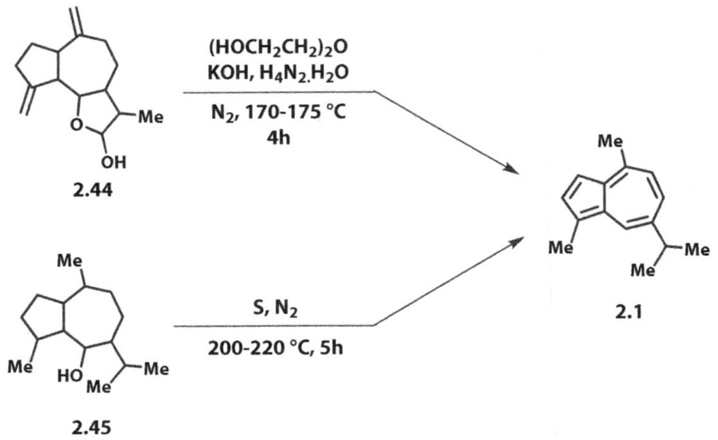

Scheme 2.6 Synthesis of guaiazulene from lactol and monol.

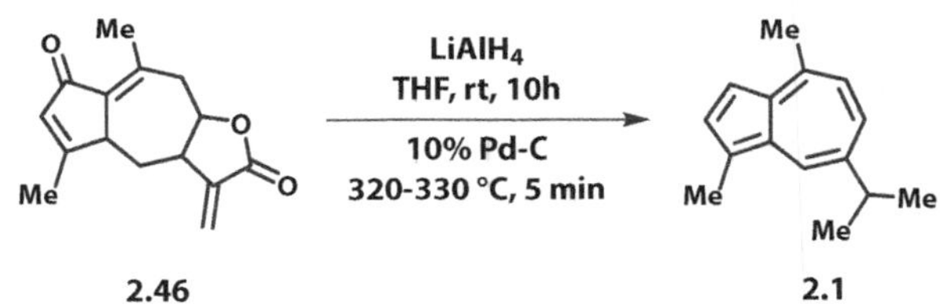

Scheme 2.7 Synthesis of guaiazulene from virginolide.

2.48 using ZnO_2 as catalyst in refluxing decalin. Thereafter, compound **2.49** was treated with CH_3MgCl in Tetrahydrofuran (THF) to yield the allyl alcohol derivative (**2.50**). The double-bonded isomer (**2.51**) was then formed by dehydrating the allyl alcohol derivative (**2.50**). Finally, treatment of **2.51** with sulfur in paraffin oil at 190°C, yielded 70% of guaiazulene (Scheme 2.8) [54].

In a completely different approach, Mukherjee *et al.* described a new methodology for the synthesis of guaiazulene and chamazulene using [6+4] cycloaddition reactions [55]. In this regard, thiophene dioxide (**2.52**) undergoes [6+4] cycloaddition with 6-methylaminofulvene (**2.53**) in refluxing pyridine to give a mixture of 7-ethyl-4-methylazulene (**2.54a**) and 4-ethyl-7-methylazulene (**2.54b**). Subsequently, 7-ethyl-4-methylazulene (**2.54a**) underwent Vilsmeier formylation at the C-1 position, resulting in a mixture of 1,4,7-trisubstituted azulene (**2.55a**) and its corresponding 1,5,8 isomer (**2.55b**). Finally, Wolff–Kishner reduction of **2.55a** produced guaiazulene (**2.1**) and chamazulene (**2.2**) in 51% and 50% yield respectively (Scheme 2.9a). Similar cycloaddition reactions occur between pyrrolidine

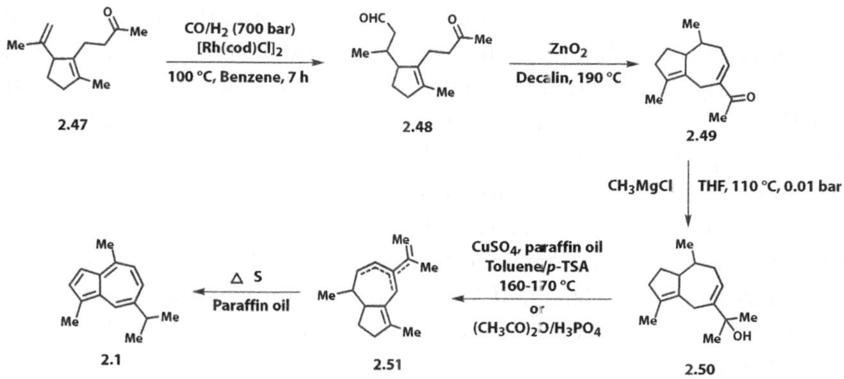

Scheme 2.8 Synthesis of guaiazulene.

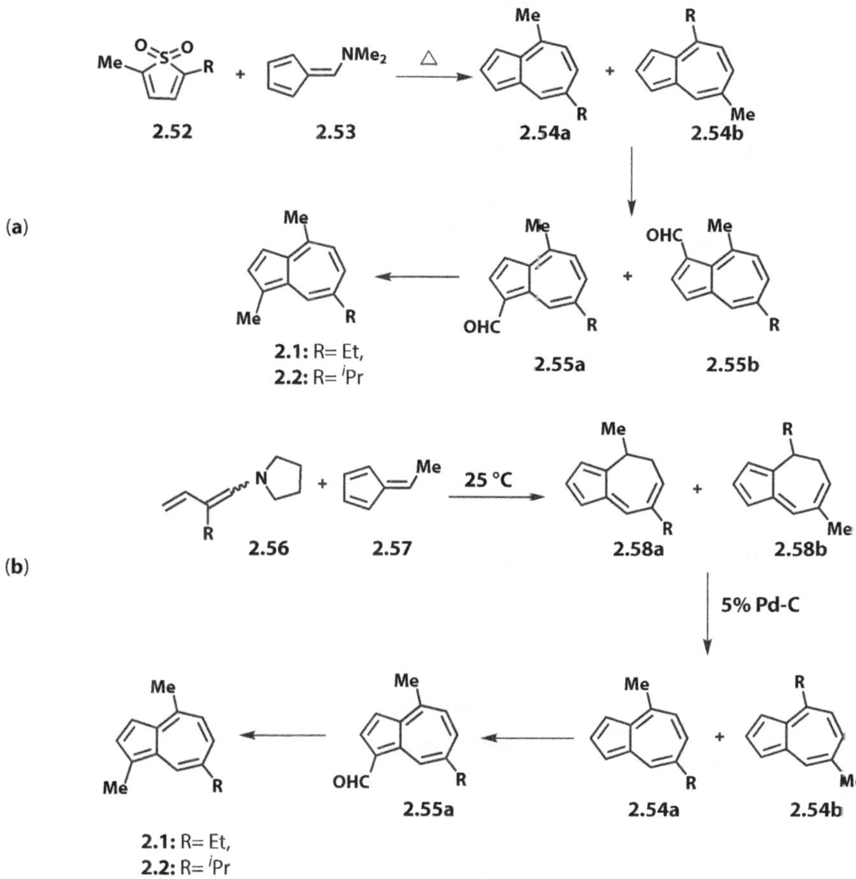

Scheme 2.9 Synthesis of guaiazulene and chamazulene *via* [6+4] cycloaddition reactions.

dienamines (**2.56**) and methylfulvene (**2.57**), yielding the dihydroazulene derivatives (**2.58a**, **2.58b**). Refluxing **2.58a** and **2.58b** with 5% Pd-C at 170°C in triglyme yielded the corresponding azulenes **2.54a** and **2.54b**. Finally, this azulene can be converted into guaiazulene and chamazulene by following the steps outlined in Scheme 2.9a.

Based on the dehydrogenation reaction strategy for the synthesis of azulene derivatives, Oshima *et al.* reported two new sesquiterpenoids, alismol and alisoxide. These are obtained from the drug takusha for the synthesis of guaiazulene [56]. Here, a 10% Pd-C is used to dehydrate both alismol (**2.59**) and alisoxide (**2.60**) to produce the desired guaiazulene (**2.1**) as blue oil (Scheme 2.10).

Sullivan *et al.* reported an alternative method that involves dehydrogenating guaia-1(5),6-diene, a sesquiterpene hydrocarbon derived from sponge *Halicondria* sp [57]. The dehydrogenation of **2.61** with a 10% Pd-C in xylene under reflux conditions led to the generation of guaiazulene (**2.1**) (Scheme 2.11).

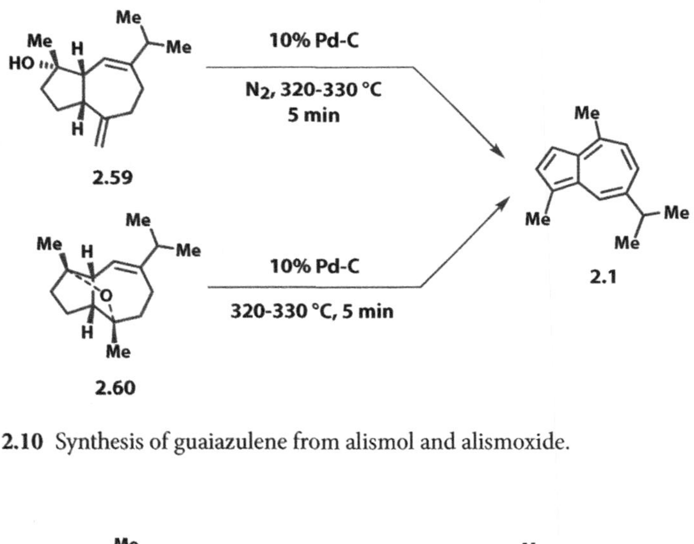

Scheme 2.10 Synthesis of guaiazulene from alismol and alismoxide.

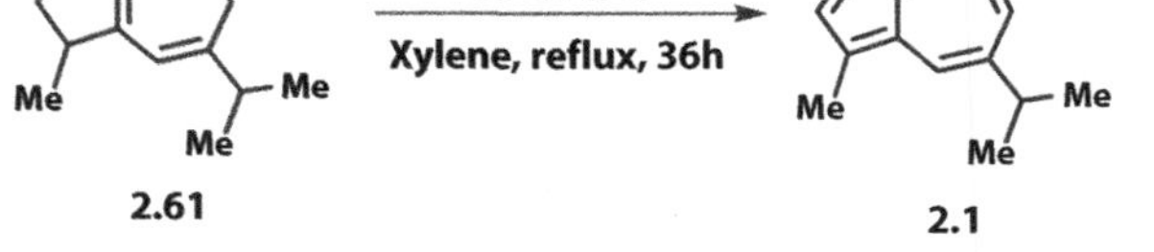

Scheme 2.11 Synthesis of guaiazulene from guaia-1(5),6-diene.

Later, Shono *et al.* proposed an intriguing electroreduction reaction employing γ-cyano ketones as a starting material to produce guaiazulene [58]. γ-cyano ketone (**2.64**) was produced *via* cyanoalkylation of β-keto ester (**2.63**) obtained from treatment of carvone (**2.62**) with allylcyanide followed by de-carbethoxylation. The electroreduction of **2.64** in the presence of isopropyl alcohol at 25°C gave α-hydroxy ketone (**2.65**) along with cyclopentanone derivative (**2.66**). Finally, dehydration of **2.65** followed by reduction with $LiAlH_4$ produces α,β-unsaturated ketone intermediate (**2.67**), upon treatment with *p*-TsOH and subsequent dehydrogenation with sulfur, provided guaiazulene (**2.1**) in 35% yield (Scheme 2.12).

The acid-catalyzed ethanolysis of di-1-azulenyl ketone entailed an entirely new synthesis of guaiazulenes. Herein, Saitoh *et al.* constructed the guaiazulene framework by treating di-1-azulenyl ketone (**2.68**) with *p*-toluenesulfonic acid in refluxing ethanol under an inert atmosphere of nitrogen along with the generation of ethyl-5-isopropyl-3,8-dimethylazulene-1-carboxylate (**2.69**) as a by-product (Scheme 2.13) [59].

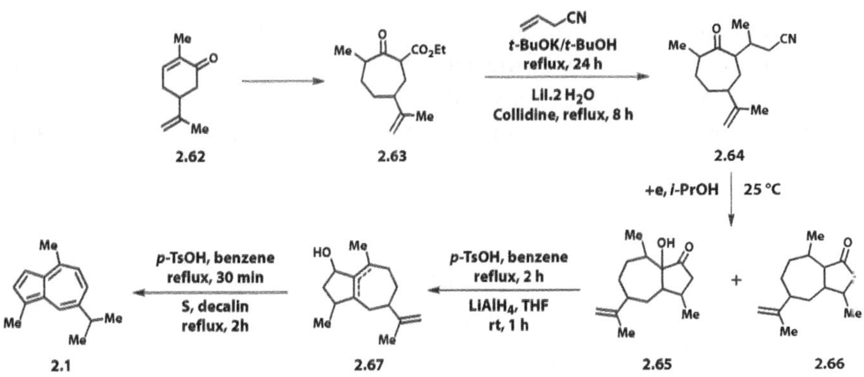

Scheme 2.12 Synthesis of guaiazulene from γ-cyano ketone.

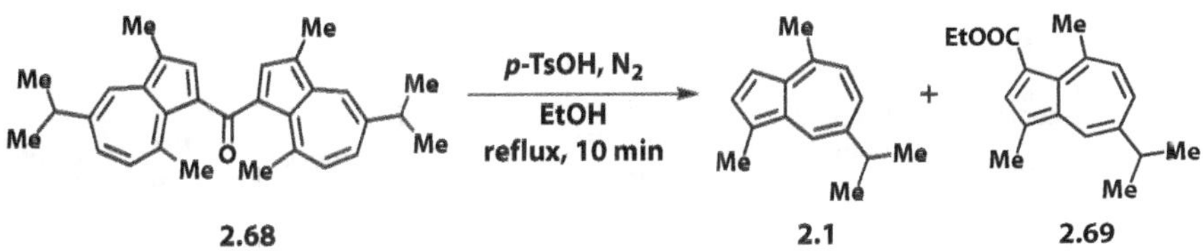

Scheme 2.13 Acid-catalyzed synthesis of guaiazulene.

Furthermore, Matsubara *et al.* reported a one-pot synthesis of guaiazulene from guaiol, an essential oil found in the wood of *Guaiacum officinalis* L [60]. Guaiol (**2.70**) undergoes dehydration-dehydrogenation in the presence of sulfur at 200 – 220°C to afford the desired guaiazulene (**2.1**) (Scheme 2.14).

As a result of discovering new guaiazulene synthesis techniques, Hansen *et al.* reported a thermolysis reaction of tricyclic compounds in polar aprotic solvent to afford guaiazulene [61]. Tricyclic dicarboxylates (**2.71**) undergo this transformation by treatment with 1,8-bis(dimethylamino) naphthalene in CH_3CN at 100°C. This results in the formation of guaiazulene (**2.1**), heptalene-1,2-dicarboxylate (**2.72**), and tricyclic compound (**2.73**) as red oil, with an overall yield of 28% (Scheme 2.15).

Based on the synthesis of guaiazulene from sesquiterpenes, de Groot *et al.* reported another sesquiterpene, (-)-isolenede isolated from the essential oil of *Citrus aurantifolia* and *Cistus ladaniferus* toward the synthesis of guaiazulene moiety [62]. Nonetheless, a variety of techniques based on the transformation of guaiane and aromadendrane sesquiterpenes into guaiazulene have been documented in the literature [57, 63]. However, in most cases, the yield is not provided, and the conversion is mostly utilized to demonstrate the existence of the azulene skeleton in an unidentified natural product. Additionally, a maximum yield of 6.3% has been observed for the dehydrogenation of aromadendrene to guaiazulene with sulfur [63b].

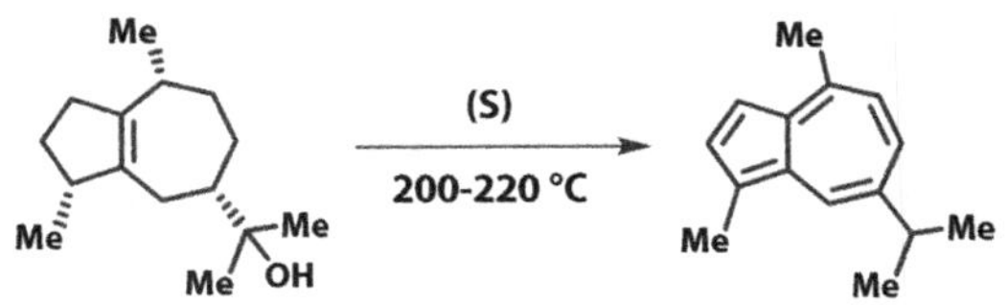

2.70 2.1

Scheme 2.14 Synthesis of guaiazulene from guaiol.

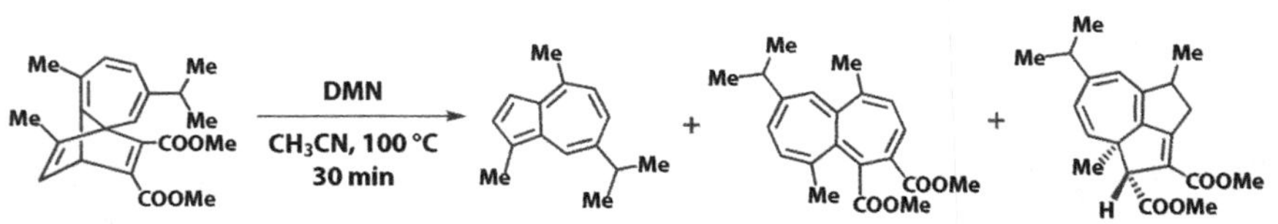

2.71 2.1 2.72 2.73

Scheme 2.15 Synthesis of guaiazulene via thermolysis of tricyclic dicarboxylates.

Interestingly, dehydrogenation of aromadendrane with dithioglycolate esters under thermal or photochemical conditions provided guaiazulene in 18% yield. Furthermore, an improved yield of 48% was observed when di-tertbutylperoxide was added to the reaction mixture [63f]. On the other hand, guaiazulene was produced in an enhanced yield of 22% by dehydrogenating a mixture of diene (**2.61, 2.75**) derived from (-)-isolenede (**2.74**) in the presence of mesitylene and sulfur at 160°C (Scheme 2.16).

In 2005, Depres *et al.* published an alternate method for the synthesis of guaiazulene from chlorotrienone [64]. In the synthesis of guaiazulene (**2.1**), 1,6-conjugate addition of chlorotrienone (**2.76**) has been exercised either by employing $i\text{Pr}_2\text{Zn}$ in the presence of CuOTf and trimethylsilyl chloride or by adding phenylselenyl chloride followed by oxidative elimination to regenerate the chlorotrienone system (**2.77**) in 62% yield. The resulting chlorotrienone (**2.77**) was then treated with $NaBH_4/CeCl_3$, the Burgess reagent, *p*-chloroanil followed by methylation with methylboronic acid, dpdb, and K_3PO_4 to yield the corresponding guaiazulene (**2.1**) in 94% yield (Scheme 2.17).

Recently, Takekuma *et al.* reported an effective method for generating guaiazulene from azulene and triethylarenes [65]. This reaction was initiated by treating azulene (**2.78**) and triethylarenes (**2.79–2.81**) in a mixed solvent of CH_3OH and CH_3CN in the presence of 36% HCl at 60°C for 3h (Scheme 2.18). Interestingly, the reaction proceeds by substituting an azulen-1-yl group for a guaiazulen-3-yl group, yielding guaiazulene (**2.1**)

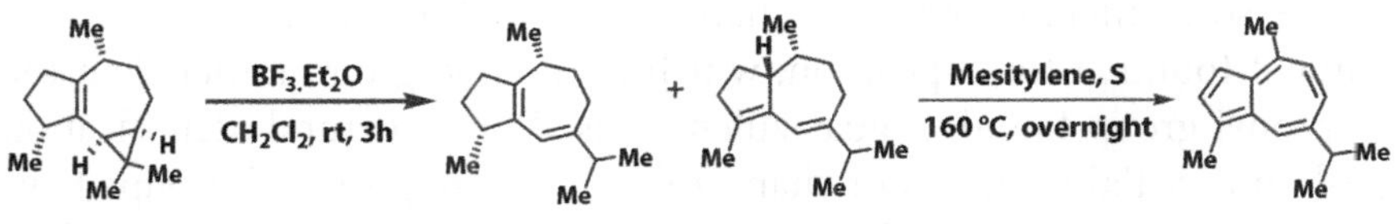

Scheme 2.16 Synthesis of guaiazulene *via* (-)-isolenede.

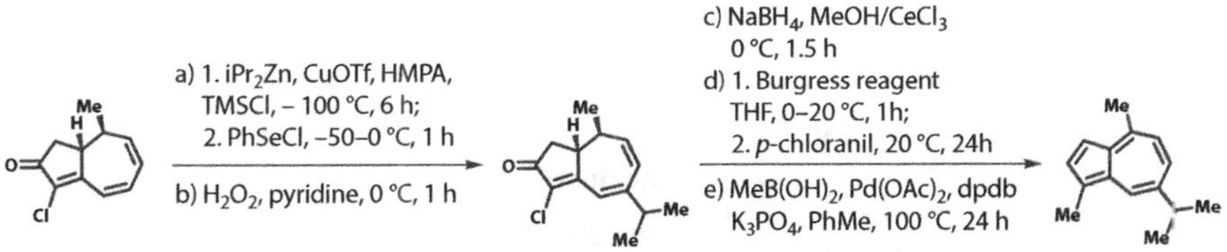

Scheme 2.17 Synthesis of guaiazulene *via* chlorotrienones.

2.78 2.1

2.79, R= $N(CH_3)_2$
2.80, R= OMe
2.81, R= $OCH_2CH_2N(CH_3)_2$

2.82, R= $N(CH_3)_2$
2.83, R= OMe
2.84, R= $OCH_2CH_2N(CH_3)_2$

Scheme 2.18 Synthesis of guaiazulene from azulene and triarylethylenes.

in excellent yield and its corresponding triarylethylene derivatives (**2.82**–**2.84**). Besides these products, the reaction between **2.78** and **2.79** provided two new triarylethylene and tetrarylethylene derivatives in 19% and 13% respectively, whereas the reaction between **2.78** and **2.80** afforded another triarylethylene in 19% yield.

2.4 Synthesis of Chamazulene

Chamazulene, a bicyclic sesquiterpene, is synthesized *via* a sequence of chemical reactions that provide the carbon framework and functional groups needed for the molecule's structure. Each step in the synthesis is designed to sequentially modify the molecule until the final product, chamazulene, is obtained.

The sole difference between chamazulene and guaiazulene is the substituent found at the 7-position, which is an ethyl group rather than the isopropyl group found in guaiazulene. Based on structural identification [66] and partial synthesis of chamazulene [66b, 67], several groups have devised alternative methods for the synthesis of chamazulenes by employing a reduction-dehydrogenation reaction strategy.

Initially, Takeda *et al.* reported a zinc-dust distillation of linderene for the synthesis of chamazulene (Scheme 2.19) [68]. In this strategy, linderene

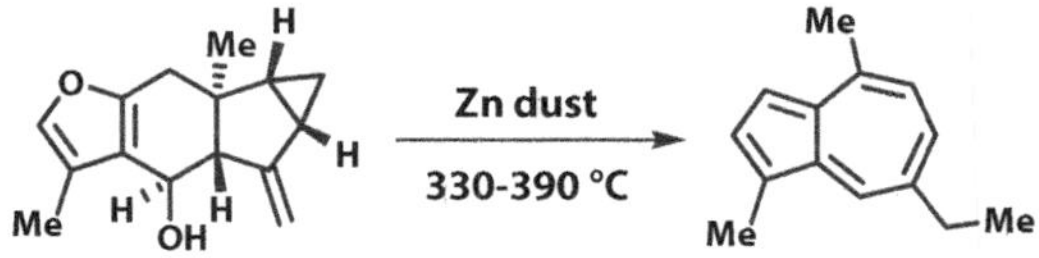

2.85 2.2

Scheme 2.19 Synthesis of chamazulene *via* zinc-dust distillation of linderene.

(**2.85**) was heated with zinc dust at 330 – 390°C for the synthesis of lindazulene as trinitrobenzene and picrate complex. According to spectroscopic data and other experimental findings, it was assumed that the structure of lindazulene corresponds to that of chamazulene (**2.2**).

Meanwhile, the first comprehensive synthesis of chamazulene (**Scheme 2.20**) was performed in 1954 by Novak *et al.* [69]. The reaction commences with the treatment of 2,3-dimethoxybenzoyl chloride (**2.86**) and di-*n*-propyl cadmium, which leads to the formation of 2,3-dimethoxybutyrophenone (**2.87**) in 72% yield. Next, **2.87** was treated with dimethylamine hydrochloride to afford the Mannich base (**2.88**), upon which exposure to malonic ester yields diester (**2.89**). Subsequently, hydrolysis and decarboxylation of **2.89** were performed to introduce the carboxylic functionality required to assemble the ring. This task was achieved by using NaOH and heat, thus allowing for the conversion of diester (**2.89**) to dimethoxybenzoyl caproic acid (**2.90**) in 67% yield. Reduction of caproic acid (**2.90**) with Pd/C furnishes the corresponding dimethoxybenzyl caproic acid (**2.91**), which is then subjected to polyphosphoric acid to facilitate the cyclization reaction to form the benzsuberane derivative (**2.92**).

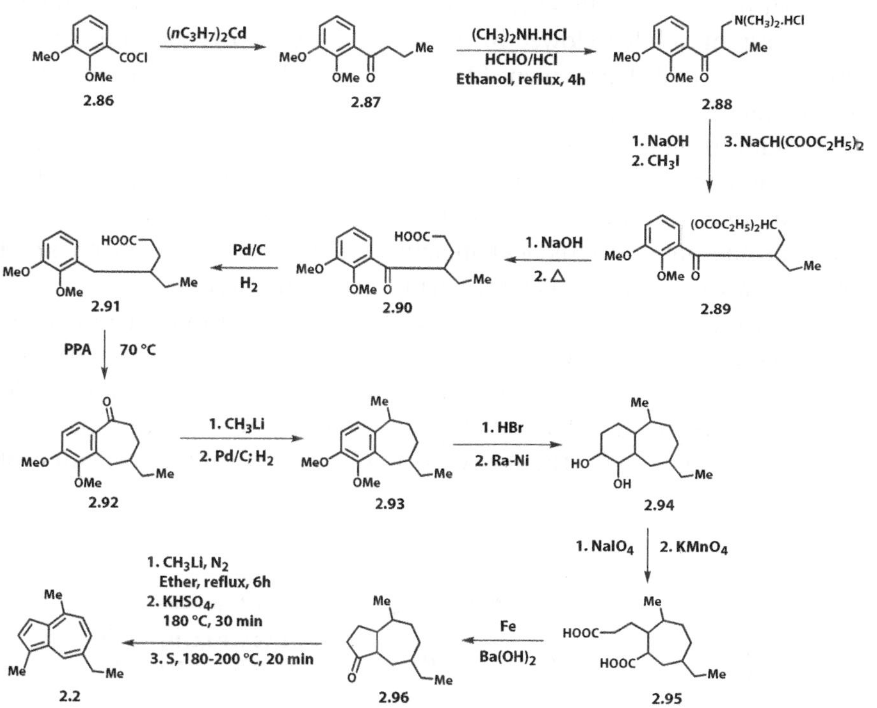

Scheme 2.20 Total synthesis of chamazulene.

Next, the introduction of a methyl group into the seven-membered ring of benzsuberane derivative (**2.92**) followed by hydrogenolysis gave cycloheptene derivative (**2.93**). Subsequently, the cycloheptene derivative (**2.93**) was demethylated using HBr, followed by hydrogenation in the presence of Raney-Nickel to produce the required diol intermediate (**2.94**). Thereafter, $NaIO_4$ and $KMnO_4$ are used to oxidize the diol intermediate, resulting in cycloheptyl carboxylic acid (**2.95**). This compound then goes through ring closure, most likely with the help of Fe and $Ba(OH)_2$, to produce bicycle-(5.3.0)-decanone (**2.96**) in 86% yield. Finally, **2.96** was transformed into carbinol by treatment with CH_3Li, followed by dehydration-dehydrogenation with $KHSO_4$ and sulfur-delivered chamazulene (**2.2**) as trinitrobenzene adduct. This comprehensive procedure describes how the benzsuberane derivative is changed into chamazulene by ring closures and functional group alterations. As a result, each step in the synthesis is crucial for building the complex structure of chamazulene from simpler starting materials.

In 1965, Minato *et al.* disclosed another reaction strategy by employing prehydro azulene derivative obtained from guaiol (**2.70**) as a starting material for the synthesis of chamazulene [70]. This reaction proceeds *via* the treatment of 6,10-dimethylbicyclo-(5.3.0)-decan-3-one **2.97** with ethylmagnesium iodide followed by dehydration-dehydrogenation of **2.98** to obtain the desired chamazulene (**2.2**) in 5.6% yield as 2,4,6-trinitrobenzene adduct (Scheme 2.21).

Sesquiterpene lactones are an important class of natural products with diverse biological activities and potential medicinal uses. The dehydrogenation of a wide variety of sesquiterpene lactones has been extensively used for one-pot synthesis of chamazulenes. However, all the synthesized products are characterized by their trinitrobenzene adduct.

In this regard, Mathur *et al.* reported a dehydrogenation reaction strategy by employing dihydrocostus lactone (**2.99** of Figure 2.4), isolated from costus root oil [52]. The desired chamazulene was produced in 8% yield by heating **2.99** with KOH and ethylene glycol in an inert atmosphere of nitrogen at 180–200°C. The low yield was attributed to the formation of by-products.

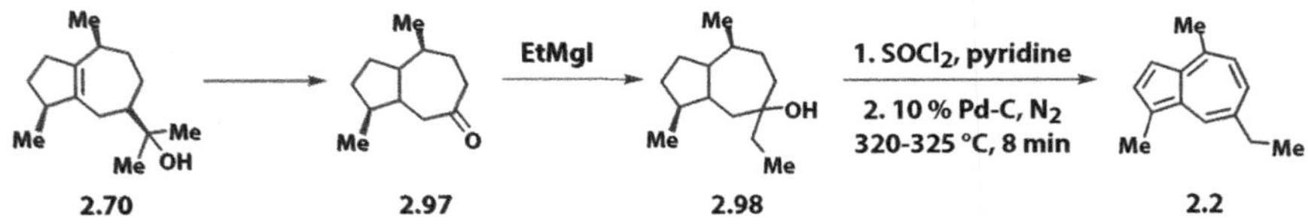

Scheme 2.21 Synthesis of chamazulene from perhydro azulene.

Similarly, Virginolide (**2.46** of Figure 2.4), a novel guaianolide derived from *Helenium virginicum* Black, was identified in 1967 by Herz and Santhanam. They reduced virginolide with $NaBH_4$ in THF at room temperature for 3h, followed by dehydrogenation in presence of a 10% Pd over charcoal, which led to the formation of chamazulene [53].

However, the direct synthesis of chamazulene *via* dehydrogenation reactions of a wide variety of sesquiterpene lactones isolated from nature has been reported in the literature. In most cases, chamazulenes are formed through the significant usage of Se and Pd catalysts supported over activated charcoal. In this regard, Pd-C catalyzed the dehydrogenation of four distinct guaianolides: eremanthine [71], carolenalin [72], isodehydrocostus lactone, and isozaluzanin C [73] (**2.100–2.103** of Figure 2.4) to chamazulenes has been reported by three groups. Similarly, chamazulene was produced in a single step by dehydrogenating olgin [74], grilactone [75], anobin [76], artausin [77], and atrenolide [78] over Se (**2.104–2.108** of Figure 2.4).

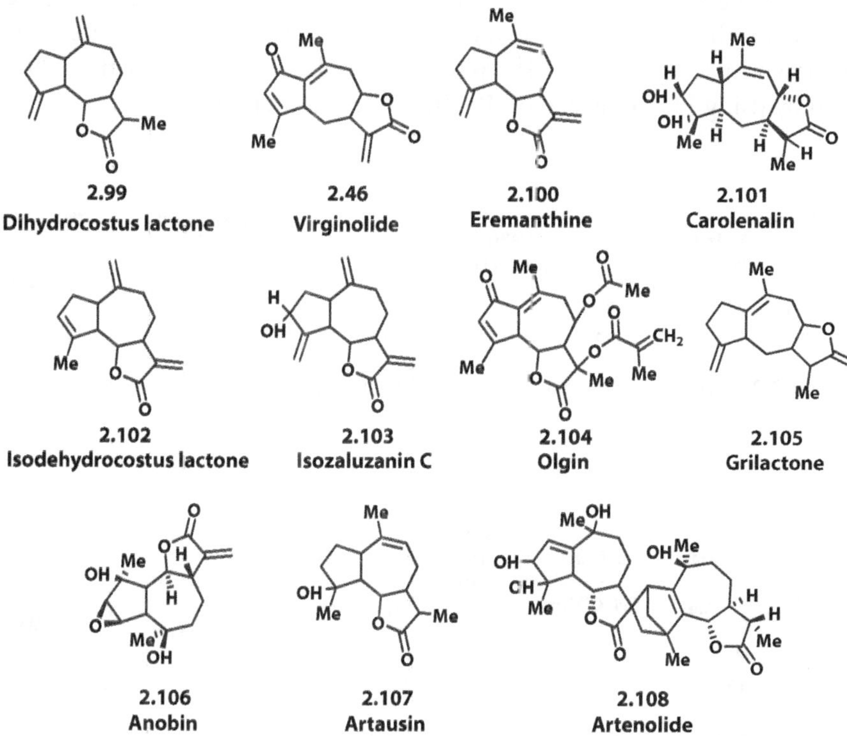

Figure 2.4 Sesquiterpene lactones employed for the synthesis of chamazulene.

In contrast to the direct synthesis of chamazulene from sesquiterpene lactones, modification of sesquiterpene and sesquiterpene lactones through sequential reaction strategy has also led to the generation of chamazulenes. In this regard, Hufford *et al.* synthesized chamazulenes from damsinic acid isolated from *Ambrosia arnbrosioides* (Cav.) Payne [79]. The reaction proceeds *via* the hydrogenation of damsinic acid (**2.109**) to its corresponding dihydrodamsinic acid (**2.110**) using 5% Pd-C in ethanol. Subsequently, $LiAlH_4$ reduction of **2.110** results in the formation of diol (**2.111**), which then undergoes dehydration followed by dehydrogenation to afford chamazulene (**2.2**) in a low yield (Scheme 2.22).

Thereafter, Shafizadeh *et al.* reported another route for synthesizing chamazulene *via* modification of viscidulin B, guaianolides isolated from *Artemisia cana* subspecies *viscidula* [80]. The reported synthetic methods involve two steps. Initially, viscidulin B (**2.112**) was treated with *p*-toluenesulfonic acid in the presence of acetic acid to yield a lactone derivative (**2.113**). Finally, chamazulene (**2.2**) was accomplished after heating lactone derivative (**2.113**) at elevated temperature (**Scheme 2.23**).

Based on the synthesis of guaiazulene from chlorotrienone, Depres *et al.* reported a similar reaction strategy for the synthesis of chamazulene (**2.2**) from the chlorotrienone system (**2.114**) by following the same steps as shown in scheme 2.17 [64]. Here, chamazulene (**2.2**) obtained an 89% yield by using Et_2Zn instead of iPr_2Zn in the first step (Scheme 2.24).

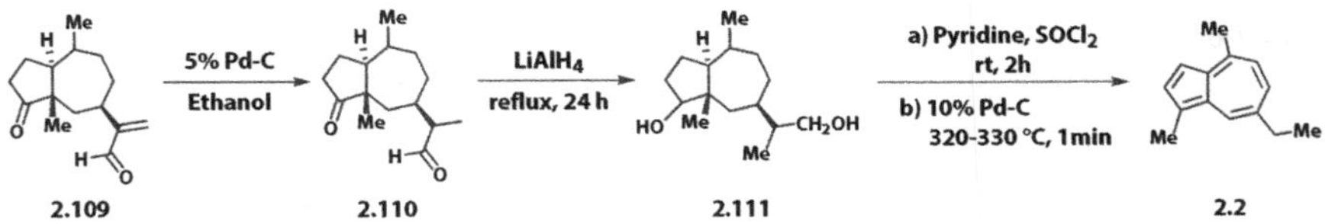

Scheme 2.22 Synthesis of chamazulene from damsinic acid.

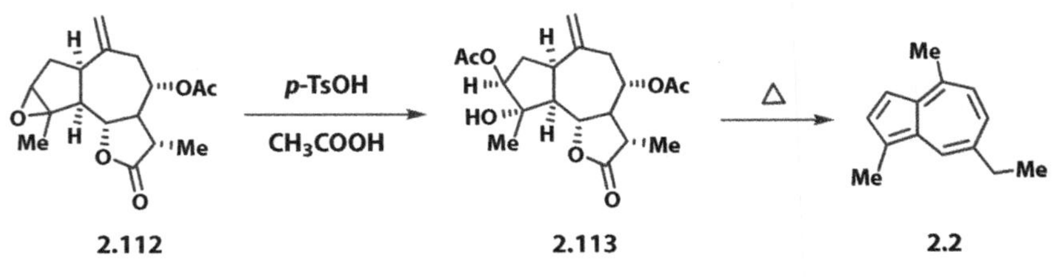

Scheme 2.23 Synthesis of chamazulene from viscidulin B.

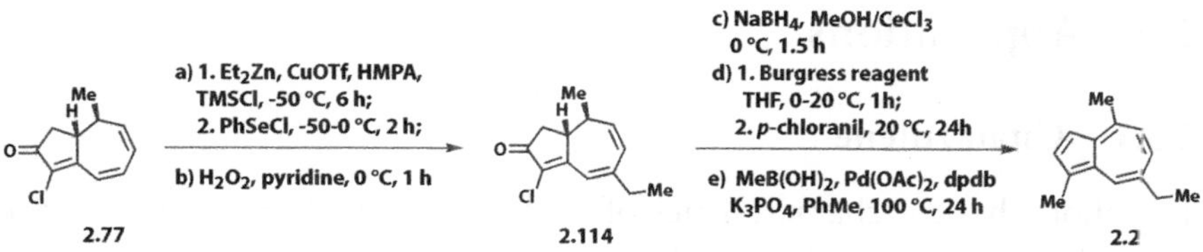

Scheme 2.24 Chamazulene synthesis from chlorotrienones.

Another proazulenic sesquiterpene lactone that acts as a precursor to chamazulenes is called matricin. In this regard, Ramadan *et al.* described the synthesis of chamazulene, which proceeds *via* hydrolytic degradation of matricin (**2.115**) to chamazulene carboxylic acid (**2.116**), followed by decarboxylation to afford the desired azulene derivative (Scheme 2.25) [81].

On the other hand, guaiane-type sesquiterpene, α-methylidene-γ-lactones can also serve as a useful scaffold for the synthesis of chamazulenes. In this context, Adekenov *et al.* employed the Heck cross-coupling reaction to provide a flexible route to chamazulene synthesis (Scheme 2.26) [82]. Here, the $Pd(OAc)_2$–(*o*-Tol)$_3$P system catalyzed the reaction between ludartin (**2.117**) and 2-bromotoluene (**2.118**) in the presence of Et_3N and Tetrabutylammonium bromide (TBAB) in Dimethylformamide (DMF) to produce the desired chamazulene (**2.2**) and its corresponding arylated product (**2.119**) in yields of 9% and 4%, respectively. Remarkably, when the reaction was performed without TBAB, only chamazulene was generated, albeit in a 9% yield.

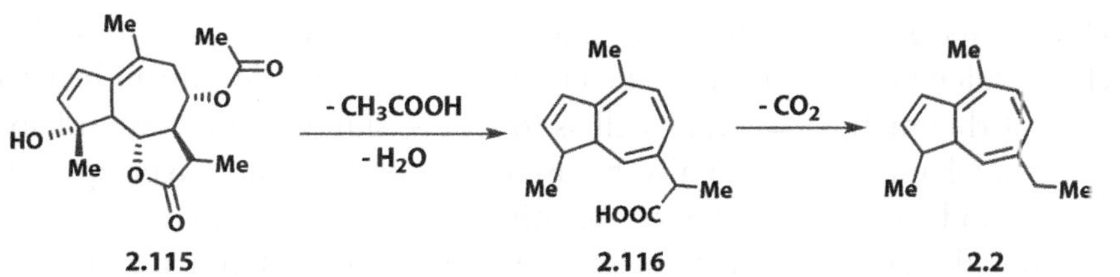

Scheme 2.25 Synthesis of chamazulene from matricin.

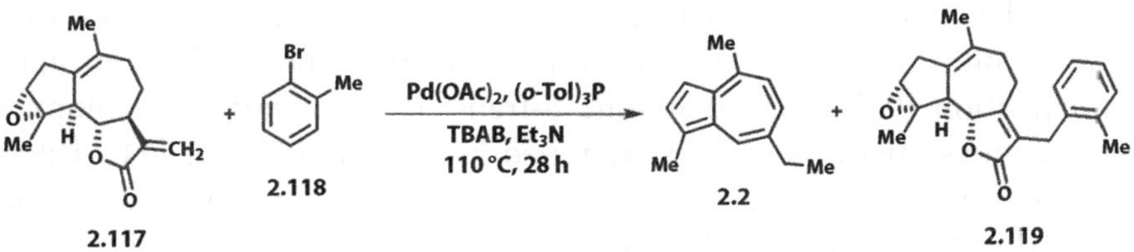

Scheme 2.26 Synthesis of chamazulene *via* Heck cross-coupling of ludartin and 2-bromotoluene.

2.5 Applications

2.5.1 Guaiazulene

The photophysical characteristics of azulene and its derivatives give rise to intriguing biological characteristics. Among these, guaiazulene was well-researched by numerous scientific groups, exhibiting diverse biological effects. In this regard, numerous biologically active compounds bearing guaiazulene have been synthesized *via* modification or conjugation of guaiazulene with other bioactive components. Anti-inflammatory, antispasmodic, and antibacterial properties of guaiazulene have been documented by Chen and Fiori's groups [31, 83]. Subsequently, Li's group has synthesized four series of guaiazulene derivatives that show remarkable anti-inflammatory, antiproliferative, and antiviral activities [84]. Furthermore, guaiazulene derivatives exhibit anticancer activities in addition to their notable anti-inflammatory capabilities [8]. Nevertheless, guaiazulene derivatives' antiviral activity [85] was not as widely recognized as azulene's antiviral activity. Additionally, the lipid peroxidation-inhibiting activities of guaiazulene was observed by Kourounakis *et al.* [86]. After that, guaiazulene's cosmetic properties and ability to scavenge hydroxyl radicals were reported by Zheng *et al.* [87] and Franchi *et al.* [88].

2.5.2 Chamazulene

Due to its excellent anti-inflammatory and antioxidant qualities [89], chamazulene is used in medicine and cosmetic industries. It can be used to treat rheumatic disease, as well as to reduce skin irritation and inflammation. Chamazulene can also be used as a coloring and flavoring agent in the food packaging industry. Additionally, chamazulene is widely used in aromatherapy products to help relieve headaches, insomnia, and stress. Safayhi *et al.* later revealed the antiallergic activities of chamomile oil [90], building on their discovery of the pharmacological properties of chamazulenes. Additionally, Rekka *et al.* [91] and Burits *et al.* [92] reported the antioxidant activity of chamazulene. Owlia *et al.* subsequently found that Iranian blue chamomile oil functions as an excellent regulator of lipid peroxidation [93]. The glutathione S-transferase activity of chamomile oil on mouse tissue was then examined by Lam and Zheng [94]. Later, Hernandez-Ceruelos *et al.* demonstrated chamomile oil's anti-mutagenic properties [95]. Recently, the cytotoxic impact of blue chamomile oil on human lungs, breasts, and prostate cancer cells was reported by Zu *et al.* [96].

2.6 Conclusion

In this chapter, we highlighted an updated perspective on naturally occurring azulene derivatives, specifically focusing on guaiazulenes and chamazulenes. The occurrence and isolation methods discussed in this chapter barely scratch the surface of the most remarkable developments in this important field of research. Moreover, it provides a comprehensive overview of different synthetic strategies and methodologies for isolating natural compounds containing azulene. Considering the importance of guaiazulene and chamazulene in diverging biological applications, we briefly discussed their biological relevance in pharmaceutical industry. In our view, this chapter will stimulate further advancements and improvements in this fascinating and dynamic field of natural product chemistry.

AI disclosure statement: The author confirms that no artificial intelligence (AI)-generated content or AI-assisted writing tools were used in the preparation of this chapter (including summaries and the abstract). All text, analysis, and conclusions are the result of the author's original work.

References

1. Anderson Jr., A.G. and Steckler, B.M., A Study of the visible absorption spectra and dipole moments of some 1- and 1,3-substituted azulenes. *J. Am. Chem. Soc.*, 81, 4941, 1959.
2. Bakun, B., Czarczynska-Goslinska, B., Goslinski, T., Lijewski, S., *In vitro* and *in vivo* biological activities of azulene derivatives with potential applications in medicine. *Med. Chem. Res.*, 30, 834, 2021.
3. Doimo, L., Azulenes, Costols and Γ-lactones from cypress-pines (*Callitris columellaris, C. glaucophylla* and *C. intratropica*) distilled oils and methanol extracts. *J. Essent. Oil Res.*, 13, 25, 2001.
4. Steer, R.P., Photophysics of molecules containing multiples of the azulene carbon framework. *J. Photochem. Photobiol. C*, 40, 68, 2019.
5. Xin, H. and Gao, X., Application of azulene in constructing organic optoelectronic materials: New tricks for an old dog. *ChemPlusChem*, 82, 945, 2017.
6. Lemal, D.M. and Goldman, G.D., Synthesis of azulene, a blue hydrocarbon. *J. Chem. Educ.*, 65, 923, 1988.
7. Zhang, L.-Y., Yang, F., Shi, W.-Q., Zhang, P., Li, Y., Yin, S.-F., Synthesis and antigastric ulcer activity of novel 5-isoproyl-3,8-dimethylazulene derivatives. *Bioorg. Med. Chem. Lett.*, 21, 5722, 2011.

8. Imanari, K., Hashimoto, M., Wakabayashi, H., Okudaira, N., Bandow, K., Nagai, J., Tomomura, M., Tomomura, A., Uesawa, Y., Sakagami, H., Quantitative structure-cytotoxicity relationship of zzulene amide derivatives. *Anticancer Res.*, 39, 3507, 2019.
9. Yang, M., Luo, J., Li, K., Hu, S., Ding, T., Liu, H., Sheng, P., Yang, M., Determination and pharmacokinetic study of guaiol in rat plasma by gas chromatography-mass spectrometry with selected ion monitoring. *J. Chromatogr. B Analyt. Technol. Biomed Life Sci.*, 1085, 30, 2018.
10. Souza, M.A., Scapinello, J., Guzatti, J.G.G., Scatolin, M., Martello, R., Schindler, M.S.Z., Calisto, J.F.F., Alves, B., Morgan, L.V., Oliveira, J.V., Magro, J.D., Muller, L.G., Antinociceptive effect and mechanism of supercritical carbon dioxide extract of *Aloysia gratissima* leaves in mice. *Biomed. J.*, 44, S63, 2021.
11. Newall, C.A., Anderson, L.A., Phillipson, J.D., *Herbal medicines: a guide for health-care professionals*, Pharmaceutical Press, 1996.
12. (a) Shaath, N.A. and Azzo, N.R., Essential oils of Egypt, in: *Food Flavors, Ingredients and Composition*, G. Charalambous (Ed.), pp. 591–603, Elsevier, Amsterdam, 1993; (b) Piccaglia, R. and Marotti, M., Characterization of several aromatic plants grown in northern Italy. *Flavour Fragance J.*, 8, 115–122, 1993; (c) Matos, F.J.A., Machado, M.I.L., Alencar, J.W., Craveiro, A.A., Constituents of brazilian Chamomile Oil. *J. Essent. Oil Res.*, 5, 337, 1993.
13. (a) Chialva, F., Gabri, G., Liddle, P.A.P., Ulian, F., Qualitative evaluation of aromatic herbs by direct headspace GC analysis. Applications of the method and comparison with the traditional analysis of essential oils. *J. High. Resolut. Chromatogr.*, 5, 182, 1982; (b) Srinivas, S.R., *Atlas of essential oils*, Srinivas, New York, 1986; (c) Zani, F., Massimo, G., Benvenuti, S., Bianchi, A., Albasini, A., Melegari, M., Vampa, G., Bellotti, A., Mazza, P., Studies on the genotoxic properties of essential oils with *Bacillus subtilis* rec-assay and *Salmonella*/microsome reversion assay. *Planta Med.*, 57, 237, 1991; (d) Lawrence, B.M., Progress in essential oils, Perfum. Flavor., p. 23, 1998.
14. Greche, H., Ismaili-Alaoui, M., Zrira, S., Benjilali, B., Belangar, B., Hajjaji, N., Composition of *Tanacetum annuum* L. Oil from Morocco. *J. Essent. Oil Res.*, 11, 343, 1999.
15. (a) Tucker, A.O., Maciarello, M.J., Sturtz, G., The essential oils of *Artemisia* 'Powis Castle' and its putative parents, *A. absinthium* and *A. arborescens*. *J. Essent. Oil Res.*, 5, 239, 1993; (b) Abderrahim, A., Belhamel, K., Chalchat, J.C., Figueredo, G., Chemical composition of the essential oil from *Artemisia arborescens* L. growing wild in algeria. *Rec. Nat. Prod.*, 4, 87, 2010.
16. Baczek, K., Kosakowska, O., Przybyl, J.L., Kuzma, P., Ejdys, M., Obiedzinski, M., Weglarz, Z., Intraspecific variability of yarrow (*Achillea millefolium* L. s.l.) in respect of developmental and chemical traits. *Herba Pol.*, 61, 37, 2015.
17. (a) Leung, A.Y. and Foster, S., *Encyclopedia of Common Natural Ingradients Used in Food, Drugs, and Cosmetics*, 2nd Ed., John Wiley & Sons Inc., New York, 1996; (b) Lawrence, B.M., *Essential oils 1988-1991*, Allured Publishing,

Wheaton, 1993; (c) Orav, A., Kalias, T., Liiv, M., Analysis of terpenoic composition of conifer needle oils by steam distillation/extraction, gas chromatography and gas chromatography-mass spectrometry. *Chromatographia*, 43, 215, 1996.
18. Fusetani, N., Matsunaga, S., Konosu, S., Bioactive marine metabolites I. Isolation of guaiazulene from the gorgonian *Euplexaura erecta*. *Experientia*, 37, 680, 1981.
19. Imre, S., Thomson, R.H., Yalhi, B., Linderazulene, a new naturally occurring pigment from the gorgonian. *Paramuricea chamaeleon*. *Experientia*, 37, 442, 1981.
20. Okuda, R.K., Klein, D., Kinnel, R.B., Li, M., Scheuer, P.J., Marine natural products: the past twenty years and beyond. *Pure Appl. Chem.*, 54, 1907, 1982.
21. Li, M.K.W. and Scheuer, P.J., A guaianolide pigment from a deep-sea gorgonian. *Tetrahedron Lett.*, 25, 2109, 1984.
22. Kapustina, I.I., Kalinovskii, A.I., Grebnev, B.B., Savina, A.S., Stonik, V.A., Sesquiterpenoids from the far-eastern gorgonaria *Stenogorgia* sp. *Chem. Nat. Compd.*, 45, 916, 2009.
23. Li, M.K.W. and Scheuer, P.J., Halogenated blue pigments of a deep-sea gorgonian. *Tetrahedron Lett.*, 25, 587, 1984.
24. Sakemi, S. and Higa, T., 2,3-Dihydrolinderazulene, a new bioactive azulene pigment from the gorgonian Acalycigorgia sp. *Experientia*, 43, 624, 1987.
25. Tanaka, J.I., Miki, H., Higa, T., Echinofuran, a new furanosesquiterpene from the gorgonian Echinogorgza praelonga. *J. Nat. Prod.*, 55, 1522, 1992.
26. Ochi, M., Kataoka, K., Tatsukawa, A., Kotsuhi, H., Shibata, K., Gorgiabisazulene and gorgiagallylazulene, two new guaiazulenoid pigments from a gorgonian *Acalycigorgia* sp. *Chem. Lett.*, 22, 2003, 1993.
27. Aknin, M., Rudi, A., Kashman, Y., Gaydou, E.M., Bebryazulene, a new guaiane metabolite from the Indian ocean gorgonian coral, *bebryce grandicalyx*. *J. Nat. Prod.*, 61, 1286, 1998.
28. Seo, Y., Rho, J.-R., Geum, N., Yoon, J.B., Shin, J., Isolation of guaianoid pigments from the gorgonian *Calicogorgia granulosa*. *J. Nat. Prod.*, 59, 985, 1996.
29. Reddy, N.S., Reed, J.K., Longley, R.E., Wright, A.E., Two new cytotoxic linderazulenes from a deep-sea gorgonian of the genus *Paramuricea*. *J. Nat. Prod.*, 68, 248, 2005.
30. Manzo, E., Ciavatta, M.L., Gresa, M.P.L., Gavagnin, M., Villani, G., Naik, C.G., Cimino, G., New bioactive hydrogenated linderazulene-derivatives from the gorgonian *Echinogorgia complexa*. *Tetrahedron Lett.*, 48, 2569, 2008.
31. Chen, D., Yu, S., van Ofwegen, L., Proksch, P., Lin, W., Anthogorgienes A-O, new guaiazulene-derived terpenoids from a Chinese gorgonian *Anthogorgia* species, and their antifouling and antibiotic activities. *J. Agric. Food. Chem.*, 60, 112, 2012.

32. Li, P., Liu, X., Zhu, H., Tang, X., Shi, X., Liu, Y., Li, G., Unusual inner-salt guaiazulene alkaloids and *bis*-sesquiterpene from the south china sea gorgonian *Muriceides collaris*. *Sci. Rep.*, 7, 7697, 2017.
33. Ahmad, V.U., Yasmeen, S., Ali, Z., Khan, M.A., Choudhary, M.I., Akhtar, F., Miana, G.A., Zahid, M., Taraxacin, a new guaianolide from *Taraxacum wallichii*. *J. Nat. Prod.*, 63, 1010, 2000.
34. Kawazoe, K., Tsubouchi, Y., Abdullah, N., Takaishi, Y., Shibata, H., Higuti, T., Hori, H., Ogawa, M., Sesquiterpenoids from *Artemisia gilvescens* and an anti-MRSA compound. *J. Nat. Prod.*, 66, 538, 2003.
35. Zhang, C.M., Fan, P.H., Li, M., Lou, H.X., Two new sesquiterpenoids from the rhizomes of *Curcuma xanthorrhiza*. *Helv. Chim. Acta*, 97, 1295, 2014.
36. Phan, M.G., Tran, T.T.N., Phan, T.S., Matsunami, K., Otsuka, H., Guaianolides from Curcuma kwangsiensis. *Phytochem. Lett.*, 9, 137, 2014.
37. (a) Vuorela, H., Holm, Y., Hiltunen, R., Harvala, T., Laitinen, A., Extraction of the volatile oil in chamomile flowerheads using supercritical carbon dioxide. *Flavour Fragr. J.*, 5, 81, 1990; (b) Zekovic, Z.P., Chamomile ligulate flowers in supercritical CO_2-extraction. *J. Essent. Oil Res.*, 12, 85, 2000; (c) Povh, N.P., Marques, M.O.M., Meireles, M.A.A., Supercritical CO_2 extraction of essential oil and oleoresin from chamomile (*Chamomilla recutita* [L.] Rauschert). *J. Supercrit. Fluids*, 21, 245, 2001.
38. Kotnik, P., Skerget, M., Knez, Z., Supercritical fluid extraction of chamomile flower heads: Comparison with conventional extraction, kinetics and scale-up. *J. Supercrit. Fluids*, 43, 192, 2007.
39. (a) Mwaniki, J.M., Mwazighe, F.M., Kamau, G.N., Analysis of blue chamomile essential oil produced by multi-solvent solvent extraction clevenger distillation method. *Afr. J. Phys. Sci.*, 2, 2313, 2015; (b) Ghasemi, M., Babaeian Jelodar, N., Modarresi, M., Bagheri, N., Jamali, A., Increase of chamazulene and α-bisabolol contents of the essential oil of german chamomile (*Matricaria chamomilla* L.) using salicylic ccid treatments under normal and heat stress conditions. *Foods*, 5, 56, 2006.
40. Tschiggerl, C. and Bucar, F., Guaianolides and volatile compounds in hamomile tea. *Plant Foods Hum. Nutr.*, 67, 129, 2012.
41. El-Assri, E.-M., Eloutassi, N., El Barnossi, A., Bakkari, F., Hmamou, A., Bouia, A., Wild Chamomile *(Matricaria recutita* L*)* from the Taounate Province, Morocco: Extraction and valorisation by evaluation of the antibacterial activity of its essential oils. *Trop. J. Nat. Prod. Res.*, 5, 883, 2021.
42. Salamon, I., Ibraliu, A., Kryvtsova, M., Essential Oil Content and Composition of the Chamomile Inflorescences (*Matricaria recutita* L.) Belonging to Central Albania. *Horticulturae*, 9, 47, 2023.
43. Capuzzo, A., Occhipinti, A., Maffei, M.E., Antioxidant and radical scavenging activities of chamazulene. *Nat. Prod. Res.*, 28, 2321, 2014.
44. Abdelhameed, R.M., Abdel-Gawad, H., Taha, M., Hegazi, B., Separation of bioactive chamazulene from chamomile extract using metal-organic framework. *J. Pharm. Biomed. Anal.*, 146, 126, 2017.

45. Ornano, L., Venditti, A., Ballero, M., Sanna, C., Quassinti, L., Bramucci, M., Lupidi, G., Papa, F., Vittori, S., Maggi, F., Bianco, A., Chemopreventive and antioxidant activity of the chamazulene-rich essential oil obtained from *Artemisia arborescens* L. growing on the Isle of La Maddalena, Sardinia, Italy. *Chem. Biodivers.*, 10, 1464, 2013.
46. (a) Plattner, P.A. and Lemay, L., Zur Kenntnis der Sesquiterpene. (44. Mitteilung). Über das Kohlenstoff-Gerüst des Guajols und des Guaj-azulens. *Helv. Chim. Acta*, 23, 897, 1940; (b) Plattner, P.A. and Magyar, G., Zur Kenntnis der Sesquiterpene. (52. Mitteilung). Abbau des Dihydro-guajols mit Chromsäure. Bereitung des 1,4,7-Trimethyl-azulens. *Helv. Chim. Acta*, 25, 581, 1942; (c) Plattner, P.A., Fürst, A., Marti, L., Schmid, H., Zur Kenntnis der Sesquiterpene und Azulene. 87. Mitteilung. Synthese des Guaj-azulens I. *Helv. Chim. Acta*, 32, 2137, 1949; (d) Plattner, P.A., Fürst, A., Marti, L., Zur Kenntnis der sesquiterpene und azulene. 88. Mitteilung. Synthese des Guaj-azulens II. *Helv. Chim. Acta*, 32, 2452, 1949; (e) Günthard, H.H. and Plattner, P.A., Zur Kenntnis der sesquiterpene und azulene. 84. Mitteilung. infrarotspektren der azulene. *Helv. Chim. Acta*, 32, 284, 1949.
47. Adams, R. and Herz, W., Helenalin. III. Reduction and dehydrogenation. *J. Am. Chem. Soc.*, 71, 2554, 1949.
48. Sorm, F., Gut, J., Hlavnička, J., Kučera, J., Šedivý, L., On Terpenoids. XXV. The total dynthesis of S-guaiazulene. *Collect. Czech. Chem. Commun.*, 16, 168, 1951.
49. Ukita, T., Watanabe, H., Miyazaki, M., Studies on azulenes. I. 2- and 3-substituted S-Guaiazulene. *J. Am. Chem. Soc.*, 76, 4584, 1954.
50. Herout, V., Romanuk, M., Sorm, F., On Terpenoids. LXXI*. Helenalin: A future lactone of the guaianolide type. *Collect. Czech. Chem. Commun.*, 21, 1359, 1956.
51. Jacob, T.M., Vatakencherry, P.A., Dev, S., Azulenes and related substances–XII: A new azulene synthesis-guaiazulene and Se-guaiazulene. *Tetrahedron*, 20, 2821, 1964.
52. Mathur, S.B., Hiremath, S.V., Kulkarni, G.H., Kelkar, G.R., Bhattacharyya, S.C., Simonovic, D., Rao, A.S., Terpenoids-LXX: Structure of dehyrocostus lactone. *Tetrahedron*, 21, 3575, 1965.
53. Herz, W. and Santhanam, P., Constituents of Helenium species. XX. Virginolide, a new guaianolide from *Helenium virginicum*. *J. Org. Chem.*, 32, 507, 1967.
54. Hoffmann, W. and Siegel, H., 7-Acylhexahydroazulene eine neue synthese des guajazulens. *Tetrahedron Lett.*, 16, 533, 1975.
55. Mukherjee, D., Dunn, L.C., Houk, K.N., Efficient guaiazulene and chamazulene syntheses involving [6 + 4] cycloadditions. *J. Am. Chem. Soc.*, 101, 251, 1979.
56. Oshima, Y., Iwakawa, T., Hikino, H., Alismol and alismoxide, sesquiterpenoids of *Alisma* rhizomes. *Phytochemistry*, 22, 183, 1983.

57. Sullivan, B.W., Faulkner, D.J., Okamoto, K.T., Chen, M.H.M., Clardy, J., (6R,7S)-7-amino-7,8-dihydro-α-bisabolene, an antimicrobial metabolite from the marine sponge *Halichondria* sp. *J. Org. Chem.*, 51, 5134, 1986.
58. Shono, T., Kise, N., Fujimoto, T., Tominaga, N., Morita, H., Electroorganic chemistry. 140. Electroreductively promoted intra- and intermolecular couplings of ketones with nitriles. *J. Org. Chem.*, 57, 7175, 1992.
59. Saitoh, M., Hashimoto, K., Nakazawa, T., Sugihara, Y., Acid-catalyzed ethanolysis of di-l-azulenyl ketones. *Tetrahedron Lett.*, 34, 3563, 1993.
60. Matsubara, Y., Yamamoto, H., Nozoe, T., Oxidation products of Guaiazulene and other azulenic hydrocarbons, in: *Studies in Natural Products Chemistry*, vol. 14, A. Rahman (Ed.), pp. 313–354, Elsevier, 1994.
61. Fallahpour, R.-A. and Hansen, H.-J., Formation and thermal rearrangement of dimethyl Tricyclo[6.2.2.0]dodeca-2,4,6,9,11-pentaene-9,10-dicarboxylates. *Helv. Chim. Acta*, 78, 1933, 1995.
62. Moreno-Dorado, F.J., Lamers, Y.M.A.W., Mironov, G., Joannes, B.P.A., Wijnberg, J.B.P.A., de Groot, A., Chemistry of (+)-aromadendrene. Part 6: Rearrangement reactions of ledene, isoledene and their epoxides. *Tetrahedron*, 59, 7743, 2003.
63. (a) Briggs, L.H. and Short, W.F., The chemistry of aromadendrene. Part I. *J. Chem. Soc.*, 2524, 1928; (b) Radcliffe, C.B. and Short, W.F., Aromadendrene. Part II. *J. Chem. Soc.*, 1200, 1938; (c) Treibs, W., Über bi- und polycyclische Azulene. III. Mitteil.: Der Patchouli-alkohol, ein tricyclischer Azulen-bildner. *Liebigs Ann. Chem.*, 564, 141, 1949; (d) Treibs, W. and Merkel, D., Über die Gurjunene, II. α-Gurjunen. *Liebigs Ann. Chem.*, 617, 129, 1958; (e) Ehret, C. and Ourisson, G., Le γ-gurjunene, structure et configuration: Isomerisation de l'α-gurjunene. *Tetrahedron*, 25, 1785, 1969; (f) Yamin, H. and Pallaud, R., *C. R. Hebd. Seances Acad. Sci. Ser. C*, 268, 1881, 1969; (g) Batey, I.L., Hellyer, R.O., Pinhey, J.T., The structure of squamulosone, a new sesquiterpene ketone from *Phebalium squamulosum*. *Aust. J. Chem.*, 24, 2173, 1971.
64. Carret, S., Blanc, A., Coquerel, Y., Berthod, M., Greene, A.E., Deprés, J.P., Approach to the Blues: A highly flexible route to the Azulenes. *Angew. Chem. Int. Ed.*, 44, 5130, 2005.
65. Takekuma, S., Kaibara, M., Minematsu, T., Takekuma, H., Reaction of azulene with 1,2-bis[4-(dimethylamino)phenyl]-1,2-ethanediol in a mixed solvent of methanol and acetonitrile in the presence of hydrochloric acid: a facile one-pot synthesis and properties of new triarylethylenes possessing an azulen-1-yl group. *Tetrahedron*, 67, 4780, 2011.
66. (a) Sorm, F., Novák, J., Herout, V., On terpenes., L.I., The composition of chamazulene. *Collect. Czech. Chem. Commun.*, 18, 527, 1953; (b) Meisels, A. and Weizmann, A., The Structure of Chamazulene. *J. Am. Chem. Soc.*, 75, 3865, 1953; (c) Sorm, F., Herout, V., Takeda, K., On terpenes. LVI. On the identity of lindazulene and chamazulene. *Collect. Czech. Chem. Commun.*, 19, 186, 1954.

67. Takeda, K., Kubota, T., Nagata, W., Studies on seven-membered ring derivatives. III: Partial synthesis of 1, 4-Dimethyl-7-ethylazulene (Lindazulene). *Pharm. Bull.*, 1, 241, 1953.
68. Takeda, K. and Nagata, W., Components of the root of *Lindera strychnifolia* Vill. V.: Azulenes isolated from Linderene by zinc-dust distillation. *Pharm. Bull.*, 1, 164, 1953.
69. Novak, J., Sorm, F., Sicher, J., On Terpenoids. LXIII*. Total synthesis of chamazulene. A simple general synthesis of 1,4,7-trisubstituted azulenes. *Collect. Czech. Chem. Commun.*, 19, 1264, 1954.
70. Minato, H., Ishikawa, M., Nagasaki, T., Studies on Sesquiterpenoids. X*. Small scale dehydrogenation of some perhydro-azulene and -naphthalene derivatives. *Chem. Pharm. Bull.*, 13, 717, 1965.
71. Vichnewski, W. and Gilbert, B., Schistosomicidal sesquiterpene lactone from *Eremanthus elaeagnus*. *Phytochemistry*, 11, 2563, 1972.
72. Furukawa, H., Lee, K.-H., Shingu, T., Meck, R., Piantadosi, C., Carolenin and Carolenalin, two new guaianolides in *Helenium autumnale* from North Carolina. *J. Org. Chem.*, 38, 1722, 1973.
73. Kalsi, P.S., Sharma, S., Kaur, G., Isodehydrocostus lactone and isozaluzanin C, two guaianolides from *Saussurea lappa*. *Phytochemistry*, 22, 1993, 1983.
74. Konovalova, O.A., Ban'kovskii, A.I., Rybalko, K.S., Sheichenko, V.I., Zakharov, P.I., Pimenov, M.G., Sesquiterpene lactones from Ferula olgae. *Chem. Nat. Compd.*, 8, 623, 1972.
75. Nurmukhamedova, M.R., Kasymov, S.Z., Melibaev, S., Grilactone from *Ferula penninervis*. *Chem. Nat. Compd.*, 18, 244, 1982.
76. Adekenov, S.M., Mukhametzhanov, M.N., Kagarlitskii, A.D., Turmukhambetov, A.Z., A chemical investigation of *Achillea nobilis*. *Chem. Nat. Compd.*, 20, 568, 1984.
77. Adekenov, S.M., Aituganov, K.A., Golovtsov, N.I., Artausin, a new sesquiterpene lactone from *Artemesia austriaca*. *Chem. Nat. Compd.*, 23, 127, 1987.
78. Ovezdurdyev, A., Abdullaev, N.D., Yusupov, M.I., Kasymov, S.Z., Artenolide-A new disesquiterpenoid from *Artemisia absinthium*. *Chem. Nat. Compd.*, 23, 553, 1987.
79. Doskotch, R.W. and Hufford, C.D., Antitumor agents. III. Structure of damsinic acid, a new sesquiterpene from Ambrosia ambros (Cav.) Paynelioides. *J. Org. Chem.*, 35, 486, 1970.
80. Shafizadeh, F. and Bhadane, N.R., Chemical constituents of sagebrush. V. Sesquiterpene lactones of sagebrush. New guaianolides from *Artemisia cana* subspecies *viscidula*. *J. Org. Chem.*, 37, 3168, 1972.
81. Ramadan, M., Goeters, S., Watzer, B., Krause, E., Lohmann, K., Bauer, R., Hempel, B., Imming, P., Chamazulene carboxylic acid and matricin: A natural profen and its natural prodrug, identified through similarity to synthetic drug substances. *J. Nat. Prod.*, 69, 1041, 2006.
82. Kihkentayeva, A.S., Shults, E.E., Gatilov, Y.V., Patrushev, S.S., Karim, S., Atazhanova, G.A., Adekenov, S.M., Synthetic transformations of

sesquiterpene lactones 10*. Synthesis of 13-Arylguaianolides. *Chem. Heterocycl. Compd.*, 52, 788, 2016.
83. Fiori, J., Teti, G., Gotti, R., Mazzotti, G., Falconi, M., Cytotoxic activity of guaiazulene on gingival fibroblasts and the influence of light exposure on guaiazulene-induced cell death. *Toxicol. Vitro*, 25, 64, 2011.
84. Ma, Z., Han, X., Ren, J., Liu, K., Zhang, W., Li, G., Design, synthesis, and biological activity of guaiazulene derivatives. *Chem. Biodivers.*, 20, e202201174, 2023.
85. Iuliano, A.D., Roguski, K.M., Chang, H.H., Muscatello, D.J., Palekar, R., Tempia, S., Cohen, C., Gran, J.M., Schanzer, D., Cowling, B.J., Wu, P., Kyncl, J., Ang, L.W., Park, M., Redlberger- Fritz, M., Yu, H., Espenhain, L., Krishnan, A., Emukule, G., van Asten, L., Pereira da Silva, S., Aungkulanon, S., Buchholz, U., Widdowson, M.-A., Bresee, J.S., Estimates of global seasonal influenza-associated respiratory mortality: a modelling study. *Lancet*, 391, 1285, 2018.
86. Kourounakis, A.P., Rekka, E.A., Kourounakis, P.N., Antioxidant activity of guaiazulene and protection against paracetamol hepatoxicity in rats. *J. Pharm. Pharmacol.*, 49, 938, 1997.
87. Zheng, Z.F., Development and application of S-guaiazulene and Na S-guaiazulene-3-sulfonate in Japan. *Flavour Fragance Cosmet.*, 5, 18, 2001.
88. Franchi, E., Ingrosso, G., Marchetti, F., Pinzino, C., Guaiazulene-based phenolic radical scavengers: Synthesis, properties, and EPR studies of their reaction with oxygen- centred radicals. *Tetrahedron*, 59, 5003, 2003.
89. Sizova, N.V., Composition and antioxidant activity of essential oils containing azulene derivatives. *Pharm. Chem. J.*, 46, 369, 2012.
90. Safayhi, H., Sabieraj, J., Sailer, E.R., Ammon, H.P., Chamazulene: An antioxidant-type inhibitor of leukotriene B_4 formation. *Planta Med.*, 60, 410, 1994.
91. Rekka, E.A., Kourounakis, A.P., Kourounakis, P.N., Investigation of the effect of chamazulene on lipid peroxidation and free radical processes. *Res. Commun. Mol. Pathol. Pharmacol.*, 92, 361, 1996.
92. Burits, M., Asres, K., Bucar, F., The antioxidant activity of the essential oils of *Artemisia afra*, *Artemisia abyssinica* and *Juniperus procera*. *Phytother. Res.*, 15, 103, 2001.
93. Owlia, P., Rasooli, I., Saderi, H., Antistreptococcal and antioxidant activity of essential oil from *Matricaria chamomilla* L. *Res. J. Biol. Sci.*, 2, 155, 2007.
94. Lam, L.K. and Zheng, B., Effects of essential oils on glutathione S-transferase activity in mice. *J. Agric. Food. Chem.*, 39, 660, 1991.
95. Hernandez-Ceruelos, A., Madrigal-Bujaidar, E., de la Cruz, C., Inhibitory effect of chamomile essential oil on the sister chromatid exchanges induced by daunorubicin and methyl methanesulfonate in mouse bone marrow. *Toxicol. Lett.*, 135, 103, 2002.
96. Zu, Y., Yu, H., Liang, L., Fu, Y., Efferth, T., Liu, X., Wu, N., Activities of ten essential oils towards *Propionibacterium* acnes and PC-3, A-549 and MCF-7 cancer cells. *Molecules*, 15, 3200, 2010.

3

Synthesis of Azulene and Its Functionalized Derivatives

G.B. Dharma Rao[1], Aditi Tiwari[2], A. Udayasri[3] and Anjaneyulu Bendi[4*]

[1]Department of Chemistry, Kommuri Pratap Reddy Institute of Technology (Autonomous), Hyderabad, Telangana, India
[2]Intertek India, Udyog Vihar, Gurugram, Haryana, India
[3]Department of Chemistry, Aditya Institute of Technology and Management, Tekkali, Andhra Pradesh, India
[4]Department of Chemistry, Kalinga University, Kotni, Near Mantralaya, New Raipur, Chhattisgarh, India

Abstract
Azulene is a non-alternant aromatic scaffold having a 10π Huckel electron system and is an isomer of naphthalene. Azulene and its derivatives have garnered significant attention from academic and industrial researchers due to their unique electronic properties, unusual chemical reactivity, and potential applications in optoelectronics. The research on azulene-hybrid polycyclic aromatic hydrocarbons (PAHs) has notably increased in recent years; however, the absence of efficient synthetic strategies has been a hindrance to detailed insights into the structure-property relationships of azulene derivatives. The application of this aromatic chameleon is dependent on the synthetic methods available. This chapter focuses on the different synthetic approaches of azulene, as well as its important functionalized derivatives. This research will be of immense significance tc organic chemists and material scientists, as well.

***Keywords*:** Azulene, polyaromatic hydrocarbons (PAHs), non-benzenoid hydrocarbons, Suzuki cross-coupling reaction, nucleophilic substitution, troponoids.

**Corresponding author*: anjaneyulu.bendi@gmail.com

Neha Rani Kumar and Abhijeet R. Agrawal (eds.) Azulene-Based Materials: Design, Synthesis, and Applications, (53–80) © 2026 Scrivener Publishing LLC

3.1 Introduction

The azulene unit consists of a five- and seven-membered fused ring structure and is deep blue in color. The word azulene found its origin from the words "azul" and "azur," which mean "blue" in Spanish and Arabic, correspondingly. The last two decades have witnessed an upsurge in the research focusing on non-alternant non-benzenoid aromatic hydrocarbons, owing to their interesting optoelectronic applications, unique chemical reactivity, and significance in organic electronics [1]. A special mention in this regard can be made for the azulene unit, which consists of a fused pentagon and heptagon (odd-membered shapes). It is aromatic and also exhibits unique photophysical properties and reactivity [2]. This push-pull nature endows azulene with a dipole moment of 1.08D and the positions, one- and three- of the five-membered ring in azulene, are electron-rich sites, whereas the four-, six-, and eight- positions of the seven-membered ring in azulene are electron deficient sites. The growing fascination with azulene stems from its inherent donor-acceptor/push-pull character, which arises from the combination of electron-rich/donor five-membered ring and electron-deficient/acceptor seven-membered ring. Azulene possesses unique photophysical properties arising from the small energy gap between the highest occupied molecular orbital and the lowest unoccupied molecular orbital and the anomalous fluorescence from the second excited state to the ground state that violates Kasha's rule (Figure 3.1) [3]. Azulene that contains small molecules and polymeric systems find wide applications in several domains that encompass pharmaceutical applications, optoelectronics, stimuli-responsive materials, etc. The progress in azulene chemistry is significantly dependent on the synthetic strategies available to design multi-substituted azulene and its derivatives. Over the years, several synthetic methodologies of azulene have emerged with each having its unique set of advantages and disadvantages [4]. In this chapter, we briefly outlined important methods for synthesizing azulene and its functionalized derivatives. It is envisioned that this chapter will be of immense significance to organic synthetic chemists, material scientists, and researchers working in the domains of organic electronics.

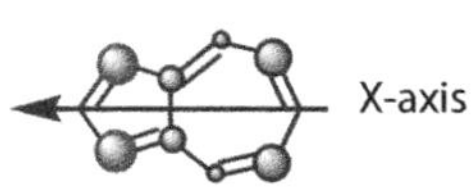

3.1

Figure 3.1 HOMO of azulene only and the X-axis symmetry.

3.2 Traditional Process for the Synthesis of Azulene

The research interest in the synthesis of azulenes and azulene derivatives has improved extensively in recent years, however, the need for these synthetic protocols to be curated in an organized manner remains. More insight into azulene chemistry is required as detailed insights into their structure-property relationships and biological applications are still limited. In the section below, we give insights into the most popular azulene synthesis that has evolved since its first synthesis in 1937.

The first synthetic method for preparing azulenes was introduced by Plattner and Pfau [5] in 1937 as shown in Scheme 3.1. Though the method provided an efficient way to synthesize the "aromatic chameleon", the robust dehydrogenation step limited its wide applications. And over the years, better synthetic methods evolved.

In the 1950s, Ziegler–Hafner provided another methodology for azulene synthesis from pyrylium salts and pyridinium salts [6] as depicted in Scheme 3.2. The efficiency of this method was more than that of Plattner and Pfau's method, but it provided a way to introduce groups only on the seven-membered ring.

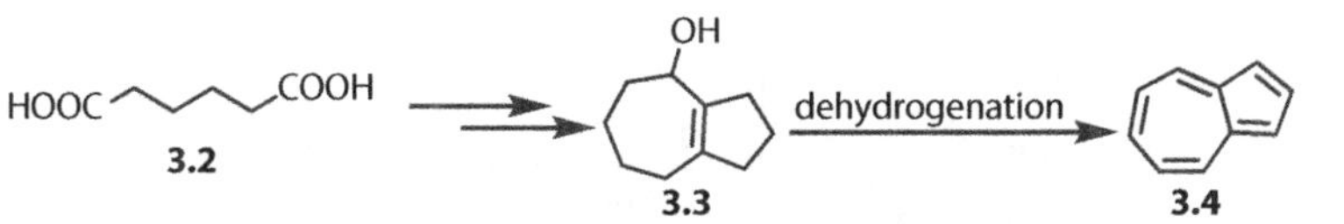

Scheme 3.1 Plattner and Pfau's synthesis of azulene.

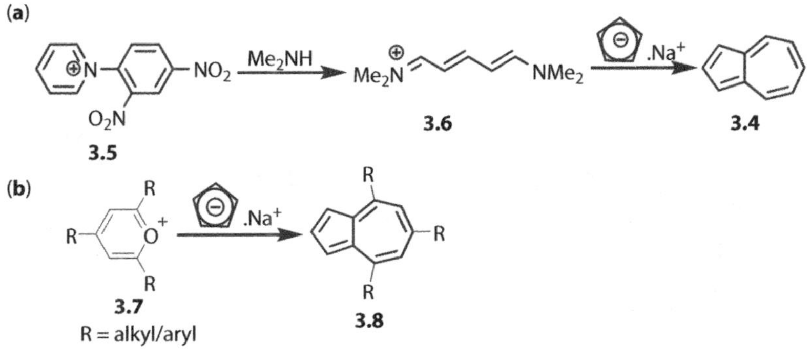

Scheme 3.2 Ziegler and Hafner's synthesis of azulene from (a) pyridinium salts (b) pyrylium salts.

In 1962, Nozoe *et al.* [7] reported the synthesis of azulenes from tropolones (**3.9**) as shown in Scheme 3.3. Till date, this approach is considered the most useful approach as it provides an easy way to introduce functional group on the five-membered ring (**3.11**). And by making use of abnormal nucleophilic substitution, even substituents can be introduced on the seven-membered ring, as well.

In 2021, Yasunami and Takase reported the synthesis of azulenes from enamines and 2H-cyclohepta-[b]furan-2-ones by employing [8+2] cycloaddition as depicted in Scheme 3.4 [8]. This approach again gave the possibility of introducing functional groups on the five-membered ring.

In 1977, Reiter *et al.* introduced another approach that involved [6+4] cycloaddition reactions leading to multisubstituted azulenes [9]. The starting materials used in this case were the electron acceptor thiophene dioxide (**3.16**) molecules and fulvene (**3.19**) as an electron acceptor unit. This method permits the synthesis of azulene derivatives (**3.20**) with diverse functional groups shown in Scheme 3.5.

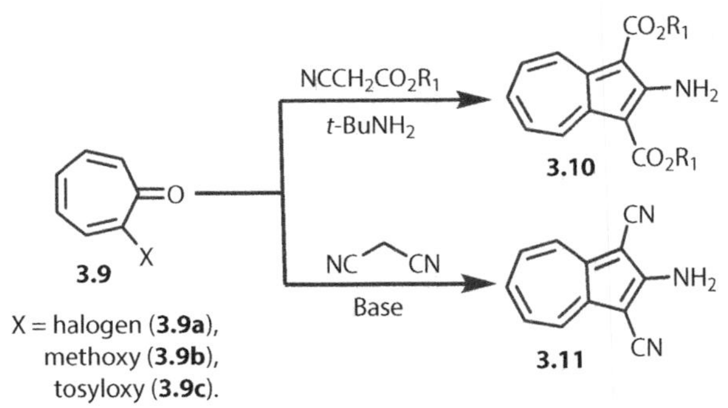

Scheme 3.3 Nozoe's azulene synthesis from tropolones.

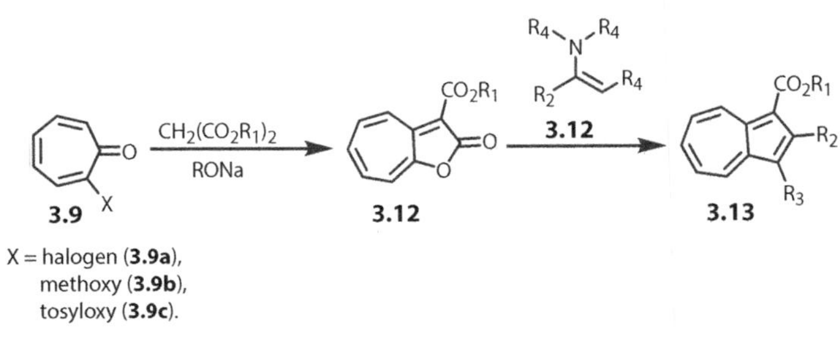

Scheme 3.4 Yasunami and Takase approach for the synthesis of azulenes from enamines.

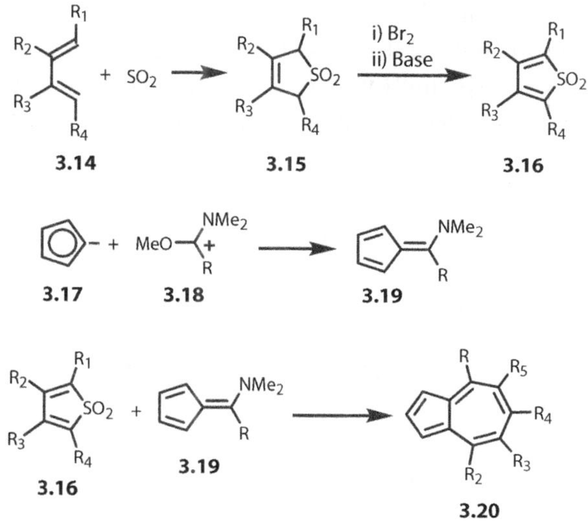

Scheme 3.5 Fulvenes undergo [6+4] cycloaddition to give azulenes.

3.3 Synthesis of Azulene Derivatives

The above section in this chapter sheds light on some of the earlier approaches that have been widely used for the synthesis of azulene derivatives. However, azulene chemistry is now emerging as a broad field, and researchers are actively working in this direction to synthesize azulene derivatives of diverse classes and exploring their photophysical properties. The section below discusses the synthesis of important derivatives of azulene that have evolved over time.

Leino *et al.* reported the synthesis of 3-methyl-N-[2-[[3-[[6-methyl-3-(2,2,2-trifluoroacetyl)azulene]-1-sulfonamido]phenyl]amino]ethyl] benzamide derivatives (**3.29**) in three steps as shown in Scheme 3.6. The first step in this approach involved the synthesis of sodium 6-methyl-3-(2,2,2-trifluoroacetyl)azulene-1-sulfonate (**3.23**) by the reaction between N,N-dimethylazulene-6-carboxamide (**3.21**) and 2,2,2-trifluoro-1-(6-methylazulen-1-yl)ethan-1-one (**3.22**) using sulfur trioxide pyridine complex as a reagent in benzene as solvent. In the second step, they synthesized the N-[2-[N-(3-aminophenyl)-2,2,2-trifluoroacetamido]ethyl]-3-methylbenzamide (**3.27**) using 1-fluoro-3-nitrobenzene (**3.24**) as a precursor and the reaction conditions as shown in Scheme 3.6. Finally in the third step, the reaction between appropriate azulene derivatives (**3.23**) and N-[2-[N-(3-aminophenyl)-2,2,2-trifluoroacetamido] ethyl]-3-methylbenzamide (**3.27**) in the presence of Et_3N for around 19 – 22 h at room temperature led to forming the final target product 3-Methyl-N-[2-[[3-[[6-methyl-3-(2,2,2-trifluoroacetyl)azulene]-1-sulfonamido]phenyl]amino]ethyl]benzamide derivatives (**3.29**) as depicted in Scheme 3.6 [10].

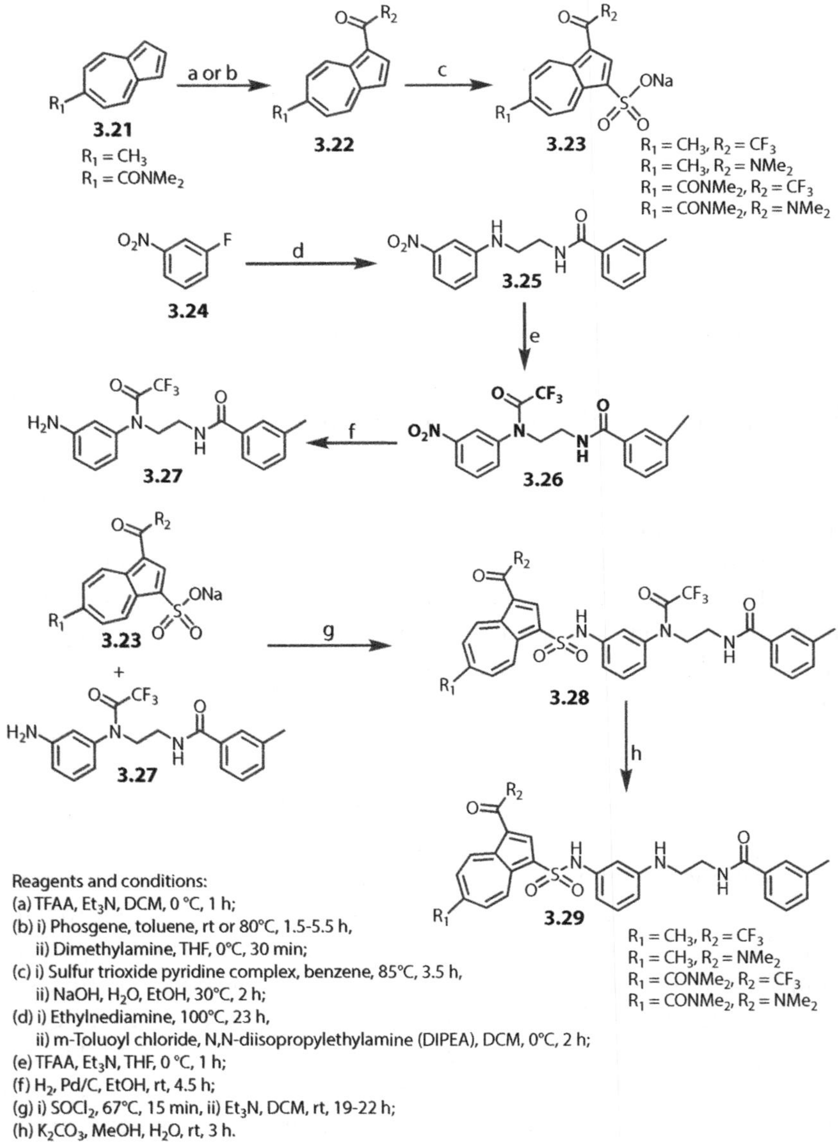

Scheme 3.6 Synthesis of azulene derivative to **3.29**.

García-Martínez *et al.* proposed the synthesis of conjugates of phenethylamine and azulene derivatives (**3.31/3.34**) by the reaction between azulene (**3.4**) and N-protected aziridine (**3.30**) in the presence of BF_3.OEt_2 and CH_2Cl_2 as a catalyst involving the SN_2-type mechanism controlling the aziridine's ring-opening as shown in Scheme 3.7.

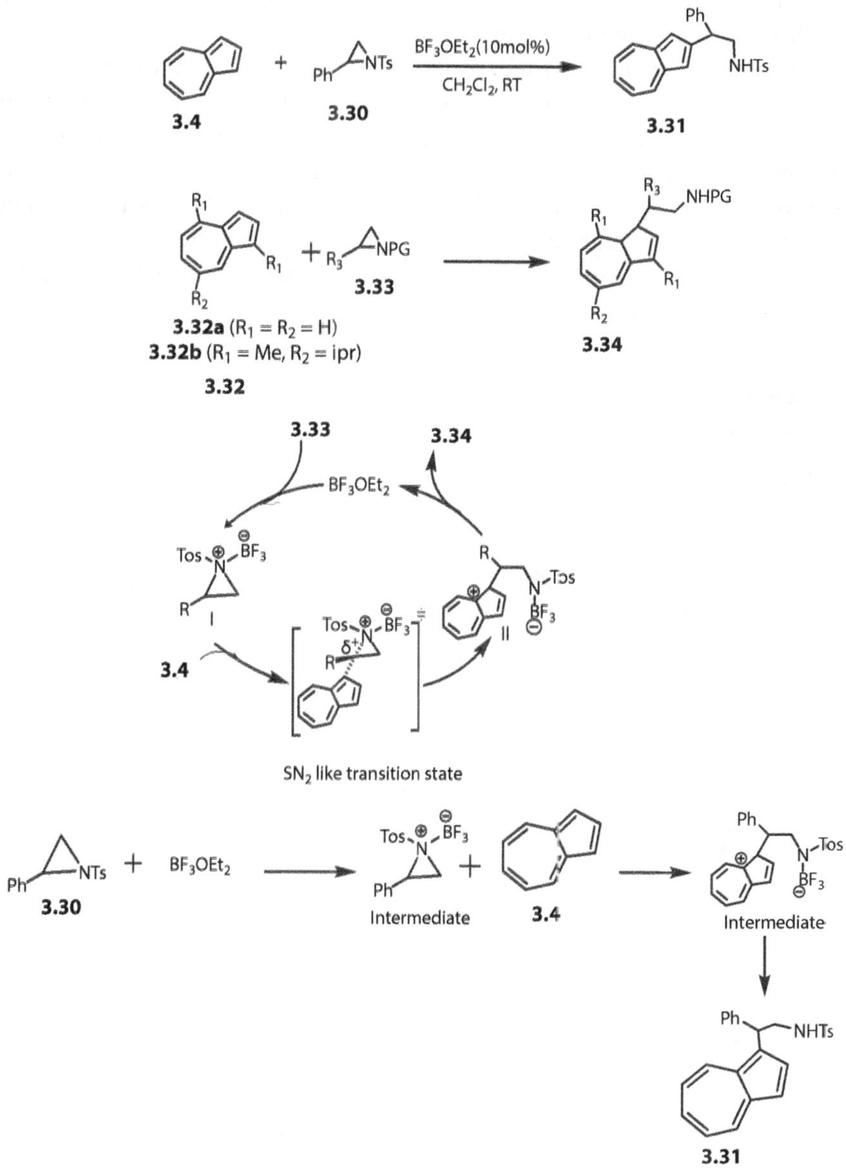

Scheme 3.7 Synthesis of azulene analogues using N-protected aziridines.

The first step is the formation of intermediate **I** from the reaction between aziridine and Lewis acid. This leads to the development of a slight positive charge on the substituted caron and further activates the aziridine system. These conditions make the attack of azulene at this position favorable [11]. The two steps involved in the formation of final product are the nucleophilic attack of the azulene to the positively charged carbon of aziridine, and the second is the elimination of proton to achieve aromatic stability leading to the formation of final product.

Mackenroth *et al.* reported the synthesis of 11H indeno-[2,1-a]azulenes (**3.37**) by using cyclohepta trienyl-substituted 1,5-diynes (**3.35**) in the presence of a gold catalyst through 5-endo-dig cyclization shown in Scheme 3.8. This methodology has a broad synthetic scope as the reaction takes place with both terminal alkynes and substituted alkynes as shown in Scheme 3.8 [12], though the reaction conditions have to be altered slightly. The synthesized indenoazulenes can also be further functionalized to get a new class of materials.

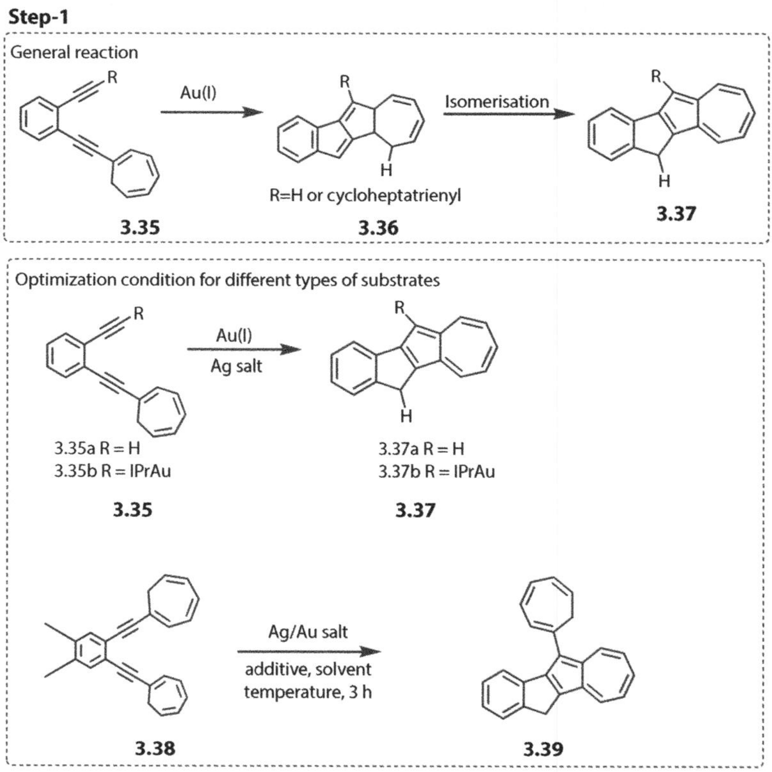

Scheme 3.8 Synthesis of benzofulvene derivatives using a gold catalyst.

Liu *et al.* presented a new synthetic protocol for the preparation of azulene-embedded polycyclic aromatic hydrocarbons (PAHs) by making use of Suzuki coupling and base-promoted Knoevenagel condensation reactions. The Suzuki coupled products (**3.44**) were obtained in approximately 57 – 93% yield and the final azulene containing embedded PAHs (**3.45**) in 82-96% yield as depicted in Scheme 3.9 [13].

Mathey *et al.* designed a synthetic procedure for the preparation of annulated azuleno-azulene derivatives (**3.49**). The synthesis step involves fourfold aldol condensation among aromatic dialdehydes (**3.47**) and tetrahydropentalene-2,5-(1H, 3H)—dione (**3.46**) and phthaldialdehyde or 2,3-thiophenediacarboxyaldehyde. This is followed by a nucleophilic addition with lithium TIPS-acetylide to give an intermediate, which was further reduced with $SnCl_2$ as depicted in Scheme 3.10 [14].

Liang *et al.* reported a highly efficient one-pot synthetic protocol for the synthesis of truxene-based benzotriazulene that consists of three azulene units. The target compound, benzo-[1,2-a:3-4-a′:5-6-a″]triazulene (BTA) (**3.55**), was synthesized by employing two key reactions that is Suzuki

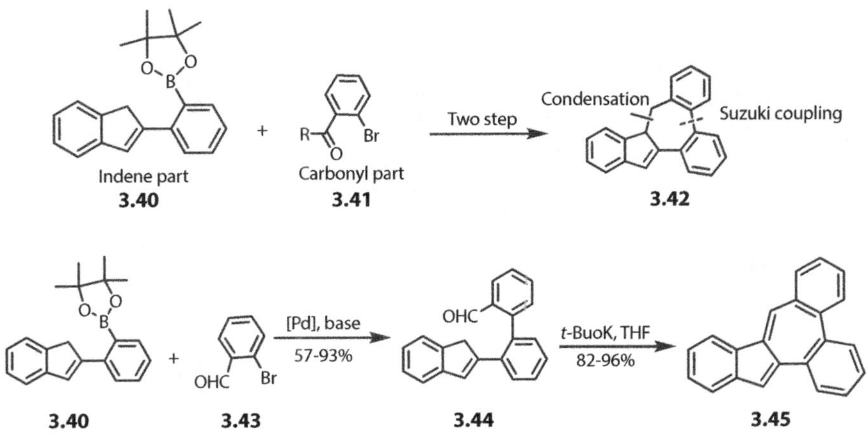

Scheme 3.9 Synthesis of azulene-incorporated PAHs.

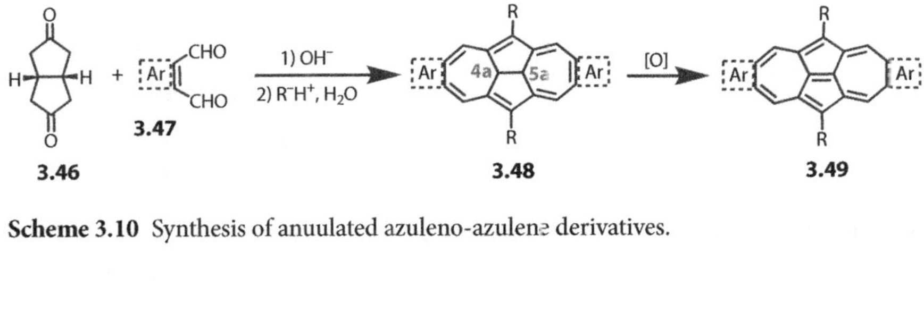

Scheme 3.10 Synthesis of anuulated azuleno-azulene derivatives.

coupling reaction and Knoevenagel condensation between triborylated truxene (**3.53**) and 8-bromo-1-napthaldehyde (**3.54**) as the starting material. The targeted compound (**3.55**) can also be trinitrated regioselectively to give 3.56 as shown in Scheme 3.11. The synthesized final trinitrated derivative (**3.56**) was found to be stable and exhibited good solubilities and fluorescence properties in solution [15].

Murai *et al.* reported the synthesis of azulene containing linear PAHs. The preparation of the targeted compounds involved several key reactions like selective borylation of azulene at position 2 to produce **3.58**, followed by Suzuki coupling to produce **3.59**. The biazulene derivative (**3.59**) then underwent Vilsmeier reaction, followed by reaction with p-toluenesulfonyl hydrazine and rhodium catalyst to give the final compound (**3.61**) as depicted Scheme 3.12 [16].

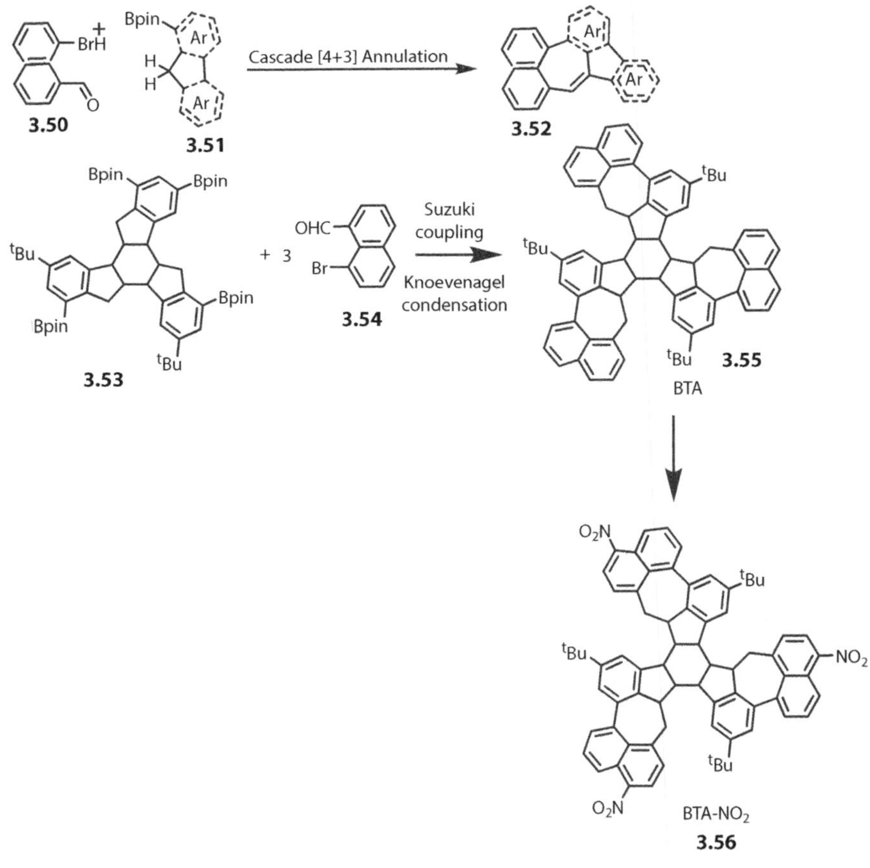

Scheme 3.11 Preparation of azulene embedded polycyclic systems.

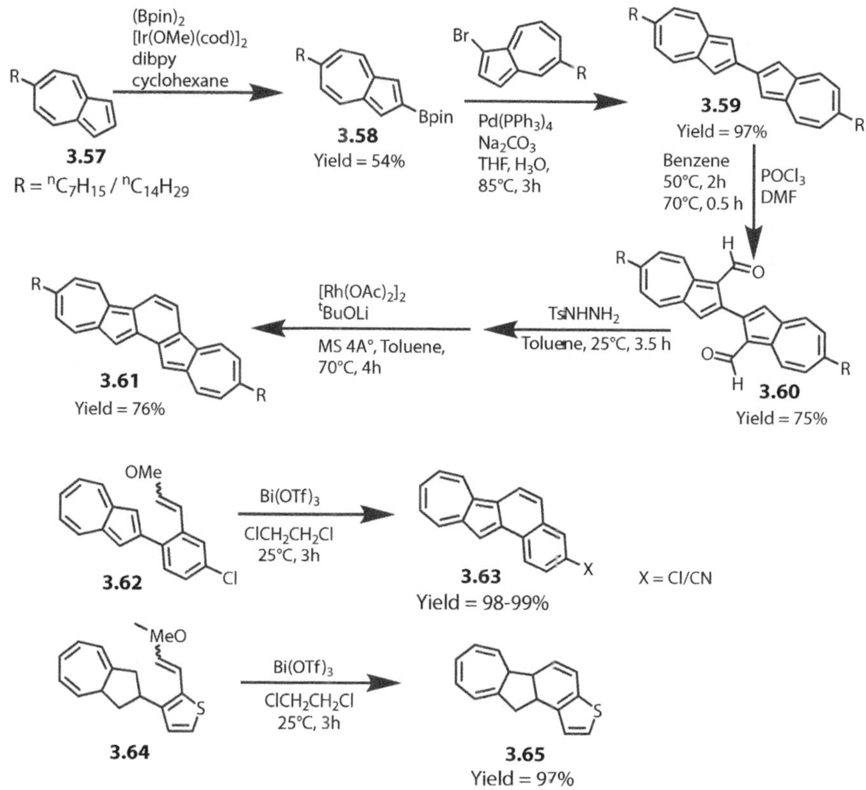

Scheme 3.12 Preparation of azoluene-fused PAHs.

Kirschbaum *et al.* reported a synthetic procedure for the synthesis of benzotrisazulene from tris(formylnaphthyl)truxene (**3.67**) as shown in Scheme 3.13. The synthesis began with a Suzuki–Miyaura cross-coupling reaction of 3.66 with 8-bromo-2-napthaldehyde using dimethoxyethane (DME) as a solvent, K_2CO_3 as base, and $Pd(PPh)_3$ as a catalyst. The formed crude product was then purified using column chromatography. This approach yielded 3.68 in only a 9% yield, and hence an alternative route was adopted to synthesize the target compound involving the set of reaction conditions shown in Scheme 3.13 [17].

Tsuchiya *et al.* designed and reported the synthesis of three different types of terazulenes molecules that included a planar structure (**3.74**), a bent structure (**3.75**), and a helical structure (**3.76**), as shown in Scheme 3.14 [18]. The bent terazulene molecule displayed anti-aromatic properties.

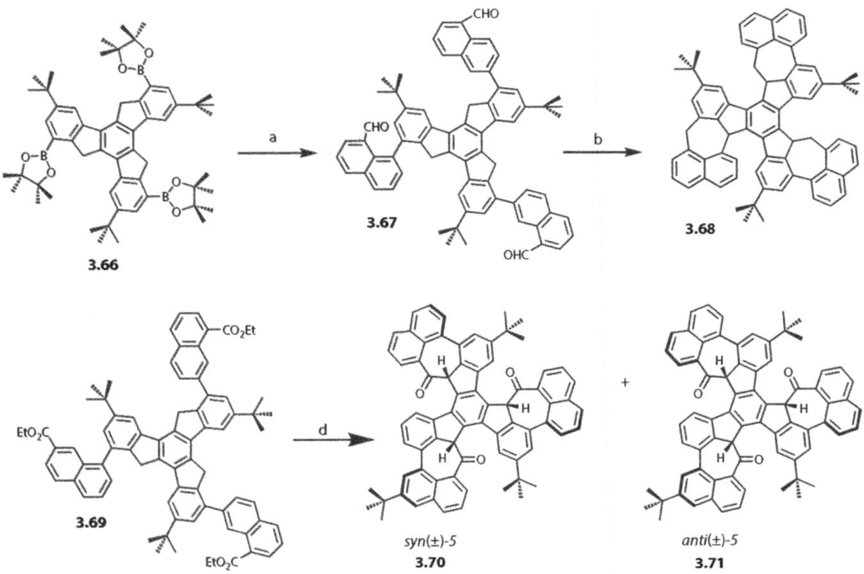

Reaction conditions:
a) 8-bromo-2-napthaldehyde (10.5 equiv), $Pd(PPh_3)_4$ (5 mol%), K_2CO_3 (2 m), DME, 90 °C, 24 h;
b) KOH (9 equiv), THF, 60°C, 24 h, 9% over two steps.
c) ethyl 8-bromo-2-naphthoate (6 equiv), Pd_2dba_3 (5 mol%), tBu_3PHBF_4 (20 mol%), K_2CO_3 (2 m), THF, 80 °C, 48 h, 85%;
d) KOtBu (9 equiv), THF, 75 °C, 72 h, 92% (anti-syn: 5:1).;
e) i) $LiAlH_4$ (4.5 equiv), THF, 0°C!r.t., 18 h;
ii) E_3tN (90 equiv), then MsCl (45 equiv), DCMabs, 0°C ⟶ 80°C, 96 h, 29% over two steps.

Scheme 3.13 Preparation of benzotrisazulene derivatives.

Li *et al.* reported the first gas phase synthesis of PAHs consisting of azulene by the reaction between stable radicals, fulvenallenyl, and propargyl guided by the novel fulvenallenyl addition cyclization aromatization mechanism. The reaction mechanism is outlined in Scheme 3.15 [19].

Mengjia *et al.* explored the anti-inflammatory properties of azulene compounds containing benzimidazole units, which were synthesized by using the Buchwald–Hartwid coupling method as shown in Scheme 3.16. Guaiazulene benzimidazole (GABI-1/**3.84**) derivatives were shown to have minimal toxicity and significant anti-inflammatory efficacy [20].

Gorgensen *et al.* reported the synthesis of an array of polysubstituted azulene compounds (for example, **3.87**) involving a (10+4) cycloaddition reaction between indene-2-carbaldehyde (**3.85**) and chromen-4-one (**3.86**) as shown in Scheme 3.17 [21]. These types of compounds find wide applications as indicators, dyes, and also in imaging.

Shoji *et al.* synthesized azuleno[2,1-*b*]quinolones by using an aromatic nucleophilic substitution reaction of 2-chloroazulenes and arylamines [22].

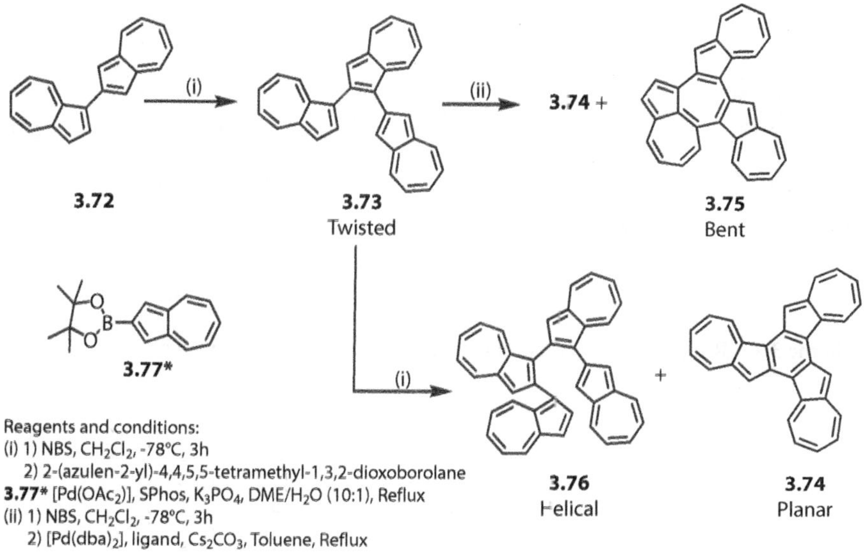

Scheme 3.14 Synthesis of three different types of azulenes by Tsuchiya *et al.*

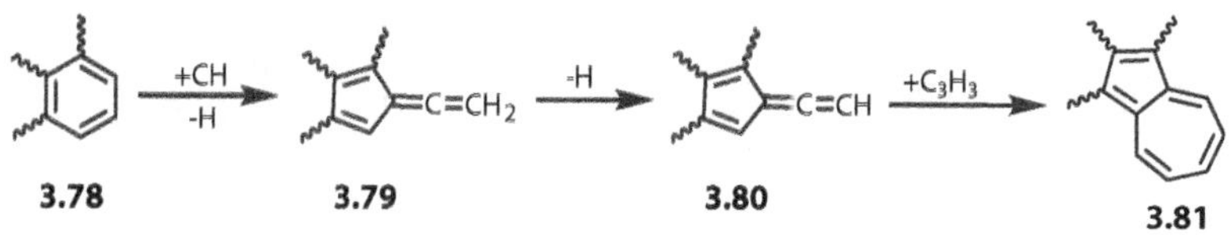

Scheme 3.15 Gas-phase synthesis of PAHs containing azulene.

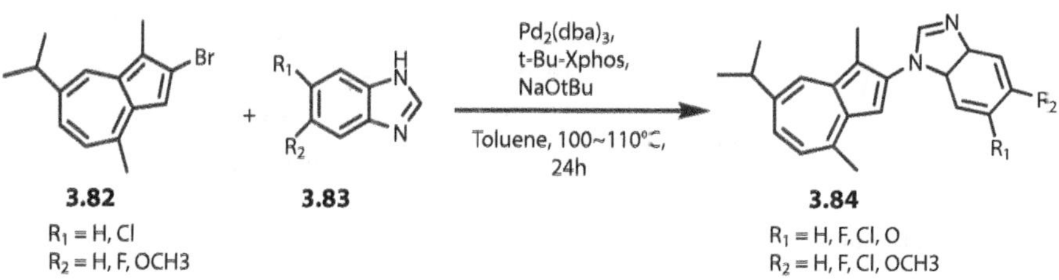

Scheme 3.16 Synthesis of azulene derivatives with benzimidazole moiety.

The coupling of haloazulene with arylamine resulted in the formation of **3.89**, which then underwent Bronsted acid-catalyzed intramolecular cyclization to give **3.90**. The final step involved the halogenative aromatization of **3.90** to give azuleno[2,1-b] quinolones (**3.91**) as shown in Scheme 3.18.

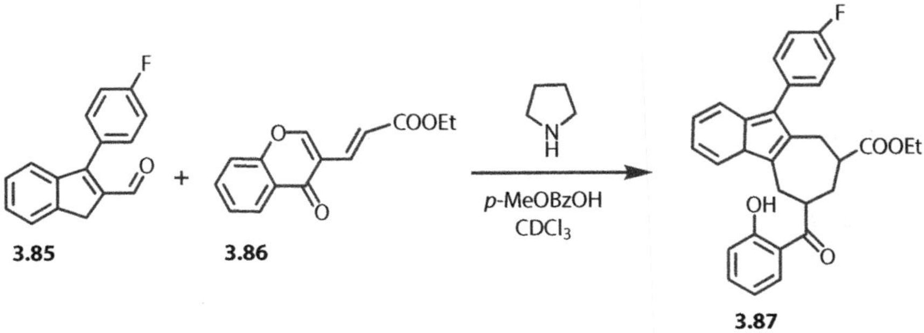

Scheme 3.17 Ring expansion reaction for formation of azulene derivatives.

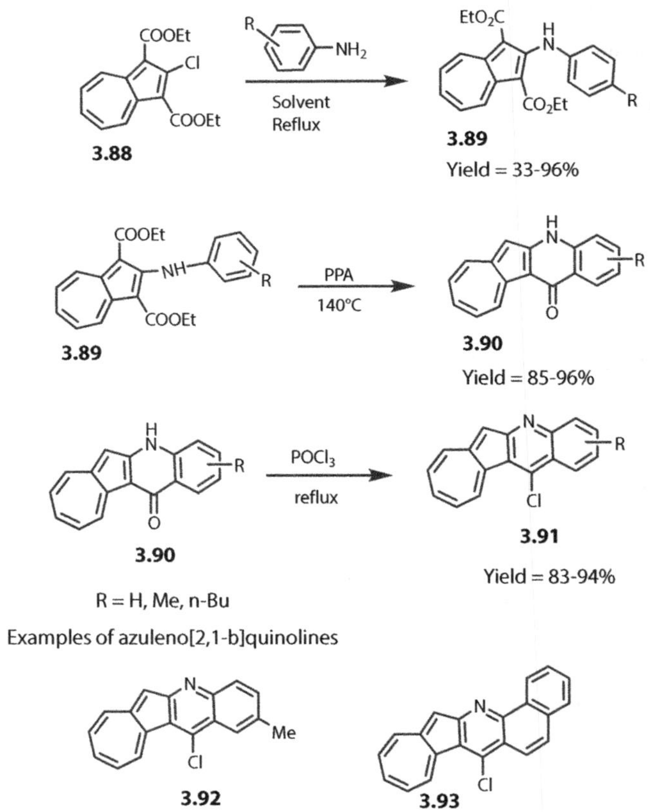

Scheme 3.18 Synthesis of azuleno-quinonlones.

Akram *et al.* reported the synthesis of Guaiazulene (**3.95**) from sodium guaiazulene sulfonate (Scheme 3.19). Guaiazulene is an azulene containing a natural product with immense physiological significance in bacterial infections, immune-modulation, tumors, expectorants, diuretics, ulcers, diaphoresis, dermatitis, gastritis, and proliferation [23].

Manea-Saghin *et al.* reported the synthesis, as well non-linear optical properties of azulene containing compounds such as **3.100a–d** [24]. All the compounds shown in Figure 3.2 exhibit non-linear optical properties. However, among all compounds, **3.100a–d** were found to be excellent candidates. They can be obtained by coupling azulenyl pyridine with diazonium salts of substituted anilines.

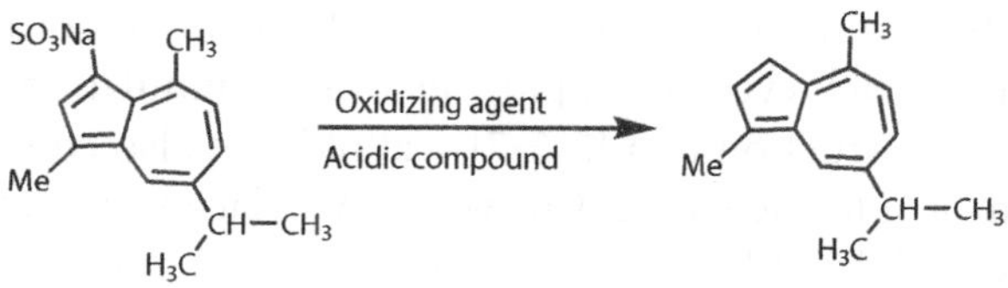

3.94 **3.95**

Scheme 3.19 Synthesis of Guaiazulene.

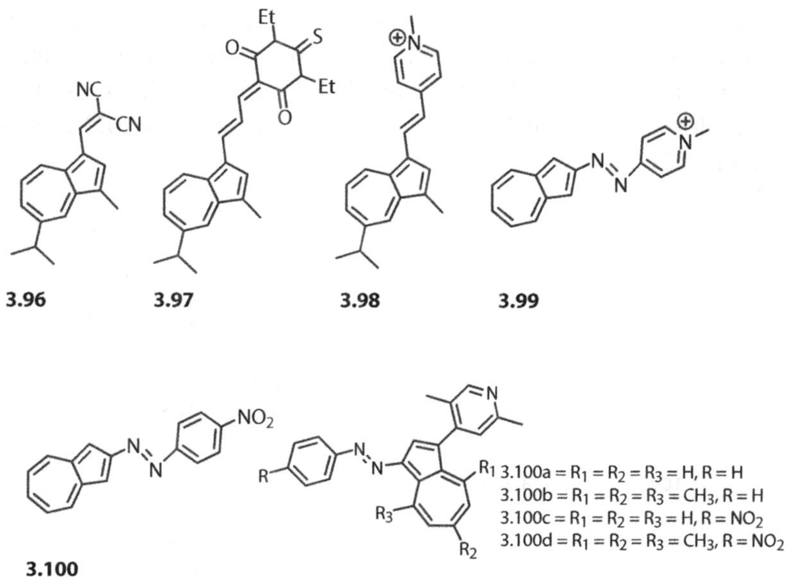

Figure 3.2 Structures of azulene-based derivatives showing non-linear optical properties.

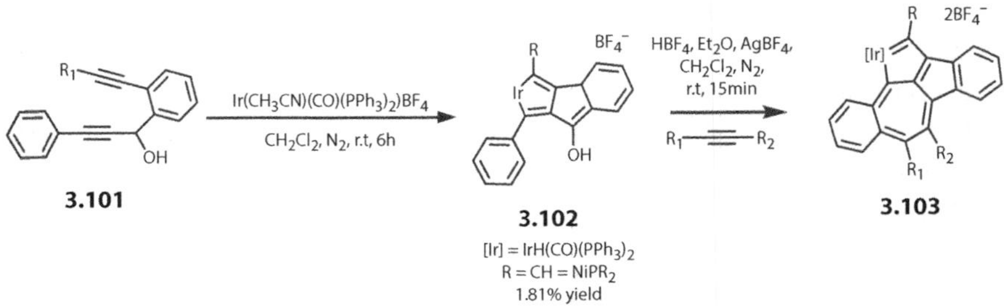

Scheme 3.20 Synthesis of metalla-dual-azulenes.

In 2013, Liu *et al.* [25] reported the synthesis of metalla-dual-azulene that comprises metallaazulene and a metal-free organic azulene, both intertwined by a common tropylium unit. The synthesis of this metalla-dual-azulene was achieved by employing a [5+2] annulation strategy followed by the construction of a [5-5-7] scaffold that also contained metal. The reaction of intermediate **3.102** with phenylacetylene and $AgBF_4$ in the presence of $HBF_4 \cdot Et_2O$ at r.t. in CH_2Cl_2 for 15 min resulted in the production of derivative **3.103** in approximately 87% yield as shown in Scheme 3.20.

Park *et al.* reported the ring expansion of an azulene system by making use of transition metal catalyst [26]. The catalyst employed for this purpose was $Cu(hfacac)_2$ (hfacac: hexafluoroacetylacetonate) along with a diazo derivative. However, when the copper catalyst was replaced by AgOTf, interestingly it led to C-H alkylation. This provided an efficient approach for synthesis of C-H functionalized azulenes (**3.107**) as shown in Scheme 3.21.

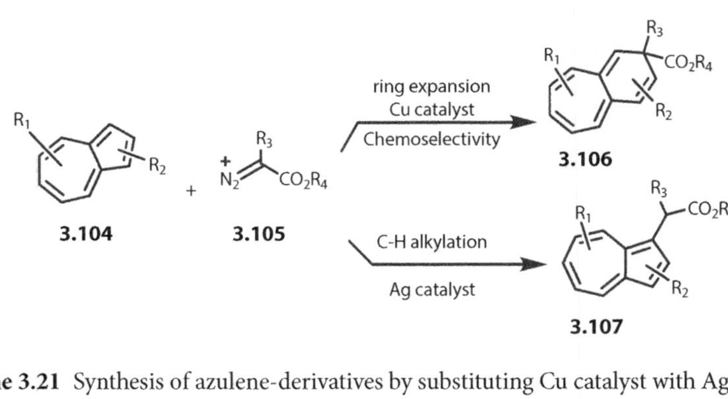

Scheme 3.21 Synthesis of azulene-derivatives by substituting Cu catalyst with AgOTf.

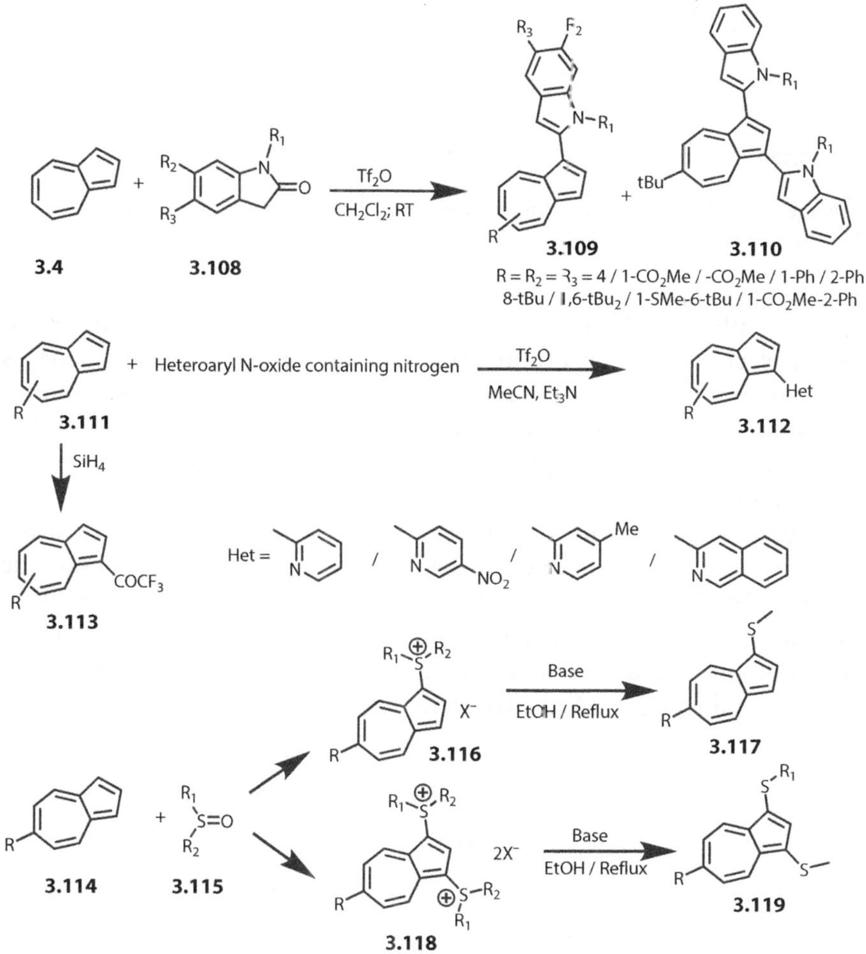

Scheme 3.22 Synthesis of functionalised azulene derivatives.

Vilmeier–Haack type arylation reaction has also been employed for the synthesis of substituted azulene compounds. The synthesis of 1-(indol-2-yl)azulene derivatives (**3.109**, **3.110**) was achieved by treating azulene and indole derivative (**3.108**) with triflic anhydride in dicholormethane. A series of heteroaryl substituted azulenes (**3.112**) were synthesized by using Reissert–Henze reaction as shown in Scheme 3.22. The synthesis of azulenyl sulphide derivatives (**3.117**, **3.119**) can be obtained by treatment of azulene (**3.114**) with sulfoxide derivatives (**3.115**) as shown in Scheme 3.22 [27].

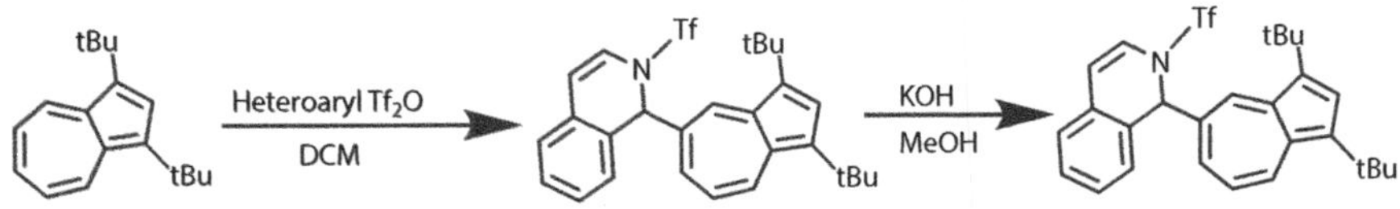

3.120 **3.121** **3.122**

Scheme 3.23 Electrophilic substitution reaction of 1,,3-di-tert-butylazulene to give N-containing heteroarylazulenes.

The synthesis of N-containing heteroarylazulenes can be accomplished by an electrophilic substitution reaction on 1,3-di-*tert*-butylazulene as depicted in Scheme 3.23. This process was capable of affording both mono- and di-substituted products depending on the condition employed [28].

Azulene derivatized at position 2- is one among the unique substitution patterns that researchers look forward to using in the synthesis of conjugated systems with narrow energy gaps. Kurotobi *et al.* reported the synthesis of -(2-Azulenyl)-4,4,5,5-tetramethyl-1,3,2-dioxaborolane (**3.123**) [29]. For the direct borylation of azulene at position 2-, a mixture of chloro(1,5-cyclooctadiene)iridium(I) dimer, 2,2-bipyridine, bis(pinacolato)diboron, and azulene are refluxed for around 14 hr to afford the target product 2-(2-Azulenyl)-4,4,5,5-tetramethyl-1,3,2-dioxaborolane

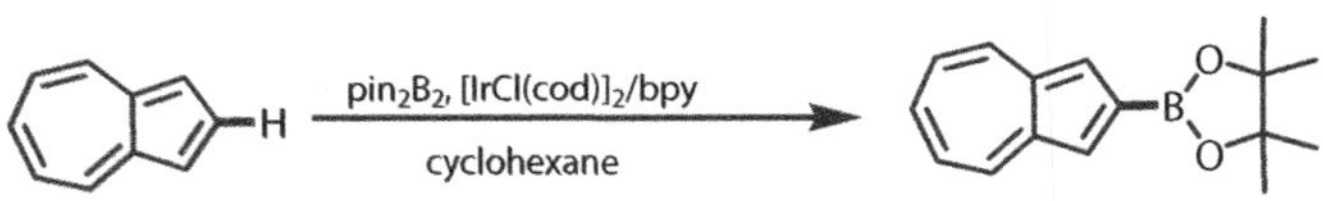

3.4 **3.123**

Scheme 3.24 Synthesis of 2-(2-Azulenyl)-4,4,5,5-tetramethyl-1,3,2-dioxaborolane.

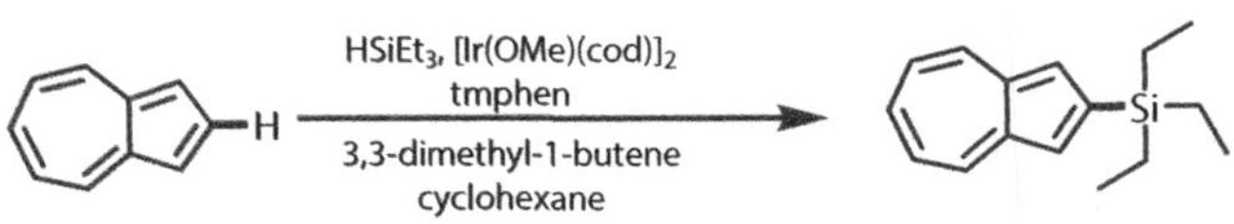

3.4 **3.124**

Scheme 3.25 Synthesis of 2-silylated azulene molecule.

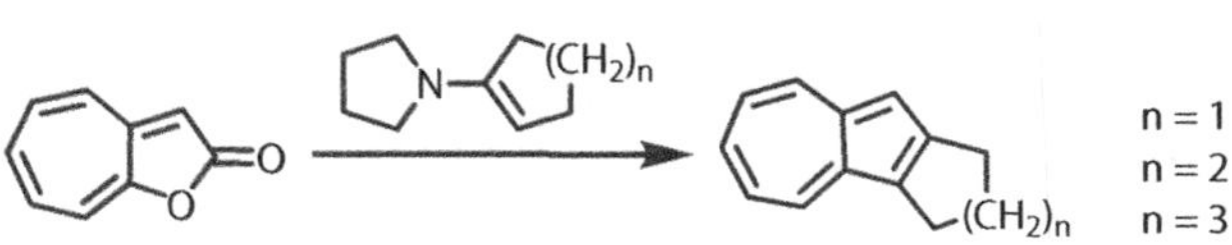

3.125 **3.126**

Scheme 3.26 Synthesis of azulene derivatives from 2H-cyclohepta[b]furan-2-ones.

as illustrated in Scheme 3.24. Similarly, Murai *et al.* reported the synthesis of 2-(triethylsilyl)azulene (**3.124**) as shown in Scheme 3.25 [30]. To achieve the target compound, a mixture of $[Ir(OMe)(cod)]_2$ and tmphen (tetramethyl phenanthroline) was added to the azulene followed by the addition of 3,3-dimethyl-1-butene and triethylsilane and then stirred for 4 hours at 100°C.

The synthesis of azulene fused with rings of different sizes (**3.126**) as shown in Scheme 3.26 was reported by Yang *et al.* [31]. This method involved the reaction between 1-N-pyrrolidinyl-1-cyclo-heptene and 2H-cyclohepta[b]furan-2-ones (**3.125**) to give fused azulene derivatives.

The synthesis of conjugated oligomers and polymers is largely dependent on the availability and easy synthesis of key precursors that can facilitate the coupling reactions. In such a scenario, halogenated precursors are considered very important and are capable of undergoing a wide array of coupling reactions. While most of the halogenated derivative of azulene can be obtained by Nozoe's approach, synthesis of 4,7-dihalogenated azulene becomes difficult. This unique substitution can be obtained by Hawker's approach to synthesize azulene [32]. This method involves initial treatment of 2,5-dibromothiophene derivative (**3.127**) with acetonitrile complex of hypofluorous acid to give alkyl substituted 2,5-dibromothiophene-S,S-dioxide (**3.128**) followed by the reaction with dimethylaminofulvene at room temperature to afford the target compound 4,7-dibromoazulene (**3.129**) as discussed in Scheme 3.27.

Xin *et al.* reported the synthesis of biazulene diimides (**3.140**) as depicted in Scheme 3.28 [33]. Initially, 2-hydroxy tropolone (**3.130**) is converted into 2-chloro tropolone (**3.9a**) by reaction with $SOCl_2$ followed by the nucleophilic substitution reaction with ethyl cyanoacetate to obtain the dicarboxylate aminoazulene compound **3.131**. The azulene compound undergoes bromination along with the reaction with bis(pinacolato)diboron to afford 6-boryl substituted azulene analogue **3.133**. The chlorination of 6-boryl substituted azulene analog followed by the Suzuki cross-coupling reaction with 2-bromoazulene leads to the formation of

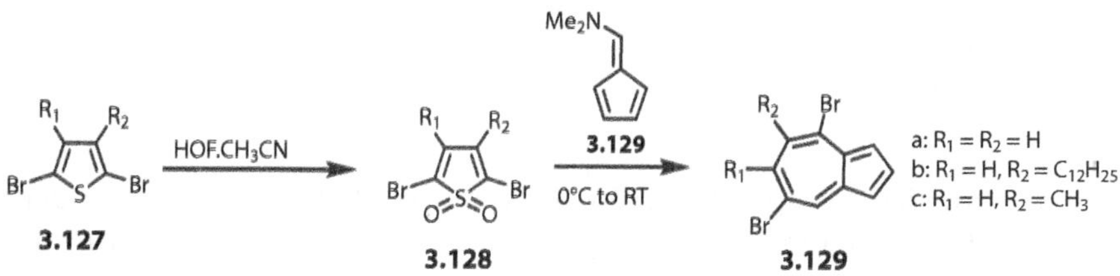

Scheme 3.27 Synthesis of substituted azulene scaffolds.

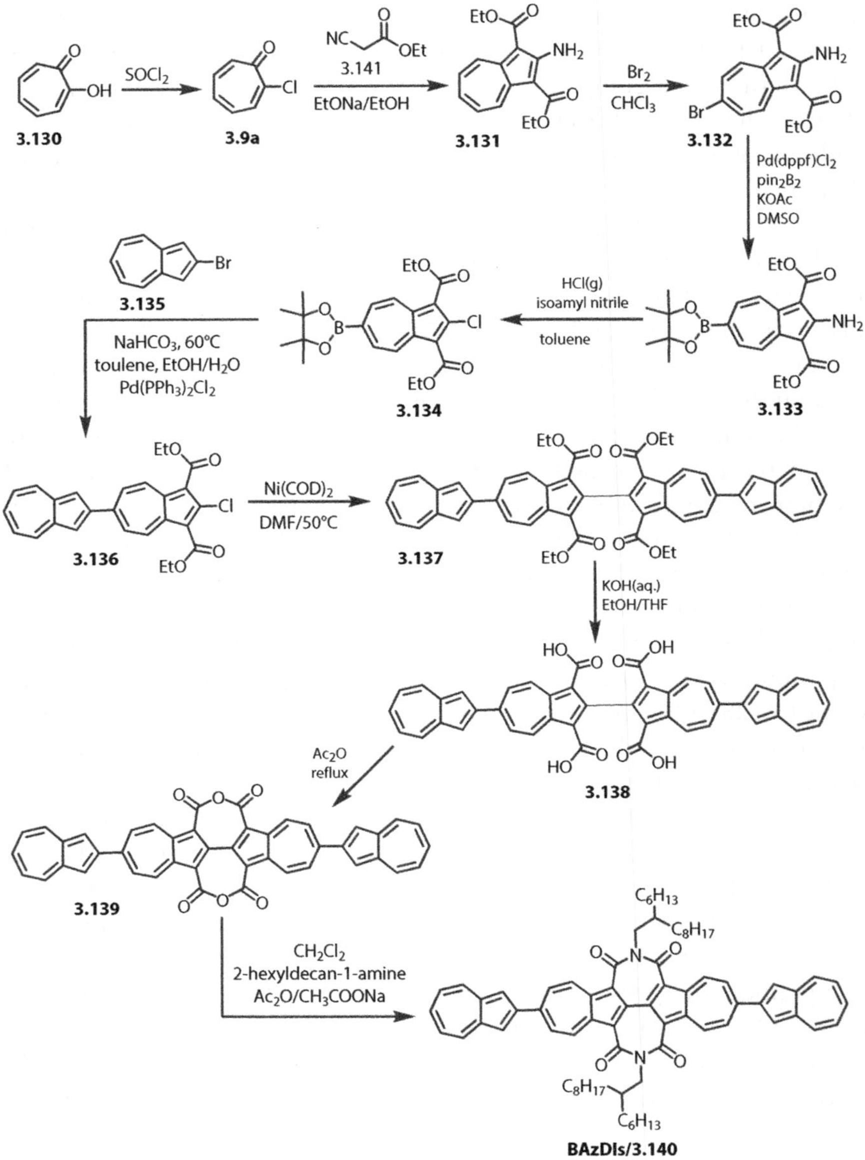

Scheme 3.28 Synthesis of N,N'-bis(2-hexyldecyl)-2,6':2',2":6",2'"-quaterazulene-1',1",3',3"-tetracarboxdiimide (BAzDIs).

key intermediate compound **3.136**. This intermediate compound undergoes a self-coupling reaction in the presence of $Ni(COD)_2$ to result in the formation of **3.137**. The ester hydrolysis of this compound followed by acetic anhydride-assisted dehydration leads to the formation of azulene-containing dianhydride (**3.139**). Finally, the dianhydride compound on being treated with 2-hexyldecylamine in DCM affords the title compound N,N'-bis(2-hexyldecyl)-2,6':2',2'':6'',2'''-quaterazulene-1',1'',3',3''-tetracarboxdiimide (BAzDIs/**3.140**). The synthesized compound exhibits unique photophysical properties and organic electronic properties.

As discussed earlier, Nozoe's azulene synthesis is considered the most significant azulene synthesis [34–37]. This method is one of the simplest methods available to achieve the synthesis of multifunctional azulenes in one step from active troponoid derivatives (**3.9**). Its significance lies in the fact that it permits a wide variety of azulene derivatives to be synthesized in high yields (**3.11**, **3.131–3.144**). Besides that, these functional groups can be easily removed or converted at will to synthesize key azulene precursors, such as azulene itself, its halogenated derivatives, and 2,6-functionalized azulenes (**3.145–3.147**). The wide variety of azulene derivatives that can be synthesized using this approach are outlined in Scheme 3.29.

Yao *et al.* [38] reported the synthesis of dithienyldiketopyrrolopyrrole–azulene conjugated systems as depicted in Scheme 3.30. The 4,7-dibromoazulene (**3.148**) compound reacts with thiophene-derivatized

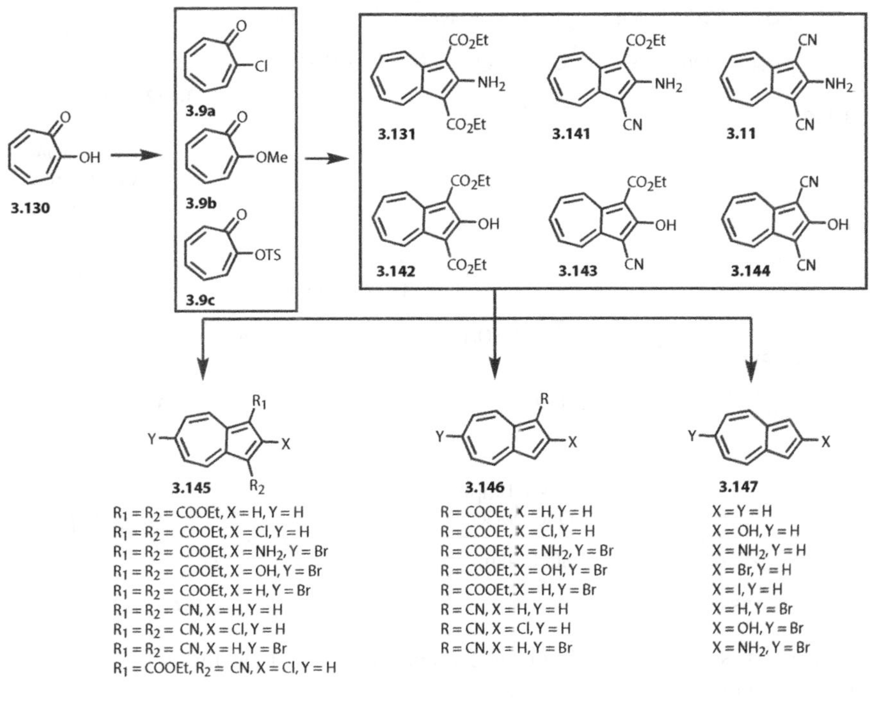

Scheme 3.29 Synthesis of multifunctional azulenes by using Nozoe's approach.

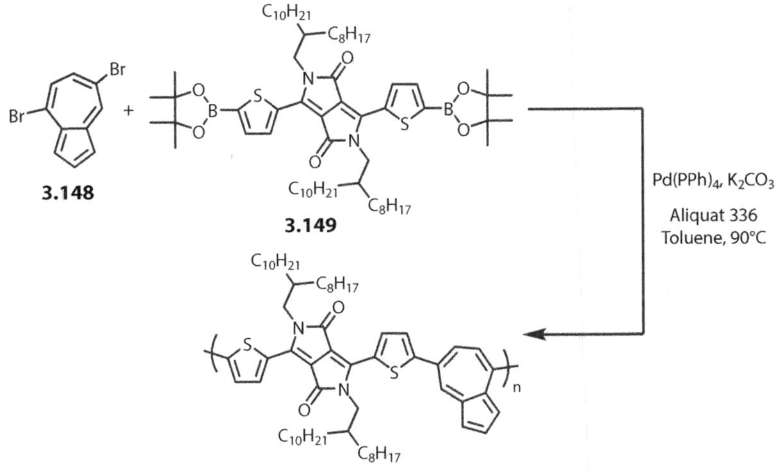

Scheme 3.30 Synthesis of dithienyldiketopyrrolopyrrole–azulene conjugated system.

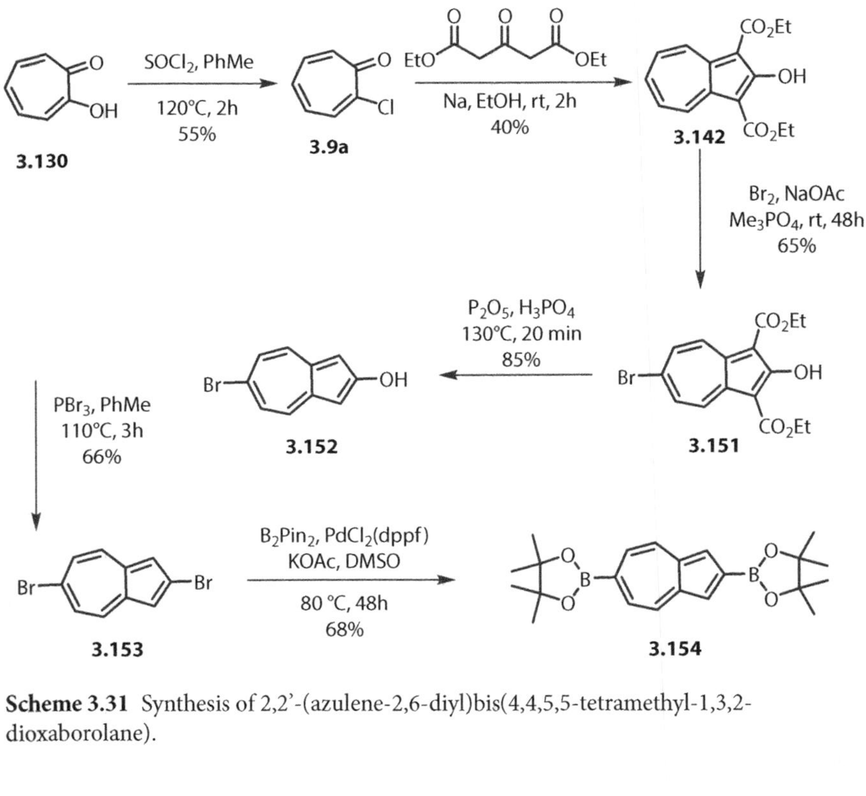

Scheme 3.31 Synthesis of 2,2'-(azulene-2,6-diyl)bis(4,4,5,5-tetramethyl-1,3,2-dioxaborolane).

diketopyrrolopyrrole (**3.149**) using the Suzuki cross-coupling reaction to afford the corresponding dithienyldiketopyrrolopyrrole–azulene conjugated polymer (**3.150**) involving connectivity *via* the seven-membered ring.

Morin *et al.* reported the synthesis of conjugated polymers that involved 2,6-connectivity of azulene. The 2,6-connectivity of azulene is considered the most advantageous as it involves the maximum utilization of the resonance structure of azulene and ensures effective delocalization of electrons around the C_{2v} dipolar axis. 2,2'-(azulene-2,6-diyl)bis(4,4,5,5-tetramethyl-1,3,2-dioxaborolane) (**3.154**), as well as 2,6-dibromoazulene (**3.153**) are some of the most important key precursors that can be utilized for the synthesis of an excellent class of conjugated systems with very high efficiency

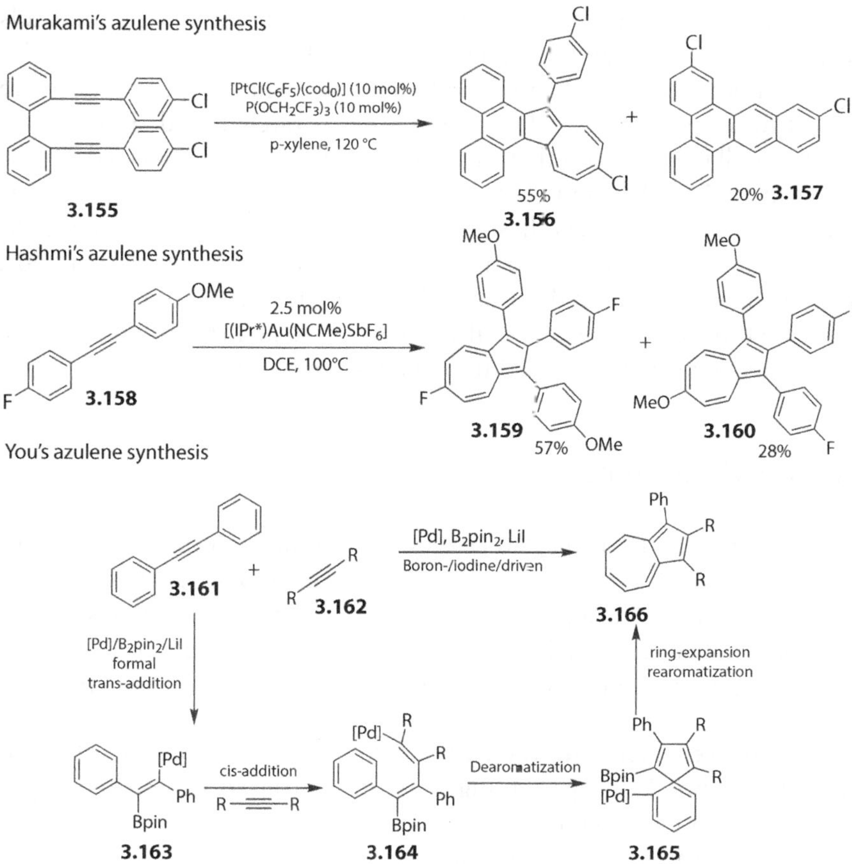

Scheme 3.32 Synthesis azulene derivatives from alkyne precursors.

and device performance. The synthesis of these precursors starting from tropolone (**3.130**) *via* the intermediate 6-bromo-2-hydroxy azulene (**3.152**) is outlined in Scheme 3.31 [39].

As discussed previously, alkynes have been emerging as an important class of precursors that can lead to the synthesis of different types of azulene derivatives (**3.156, 3.157, 3.159, 3.160, 3.166**). Some recent approaches as reported by Murakami *et al.* [40], Hashmi *et al.* [41], and Zhou *et al.* [42] are shown in Scheme 3.32.

3.4 Conclusion

In summary, this book chapter has been written to present an account of the developments that have taken place in the major synthetic protocols available for azulene and its derivatives. Though azulene is now a popular moiety among organic chemists and material scientists, a comprehensive report focusing on the synthetic approaches is rarely found. This chapter discusses all the major synthetic approaches of azulene from its oldest synthesis in 1937 by Pfau and Plattner to its most widely recognized synthesis that is Nozoe's azulene synthesis. The synthesis of key azulene intermediates that act as the precursor for the synthesis of larger conjugated scaffolds, such as halo azulenes and boronated azulene derivatives, have also been discussed. In the end, we have also briefly presented the synthesis of the most interesting azulene substitution pattern that is the 2,6-substitution.

Acknowledgment

The author G. B. Dharma Rao conveys his cordial gratitude to the management of KPRIT (Autonomous), Hyderabad, India. Aditi is thankful to Intertek India, Gurugram, Haryana, India. Udayasri expresses special thanks to management, AITAM, Tekkali, India. Anjaneyulu is thankful to the management of Presidency University, Bangalore, Karnataka, India.

AI disclosure statement: The author acknowledges the use of AI-based tools for minor language editing or grammatical correction. However, all intellectual contributions, analyses, and conclusions presented in this chapter are entirely the author's own.

References

1. Chaolumen, Stepek, I.A., Yamada, K.E., Ito, H., Itami, K., Construction of heptagon-containing molecular nanocarbons. *Angew. Chem. Int. Ed.*, 60, 23508, 2021.
2. Konishi, A. and Yasuda, M., Breathing new life into nonalternant hydrocarbon chemistry: Syntheses and properties of polycyclic hydrocarbons containing azulene, pentalene, and heptalene frameworks. *Chem. Lett.*, 50, 195, 2021.
3. Dunlop, D., Ludvíková, L., Banerjee, A., Ottosson, H., Slanina, T., Excited-State (Anti) aromaticity explains why azulene disobeys Kasha's rule. *J. Am. Chem. Soc.*, 145, 21569, 2023.
4. (a) Homocianu, M., Airinei, A., Matica, O.T., Cristea, M., Ungureanu, E.M., Solvent effects and metal ion recognition in several azulenyl-vinyl-oxazolones. *Symmetry*, 15, 327, 2023. (b) Xin, H. and Gao, X., Application of azulene in constructing organic optoelectronic materials: New tricks for an old dog. *ChemPlusChem*, 82, 7, 945, 2017. (c) Xin, H., Hou, B., Gao, X., Azulene-based π-functional materials: Design, synthesis and applications. *Acc. Chem. Res.*, 54, 7, 1737, 2021.
5. St. Pfau, A. and Plattner., P.I.A., Zur kenntnis der flüchtigen pflanzenstoffe VIII. Synthese des vetivazulens. *Helv. Chim. Acta*, 22, 202, 1939.
6. Ziegler, E.H.K. and Hafner, K., Rational Synthesis of Azulens. *Angew. Chem. Int. Ed.*, 67, 301, 1955.
7. (a) Nozoe, T., Seto, S., Matsumura, S., Murase, Y., The synthesis of azulene derivatives from troponoids. *Bull. Chem. Soc. Jpn.*, 35, 1179, 1962. (b) Kumar, N.R., Agrawal, A.R., Zade, S.S., Abnormal nucleophlic susbtitution on methoxytropone derivatives: Steric strategy to synthesize 5-substituted azulenes. *Chem. Eur. J.*, 25, 62, 14064, 2019. (c) Kumar, N.R., Agrawal, A.R., Choudhury, A., Zade, S.S., The effect of base and nucleophile on the nucleophilic susbtitution on methoxytropone derivatives: An easy access to 4- and 5-substituted multifunctional azulenes. *J. Org. Chem.*, 85, 14, 9029, 2020.
8. Shoji, T., Ito, S., Yasunami, M., Synthesis of azulene derivatives from 2 H-Cyclohepta [b] furan-2-ones as starting materials: Their reactivity and properties. *Int. J. Mol. Sci.*, 22, 10686, 2021.
9. Reiter, S.E., Dunn, L.C., Houk, K.N., Synthesis of azulenes by the [6+4] cycloadditions of 6-aminofulvenes to thiophene S, S-dioxides. *J. Am. Chem. Soc.*, 99, 4199, 1977.
10. Leino, T.O., Turku, A., Urvas, L., Adhikari, K., Oksanen, J., Steynen, Y., Yli-Kauhaluoma, J., Xhaard, H., Kukkonen, J.P., Wallén, E.A., Azulene as a biphenyl mimetic in orexin/hypocretin receptor agonists. *Bioorg. Med. Chem.*, 88, 117325, 2023.
11. García-Martínez, P., Bernardo, O., Borge, J., González, J., López, L.A., Synthesis of phenethylamine-azulene conjugates enabled by regioselective ring opening of aziridines. *Adv. Synth. Catal.*, 366, 1007, 2024.

12. Mackenroth, A.V., Ahrens, A., Wunsch, J.F., Berger, R., Rominger, F., Rudolph, M., Hashmi, A.S.K., Accessing indenoazulenes *via* a gold-catalysed cyclisation of cycloheptatrienyl-substituted 1, 5-diynes. *Adv. Synth. Catal.*, 366, 1331, 2024.
13. Liu, R., Fu, Y., Wu, F., Liu, F., Zhang, J.J., Yang, L., Popov, A.A., Ma, J., Feng, X., Modular synthesis of structurally diverse azulene-embedded polycyclic aromatic hydrocarbons by Knoevenagel-type condensation. *Angew. Chem. Int. Ed.*, 62, e202219091, 2023.
14. Mathey, P., Lirette, F., Fernández, I., Renn, L., Weitz, R.T., Morin, J., Annulated azuleno[2,1,8-*ija*]azulenes: Synthesis and properties. *Angew. Chem. Int. Ed.*, 135, e202216281, 2023.
15. Liang, Y., Wang, S., Tang, M., Wu, L., Bian, L., Jiang, L., Tang, Z.B., Liu, J., Guan, A., Liu, Z., Cascade synthesis of benzotriazulene with three embedded azulene units and large Stokes shifts. *Angew. Chem. Int. Ed.*, 135, e202218839, 2023.
16. Murai, M., Iba, S., Hamao, S., Kubozono, Y., Ota, H., Takai, K., Azulene-fused linearly π-extended polycyclic aromatic compounds: Synthesis, photophysical properties, and OFETs applications. *Bull. Chem. Soc. Jpn.*, 96, 1077, 2023.
17. Kirschbaum, T., Rominger, F., Mastalerz, M., Synthesis of a benzotrisazulene *via* trioxobenzotrisazulene. *Chem. Eur. J.*, 29, e202301470, 2023.
18. Tsuchiya, T., Higashibeppu, M., Mazaki, Y., Synthesis and properties of twisted and helical azulene oligomers and azulene-based polycyclic hydrocarbons. *ChemistryOpen*, 12, e202100298, 2023.
19. Li, W., Yang, J., Zhao, L., Couch, D., San Marchi, M., Hansen, N., Morozov, A.N., Mebel, A.M., Kaiser, R.I., Gas-phase preparation of azulene ($C_{10}H_8$) and naphthalene ($C_{10}H_8$) *via* the reaction of the resonantly stabilized fulvenallenyl ($C_7H_5^{\cdot}$) and propargyl ($C_3H_3^{\cdot}$) radicals. *Chem. Sci.*, 14, 9795, 2023.
20. Xiao, M. and Gao, X., Design, synthesis and anti-inflammatory activity of azulene derivatives containing benzimidazole unit. *Chin. J. Org. Chem.*, 43, 3246, 2023.
21. Noor, R., Zahoor, A.F., Mansha, A., Khan, S.G., Haq, A.U., Ahmad, S., Al-Hussain, S.A., Irfan, A., Zaki, M.E., Synthetic potential of regio-and stereoselective ring expansion reactions of six-membered carbo-and heterocyclic ring systems: A review. *Int. J. Mol. Sci.*, 24, 6692, 2023.
22. Shoji, T., Takeuchi, M., Uda, M., Ariga, Y., Yamazaki, A., Sekiguchi, R., Ito, S., Synthesis of azuleno [2, 1-b] quinolones and quinolines *via* Brønsted acid-catalyzed cyclization of 2-arylaminoazulenes. *Molecules*, 28, 5785, 2023.
23. Akram, W., Tagde, P., Ahmed, S., Arora, S., Emran, T.B., Babalghith, A.O., Sweilam, S.H., Simal-Gandara, J., Guaiazulene and related compounds: A review of current perspective on biomedical applications. *Life Sci.*, 316, 121389, 2023.

24. Manea-Saghin, A.M., Ion, A.E., Kajzar, F., Nica, S., Second order nonlinear optical properties of poled films containing azobenzenes tailored with azulen-1-yl-pyridine. *Heliyon*, 9, e17360, 2023.
25. Liu, H.C., Ruan, K., Ma, K., Fei, J., Lin, Y.M., Xia, H., Synthesis of metalla-dual-azulenes with fluoride ion recognition properties. *Nat. Commun.*, 14, 5583, 2023.
26. Park, S., Kim, C.E., Jeong, J., Ryu, H., Maeng, C., Kim, D., Baik, M.H., Lee, P.H., Selective ring expansion and C– H functionalization of azulenes. *Nat. Commun.*, 14, 7936, 2023.
27. (a) Shoji, T., Inoue, Y., Ito, S., First synthesis of 1-(indol-2-yl)azulenes by the Vilsmeier-Haack type arylation with triflic anhydride as an activating reagent. *Tetrahedron Lett.*, 53, 1493, 2012. (b) Shoji, T., Okada, K., Ito, S., Toyota, K., Morita, N., Synthesis of 1-(pyridyl, quinolyl, and isoquinolyl) azulenes by Reissert-Henze type reaction. *Tetrahedron Lett.*, 51, 5127, 2010. (c) Shoji, T., Higashi, J., Ito, S., Toyota, K., Asao, T., Yasunami, M., Morita, N., Synthesis and redox behavior of 1-azulenyl sulfides and efficient synthesis of 1,1'-biazulenes. *Eur. J. Org. Chem.*, 2008, 7, 1242, 2008.
28. Shoji, T., Ito, S., Toyota, K., Morita, N., Synthesis of 5-heteroaryl- and 5,7-bis(heteroaryl)azulenes by electrophilic susbtitution of 1,3-di-tert-butyl-azulene with triflates of N-containing heterocycles. *Eur. J. Org. Chem.*, 80, 1059, 2010.
29. Kurotobi, K., Miyauchi, M., Takakura, K., Murafuji, T., Sugihara, Y., Direct introduction of a boryl substituent into the 2-position of azulene: Application of the Miyaura and Smith mthods to azulene. *Eur. J. Org. Chem.*, 2003, 18, 3663, 2003.
30. Murai, M., Takami, K., Takeshima, H., Takai, K., Iridium-catalyzed dehydrogenative silylation of azulenes based on regioselective C–H bond activation. *Org. Lett.*, 17, 1798, 2015.
31. Yang, P.-W., Yasunami, M., Takase, K., The formation of azulene derivatives by the reaction of 2h-cyclohepta[b]furan-2-ones with enamines. *Tetrahedron Lett.*, 12, 4275, 1971.
32. Amir, E., Amir, R.J., Campos, L.M., Hawker, C.J., Stimuli-responsive azulene-based conjugated oligomers with polyaniline-like properties. *J. Am. Chem. Soc.*, 133, 10046, 2011.
33. Xin, H., Ge, C., Yang, X., Gao, H., Yang, X., Gao, X., A rational synthesis of azulene. *Chem. Sci.*, 7, 6701, 2016.
34. Nozoe, T., Seto, S., Matsumura, S., Tropone and its derivatives. VIII On the Reaction of Halotropones. *Proc. Jpn. Acad.*, 28, 483, 1952.
35. Nozoe, T., Seto, S., Matsumura, S., Asano, T., Synthesis of azulene derivatives from troponoids and cyanoacetic ester. *Proc. Jpn. Acad.*, 32, 339, 1956.
36. Nozoe, T., Seto, S., Matsumura, S., Synthesis of 2-substituted azulenes by nucleophilic substitution reactions of 2-haloazulene derivatives. *Bull. Chem. Soc. Jpn.*, 35, 1990, 1962.
37. In Memorian: Tetsuo Nozoe (1902-1996). *Eur. J. Org. Chem.*, 2004, 4, 899, 2004.

38. Yao, J., Cai, Z., Liu, Z., Yu, C., Luo, H., Yang, Y., Yang, S., Zhang, G., Zhang, D., Tuning the semiconducting behaviors of new alternating dithienyldiketopyrrolopyrrole–azulene conjugated polymers by varying the linking positions of azulene. *Macromolecules*, 48, 2039, 2015.
39. Chamelot, G., Idir, M., Leclerc, M., Morin, J.-F., Synthesis and properties of 2,6-azulene-based conjugated polymers and their applications in dispersing single-walled carbon nanotubes. *Polym. Chem.*, 14, 1206, 2023.
40. Matsuda, T., Goya, T., Liu, L., Sakurai, Y., Watanuki, S., Ishida, N., Murakami, M., Azulenophenanthrenes from 2,2'-di(arylethynyl)biphenyls through C-C bond cleavage of a benzene ring. *Angew. Chem. Int. Ed.*, 52, 6492, 2013.
41. Claus, V., Schukin, M., Harrer, S., Rudolph, M., Rominger, F., Asiri, A.M., Hashmi, A.S.K., Gold-catalyzed dimerization of diarylalkynes: Direct access to azulenes. *Angew. Chem. Int. Ed.*, 57, 12966, 2018.
42. Zhou, F., Shi, W., Liao, X., Yang, Y., Yu, Z.-X., You, J., Palladium-catalyzed [3+2] annulation of alkynes with concomitant aromatic ring expansion: A concise approach to (pseudo)azulenes. *ACS Catal.*, 12, 676, 2022.

4

Azaazulenes: Synthesis, Reactivity, and Applications

R. Balavardhana Rao Adapa

Department of Chemistry, Vallurupalli Nageswara Rao Vignana Jyothi Institute of Engineering & Technology, Hyderabad, India

Abstract

Azaazulenes, nitrogen-containing analogs of azulene, are fascinating heterocyclic compounds with distinctive structural, electronic, and biological properties. This chapter explores their chemistry, beginning with an overview of their structural diversity and various isomeric forms. Detailed synthetic methodologies are presented, including the amino-Heck reaction using oximes, transformations of azaheptafulvenes, and reactions involving 5-nitrosotropolone and pyridine. Classical approaches such as Buchner ring expansion, Bischler–Napieralski reactions, and intramolecular cyclizations are also highlighted. The chapter emphasizes the reactivity of azaazulenes, focusing on their participation in key coupling reactions such as the Suzuki, Negishi, Sonogashira–Hagihara, and Hartwig–Buchwald reactions, which demonstrate their utility in the construction of complex molecules. Additionally, the applications of azaazulenes are discussed, particularly their roles as phosphodiesterase 2 (PDE2) inhibitors, antiviral agents, and epigenetic modulators, showcasing their therapeutic potential. By bridging fundamental azaazulene chemistry and cutting-edge applications, this chapter provides a comprehensive guide to their synthesis, reactivity, and functionality, serving as a valuable resource for researchers exploring this versatile class of heterocycles.

Keywords: Azaazulenes, troponoids, amino-Heck reactions, Buchner ring expansion, fuzed azaazulenes

Email: balavardhanarao_ar@vnrvjiet.in

Neha Rani Kumar and Abhijeet R. Agrawal (eds.) Azulene-Based Materials: Design, Synthesis, and Applications, (81–116) © 2026 Scrivener Publishing LLC

4.1 Introduction

Azaazulenes, a class of bicyclic aromatic hydrocarbons derived from azulene, have garnered significant attention due to their unique properties and potential applications. Azulene itself, a non-benzenoid aromatic hydrocarbon, exhibits a distinctive blue color and a dipole moment, distinguishing it from benzene. Incorporating nitrogen atoms into the azulene framework leads to a diverse array of azaazulenes, each with its own characteristic electronic and structural features [1–6].

The positioning of nitrogen atoms within the azulene ring system significantly influences the compound's properties. For instance, nitrogen atoms in the five-membered ring can enhance the electron-withdrawing character, while those in the seven-membered ring may exhibit electron-donating behavior. This interplay of electronic effects can modulate the reactivity, stability, and spectroscopic properties of azaazulenes.

Azaazulenes have been extensively studied for their potential applications in various fields. Their unique electronic properties make them promising candidates for materials science, such as organic light-emitting diodes and organic field-effect transistors. The ability of azaazulenes to form stable radical cations has also attracted interest in the context of energy storage and redox-active materials. Moreover, certain azaazulenes exhibit biological activity, suggesting potential applications in medicinal chemistry [7].

4.2 Types of Azaazulene Isomers

Azaazulenes are formed by replacing one or more carbon atoms in the azulene ring with nitrogen atoms. This substitution leads to a variety of isomers, each with distinct chemical and physical properties.

Azaazulene can result in the replacement of one or more carbon atoms by nitrogen in five- or seven-membered rings of azulene leading to the formation of different isomers [8]. Replacing one carbon with nitrogen results in the formation of monoazaazulenes. More common monoazaazulenes researchers studied are the 1-azaazulene and 2-azaazulenes (Figure 4.1).

4.2.1 Monoazaazulenes

More isomers of monoazaazulenes can be formed by replacement of carbon atoms in seven membered rings like 4-azaazulenes, 5-azaazulenes and 6-azaazulenes (Figure 4.2).

cyclohepta[b]pyrrole cyclohepta[c]pyrrole

Figure 4.1 Monoazaazulenes.

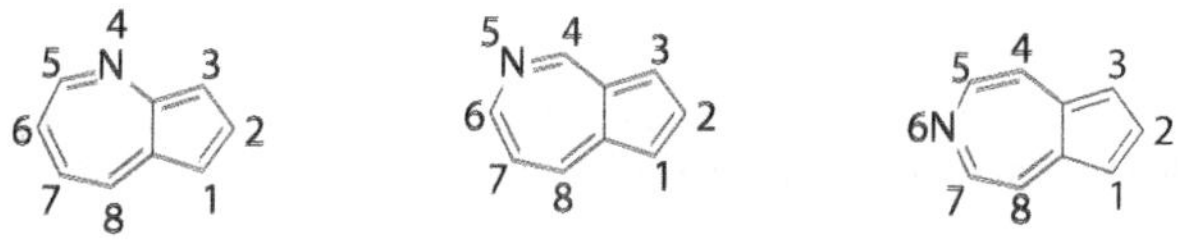

4-Azaazulne (or) cyclopenta[b]azepine

5-Azaazulne (or) cyclopenta[c]azepine

6-Azaazulne (or) cyclopenta[d]azepine

Figure 4.2 Isomers of monoazaazulenes.

4.2.2 Diazaazulenes

Diazaazulenes are a group of organic compounds derived from azulene, a bicyclic aromatic hydrocarbon. They are characterized by the presence of two nitrogen atoms within their molecular structure. These nitrogen atoms can be located in either the seven-membered ring, the five-membered ring, or both.

The substitution of carbon atoms with nitrogen atoms in the azulene framework leads to a variety of diazaazulene isomers, each with distinct chemical and physical properties. These properties are influenced by the position of the nitrogen atoms, as well as other factors such as substituents on the molecule. Diazaazulenes have attracted significant interest in recent years due to their potential applications in various fields. Plausible diazaazulenes are shown in Figure 4.3 some of which are synthesized and theoretically reported.

4.2.3 Triazaazulenes

Triazaazulenes are a unique class of heterocyclic compounds featuring a seven-membered ring structure adorned with three nitrogen atoms. Their aromatic nature and electron-deficient centers contribute to their distinctive chemical and physical properties. These compounds have shown promise in materials science, pharmaceuticals, and organic synthesis, with potential applications in areas such as organic light-emitting diodes, drug development, and the preparation of complex molecules (Figure 4.4).

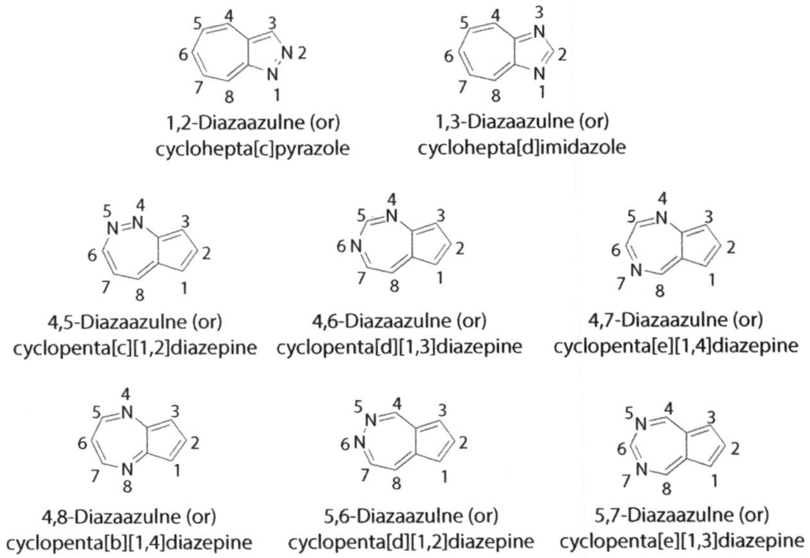

Figure 4.3 Isomers of diazaazulenes.

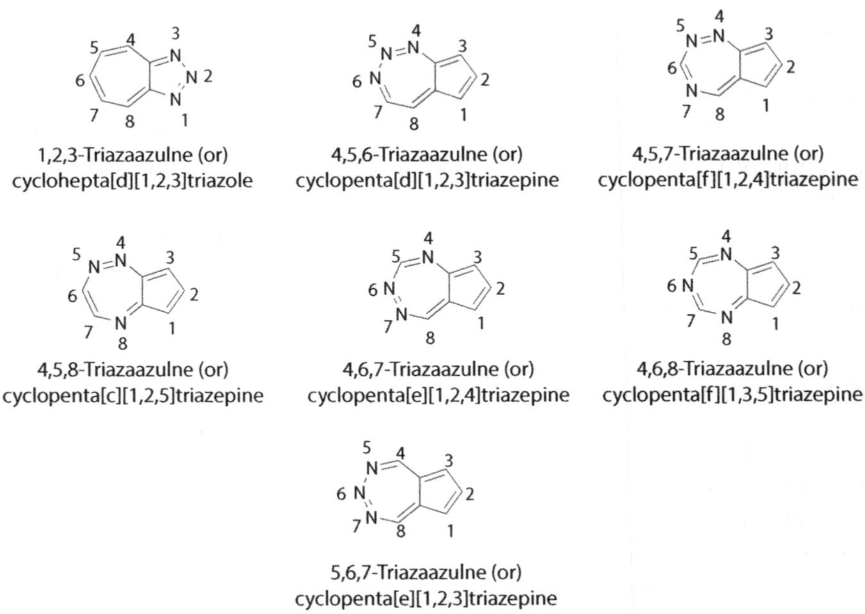

Figure 4.4 Isomers of triazaazulenes.

4.3 Synthesis of Azaazulenes

Azaazulenes were first reported in the mid-20th century [9, 10]. The synthesis of azaazulenes has been a subject of ongoing research. Various synthetic strategies, including cycloadditions, ring-opening reactions, and rearrangements, have been employed to access these compounds. The development of efficient and selective synthetic methods is crucial for further exploration and application of azaazulenes.

Azaazulenes represent a fascinating class of heterocyclic compounds with diverse properties and potential applications. Their unique electronic characteristics, coupled with their structural diversity, make them attractive targets for both fundamental and applied research. As synthetic methodologies and understanding of their properties continue to evolve, azaazulenes are poised to play an increasingly important role in various scientific disciplines.

4.3.1 Syntheses from Reactive Troponoids

Azaazulenes, especially those containing nitrogen atoms in the five-membered ring, can be synthesized efficiently from reactive troponoids. These troponoids include derivatives like 2-amino-, 2-methoxy-, and 2-chlorotropones. In the 1950s, several annelation reactions were developed, using these reactive troponoids to form azaazulenes such as cyclohepta[h] pyrroles (1-azaazulenes) and cycloheptimidazoles (1,3-diazaazulenes).

The typical approach involves cyclization reactions where the substituents at C-2 on tropones guide the regiospecific formation of these azaazulenes. For instance, electron-withdrawing groups at C-2 favor cyclization at C(7)-C(1), while electron-donating groups (EDGs) like methoxy induce cyclization at C(2)-C(1). These cyclization processes are influenced by the nature of the substituents, leading to different azaazulene structures.

One-step synthesis methods have also been reported, such as the reaction between tropone and pyridinium ylide in the presence of ammonium bromide, though these often suffer from low yields. Various tropones substituted with groups such as methoxy, dimethylamino, or tosylates have expanded the synthetic versatility of azaazulenes, allowing the construction of more complex structures through advanced cyclization and condensation techniques. Recent developments include the use of reagents like acylhydrazines, ketene dimer, and pyridinium ylides, further broadening the scope of azaazulene synthesis. These methods remain pivotal in producing azaazulene derivatives, which are valuable in medicinal chemistry due to their structural similarity to biologically active compounds.

4.3.1.1 *Synthesis from 2-Aminotropones*

The synthesis of 1-azaazulene involves a multi-step process that begins with the modification of the starting material, 2-aminocyclohepta-2,4,6-trienone (**4.1**). This compound is treated with diethyl malonate in the presence of a base to introduce a new carbon-carbon and carbon-nitrogen bond. The resulting intermediate (**4.2**) is then subjected to hydrolysis and decarboxylation to convert into a ketone (**4.3**).

In the final steps, the ketone is transformed into a chloroimine (**4.4**) using phosphorus oxychloride. This chloroimine is then reduced using hydrazine, copper sulfate, and acetic acid to form the desired 1-azaazulene product (**4.5**). This compound is characterized by its seven-membered ring fused to a five-membered ring containing a nitrogen atom (Scheme 4.1).

4.3.1.2 *Synthesis from 2-Aminotropones and Ketene Dimer*

The synthesis of 3-acetyl-2-chloro-1-azaazulene (**4.8**) involves a three-step process. First, 2-aminotropone (**4.1**) and diketene undergo a [4+2] cycloaddition to form 2-acetoacetamidotropone (**4.6**). Second, a condensation reaction between the acetyl and amine groups within this intermediate leads to the formation of 3-acetyl-1-azaazulan-2-one (**4.7**) (Scheme 4.2).

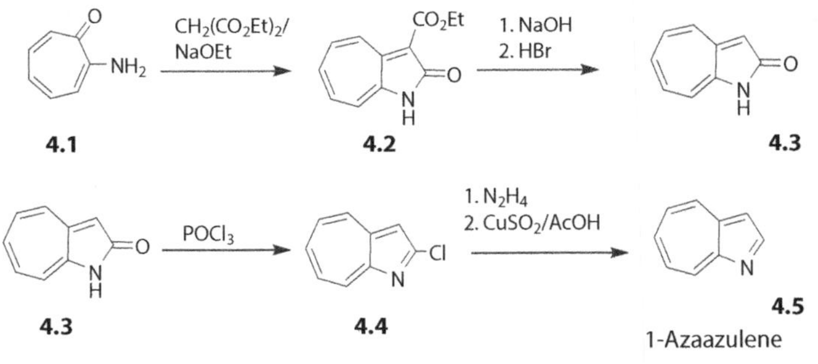

Scheme 4.1 Synthesis of 1-azaazulene from amino tropolone.

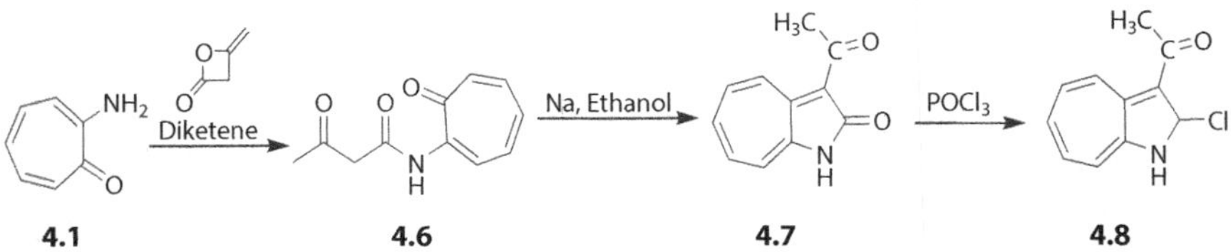

Scheme 4.2 Synthesis of azaazulene from 2- aminotropones and ketene dimer.

Finally, chlorination with phosphorus oxychloride introduces a chlorine atom at the C-2 position, yielding the desired product (**4.8**). This synthetic route provides a versatile approach to accessing azaazulene derivatives [11].

4.3.1.3 Synthesis from 2-Aminotropone and Heterocumulenes

Nitta group [12] reported a straightforward method for synthesizing heteroazulenes through the reaction of 2-aminotropone (**4.1**) with heterocumulenes. When **1** was treated with N,N'-diphenylcarbodiimide (**4.9a**) in the presence of t-BuOK, a 1,3-diazaazulene derivative (**4.10a**) was obtained. Similarly, the reaction of **1** with heterocumulenes like phenyl isothiocyanate (**4.9b**), phenyl isocyanate (**4.9c**), and carbon disulfide (**4.11d**), also in the presence of t-BuOK, led to the formation of heteroazulenes (**4.12b-f**) (Scheme 4.3).

In further investigations, Nitta group [13–15] described the reactions of (tropon-2-ylimino)pnictoranes (**4.11a–d**) with heterocumulenes (**4.12a–d**), presenting an efficient approach for the construction of azaazulene frameworks (**4.13a–d**). By optimizing reaction conditions and exploring different substrates, the Nitta group demonstrated the broad applicability of this approach, thereby providing a valuable synthetic route for the formation of complex heterocyclic structures. The reaction pathway and structural transformations are outlined in Scheme 4.4.

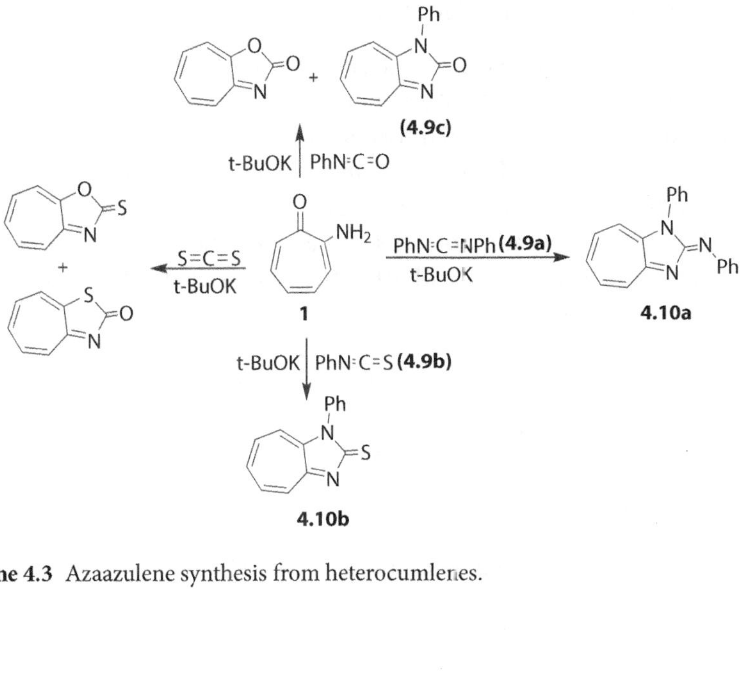

Scheme 4.3 Azaazulene synthesis from heterocumlenes.

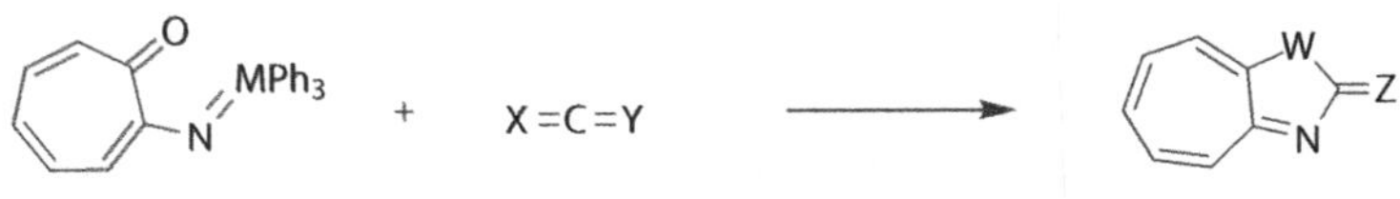

4.11a : M = P
4.11b : M = As
4.11c : M = Sb
4.11d : M = Bi

4.12a : X = Y = NPh
4.12b : X = S, Y = NPh
4.12c : X = O, Y = NPh
4.12d : X = Y = S

4.13a : W = Z = NPh
4.13b : W = O, Z = NPh
4.13c : W = Z = O
4.13d : W = O, Z = S

Scheme 4.4 Synthesis of azaazulenes from (tropon-2-ylimino)pnictoranes with heterocumulenes.

4.3.1.4 Synthesis from 2-Troponylhydrazides

The use of 2-troponylhydrazides as starting materials provides a versatile approach to the synthesis of 1-azaazulene derivatives. Under acidic conditions, 2-troponylhydrazides (**4.14**) undergoes cyclization to form a bicyclic intermediate (**4.15**). In the final step, this intermediate (**4.15**) further undergoes modifications in presence of phosphorus oxychloride and hydrazine and results in 1-azaazulene derivatives (**4.16**) Chemists can selectively modify the 2- and 3- positions of the 1-azaazulene ring. This flexibility in derivatization allows for the introduction of a wide range of functional groups, including alkyl, aryl, and heteroaryl substituents. These modifications can significantly influence the properties of the 1-azaazulene derivatives (Scheme 4.5) [16].

4.3.1.5 Synthesis from 2-Methoxytropones

The synthesis of a diazaazulene derivative (**4.19**) begins with reacting methoxy-substituted cycloheptatrienone (**4.17**) with thiourea in the presence of sodium ethoxide, which forms a thioamide derivative featuring a fused 1,3-thiazine ring (**4.18**). This intermediate is then subjected to oxidative desulfurization using either dilute nitric acid or hydrogen peroxide, leading to the formation of a fused diazaazulene structure, specifically a 1,3-diazaazulene derivative (**4.19**). This two-step synthesis efficiently

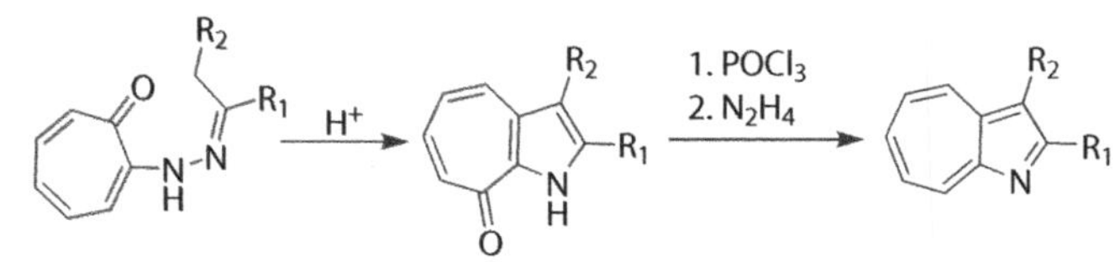

4.14 **4.15** **4.16**

Scheme 4.5 Azaazulene synthesis from 2-troponylhydrazides.

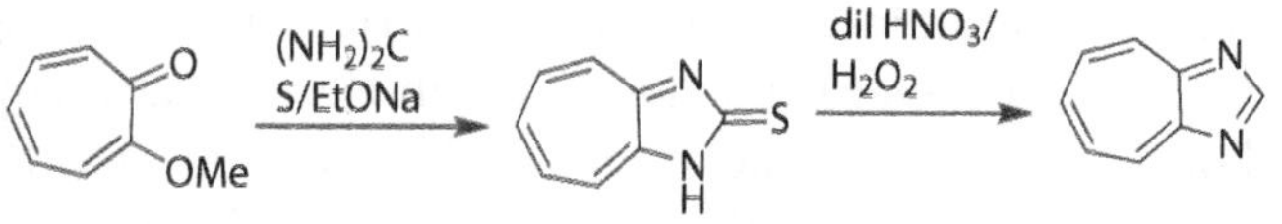

4.17 4.18 4.19

Scheme 4.6 Azaazulene synthesis from 2-methoxytropones.

converts the starting material into the desired diazaazulene compound (Scheme 4.6) [16].

4.3.1.6 Regiospecific Synthesis of 1,3-Diazaazulene

The regioselectivity of cyclization reactions in organic chemistry is a cornerstone of molecular design and synthesis. Tropones have garnered significant attention due to their propensity for regioselective cyclizations. This study delves into the mechanistic underpinnings of these transformations, focusing on the influence of substituents on the C-2 position. The electronic properties of substituents at the C-2 position of tropones play a pivotal role in determining the regioselectivity of cyclization. Electron-withdrawing groups, such as halogens or nitro groups, increase the electrophilicity of the C-2 carbon, rendering it more susceptible to nucleophilic attack. Conversely, EDGs, such as alkoxy or amino groups, decrease the electrophilicity of C-2, redirecting the nucleophilic attack to the C-7 position.

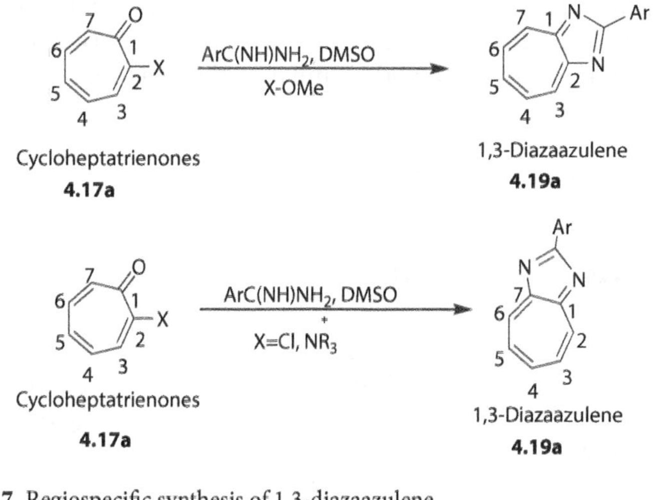

Scheme 4.7 Regiospecific synthesis of 1,3-diazaazulene.

The reaction of amidines with cycloheptatrienones carrying mobile groups at C-2 gives 1,3-diazaazulenes. Thus, an electron-attracting group, such a s Cl or R_3N^+, at C(2) induces clean cyclization at C(7)-C(l), whereas a mesomerically EDG, such as OMe, induces clean cyclization at C(2)-C(1) (Scheme 4.7) [17].

4.3.1.7 *Synthesis from 2-Chlorotropones and Acyl Hydrazides*

The reaction between 2-chlorotropone (**4.20**) and various hydrazides (**4.21**, **4.23**) catalyzed by ethanolic sodium hydroxide, provides a versatile synthetic route to 1-azaazulene derivatives (**4.22**, **4.24**). The cyclization occurs exclusively between the C-2 and C-1 positions of the tropone, demonstrating high region control. The reaction can be modified by varying the substituents on the hydrazide, allowing for the synthesis of diverse derivatives.

The mechanism of the reaction involves nucleophilic addition of the deprotonated hydrazide to the electrophilic C-2 carbon of the tropone, followed by intramolecular cyclization. The chlorine substituent on the tropone plays a crucial role in enhancing the electrophilicity of the C-2 position, facilitating the nucleophilic attack (Scheme 4.8) [18].

4.3.1.8 *Synthesis from 2-Substituted Tropones and N-Silylenamine or the Enamine*

Konakahara introduced a novel approach for synthesizing functionalized 1-azaazulenes *via* the reaction of N-silylenamines or enamines with 2-substituted tropones. The reaction between N-silylenamine (**4.25a**) and 2-chlorotropone (**4.26a**) in the presence of potassium fluoride (KF) yielded 2-substituted-azaazulene (**4.27a**). Similarly, the reactions of enamines

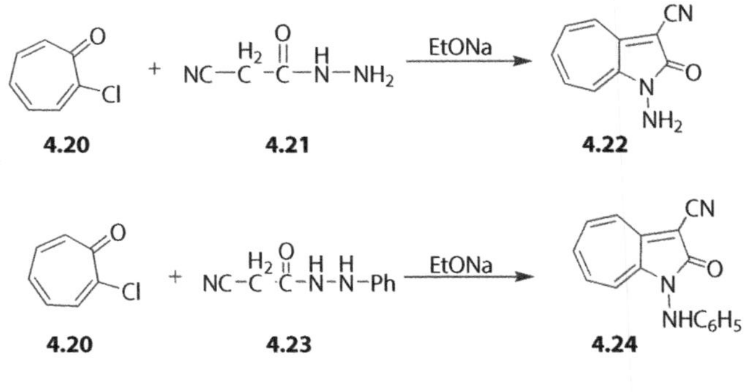

Scheme 4.8 Azaazulene synthesis from 2- chlorotropones and acyl hydrazides.

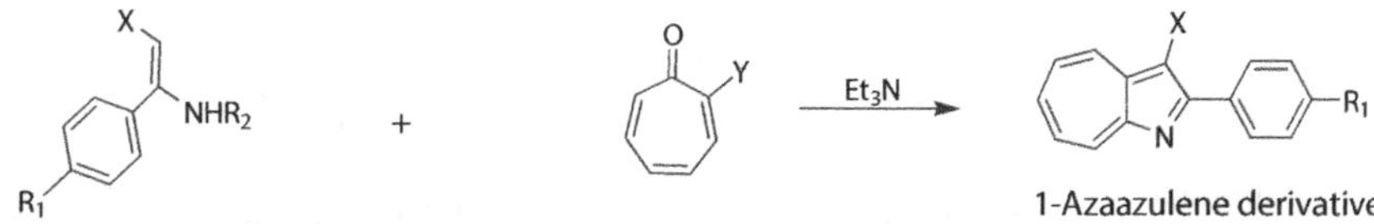

1-Azaazulene derivatives

4.25a : X = 2-pyridyl, R_1 = H, R_2 = $SiMe_3$
4.25b : X = Ar, R_1 = H, R_2 = H
4.25c : X = Ar, R_1 = OMe, R_2 = H
4.25d : X = Ar, R_1 = NMe_2, R_2 = H
4.25e : X = 2-pyridyl, R_1 = H, R_2 = H
Ar = 3-methyl-5-isoxazolyl

4.26a : Y = Cl
4.26b : Y = OTs

4.27a-e

Scheme 4.9 Azaazulene synthesis from 2-substituted tropones and N-silylenamine or the enamine.

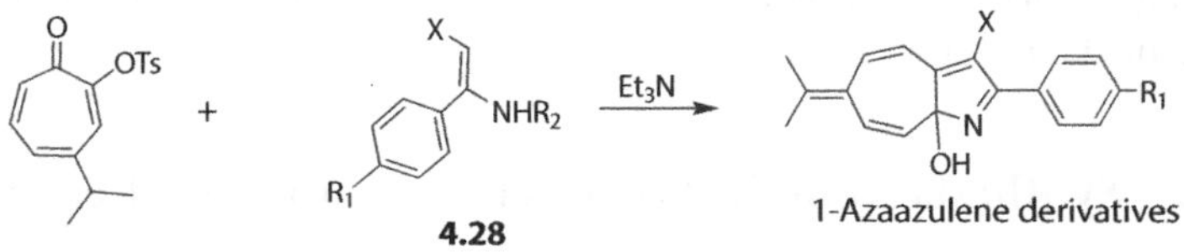

4.28

1-Azaazulene derivatives

4.25c : X = Ar, R_1 = OMe, R_2 = H
4.25d : X = Ar, R_1 = NMe_2, R_2 = H
4.29f : X = Ar, R_1 = Me, R_2 = H
Ar = 3-methyl-5-isoxazolyl

4.29c : X = Ar, R_1 = OMe
4.29d : X = Ar, R_1 = NMe_2
4.29f : X = Ar, R_1 = Me

Scheme 4.10 Reaction of enamines with 4-isopropyltropolone tosylate.

(**4.25b–e**) with either 2-chlorotropone (**4.26a**) or tropolone tosylate (**4.26b**) in the presence of triethylamine (Et_3N) resulted in the formation of the corresponding azaazulenes (**4.27b–e**) (Scheme 4.9) [19].

Notably, the reaction of enamines (**4.25c–f**) with 4-isopropyltropolone tosylate (**4.28**) led to the production of 6-isopropylidene-8a-hydroxy-6,8a-dihydroazaazulene derivatives (**4.29c–f**), respectively (Scheme 4.10) [19].

4.3.1.9 *Synthesis from Tropone and Pyridinium Ylide*

The reaction between a tropone (**4.30**) and a pyridinium ylide (**4.31a–b**), catalyzed by acetic anhydride and acetic acid, results in the formation of a 1-azaazulene derivative (**4.32a–b**). The ylide, acting as a nucleophile, attacks the electrophilic carbon of the tropone, leading to ring closure and the formation of the heterocyclic product. The electron-withdrawing effect of the pyridinium ylide enhances the reactivity of the tropone. The type of ylide used significantly influences the structure of the resulting 1-azaazulene,

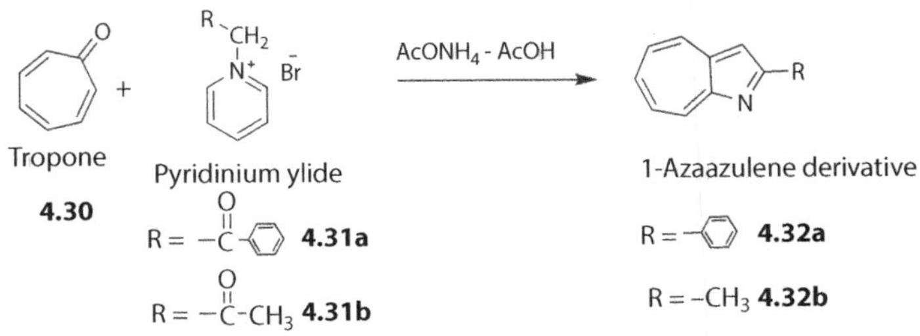

Scheme 4.11 Azaazulene synthesis from tropone and pyridinium ylide.

with alkyl-substituted ylides leading to alkyl-substituted derivatives and aryl-substituted ylides leading to aryl-substituted derivatives. Other functional groups on the ylide can also affect the properties of the final product (Scheme 4.11) [20].

4.3.2 Synthesis from the Oximes by Amino-Heck Reaction

Narasaka [21–23] reported sophisticated synthetic methods involving the amino-Heck reaction applied to cycloheptatrienylmethyl ketone oximes (**4.33a–c**). In this process, cycloheptatrienylmethyl ketone oximes (**4.33a–c**) were treated with a catalytic amount of $Pd(dba)_2$, t-Bu_3P, and triethylamine in the presence of MS 4A, followed by oxidation with MnO_2, yielding various 1-azaazulenes (**4.34a–c**) in moderate to good yields (Scheme 4.12).

This method was further utilized for the synthesis of fused azaazulene derivatives, leading to the formation of 6-azanaphth[1,2-b]azulene (**4.36**) and 11,12-dihydro-6-azanaphth[1,2-b]azulene (**4.37**) from o-pentafluorobenzoyloxime (**4.35**) (Scheme 4.13) [23].

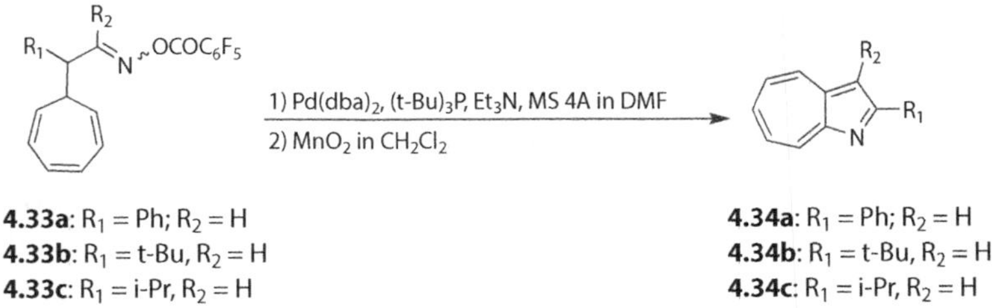

Scheme 4.12 Synthesis of azaazulene from the oximes by amino-Heck reaction.

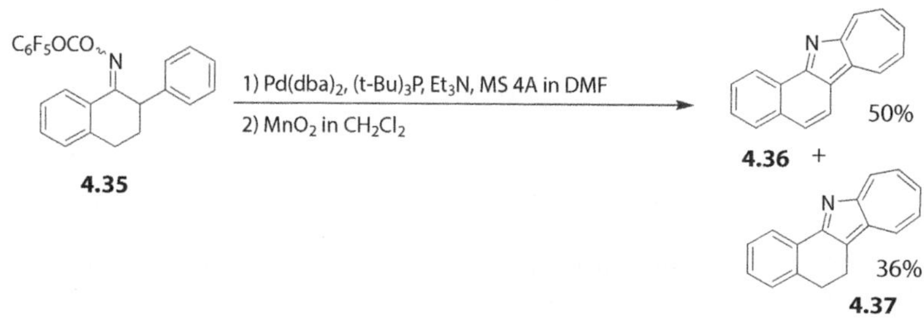

Scheme 4.13 Synthesis of fused azaazulene derivatives.

4.3.3 Synthesis from Azaheptafulvenes

The [8+2]-cycloaddition of 8-azaheptafulvenes (also known as cyclo-heptatrienylideneamines or troponeimines) offers an efficient approach for constructing the 1-azaazulene framework. The reaction of 8-aryl-8-azaheptafulvenes (**4.38a–c**) with trimethylsilylketene, followed by desilylation using TBAF and subsequent oxidation with MnO_2, produces 1-aryl-1-azaazulen-2(1H)-ones (**4.39a–c**) (Scheme 4.14) [24].

A concise and efficient method for synthesizing fully conjugated and substituted-1-azaazulenes was also developed. This approach involves treating 8-(substituted sulfonyloxy)-8-azaheptafulvenes (**4.40**) with ketonic methylene compounds (**4.41**) in the presence of a base, such as an amine. This reaction yields 1-azaazulene derivatives (**4.42**), as shown in Scheme 4.15 [25].

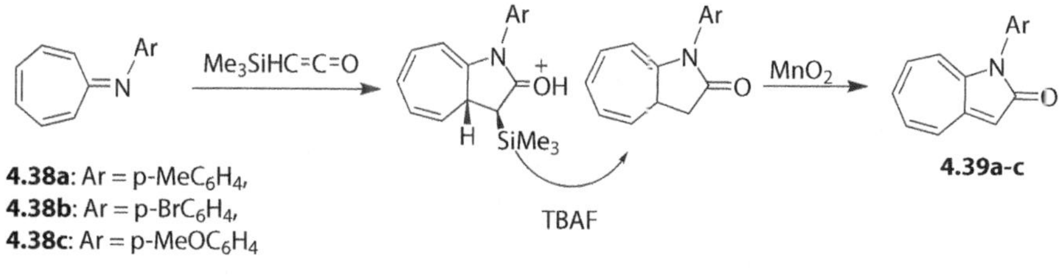

Scheme 4.14 Azaazulene synthesis from azaheptafulvenes.

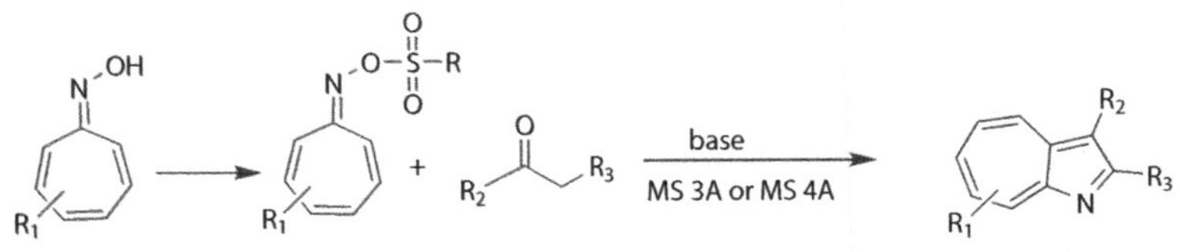

4.40 4.41 4.42

base : TEA, diisopropylethylamine, DBU, pyridine, DMAP, lutidine, ammonia sourses
R_1 = H, NH_2, OH, OR, CH_3S, CH_3SO, CH_3SO_2

Scheme 4.15 Synthesis of fully conjugated substituted-1-azaazulenes.

4.3.4 Synthesis from 5-Nitrosotropolone and Pyridine

A versatile method for constructing fused heteroaromatic rings based on 1,3-diazaazulene has been developed. The reaction of 5-nitrosotropolone (**4.43**) with various azines, including pyridine, pyridazine, isoquinoline, phthalazine, and thiazole, yields tricyclic heterocycles fused to the 1,3-diazaazulene core. The reaction proceeds through a tropoquinone intermediate (**4.46**), which was confirmed by its reaction with pyridine to produce pyrido[1',2':1,2]imidazo[4,5-e]tropolone (**4.44**) (Scheme 4.16) [26].

Nitta's pioneering work [27–31] led to the synthesis of uracil-annulated azaazulenes, which exhibit notable oxidizing properties. These compounds were prepared through two distinct routes: First is a direct coupling reaction of **4.20** with **4.47** in the presence of t-$BuNH_2$ and ethanol at room temperature to give **4.48** in high yields (Scheme 4.17). Another approach is through ring transformation reaction in which compound **4.49** was

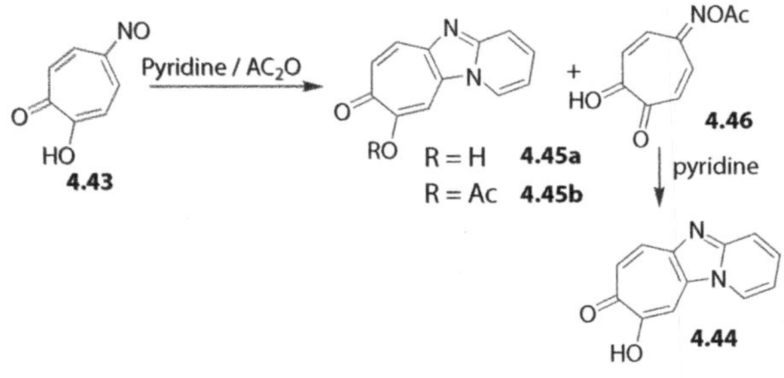

Scheme 4.16 Azaazulene synthesis from 5-nitrosotropolone and pyridine.

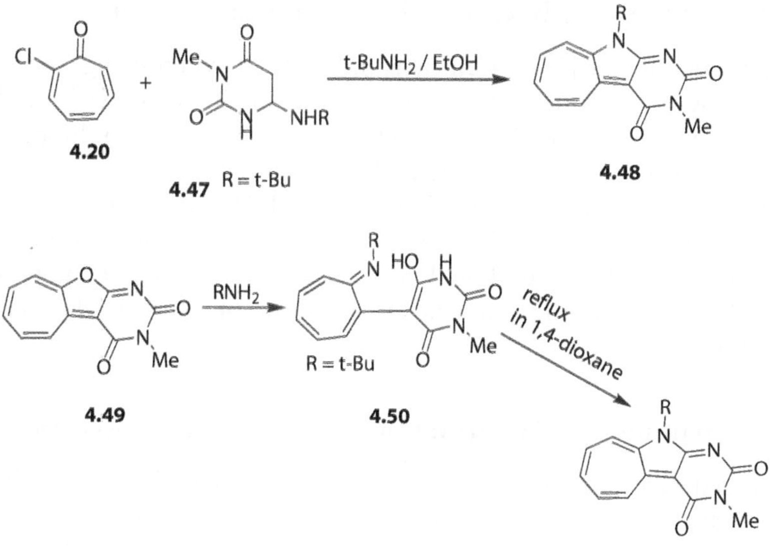

Scheme 4.17 Synthesis of uracil-annulated azaazulenes.

transformed into **4.50** by treatment with amines followed by heating at 90°C in 1,4-dioxane, also affording **4.48** in excellent yields. This versatility in the synthetic approach highlights the potential for further exploration and application of these unique azaazulene derivatives.

By using similar ring transformation, Nitta synthesized **4.53** from **4.51** with the amines *via* **4.52** (Scheme 4.18) [30].

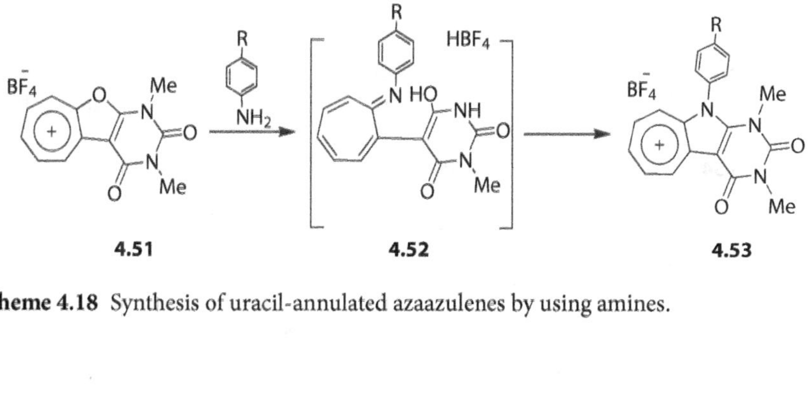

Scheme 4.18 Synthesis of uracil-annulated azaazulenes by using amines.

4.3.5 Synthesis by Buchner Ring Expansion Reaction

Aoyama developed [32] an efficient method for synthesizing 2-azaazulenes using the Buchner ring expansion reaction. This approach involves the reaction of trimethylsilyldiazomethane ($TMSCHN_2$) with thiobenzoyl isocyanate, generated from 2-phenylthiazol-4,5-dione (Scheme 4.19). The key intermediate, **4.55**, is formed by treating **4.54** with refluxing toluene. Subsequent addition of $TMSCHN_2$ followed by treatment with diisopropylamine ($i\text{-}Pr_2NH$) yields **4.56**. Finally, catalytic conversion of **4.56** using $Rh_2(OAc)_4$ produces the 2-azaazulene derivative (**4.58**) through intermediate **4.57**.

4.3.6 Synthesis by Abnormal Bischler–Napieralski Reaction

Ishikawa described an unconventional Bischler–Napieralski reaction (BNR) [33–35] that leads to an intriguing ring expansion. The treatment of **4.59** with $POCl_3$ followed by the evaporation and recrystallization resulted in the formation of **4.60**. Conversely, when **4.59** was treated with $POCl_3$ and the reaction was quenched with water and further subjected to extraction and chromatographic purification, compound **4.61** was obtained (Scheme 4.20).

4.3.7 Synthesis by Intramolecular Cyclization

4.3.7.1 Synthesis by Flash Vacuum Pyrolysis

Intramolecular cyclization was often used to construct a seven-membered ring. The formation of **4.64** was achieved by flash vacuum pyrolysis of **4.62** and **4.63** (Scheme 4.21) [36].

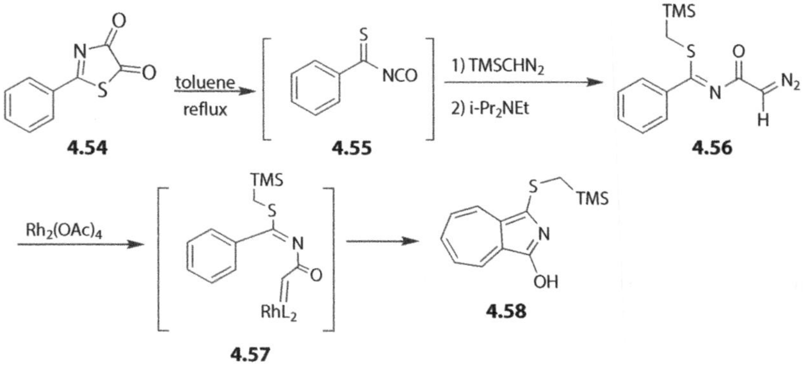

Scheme 4.19 Azaazulene synthesis by Buchner ring expansion reaction.

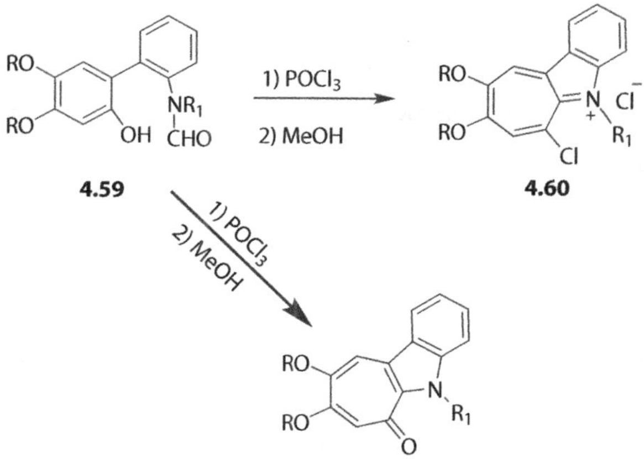

Scheme 4.20 Azaazulene synthesis by abnormal Bischler-Napieralski reaction.

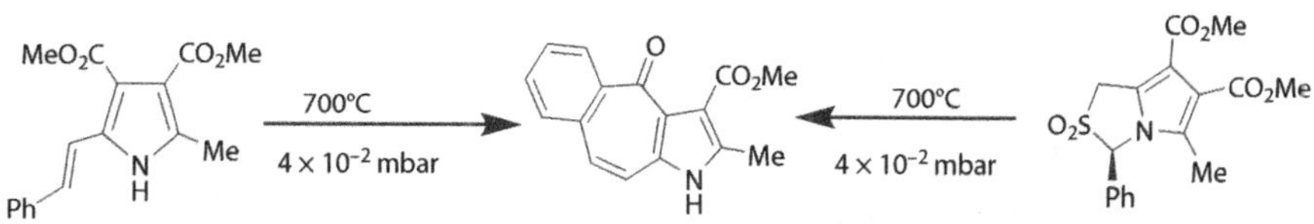

Scheme 4.21 Azaazulene synthesis by flash vacuum pyrolysis.

4.3.7.2 *Synthesis by Platinum-Catalyzed Intramolecular Cyclization of the Alkynes*

A platinum-catalyzed intramolecular cyclization of alkynes (**4.65**) provides an efficient route to the 1,5-diazaazulene skeleton. The reaction predominantly yields compound **4.66** as the major product (Scheme 4.22) [37].

4.3.8 Synthesis from Intramolecular Nucleophilic Attack by Organolithium Reagents

Pyrrole rings are prone to intramolecular nucleophilic attack by organolithium reagents. This property was exploited in the synthesis of azaazulene derivatives (**4.68**). The treatment of N-allylpyrrolocarboxamide (**4.67**) with lithium diisopropylamide induced an intramolecular cyclization, leading to the formation of **4.68** (Scheme 4.23) [38].

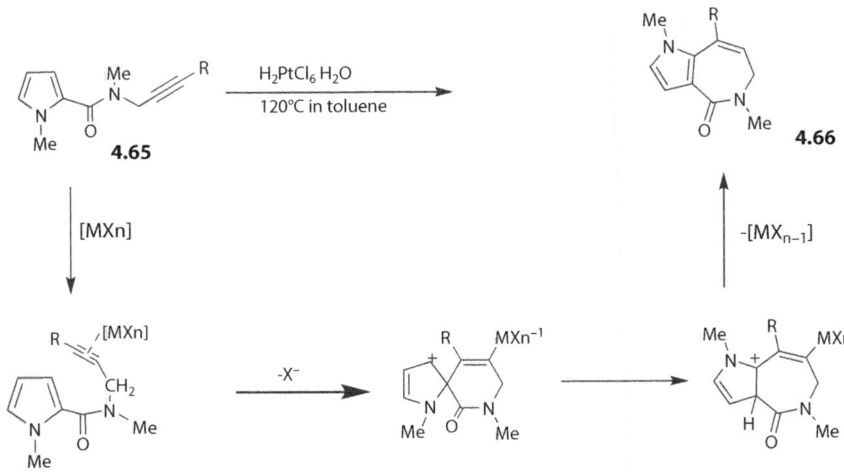

Scheme 4.22 Azaazulene synthesis by platinum-catalyzed intramolecular cyclization of the alkynes.

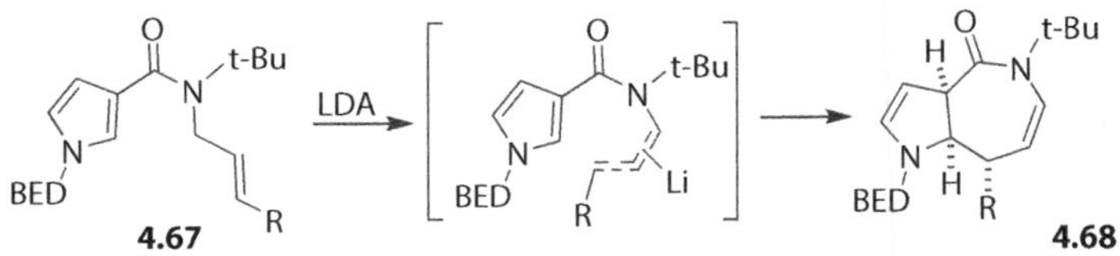

Scheme 4.23 Azaazulene synthesis from intramolecular nucleophilic attack by organolithium reagents.

4.3.9 Synthesis from Acid-Catalyzed Condensation of Dehydrotryptophan Ester with Aldehydes

Goto reported [39] that dehydrotryptophan ester (**4.69**) undergoes acid-catalyzed condensation with aldehydes to yield the corresponding 1,5-diazabenz[cd]azulene derivatives (**4.70**) (Scheme 4.24).

Khashashneh reported that 3-(2,5-diaminophenyl)indole (**4.71**) undergoes a double Pictet–Spengler condensation with aldehydes to yield compounds (**4.72**) (Scheme 4.25) [40].

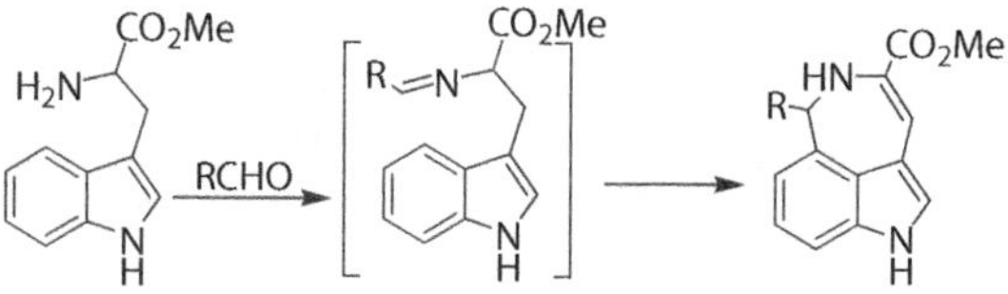

4.69 **4.70**

Scheme 4.24 Azaazulene synthesis from acid-catalyzed condensation of dehydrotryptophan ester with aldehydes.

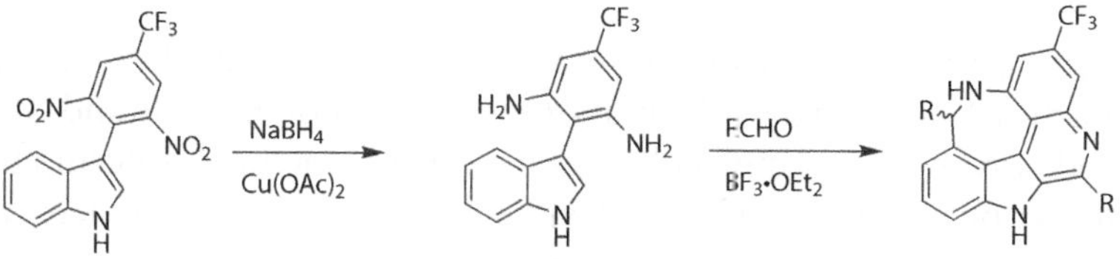

4.71 **4.72**

Scheme 4.25 Synthesis of azaazulene derivative from 3-(2,5-diaminophenyl)indole by double Pictet–Spengler condensation with aldehydes.

4.3.10 Synthesis by Bischler–Napieralski Reaction of the Carboxamides

The Bischler–Napieralski reaction of the carboxamides (**4.73**) using $POCl_3$ in MeCN under reflux gave **4.74** (Scheme 4.26) [41], and the reaction of the carboxamides (**4.75**) using polyphosphoric acid (PPA) at 140°C gave **4.76** (Scheme 4.27) [42].

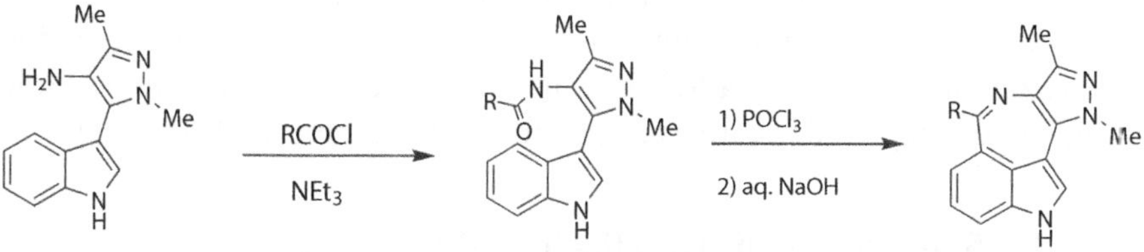

4.73 **4.74**

Scheme 4.26 Synthesis of 4.74.

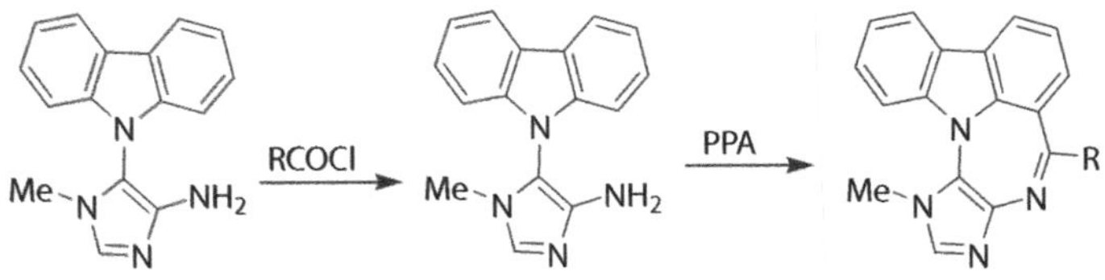

4.75 4.76

Scheme 4.27 Azaazulene synthesis by Bischler–Napieralski reaction of the carboxamides.

4.3.11 Syntheses Utilizing Intermolecular Condensations

The synthesis of azaazulenes has been extensively studied, with various approaches developed over the years. One of the most versatile and efficient methods involves intermolecular condensation reactions. Intermolecular condensation reactions offer a flexible and adaptable approach to the synthesis of azaazulenes. By carefully selecting the starting materials and reaction conditions, chemists can access a diverse array of derivatives with varying structural features and properties. This versatility has made intermolecular condensation a valuable tool for exploring the chemistry of azaazulenes and developing new synthetic routes.

A single-step annelation of a seven-membered ring onto a five-membered ring can be achieved through intermolecular condensations between 1,2-dialdehydes, diacid chloride, diamine etc., with a suitable condensation partner [43–46].

4.3.12 Synthesis from Dialdehyde and Ketone

Cycloaddition between the dialdehyde (**4.77**) and propan-2-one (**4.78**), led to the formation of a bicyclic intermediate (**4.79**), which on further rearrangement formed a new ring structure and finally underwent deprotection of the nitrogen using boron trifluoride etherate and sodium hydroxide. This synthetic route provides a versatile approach to accessing 2-azaazulene derivatives (**4.81**). By varying dialdehydes and ketones wide range of azaazulene derivatives can be formed (Scheme 4.28) [43, 44].

4.3.13 Synthesis from 1,2-Diacid Chloride and n,n'-Diphenylacetamidine

The reaction involves the condensation of 2,5-dimethyl-1-phenyl-1H-pyrrole-3,4-dicarbonyl dichloride (**4.86**) with N,N'-diphenylacetamidine (**4.87**) in the presence of triethylamine (Et_3N) to produce a polyazaazulene

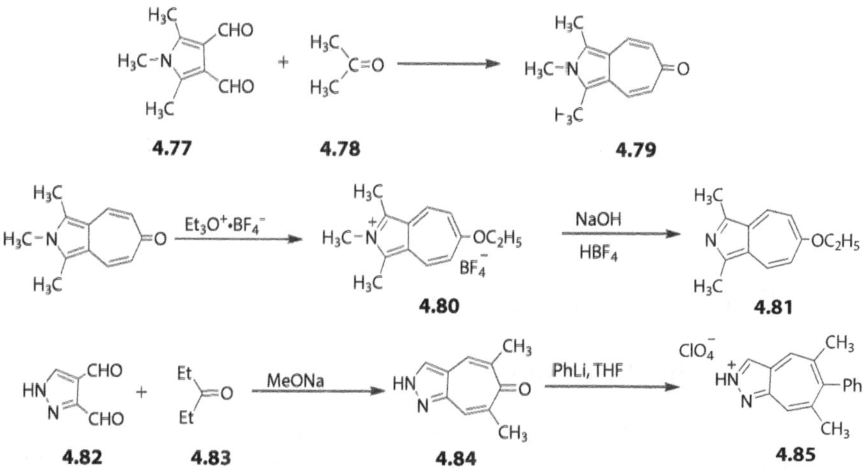

Scheme 4.28 Azaazulene synthesis from dialdehyde and ketone.

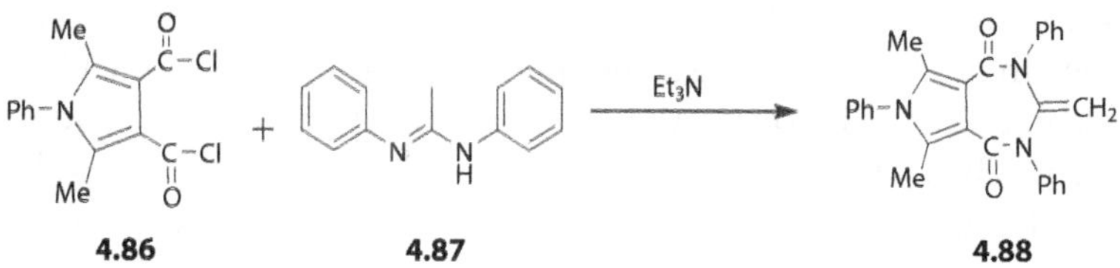

Scheme 4.29 Azaazulene synthesis from 1,2-diacid chloride and N,N'-diphenylacetamidine.

derivative (**4.88**). The process is likely initiated by a nucleophilic attack of the acetamidine nitrogen on one of the carbonyl carbons of the dicarbonyl chloride, followed by intramolecular cyclization to form the fused polyazaazulene ring system. The role of triethylamine is to deprotonate the acetamidine, increasing its nucleophilicity and facilitating the initial attack on the carbonyl carbon (Scheme 4.29) [45].

4.3.14 Synthesis from 1,2-Diamine and Ethyl Acetoacetate

The condensation reaction between 3,4-dicyano-1,2,5-triaminopyrrol (**4.89**) and ethyl acetoacetate (**4.90**) led to the creation of polyazaazulenes (**4.91**) (Scheme 4.30) [46]. Here, amino groups can be condensed with ethyl acetoacetate under the influence of p-toluenesulphonic acid intermolecular condensation of dialdehydes with suitable reagents results in the formation of mono and diazaazulenes. On the other hand, intermolecular condensation of diacid chlorides and diamines results in the formation of polyazaazulenes.

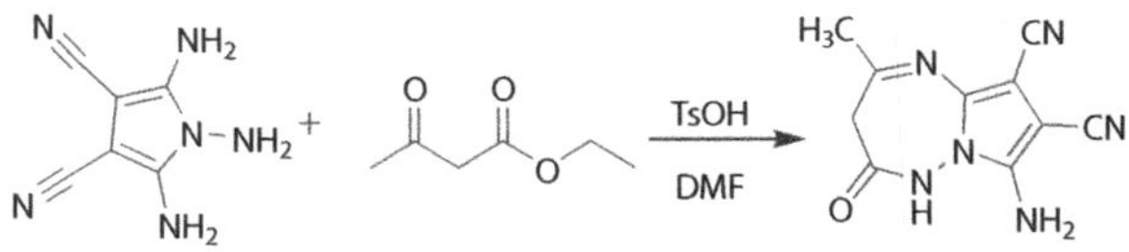

4.89 **4.90** **4.91**

Scheme 4.30 Azaazulene synthesis from 1,2-diamine and ethyl acetoacetate.

4.4 Reactions of Azaazulene

4.4.1 Coupling Reactions

Cross-coupling reactions have emerged as indispensable tools in modern organic synthesis, enabling the formation of carbon-carbon and carbon-heteroatom bonds with exceptional precision and selectivity. These reactions typically involve the coupling of a nucleophile (often an organometallic compound) with an electrophile, catalyzed by a transition metal complex. In the realm of azaazulene chemistry, the versatility and efficiency of cross-coupling reactions have been harnessed to construct a wide range of structurally diverse and biologically relevant compounds [47–49].

Several cross-coupling reactions have been extensively employed in azaazulene chemistry, each offering distinct advantages and applications. The Sonogashira–Hagihara reaction introduces alkynyl groups, while Suzuki coupling adds aryl or vinyl substituents. Negishi coupling is suitable for alkyl or alkenyl groups, and Hartwig–Buchwald coupling forms C-N bonds. These reactions enable the synthesis of diverse azaazulene derivatives with potential applications in materials science and medicinal chemistry.

4.4.2 Suzuki Coupling Reactions

The Suzuki coupling reaction between halo-1-azaazulenes and phenylboronic acid ($PhB(OH)_2$) resulted in efficient phenylation at the 2-position. Initially, it was believed that the bromine at C-2 was more reactive than the one at C-3. However, the reaction of 2,3-dibromo-1-azaazulene (**4.92b**) with $PhB(OH)_2$ yielded only the diphenylated product (**4.93b**) in 87% yield, suggesting that both bromines are equally reactive (Scheme 4.31) [50].

Interestingly, the treatment of **4.92d** with $PhB(OH)_2$ led to a mixture of products: **4.93d** (21%), 94 (31%), and unreacted starting material, **4.92d** (31%). It seems that the phenyl group hastened the reactivity of C-2. But the reaction of **4.93e** with $PhB(OH)_2$ under prolonged reaction time gave

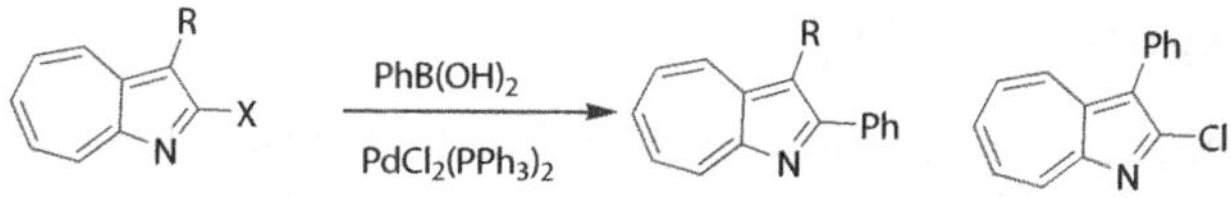

4.92a : X = Br, R = H
4.92b : X = R = Br
4.92c : X = Br, R = CO_2Et
4.92d : X = Cl, R = Br
4.92e : X = Cl, R = I

4.93a : R = H
4.93b : R = Br
4.93c : R = CO_2Et
4.93d : R = Ph

4.94

Scheme 4.31 Suzuki coupling reactions in azaazulenes.

only **4.94** in 83% yield. The result suggests that the bromine at C-3 in **4.92** hastens the reactivity of C-2.

The reaction of **4.92e** with bis(pinacolate)diboran [$(Bpin)_2$] in the presence of $PdCl_2$(dppf) yielded **4.95** (14%) and bi(1-azaazulene) **4.96** (57%) (Scheme 4.32). The intermediate product, **4.95**, was highly reactive and readily underwent further reaction with **4.92e** under the reaction conditions to give **4.96** [50].

The Suzuki coupling reaction of **4.92e** with PyB(NPDEA) resulted in the formation of **4.98a** (3%) and **4.97a** (50%). Similarly, the reaction of **4.92f** with PyB(NPDEA) produced the dimer **4.99b** (6%) and **4.97b** (84%) (Scheme 4.33) [51].

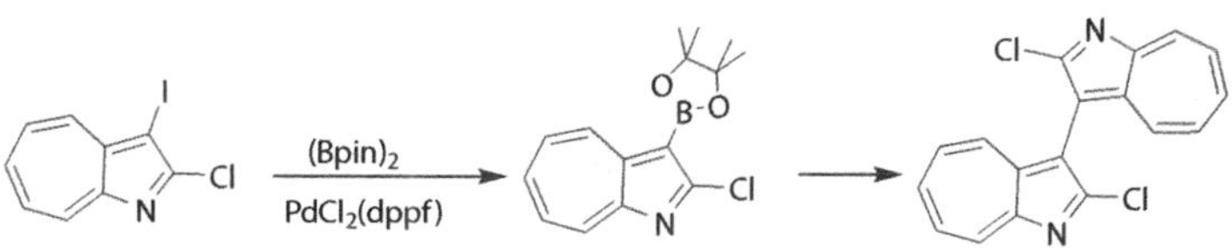

4.92e **4.95** **4.96**

Scheme 4.32 Synthesis of bi(1-azaazulene).

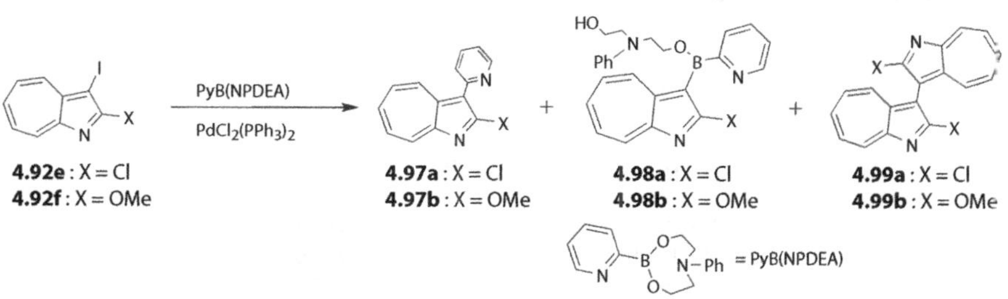

4.92e : X = Cl
4.92f : X = OMe

4.97a : X = Cl
4.97b : X = OMe

4.98a : X = Cl
4.98b : X = OMe

4.99a : X = Cl
4.99b : X = OMe

Scheme 4.33 Suzuki coupling reaction of azaazulene with PyB(NPDEA).

4.4.3 Negishi Coupling

The Negishi coupling reaction between 2-halo-1-azaazulenes and 2-pyridyl bromide, catalyzed by palladium, provides an efficient route to the synthesis of 2-(2-pyridyl)-1-azaazulenes (**4.102**). This reaction involves the formation of a carbon-carbon bond between the 2-position of 1-azaazulene and the 2-position of the pyridine ring. The palladium catalyst plays a crucial role in facilitating this coupling reaction (Scheme 4.34) [52].

The Negishi coupling reaction between **4.92a–c** and dimethylzinc ($ZnMe_2$) in the presence of $PdCl_2$(dppf) resulted in the efficient formation of 2-methyl-1-azaazulenes (**4.103a–c**). This reaction involves the coupling of the methyl group from $ZnMe_2$ to the 2-position of the 1-azaazulene ring. The palladium catalyst facilitates this coupling reaction, leading to the desired products in good yields (Scheme 4.35) [53].

The Negishi coupling reaction between 3-iodo-1-azaazulen-2(1H)-one (**4.104**) and dimethylzinc ($ZnMe_2$) yielded **4.105** (75%) and the deiodinated product **4.106** (9%) (Scheme 4.36). Subsequent treatment of **4.104** with phosphorus oxybromide ($POBr_3$) afforded **4.92g**.

In this reaction, the methyl group from $ZnMe_2$ is coupled to the 3-position of the 1-azaazulene ring, replacing the iodine atom. The deiodination side product (**4.106**) likely arises from a competing reaction pathway where the iodine is directly removed from the molecule. The conversion of **4.105** to **4.92g** involves the replacement of the hydroxyl group in **4.105** with a bromine atom using $POBr_3$.

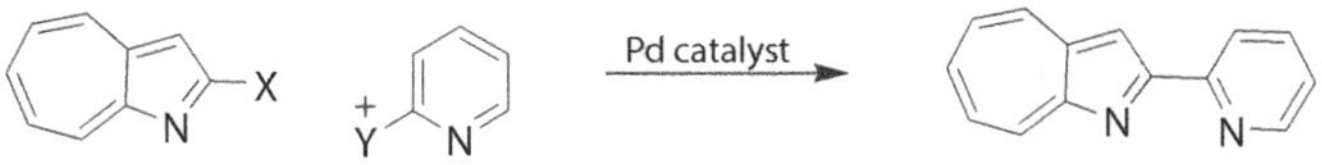

4.92a : X = Br
4.100 : X = Cl
4.101 : X = I

Y = ZnBr

4.102

Scheme 4.34 Negishi coupling reaction in azaazulenes.

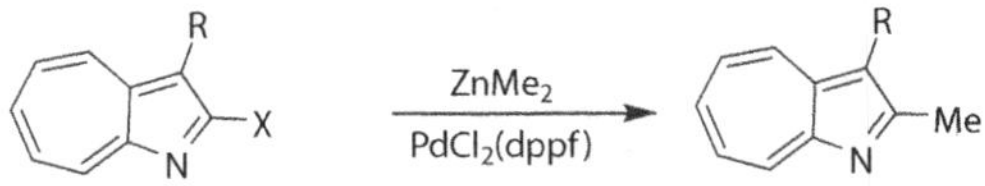

4.92a : X = Br, R = H
4.92b : X = R = Br
4.92g : X = Br, R = Me

4.103a : R = H
4.103b : R = Br
4.103g : R = Me

Scheme 4.35 Synthesis of 2-methyl-1-azaazulenes.

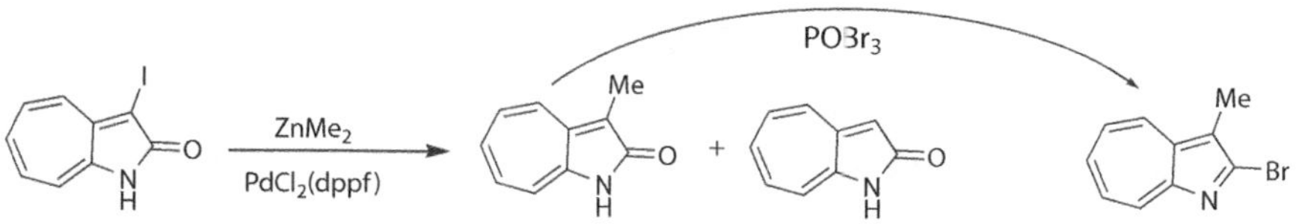

4.104 4.105 4.106 4.92g

Scheme 4.36 Negishi coupling reaction between 3-iodo-1-azaazulen-2(1H)-one and dimethylzinc.

4.4.4 Sonogashira–Hagihara Reaction

The Sonogashira coupling reaction between 2-chloro-3-iodo-1-azaazulene (**4.92e**) and trimethylsilylacetylene (**4.108a**), catalyzed by $Pd(PPh_3)_2Cl_2$ and CuI, resulted in the formation of 2-chloro-3-trimethylsilylethynyl-1-azaazulene (**4.109a**) in good yield. Similarly, the reactions of 2-amino-3-iodo-1-azaazulene (**4.107a**) with phenylacetylene or trimethylsilylacetylene yielded the **4.109b** and **4.109d**, respectively (Scheme 4.37) [54, 55].

The Sonogashira–Hagihara reaction can also be applied to seven-membered rings of azaazulenes. In this case, the reaction of 8-bromo-1-azaazulene derivative (**4.110**) with phenylacetylene in the presence of $PdCl_2(PPh_3)_2$ and CuI yielded **4.111** in good yield (Scheme 4.38) [55].

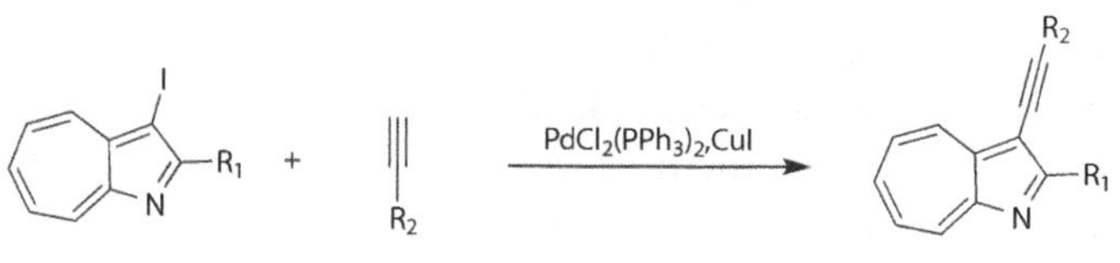

4.92e: $R_1 = Cl$
4.107a: $R_1 = NH_2$

4.108a: $R_2 = TMS$
4.108b: $R_2 = Ph$

4.109a: $R_1 = Cl$, $R_2 = TMS$
4.109b: $R_1 = NH_2$, $R_2 = Ph$
4.109c: $R_1 = Cl$, $R_2 = Ph$
4.109d: $R_1 = NH_2$, $R_2 = TMS$

Scheme 4.37 Sonogashira–Hagihara coupling reaction of azaazulene.

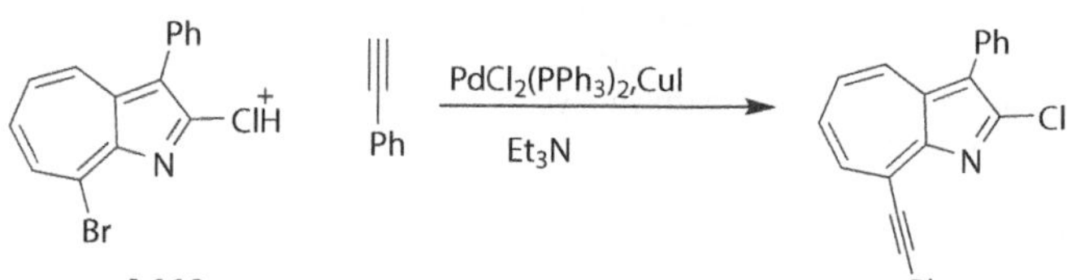

4.110 4.111

Scheme 4.38 Sonogashira–Hagihara coupling reaction in a seven-membered ring of azaazulene.

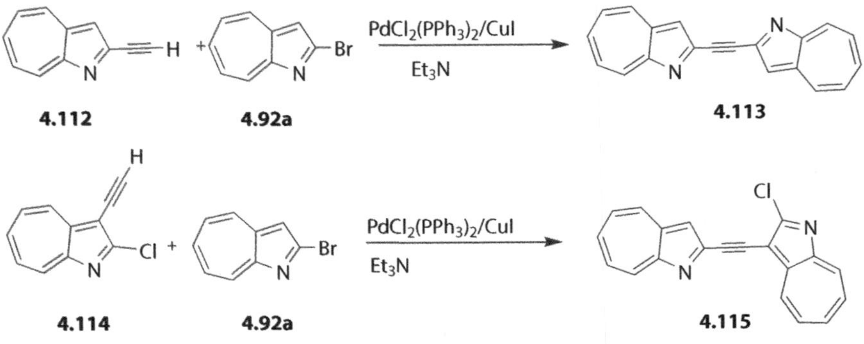

Scheme 4.39 Sonogashira–Hagihara coupling reaction of azaazulenes with their halogenated derivatives.

Sonogashira–Hagihara coupling of **4.92a** with **4.112** and **4.92a** with **4.114** gave the **4.113** and **4.115**, respectively (Scheme 4.39) [54].

4.4.5 Hartwig–Buchwald Reaction

The Hartwig–Buchwald reaction, a well-established method for the amination of aromatic compounds, was successfully applied to introduce heteroarylamino groups onto the 1-azaazulene nucleus.

By reacting **4.116** with various heteroaryl amines in the presence of $Pd_2(dba)_3$, Xantphos, and $CsCO_3$ in dioxane at 120°C, the corresponding 2-(heteroarylamino)-1-azaazulenes (**4.117**) were obtained in good yields (Scheme 4.40) [56–59]. This method was also suited for the synthesis of bis(1-azaazulenyl)amines (Scheme 4.41).

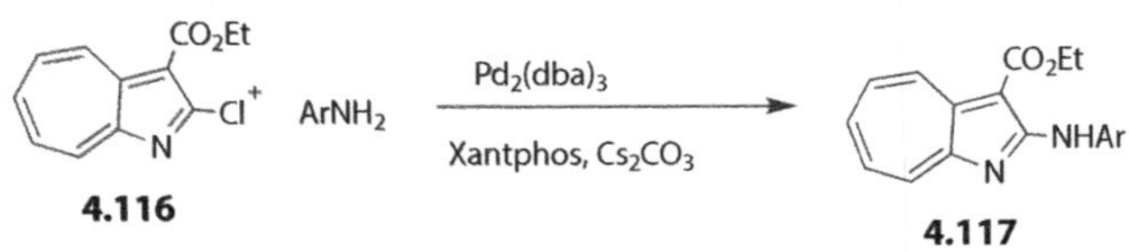

Ar=2-Py, 4-Py, 2-pyrimidinyl, 2-pyrazinyl, 2-thiazolyl, 2-benzimidazolyl, 2-benzothiazolyl, 1-azaazulen-2-yl, 1-azaazulen-8-yl, 1-isoquinoliny

Scheme 4.40 Hartwig–Buchwald reaction in azaazulenes.

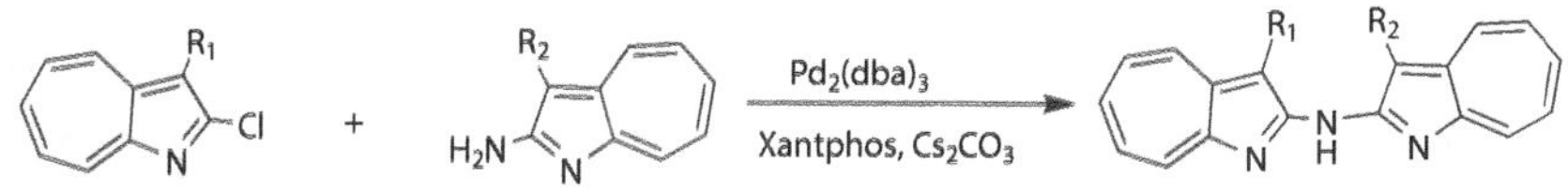

R_1 = H, CO_2Et, NO_2, Ph

4.117a

R_2 = H, Ph

4.117b

4.117c

Scheme 4.41 Synthesis of bis(1-azaazulenyl)amines by Hartwig–Buchwald reaction.

4.4.6 Cycloaddition and Annulation

Cycloadditions and annulations, two fundamental chemical transformations, have been extensively explored to exploit the reactivity of the azaazulene nucleus and create diverse heterocyclic scaffolds. Cycloadditions involve the formation of cyclic compounds through the simultaneous formation of two or more new bonds. Azaazulenes, with their electron-rich aromatic character, are prone to undergo various cycloaddition reactions, such as Diels-Alder reactions, 1,3-Dipolar cycloadditions and [2+2] cycloadditions.

The cycloaddition reaction between 2-chloro-1-azaazulene (**4.118**) and diphenylcyclopropenone (**4.119**) resulted in an intriguing outcome, yielding compounds **4.120** and **4.121** alongside unreacted **4.118** (58%) [60]. Initially, an extended dipolar cycloaddition took place between the N-1 and C-8 positions of the 1-azaazulene nucleus and diphenylcyclopropenone, forming an intermediate labeled A. Subsequently, a nucleophilic attack by water on intermediate A, accompanied by the elimination of HCl and dehydrogenation, led to the formation of compound **4.120**. Alternatively, intermediate A could undergo a hydrogen migration to generate an intermediate B. This intermediate B then underwent a cine-substitution reaction with another molecule of diphenylcyclopropenone, resulting in the formation of intermediate C. The cycloheptatriene-norcaradiene interconversion of intermediate C afforded intermediate D. Finally, a ring contraction of intermediate D, coupled with the elimination of HCl, produced compound **4.121** (Scheme 4.42). In summary, the reaction pathway involves multiple steps, including cycloaddition, nucleophilic attack, elimination, hydrogen migration, cine-substitution, interconversion, and ring contraction, ultimately leading to the formation of the desired products.

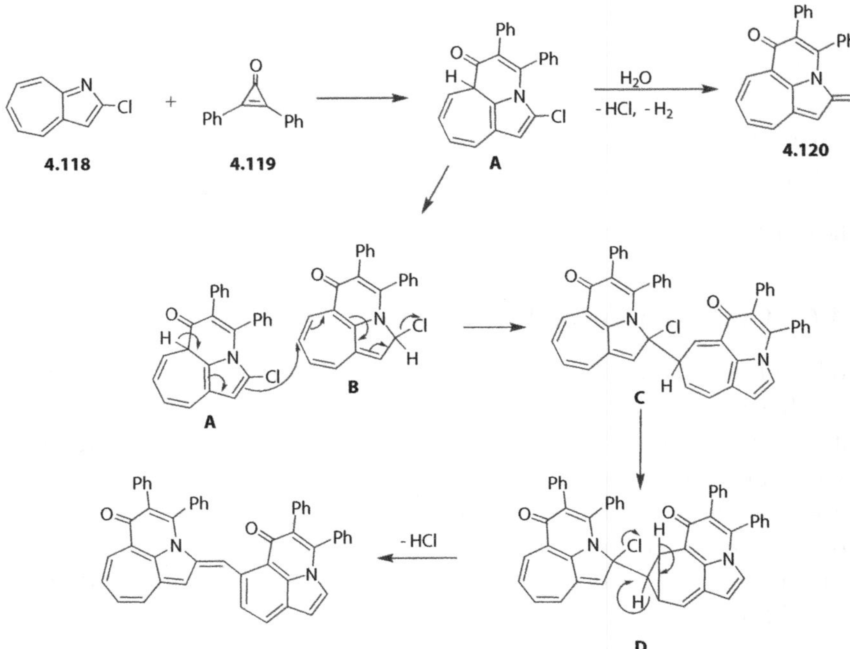

Scheme 4.42 Cycloaddition reaction between 2-chloro-1-azaazulene (4.118) and diphenylcyclopropenone.

4.4.7 Aza-Wittig Reaction

The aza-Wittig reaction, involving iminophosphoranes, is a valuable method for constructing nitrogen-containing heterocycles. Both electrochemical and photochemical approaches have been explored for this reaction. Nitta and Abe independently reported the aza-Wittig reactions of phosphoimino-azaazulene derivatives.

Nitta and his colleagues initiated their research by synthesizing 2-phosphoimino-1-azaazulene (**4.124**) from 2-azido-1-azaazulene (**4.123**), which was itself prepared from 2-hydrazino-1-azaazulene (**4.122**) using the Staudinger reaction. Subsequently, they reacted compound **4.125** with 2-bromotropone to successfully synthesize compound **4.125** (Scheme 4.43) [61–64].

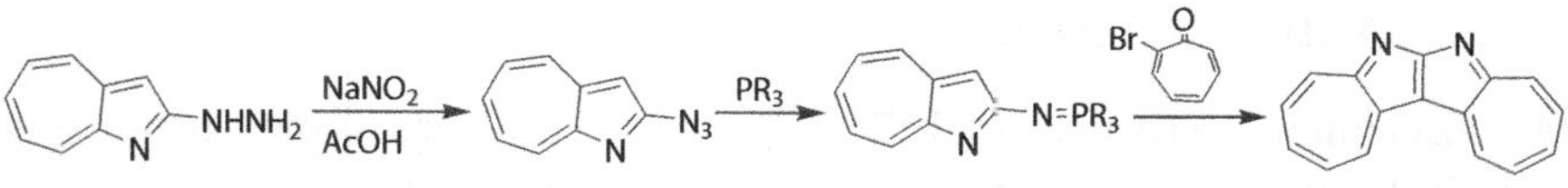

Scheme 4.43 Aza–Wittig reaction in azaazulenes.

4.5 Applications

Azaazulenes and polycyclic compounds incorporating azaazulene or azepine rings, like pyrrolobenzazepines and azepinoindole derivatives, have garnered significant interest due to their promising biological activities, particularly in the realm of antitumor antibiotics. Among these compounds, pyrrolobenzodiazepine has emerged as a particularly promising candidate for anticancer drug development.

4.5.1 Phosphodiesterase 2 Inhibitors

A novel pyrazolodiazepinone compound **4.126** shown in Figure 4.5 acts as a potent and selective inhibitor of phosphodiesterase 2 (PDE2). This compound exhibits an impressive selectivity profile, showing over 1000-fold selectivity against other phosphodiesterases. *In vivo* studies demonstrated significant analgesic effects in a rat model of osteoarthritis pain, with a half-life of 6.9 minutes and an IC50 value of 3.3 nM for PDE2, indicating its potential for pain management therapies [65].

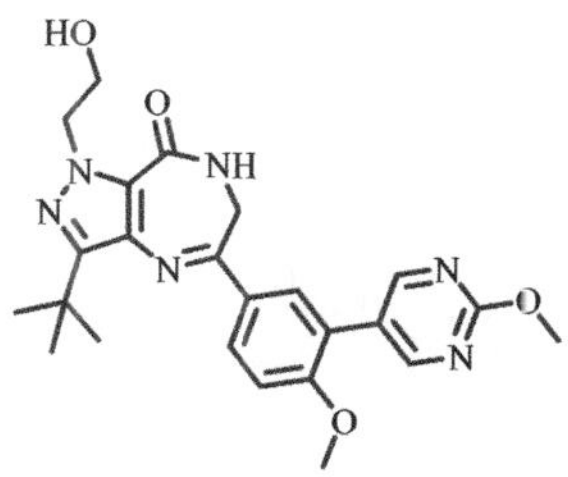

4.126

Figure 4.5 Structure of a PDE2 inhibitor.

4.5.2 Antiviral Activity

An azaazulene derivative (**4.127**) (Figure 4.6) that has shown promising antiviral activity against hepatitis C virus in a Phase Ib clinical trial is shown below. This compound was noted for its safety profile and efficacy, suggesting its potential as a therapeutic option for patients with hepatitis C [66].

4.5.3 Cholecystokinin-1 Receptor Antagonists

Azaazulene compounds have also been found to act as antagonists of the cholecystokinin-1 (CCK1) receptor. One of the compounds, CE-326597 (**4.128**) (Figure 4.7), demonstrated efficacy in regulating food intake in rodent models, indicating its potential for obesity treatment. However, despite its initial promise, it showed insufficient efficacy in treating diabetes or obesity in a Phase II clinical trial, highlighting the challenges in translating preclinical success to clinical outcomes [67].

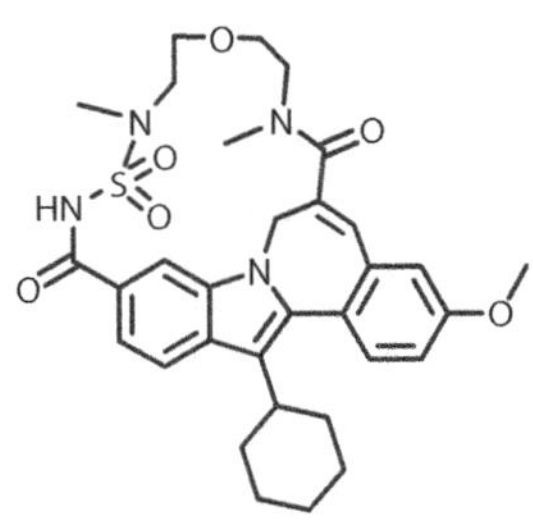

4.127

Figure 4.6 Azaazulene derivative with antiviral activity.

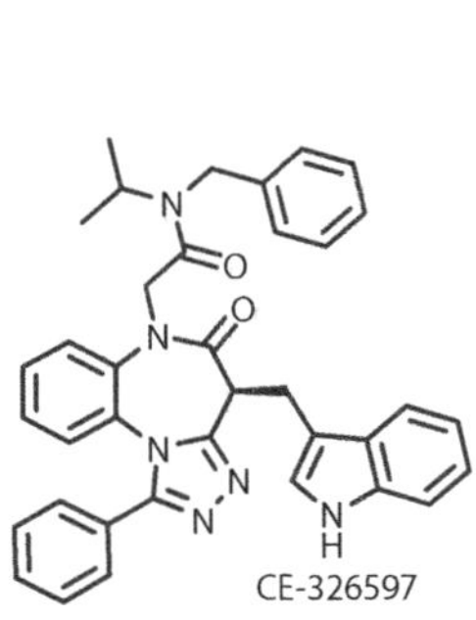

4.128

Figure 4.7 Structure of triazolobenzobenzodiazepinone.

4.5.4 Epigenetic Inhibitors

1,4-benzodiazepine scaffolds act as inhibitors of bromodomain and extra-terminal proteins. These proteins are key players in epigenetic regulation and targeting them holds significant promise for cancer therapy. By modulating gene expressions associated with tumor progression, bromodomain and extra terminal inhibitors could potentially offer novel treatment options. Additionally, the gamma-aminobutyric acid A receptors are heteropentameric chloride ion channels responsible for various brain functions. The modulation of these receptors is critical in the development of anxiolytic and sedative drugs, emphasizing their importance in pharmacology.

AI disclosure statement: The author acknowledges the use of AI-based tools for minor language editing or grammatical correction. However, all intellectual contributions, analyses, and conclusions presented in this chapter are entirely the author's own.

References

1. Nishiwaki, T. and Abe, N., Recent progress in the chemistry of azaazulenes: Synthetic methodology and chemical reactions. *Heterocycles*, 15, 1, 547, 1981.
2. Flitsch., W., The chemistry of 4-azaazulenes. *Adv. Heterocycl. Chem.*, 43, 35, 1988.
3. Abe, N. and Gunji, T., The chemistry of azaazulenes. *Heterocycles*, 82, 201, 2010.
4. Abe, N., Recent progress of the chemistry of azaazulenes and related compounds. *Heterocycles*, 97, 43, 2018.
5. Abe, N., The chemistry of azaazulenes. *Heterocycles*, 101, 33, 2020.
6. Baric, D., Utilizing the azaazulene scaffolds in the design of new organic Superbases. *ACS Omega*, 4, 12, 15197, 2019.
7. Somwar, R., Koterski, S., Sweeney, G., Sciotti, R., Djuric, S., Berg, C., Trevillyan, J., Scherer, P.E., Rondinone, C.M., Klip, A., A dominant-negative p38 MAPK mutant and novel selective inhibitors of p38 MAPK reduce insulin-stimulated glucose uptake in 3T3-L1 adipocytes without affecting GLUT4 translocation. *J. Biol. Chem.*, 277, 50386, 2002.
8. Selçuk, G., A computational study on azaazulenes. *Heterocycl. Commun.*, 19, 5, 369, 2013.
9. Nozoe, T., Kitahara, Y., Arai, T., Synthesis of 3-phenyl-1-azazulene. *Proc. Jpn. Acad.*, 30, 478–481, 1954.

10. Seto, S. and Nozoe, S., Syntheses of methyl derivatives of 1-azaazulene. *Proc. Jpn. Acad.*, 32, 765, 1956.
11. Ogura, K., Sasaki, H., Seto, S., The reaction of aminotropones with diketene. I. On the synthesis of 3-acetyl-1-azaazulenes and related reactions. *Bull. Chem. Soc. Jpn.*, 32, 306, 1965.
12. Matsumoto, Y. and Nitta, M., Convenient preparation of heteroazulenes by the reaction of 2-aminotropone with heterocumulenes in the presence of a base. *Heterocycles*, 55, 2131, 2001.
13. Yamamoto, H., Ohnuma, M., Nitta, M., Synthesis and structure of 2-(triphenylphosphoranylideneamino)tropone and its utility in the preparation of new cyclohepta-annulated heterocycles. *J. Chem. Res. (S)*, 3, 173, 1999.
14. Nitta, M. and Mitsumoto, Y., A convenient *in situ* preparation of triphenylbismuthane (tropon-2-yl)imide: Reaction with heterocumulenes and activated alcohols. *Heterocycles*, 55, 629, 2001.
15. Nitta, M., Mitsumoto, Y., Yamamoto, H., Synthesis, structure, and reactivity of (tropon-2-ylimino)arsorane and *in situ* generation of its stiborane and -bismuthorane analogues: reactions with heterocumulenes and an activated acetylene giving heteroazulenes. *J. Chem. Soc. Perkin Trans.*, 1, 1901, 2001.
16. Pietra, F., Seven-membered conjugated carbo- and heterocyclic compounds and their homoconjugated analogs and metal complexes. Synthesis, biosynthesis, structure, and reactivity. *Chem. Rev.*, 73, 293, 1973.
17. Cavazza, M., Carbino, R., Cima, F.D., Pietra, F., A route to 6-functionalised 1,3-diaza-azulenes and aminotropones *via* hydride replacement from cycloheptatrienones. *J.C.S. Perkin Trans. I*, 6, 609, 1978.
18. Sunagawa, G. and Nakao, H., Studies on seven-membered ring compounds. XIV. Reactions of 2-methoxytropone and 2-halotropone with hydrazides. *Chem. Pharm. Bull. (Japan)*, 12, 450, 1965.
19. Sakai, N., Hattori, N., Tomizawa, N., Abe, N., Konakahara, T., New approach to the synthesis of non-benzenoid aromatic compounds, functionalized 1-azaazulene derivatives: cyclization reactions of 2-substituted tropones with n-silylenamine and enamine. *Heterocycles*, 65, 2799, 2005.
20. Sugimura, Y., Soma, N., Kishida, Y., The reaction of troponoid with ylide. III. The reaction of tropone with pyridinium compounds. *Bull. Chem. Soc. Jpn.*, 42, 3174, 1972.
21. Kitamura, M., Chiba, S., Saku, O., Narasaka, K., Palladium-catalyzed synthesis of 1-azaazulenes from cycloheptatrienylmethyl ketone o-pentafluorobenzoyl oximes. *Chem. Lett.*, 31, 6, 606, 2002.
22. Kitamura, M. and Narasaka, K., Synthesis of aza-heterocycles from oximes by amino-heck reaction. *Chem. Rec.*, 2, 268, 2002.
23. Chiba, S., Kitamura, M., Saku, O., Narasaka, K., Synthesis of 1-azaazulenes from cycloheptatrienylmethyl ketone o-pentafluorobenzoyloximes by palladium-catalyzed cyclization and oxidation. *Bull. Chem. Soc. Jpn.*, 77, 785, 2004.

24. Takaoka, K., Aoyama, T., Shioiri, T., Cycloaddition reactions of trimethylsilylketene with 8-azaheptafulvenes and 6-amino-1-azafulvenes. *Heterocycles*, 54, 209, 2001.
25. Sciotti, R.J. and Wagner, R., Azaazulene inhibitors of p38 map kinase and tnf-alpha. PCT/US01/02944, assigned to Abbott laboratories, 2001.
26. Sato, O., Tsurumaki, K., Tamaru, S., Tsunetsugu, J., Novel heterocyclic troponoids: Synthesis of pyridoimidazotropolones and related compounds by the reaction of 5-nitrosotropolone with several azines. *Heterocycles*, 62, 535, 2004.
27. Nitta, M. and Tajima, Y., Studies on pyrimidine-annulated heterocycles: A short synthesis of novel 6,9-disubstituted cyclohepta[b]pyrimido[5,4-d]pyrrole-8(6H),10(9H)-diones. *Synthesis*, 5, 651, 2000.
28. Naya, S., Iida, Y., Nitta, M., Alternative synthesis and novel oxidizing ability of 6,9-disubstituted cyclohepta[b]pyrimido[5,4-d]pyrrole-8(6H),10(9H)-dione derivatives. *Tetrahedron*, 60, 459, 2004.
29. Naya, S. and Nitta, M., Synthesis, properties, and oxidizing function of 6-substituted 7,9-dimethylcyclohepta[b]pyrimido[5,4-d]pyrrole-8(7H),10(9H)-dionylium tetrafluoroborates. *Tetrahedron*, 59, 7291, 2003.
30. Naya, S. and Nitta, M., Ring transformation of cyclohepta[b]pyrimido[5,4-d]furan-8(7H),10(9H)-dionylium ion to the corresponding pyrrole derivatives *via* troponeimine intermediates: photo-induced autorecycling oxidizing reactions of some amines. *Tetrahedron*, 60, 9139, 2004.
31. Naya, S., Ohtoshi, H., Nitta, M., Synthesis, properties, and oxidizing ability of areno-annulated 1,3-dimethyl-10-phenylcyclohepta[4,5]pyrrolo[2,3-d]pyrimidine- 2,4(1,3H)-dionylium ions. *J. Org. Chem.*, 71, 176, 2006.
32. Morita, M., Hari, Y., Iguchi, T., Aoyama, T., Facile synthesis of 2-azaazulenes from thiobenzoyl isocyanates using trimethylsilyldiazomethane. *Tetrahedron*, 64, 1753, 2008.
33. Ishikawa, T., Saito, T., Noguchi, S., Ishii, H., Ito, S., Hata, T., New findings in Bischler-Napieralski reaction: Formation of 12-azonianaphth[1.2-b]azulenes from 2-aryl-1-naphthylformamides and their unexpected transformation into benz[g]indoles through hydride reduction. *Tetrahedron Lett.*, 36, 2795, 1995.
34. Ishikawa, T., Shimooka, K., Narioka, T., Noguchi, S., Saito, T., Ishikawa, A., Yamazaki, E., Harayama, T., Seki, H., Yamaguchi, N., Anomalous substituent effects in the Bischler–Napieralski Reaction of 2-aryl aromatic formamides. *J. Org. Chem.*, 65, 9143, 2000.
35. Narioka, T. and Ishikawa, T., Further approaches to anomalous carbon insertion into a benzene ring in the Bischler-Napieralski reaction of phenolic 2-phenylformanilides. *Heterocycles*, 56, 413, 2002.
36. Pinho e Melo, T.M.V.D., Soares, M.I.L., Rocha Gonsalves, A.M.A., McNab, H., Reactivity of azafulvenium methides derived from pyrrolo[1,2-c]thiazole-2,2-dioxides: synthesis of functionalised pyrroles. *Tetrahedron Lett.*, 45, 3889, 2004.

37. Gruit, M., Michalik, D., Tillack, A., Beller, M., Platinum-catalyzed intramolecular cyclizations of alkynes: efficient synthesis of pyrroloazepinone derivatives. *Angew. Chem. Int. Ed.*, 48, 7212, 2009.
38. Clayden, J., Turnbull, R., Pinto, I., Nucleophilic addition to electron-rich heteroaromatics: dearomatizing anionic cyclizations of pyrrolecarboxamides. *Org. Lett.*, 6, 609, 2004.
39. Nakatsuka, S., Miyazaki, H., Goto, T., Introduction of substituent onto 4-position of indole nucleus by intermolecular cyclization of α,β-dehydrotryptophan methyl ester with aldehyde. *Tetrahedron Lett.*, 27, 4757, 1986.
40. Al-Khashashneh, A.M., El-Abadelah, M.M., Boese, R., Novel synthesis of model dihydroazepine-fused indolo[2,3-c]quinolines. *Heterocycles*, 60, 73, 2003.
41. Gmeiner, P., Hübner, H., Abu Safieh, K.A., Fasfous, I.I., El-Abadelah, M.M., Sabri, S.S., Voelter, W., Synthesis and dopamine receptor binding of some pyrazolo[3',4':6,7]azepino[5,4,3-cd]indoles. *Heterocycles*, 60, 1339, 2003.
42. Abu Safieh, K.A., Darras, F.H., Ayoub, M.T., El-Abadelah, M.M., Sabri, S.S., Voelter, W., Synthesis of 8-substituted 12-methyl-12H-imidazo[4',5': 2,3][1,4]diazepino[6,7,1-jk]carbazoles. *Heterocycles*, 71, 2681, 2007.
43. Kreher, R., Vogt, G., Schultz, M.-L., Synthesis of 2H-cyclohepta[c]pyrrol-6-ones. A route to 2-azaazulenes. *Angew. Chem. Int. Ed.*, 14, 821, 1975.
44. Greco, C.V., Pellegrini, F.C., Pesce, N.A., An approach to the synthesis of a cycloheptapyrazole (1,2-diaza-azulene). *J.C.S. Perkin Trans.*, 1623, 1972.
45. Heine, H.W., Ludovici, D.W., Pardoen, J.A., Weber II, R.C., Bonsall, E., Osterhout, K.R., Preparation and reactions of some derivatives of 2,4-benzodiazepines and 1,3-diazepines. *J. Org. Chem.*, 44, 3843, 1979.
46. Dickinson, C.L., Middleton, W.J., Engelhardt, V.A., Cyanocarbon chemistry. XXII.1 3,4-dicyano-1,2,5-triaminopyrrole and its derivatives. *J. Org. Chem.*, 27, 2470, 1962.
47. Hassan, J., Sévignon, M., Gozzi, C., Schulz, E., Lemaire, M., Aryl–Aryl Bond Formation One Century after the Discovery of the Ullmann Reaction. *Chem. Rev.*, 102, 1359, 2002.
48. Diederich, F. and Stang, P.J., *Metal-Catalyzed Cross-Coupling Reactions*, Willey-VCH, Weinheim, 1998.
49. Miyaura, N., *Topics in Current Chemistry*, vol. 219, Springer, Heidelberg, New York, 2002.
50. Abe, N., Tanaka, M., Maeda, T., Fujii, H., Kakehi, A., Effective Methods for Introducing Some Aryl and Heteroaryl Substituent onto 1-Azaazulene Nuclei. *Heterocycles*, 66, 229, 2005.
51. Ariyoshi, T., Noda, T., Watarai, S., Tagashira, S., Murakami, Y., Fujii, H., Abe, N., Synthesis of [Poly(2-pyridyl)-Substituted]-1-azaazulenes. *Heterocycles*, 77, 565, 2009.
52. Oda, M., Ogura, K., Thanh, N.C., Kishi, S., Kuroda, S., Fujimori, K., Noda, T., Abe, N., Synthesis and properties of 2-(2-pyridyl)-1-azaazulene. *Tetrahedron Lett.*, 48, 4471, 2007.

53. Koizumi, K., Miyake, C., Ariyoshi, T., Umeda, K., Yamauchi, N., Tagashira, S., Murakami, Y., Fujii, H., Abe, N., Syntheses of 2-Formyl-, 2,3-Diformyl- and 8-Formyl-1-azaazulenes. *Heterocycles*, 73, 325, 2007.
54. Abe, N., Harada, Y., Imachi, Y., Fujii, H., Kakehi, A., Shiro, M., Synthesis of aryl conjugated (1-Azaazulenyl)acetylenes and facile synthesis of thiophene fused 1-azaazulenes. *Heterocycles*, 72, 459, 2007.
55. Fujii, H., Abe, N., Umeda, N., Kakehi, A., Synthesis and reactions of 3- and 8-Ethynyl-1-azaazulenes. *Heterocycles*, 58, 283, 2002.
56. Koizumi, K., Shimabara, K., Takemoto, A., Yamazaki, S., Yamauchi, N., Fujii, H., Kurosawa, M., Konakahara, T., Abe, N., Palladium-catalyzed heteroarylamination of Ethyl 2-Chloro-1-azaazulene-3-carboxylate and annulation of heteroarylamino-1-azaazurenes. *Heterocycles*, 79, 319, 2009.
57. Hartwig, J.F., Transition metal catalyzed synthesis of arylamines and aryl ethers from aryl halides and triflates: scope and mechanism. *Angew. Chem. Int. Ed.*, 37, 2046, 1998.
58. Wolfe, J.P., Wagaw, S., Marcoux, J.-F., Buchwald, S.L., Rational development of practical catalysts for aromatic carbon–nitrogen bond formation. *Acc. Chem. Res.*, 31, 805, 1998.
59. Prim, D., Campagne, J.-M., Joseph, D., Andrioletti, B., Palladium-catalysed reactions of aryl halides with soft, non-organometallic nucleophiles. *Tetrahedron*, 58, 2041, 2002.
60. Abe, N., Fujii, H., Kakehi, A., Reaction of 2-chloro-1-azaazulene with diphenylcyclopropenone; dimerization of the cycloadduct attended by cine-substitution and rearrangement. *Heterocycles*, 55, 1189, 2001.
61. Nitta, M., Iino, Y., Sugiyama, T., Akaogi, A., Synthetic studies of novel 6-aza- and 6,7-diazaazuleno[1,2-a]azulenes. *Tetrahedron Lett.*, 34, 831, 1993.
62. Gololobov, Y.G. and Kasukhin, L.F., Recent advances in the Staudinger reaction. *Tetrahedron*, 48, 1353, 1992.
63. Nitta, M., Ohtsuki, D., Mitsumoto, Y., Naya, S., Synthesis and properties of 3-arylcyclohepta[4,5]imidazo[1,2-a]-1,3,5-triazine-2,4(3H)-diones and related compounds: photo-induced autorecycling oxidation of some amines. *Tetrahedron*, 61, 6073, 2005.
64. Nitta, M., Morito, T., Mitsumoto, Y., Naya, S., Synthesis and properties of 3-arylcyclohepta[4,5]pyrrolo[1,2-a]-1,3,5-triazine-2,4(3h)-diones and related compounds: photo-induced autorecycling oxidation of amines. *Heterocycles*, 65, 1629, 2005.
65. Plummer, M.S., Cornicelli, J., Roark, H., Skalitzky, D.J., Stankovic, C.J., Bove, S., Pandit, J., Goodman, A., Hicks, J., Shahripour, A., Beidler, D., Lu, X.K., Sanchez, B., Whitehead, C., Sarver, R., Braden, T., Gowan, R., Shen, X.Q., Welch, K., Ogden, A., Sadagopan, N., Baum, H., Miller, H., Banotai, C., Spessard, C., Lightle, S., Discovery of potent selective bioavailable phosphodiesterase 2 (PDE2) inhibitors active in an osteoarthritis pain model. Part II: Optimization studies and demonstration of *in vivo* efficacy. *Bioorg. Med. Chem. Lett.*, 23, 3443, 2013.

66. Cummings, M.D., Lin, T., Hu, L., Tahri, A., McGowan, D., Amssoms, K., Last, S., Devogelaere, B., Rouan, M.-C., Vijgen, L., Berke, J.M., Dehertogh, P., Fransen, E., Cleiren, E., Helm, L.V.D., Fanning, G., Nyanguile, O., Simmen, K., Remoortere, P.V., Raboisson, P., Vendeville, S., Discovery and early development of TMC647055, a non-nucleoside inhibitor of the hepatitis C virus NS5B polymerase. *J. Med. Chem.*, 57, 5, 1880, 2014.
67. Elliott, R.L., Cameron, K.O., Chin, J.E., Bartlett, J.A., Beretta, E.E., Chen, Y., Jardine, P.D.S., Dubins, J.S., Gillaspy, M.L., Hargrove, D.M., Kalgutkar, A.S., LaFlamme, J.A., Lame, M.E., Martin, K.A., Maurer, T.S., Nardone, N.A., Oliver, R.M., Scott, D.O., Sun, D., Swick, A.G., Trebino, C.E., Discovery of N-benzyl-2-[(4S)-4-(1H-indol-3-ylmethyl)-5-oxo-1-phenyl-4,5-dihydro-6H-[1,2,4]triazolo[4,3-a][1,5]benzodiazepin-6-yl]-N-isopropylacetamide, an orally active, gut-selective CCK1 receptor agonist for the potential treatment of obesity. *Bioorg. Med. Chem. Lett.*, 20, 6797, 2010.

Part 2

AZULENE IN FUNCTIONAL AND BIOACTIVE MATERIALS

5

Azulene-Based System in Optoelectronics

Parth[1], Shivani[2] and Amrit Kaur[3*]

[1]Department of Chemistry, Khalsa College, Amritsar, Punjab, India
[2]Department of Chemistry, City University of Hong Kong, Hong Kong SAR, China
[3]Department of Chemistry, Thomas Jefferson University College East Fall, Philadelphia, United States

Abstract

Azulene is a non-benzenoid, non-alternant aromatic bicyclic hydrocarbon of naphthalene. This isoelectronic isomer of naphthalene, due to its unique photophysical properties including a narrow highest occupied molecular orbital (HOMO)-lowest unoccupied molecular orbital (LUMO) energy band gap and anti-Kasha fluorescence, has become a molecule of significant interest. Furthermore, it gained enormous attention due to its unique electronic structure with an electron-rich cyclopentadiene ring and electron-deficient cycloheptatriene ring, with a net dipole moment of 1.08 D, resulting in different reactivities at odd and even positions. Due to its unusual structure, azulene-based molecules are broadly used in technology, especially in optoelectronics such as nonlinear optics (NLO), molecular devices, and organic field effect transistors (OFETs), as well as medicinal chemistry. In this chapter, we have focused on the applications of azulene derivatives in varied fields of organic electronics including OFETs, organic photovoltaics, and NLOs.

***Keywords*:** Non-benzenoid, anti-Kasha, nonlinear optics (NLO), organic field effect transistors (OFETs)

**Corresponding author*: amritkaurchem92@gmail.com

Neha Rani Kumar and Abhijeet R. Agrawal (eds.) Azulene-Based Materials: Design, Synthesis, and Applications, (119–150) © 2026 Scrivener Publishing LLC

5.1 Introduction

Azulene, as an intense blue crystalline isoelectronic isomer of naphthalene, with a fused electron-rich pentagon and electron-deficient heptagon rings, contributes to an intramolecular dipole moment of 1.08 D (Figure 5.1a). Azulene was first discovered by Piesse in 1863, and its structure was first reported by Ruzicka [1]. It has become a molecule of considerable interest since its discovery [2]. In 1937, Plattner and Pfau reported the first synthesis of azulene, but their approach gave low yields [3], whereas Zeigler and Hafner articulated an extremely efficient method for synthesizing azulene and its derivatives [4]. Some of the fascinating characteristics of azulene include its polar nature, distinct energy transition, and "anomalous" fluorescence which is anti-Kasha fluorescence. Due to its non-mirrored frontier molecular orbital (FMO) geometry, where the highest occupied molecular orbital (HOMO) electron density is predominantly localized on the odd positions and the lowest unoccupied molecular orbital (LUMO) electron density on the even positions, azulene exhibits a narrow energy gap between the ground state (S_0) and the first excited state (S_1). [5] The absorption transition from $S_0 \rightarrow S_1$ in the visible range accounts for its intense crystalline blue color.

The synthesis of azulene derivatives, by functionalization at 1- and/or 3-positions of azulene, is a highly effective method [6]. However, for other positions, direct functionalization becomes a tedious job [7]. As the odd-numbered positions especially 1- and 3-positions of azulene are electron-rich due to resonance delocalization, they have the highest activity towards electrophilic reactions [8], and the electron-deficient even-numbered positions 2-, 4-, 6- and 8- prefer to undergo reactions with nucleophiles (Figure 5.1b). Another fascinating feature of azulene is its anti-Kasha fluorescence, [9] which mainly originates from S_2 state, whereas the $S_1 \rightarrow S_0$ transition is negligible [10]. Due to its relatively large energy gap

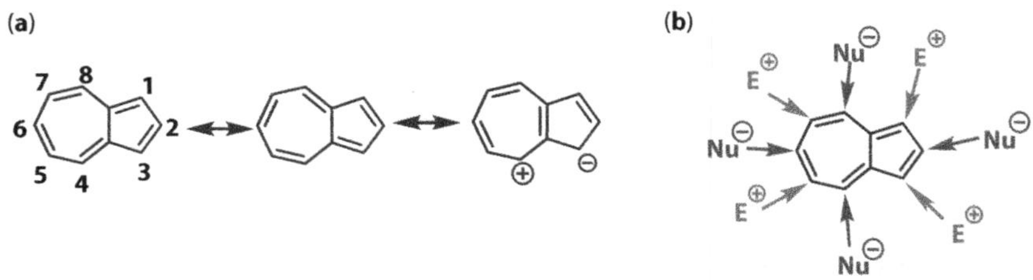

Figure 5.1 (a) Azulene and its polarized resonance structure; (b) the reactivity of different positions of azulene.

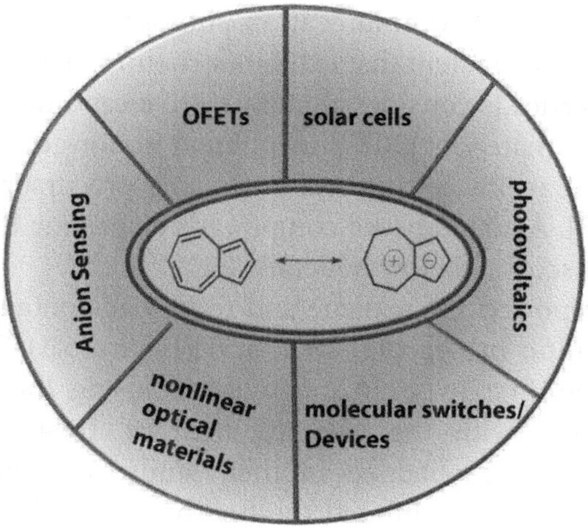

Figure 5.2 Application of azulene in various fields.

($\Delta E > 10000\ cm^{-1}$) between S_1 and S_2 states, a decrease in transition $S_2 \rightarrow S_1$ is observed, which makes the $S_2 \rightarrow S_0$ transition the dominant pathway [11].

In recent years, azulene has attracted the attention of researchers due to its unique structure, electronic and optical properties toward designing advanced π-functional organic materials, including molecular switches [12, 13], anion sensors [14], electrochromic materials [15], charge transfer complexes [16], organic conductors [17], near-infrared (NIR) resonance materials [18], nonlinear optical materials [19], organic field effect transistors (OFETs) [20], and solar cells [21] (Figure 5.2). In this chapter, we try to shine a light on the different applications of azulene-based molecules in organic photovoltaics (OPVs), nonlinear optics (NLO), OFETs, and molecular devices.

5.2 Applications in Nonlinear Optics

The last three decades have witnessed a growing interest of researchers in the organic nonlinear optical molecules [22], owing to their large application in optical communication [23], data storage devices [24], dye-sensitized solar cells [25], photodynamic therapy [26], and optoelectronic devices, etc. [27]. The organic molecules, having applications in NLOs, are generally conjugated systems comprising electron acceptor and donor units [27]. These systems are known as push-pull systems and consist of both donor and acceptor groups bridged by a π-conjugated system [28]. The successful engineering of new NLO molecules helps in the modulation of lights, and in the creation of optical communication/photonic devices [29]. Having an organic π-conjugated system that exhibits structural flexibility and tunable electronic properties is considered an attractive candidate for developing nonlinear optical materials [30]. The inherent donor-acceptor property of azulene makes it an interesting and promising precursor for application in NLOs [31–33]. Azulene-based compounds displaying nonlinear optical properties are discussed below in detail.

Youn *et al.* reported the azulene and ferrocene-containing derivatives linked by an imino bridge [34] (**5.3–5.5**) (Figure 5.3). The researchers investigated their first hyperpolarizability by using the hyper-Rayleigh scattering method with the Nd:YAG laser at 1064 nm wavelength. Excellent results were obtained in their study, and one of the compounds from this series exhibited the first hyperpolarizability at around $\beta_{333}^{1500} = 9.6\times10^{-50}\ \mathrm{Cm^3V^{-2}}$. This value was in fact close to the reference compound 4-(*N,N*-dimethylamino) cinnamic aldehyde that has a first hyperpolarizability around $\beta_{333}^{1500} = 13.0\times10^{-50}\ \mathrm{Cm^3V^{-2}}$. Delarie *et al.* also studied the NLO response of azo-azulene derivatives as shown in Figure 5.4 [35]. The NLO response of phenylazo-azulene-based compounds was investigated using the electric field-induced second harmonic (EFISH) technique. The study revealed that compound **5.6b** of this series has a static hyperpolarizability around 57 × 10^{-30} $\mathrm{cm^5\ esu^{-1}}$, which is also better than the Dispersed Red chromophore **5.7**.

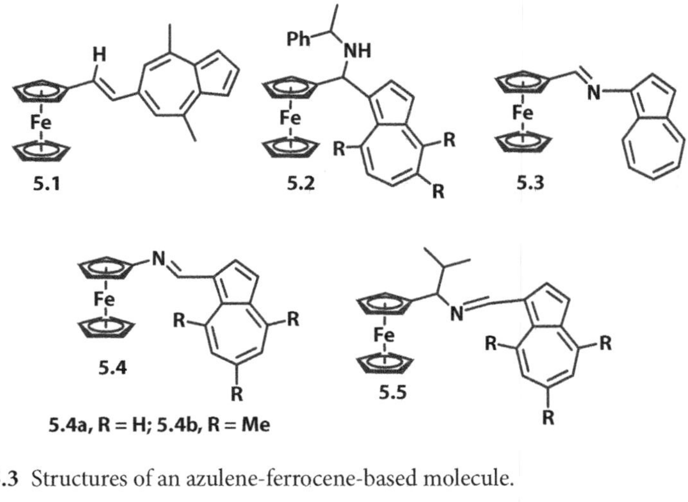

Figure 5.3 Structures of an azulene-ferrocene-based molecule.

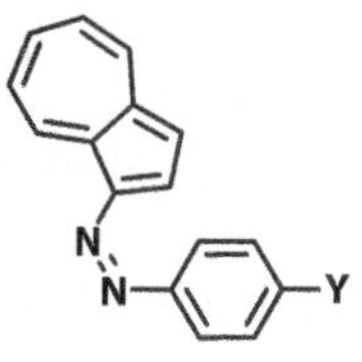

5.6

5.6a, Y = H; 5.6b, R = NO_2;
5.6c, Y = OCH_3

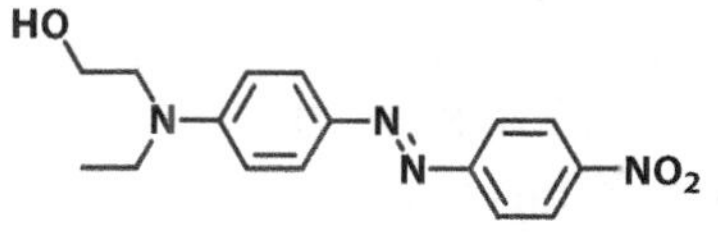

5.7

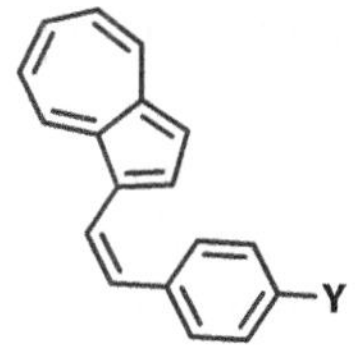

5.8

Figure 5.4 Structures of azo-azulene-based compounds exhibiting NLO properties.

Yang *et al.* employed a distinct approach in designing a novel NLO material by extending the π-conjugation using different geometric arrangements (Figure 5.5) [36]. In their approach, three different moieties were used as a functional unit, which included benzene, naphthalene, and azulene, and these were linked in classical D-π-A fashion either by extension of π-chain extended or by inclusion of a direct fused system. The inclusion of azulene plays a unique role here as a redistribution of the π-electrons in azulene-based polar carbon chains leads to lowering of the HUMO-LUNO

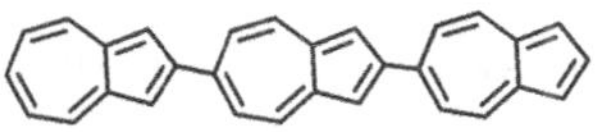

5.9a

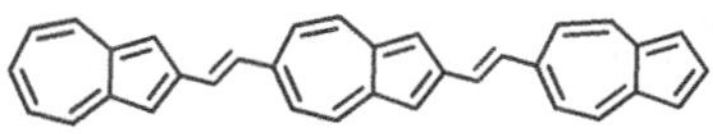

5.9b

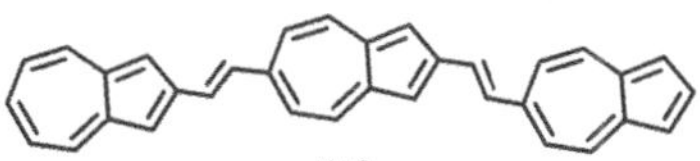

5.9c

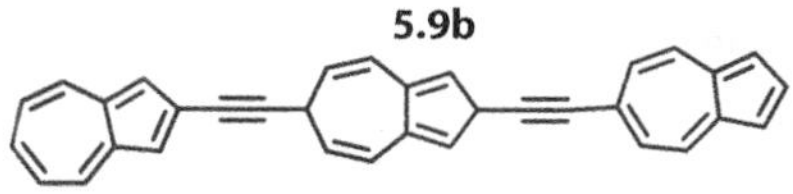

5.9d

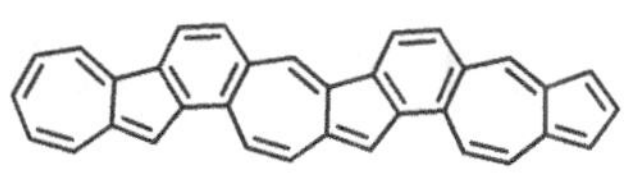

5.9e

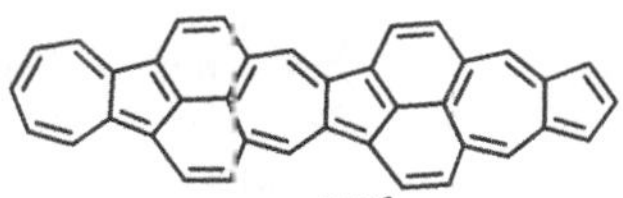

5.9f

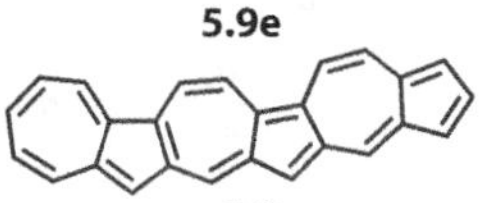

5.9g

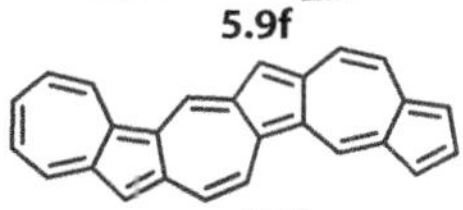

5.9h

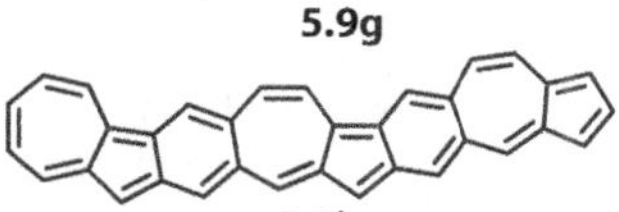

5.9i

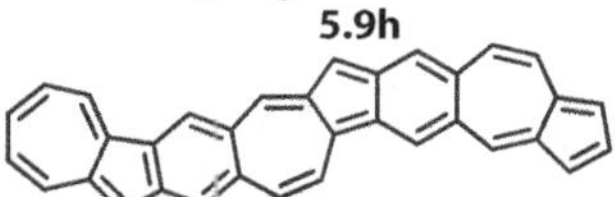

5.9j

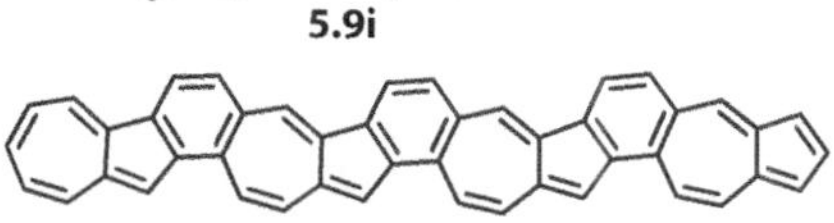

5.9k

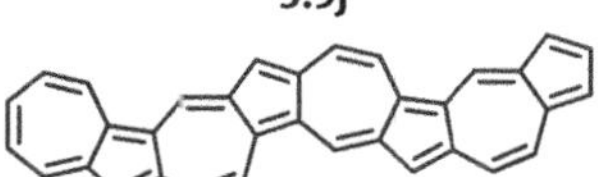

5.9l

Figure 5.5 Structures of azulene-based molecules with an extension of π-conjugation.

energy gaps and further a higher second-order NLO response. The fused azulene systems lead to a large change in the dipole moment and better NLO response as shown in compound **5.9l** that showed the highest hyperpolarizability (β_0), of around 70.85 × 10^{-30} esu per heavy atom.

In their recent report, Yang *et al.* reported some more stable NLO compounds comprising azulene and porphyrin units (**5.10a–n**) (Figure 5.6) that lead to an enhanced π-conjugation resulting in a second-order NLO response of the systems [37]. The orientation of azulene is crucial in the increment of static first hyperpolarizability from 0.31 × 10^{-30} to 9.78 × 10^{-30} esu. Further, the induction of metals to these systems lead to enhanced static first hyperpolarizability of around 41.59 × 10^{-30} esu.

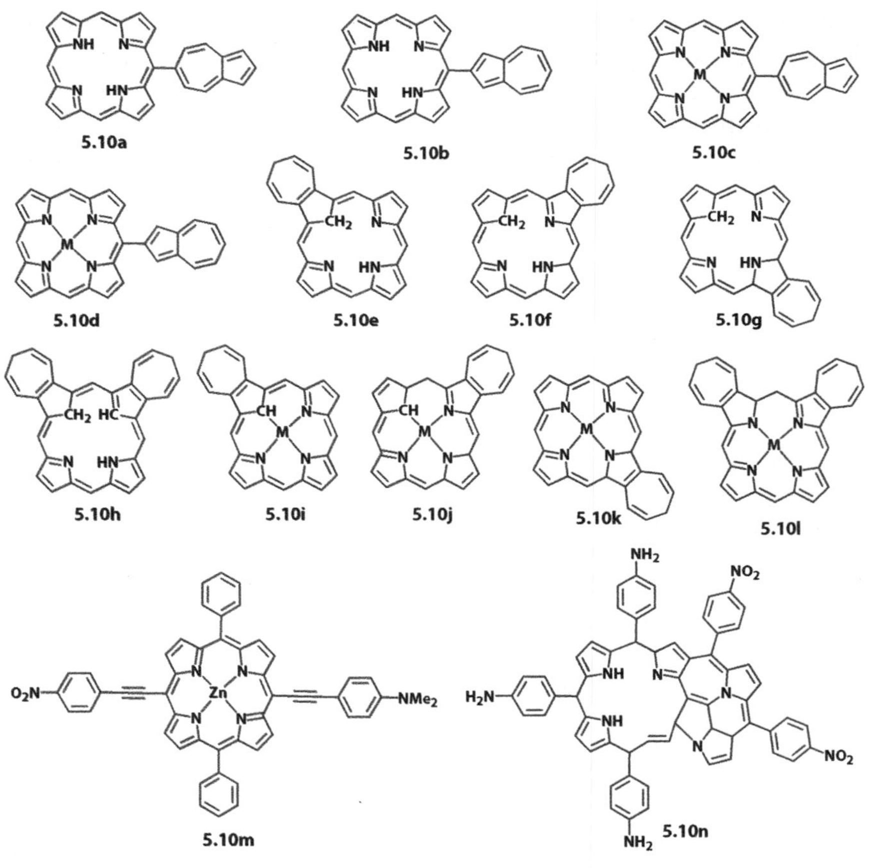

Figure 5.6 Structures of azulene-porphyrin-based molecules showing an NLO response.

5.3 Applications in Molecular Devices

Molecules that are π-conjugated are widely used in single-molecular devices, which enables the single molecule to be integrated into molecular devices. The advancement of single-molecule conductance has uplifted the development of diodes, transistors, rectifiers, and photo-voltaic cells [38]. A typical single-molecule device comprises three components viz. two electrodes, a conjugated organic molecule, and anchoring groups. The molecular conductance largely depends upon the molecular type of the electrode employed, the structure of the conjugated organic molecule, and molecule-electrode bonding [39]. In general, the high performance of a molecular device is directly related to the extent of π-delocalization of the conjugated molecule and the low HOMO-LUMO gap. Further, any modulation in the chemical structure of the organic molecule can alter the energy band gap and can control the charge transport within the single molecule, leading to device miniaturization. For example, including electronegative fragments in the molecular structure, single-molecule positive-negative (PN) junctions can be constructed, which can display the characteristics of a rectifier [40–42]. A judicious modulation in the energy band can lead to efficient positive-negative-positive (PNP) junction with wide application in molecular materials and devices, like photodetectors [43–45]. The platform for single-molecule devices offers an inherent method for examining the PNP structure from the viewpoint of a single molecule. By using single-molecule PNP junctions, new single-molecule functional devices can be designed, and important insights into molecular charge transport pathways can be gained. The most used organic conjugated systems in molecular devices include benzene, thiophene, furan, and ethylene. Azulene, a 10-π electron aromatic compound with a narrow energy gap and inherent dipole moment of 1.08D is a molecule with one of the biggest potential for investigating single-molecule conductance [46]. Below, we discuss examples of incorporating azulene-containing systems in single-molecule devices.

Guo *et al.* reported the design and synthesis of single-molecule devices (**5.11**) (Figure 5.7) with a PNP structure comprising an azulene unit

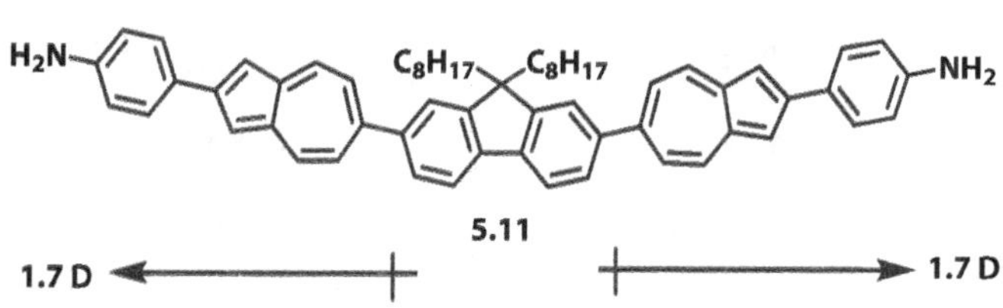

Figure 5.7 The chemical structure of derivative **5.11**.

flanked on the sides of fluorine [47]. The orientation of azulene in this system results in the opposite dipole of 1.7D, thus generating a PNP-type single-molecule junctions. The unique structure of the molecule led to the minimization of the transport barrier and the lowering of the transition temperature from thermionic emission to tunneling.

The electrical conductance of amino and cyanoazulenes sandwiched between two gold electrodes (**5.12, 5.13**) as shown in Figure 5.8 was investigated by Yoshizawa *et al.* by employing [48] non-equilibrium Green's function approach and density functional theory. The studies revealed that conductance increases when the $-NH_2$ group is positioned at C-1 and C-3 of the azulene unit whereas the –CN group is positioned at C-4 and C-8 of the same. The presence of electron donor groups such as $-NH_2$ at C-4 or C-8 also impart diode-like characteristics to 1, 3-azulene dithiolate. A similar behavior can be observed when the carbons at 2-, 4-, 6-, and 8-positions of azulene are replaced by nitrogen atoms.

Campos *et al.* [49] analyzed the use of thiochroman-coupled azulene-based molecules (**5.14a–d**) (Figure 5.9) as anchoring groups to connect Au electrodes. The distinct substitute pattern of the four molecules alters not only their optical properties but also exerts influence on their conductivity. Their study also helped in concluding that the often-used models to predict the quantum interference characteristics do not apply to azulene-like systems that are non-alternant hydrocarbons.

Chen *et al.* [50] developed a biazulene-bridged cobalt-coordinated framework on an Au rod surface by utilizing a dip-coating method (**5.14e**) (Figure 5.9). The wrapping of the coordination framework onto the gold rod results in a rectification ratio of 5:7 and excellent cycle stability. This method was also considered superior as it rules out the necessity of using expensive scanning tunneling microscopes for the fabrication step.

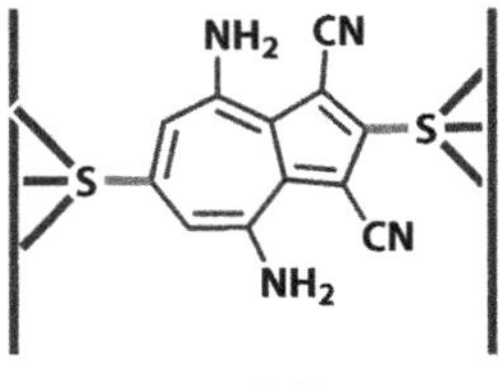

5.12

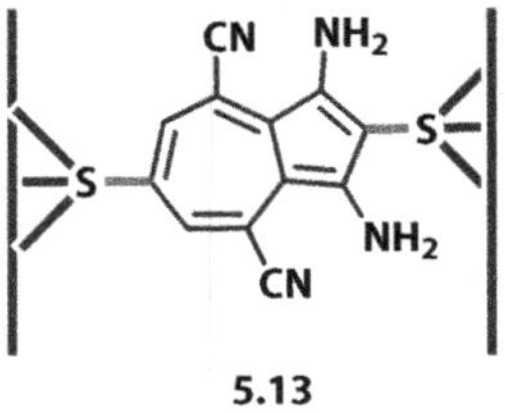

5.13

Figure 5.8 The representation of derivatives **5.12** and **5.13** sandwiched between a gold electrode.

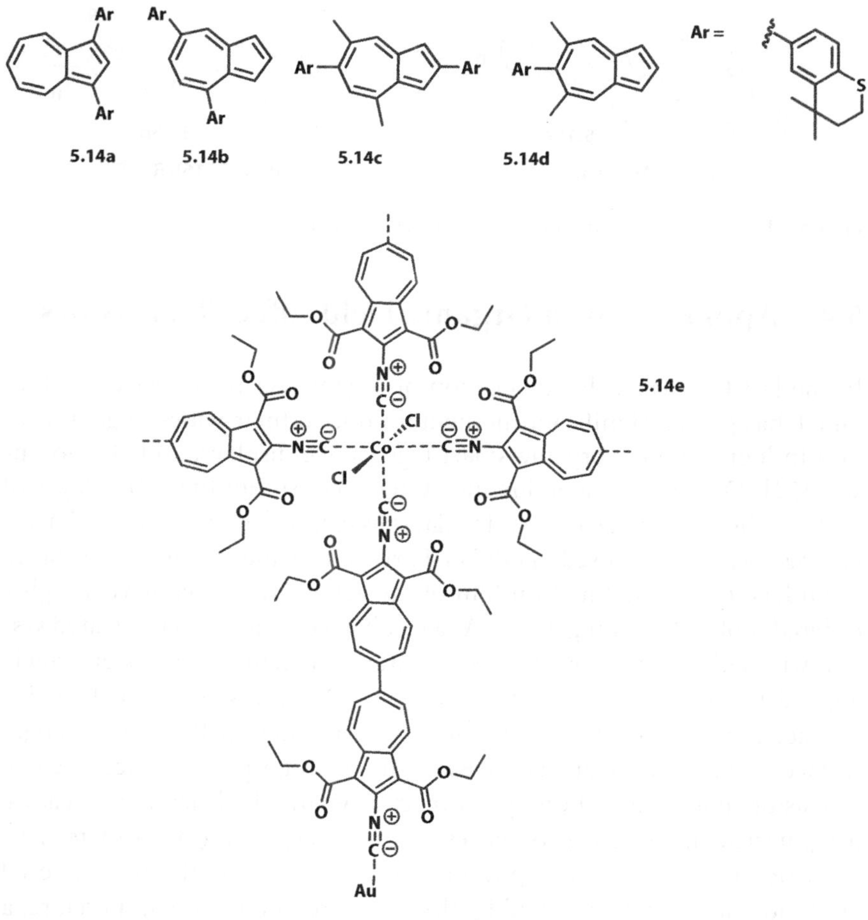

Figure 5.9 The chemical structure of derivative **5.14a–d.**

Gao *et al.* [51] reported the synthesis of two azulene polymers involving 2-, 6-connectivity, and solubilizing groups were introduced at 1-position of azulene by using Friedel–Craft's acylation reaction. Both these polymer (**5.15a**) RP(Az-AC16) and (**5.15b**) P(Az-AC16) showed good proton-responsive properties in solution states and thin films (Figure 5.10). Further, incorporating polymers in a proton-exchange membrane, such as Nafion matrices, pronouncedly enhanced its proton-conductivity properties.

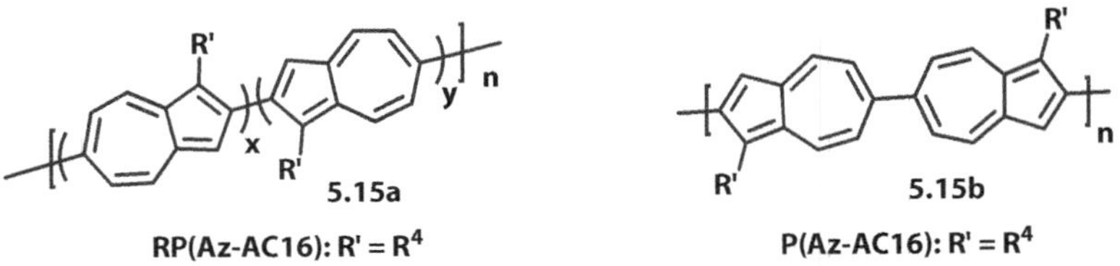

Figure 5.10 The chemical structure of derivatives **5.15a–b**.

5.4 Applications in Organic Field Effect Transistors

In the last few years, the application of organic electron devices such as OFET has gained significant momentum due to their wide-range utilization in logic circuits, organic smart tags, active displays, and chemosensors [52]. Organic semiconducting systems constitute the active layer of OFETs. The π-conjugated organic materials exhibiting crystalline behavior are the ones typically used in OFETs. Further, to achieve high performance in OFETs, the crystalline thin film of the active layer must have a highly ordered molecular arrangement. As a result, an extended π-conjugated system with either planar or linear structural arrangements has been widely applied to increase dispersion forces [53]. Organic semiconductors like pentacene and other two-dimensional herringbone patterns with edge-to-face arrangements are good examples of high-performance organic transistor materials. These polycyclic aromatic hydrocarbons feature long, narrow molecular geometries and large aspect ratios in their crystal structures [54, 55]. The optical and electronic properties of a material are fundamentally determined by the structure of a material, therefore, a synthetic chemist working in the development of organic electronic materials is keen to bring the structural innovation and synthesis of organic semiconductors. Compared to other aromatic compounds, such as thiophene and napthalene, azulene, owing to its unique molecular structure, displays distinct photophysical and chemical properties. Developing new azulene-based molecular structures presents a chance to produce unique, organic, and functional materials [56]. Nozoe's azulene synthesis provides access to multifunctional azulenes that offer the possibility of modifying the groups to other key groups, assisting in the synthesis of azulene-based conjugated systems (Figure 5.11). As discussed previously, azulene is a unique system as it offers the possibility of different connectivity patterns with each type exhibiting distinct properties. In the section below, we shine a light on azulene-based small molecules and polymers that have found applications in OFETs [57–60].

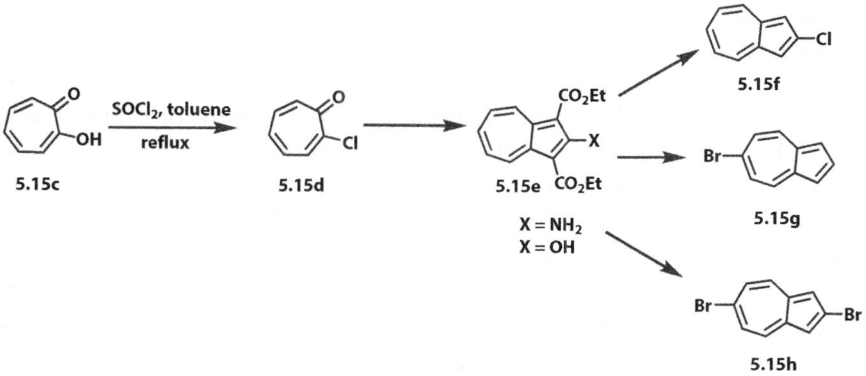

Figure 5.11 The synthetic route of the Nozoe–Seto synthesis for developing azulene derivatives.

5.4.1 Azulene-Based Polymers in Organic Field Effect Transistors

The development of azulene-based derivatives, where azulene is conjugated at the 2- and 6-positions, has paved for the creation of conjugated systems, exhibiting excellent behavior as semiconductors in OFETs [61].

Gao *et al.* [62] reported the design and synthesis of 2-, 6-functionalized azulene-based conjugated polymers **5.16a** and **5.16b**, which contain a diimide bridge and exhibit high conductivity for application in OFETs (Figure 5.12). The designed molecules were synthesized with a view that

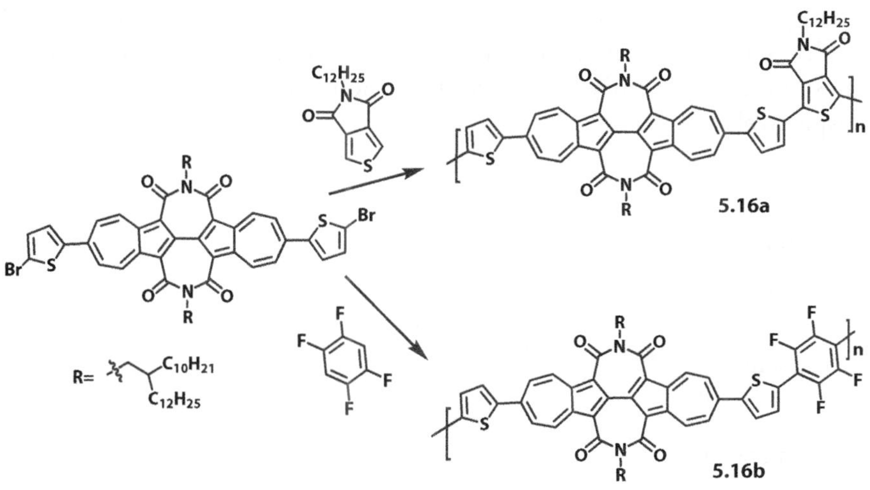

Figure 5.12 The chemical structure of derivatives **5.16a–b**.

utilizing 2- and 6- connectivity of azulene would ensure maximum utilization of its inherited D-A system and thus increase the chances of π-π stacking, which is a desirable strategy to obtain high performance OFETs. Theoretical insights revealed that the electron density of HOMO and LUMO is dispersed all over the polymeric backbone, thus enabling the FMO energy levels of the polymeric unit to be easily tuned by tuning the co-monomeric unit. Both **5.16a** and **5.16b** exhibited a dual absorption band in the visible region and a further broad and weak band in the lower-energy region, attributed to the intramolecular charge transfer absorption of the polymeric unit. Electrochemical insights into both the polymers 5.16a and 5.16b revealed an irreversible oxidation wave and multiple reduction waves, with HOMO/LUMO energy levels of **5.16a** and **5.16b** at -5.82/-3.76 eV and -5.76/-3.60 eV, respectively. The thin-film transistors developed by the bottom-gate top-contact approach using **5.16a** and **5.16b** as the active materials exhibited the n-type property. Thin films based on **5.16a** and **5.16b** displayed electron mobilities of 0.42 $cm^2V^{-1}s^{-1}$ and 0.24 $cm^2V^{-1}s^{-1}$, respectively.

Zhang *et al.* reported three dithienyldiketopyrrolopyrrole and azulene-based D-A conjugated polymers (**5.17a–c**) (Figure 5.13) [63]. The semiconducting behavior of these three polymers could be easily modulated by varying the connectivity of azulene units in the polymeric backbone. The OFET devices constructed using polymers **5.17a** and **5.17b**, wherein the connectivity of azulene was *via* a five-membered ring that exhibits p-type semiconducting behavior with hole mobilities of the order of 10^{-2} $cm^2V^{-1}s^{-1}$, respectively. On the contrary, polymer **5.17c** involving the connectivity of azulene *via* a seven-membered ring exhibits lower LUMO energy and well-diffused LUMO density. Consistent with the theoretical findings, this polymer exhibits ambipolar characteristics with hole mobilities of 6.2×10^{-2} $cm^2V^{-1}s^{-1}$ and electron mobilities of 2.1×10^{-2} $cm^2V^{-1}s^{-1}$, respectively.

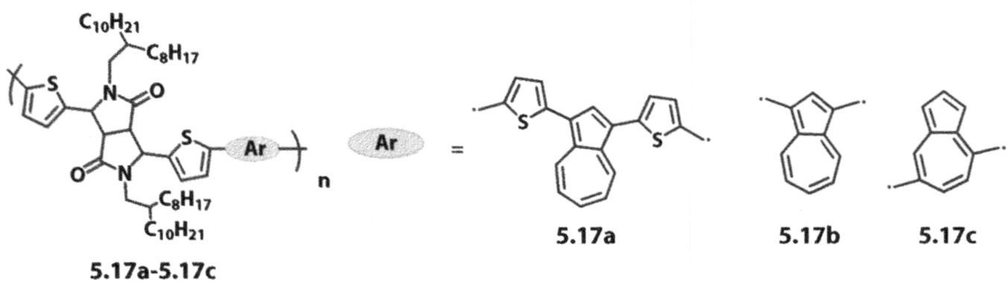

Figure 5.13 The chemical structure of derivative 5.17a–c.

Gao *et al.* reported conjugated polymers **5.18**, **5.19**, and **5.20** (Figure 5.14) comprising azulene and indacenodithiophene units and involving 2-, 6- connectivity of azulene [64]. The presence of the indacenodithiophene unit in the polymer backbone permits its easy protonation in the solution state and solid state by using acids like trifluoroacetic acid. The protonation of the polymer could be easily detected as it results in a bathochromic shift in UV-vis-NIR absorption spectra, and the same was also visible to the naked eye as protonation was accompanied by a change in color from the original blue in a neutral state to colorless when protonated. The OFET studies revealed that all the polymers exhibited p-type behavior with hole mobilities reaching as high as 0.46 $cm^2V^{-1}s^{-1}$.

The orientation of azulene in the polymer backbone can also affect its device performance. This was studied by the synthesis of three azulene and thieno[3,4-c]pyrrole-4,6-dione-containing conjugated polymers as shown

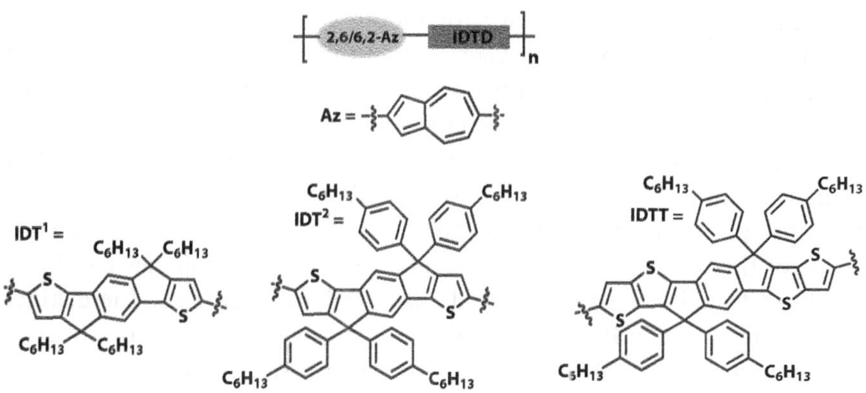

5.18 **5.19** **5.20**

Figure 5.14 Chemical structure of derivatives **5.18–20**.

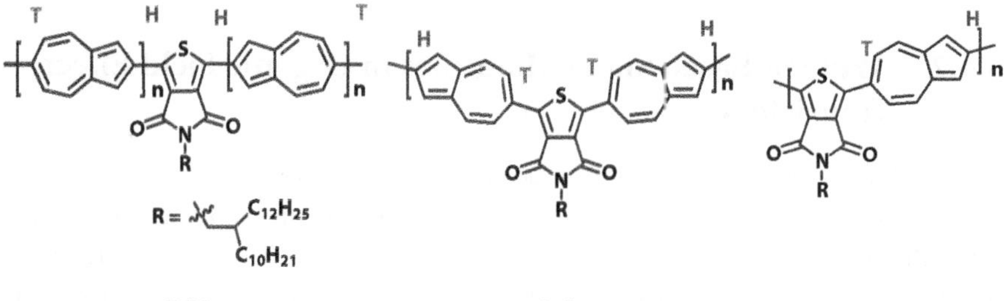

5.21a **5.21b** **5.21c**

Figure 5.15 The chemical structure of derivatives **5.21a–c**.

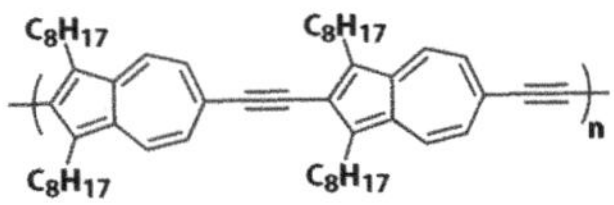

PAzE-1 5.22a

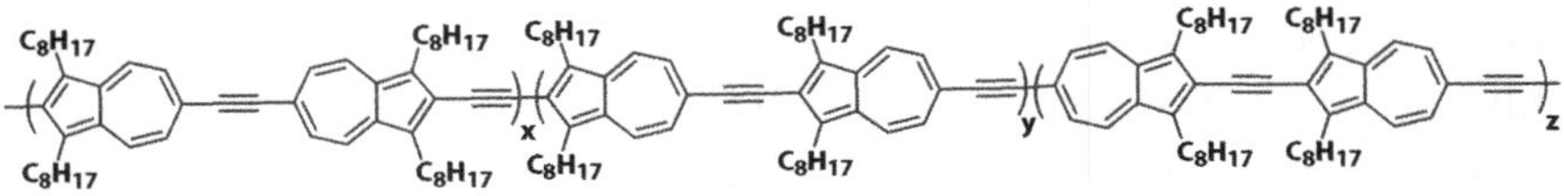

PAzE-2 5.22b

Figure 5.16 Chemical structures of PAzE-1 and PAzE-2.

in Figure 5.15 [65]. It was found that the electron mobility, crystallinity, and thermal stability of these polymers were impacted by the dipole arrangement of 2-, and 6-azulene units, and all the properties improved with enhanced regularity. The regioregular polymer **5.21c** displayed the best device performance, with an electron mobility of around 0.33 $cm^2V^{-1}s^{-1}$.

The contributions made by Gao *et al.* in the synthesis of 2,6-connected azulene polymers have been very significant [66]. In their other report, they made use of the Sonogashira coupling reaction to synthesize 2,6-azulene-based poly(aryleneethynylene)s polymers **PAzE-1** (**5.22a**) and **PAzE-2** (**5.22b**) (Figure 5.16). While the polymer **PAzE-1** contained a systematic head-to-tail arrangement of the azulene unit, the polymer **PAzE-2** consisted of three types of orientations of azulenes in the polymer backbone viz. head-to-tail, head-to-head, and tail-to-tail arrangements. The distinct dipole orientation in **PAzE-1** results in a strong aggregation in solvent even in dilute conditions. Whereas, the **PAzE-2** polymer consisting of mixed dipole orientations results in superior thin-film microstructure and surface morphology, and hence a superior device performance compared to the other polymer. The proton responsiveness of these polymers further makes them potential candidates in sensing and optoelectronics applications.

5.4.2 Azulene-Based Small Molecules in Organic Field Effect Transistors

The OFETs based on small molecule organic semiconductors are considered to be advantageous owing to their ordered molecular arrangement, reproducibility factors, and easy synthesis to achieve enhanced device performance. Push-pull or donor-acceptor types of molecules are known to exhibit high-charge carrier mobility, low-band gap, and high open-circuit voltages, thus making azulene type systems a potential candidate in the design and synthesis of small molecules for application in OFETs [67, 68]. Some of the important examples of azulene derivatives involving connectivity *via* 2- or 6- or 2,6- positions that have been applied in OFETs are discussed herewith.

Katagiri *et al.* reported the design and synthesis of azulene end-capped 1,3,4-thiadiazole molecules (**5.23a**, **5.23b**) by the Suzuki coupling reaction between a boronate derivative of azulene and dibromo derivative of 1,3,4-thiadiazole (Figure 5.17) [69]. The low-lying LUMO level in 6-azulene end-capped derivative and its efficient crystal arrangement involving herringbone type arrangement in terminal azulene and brick-like arrangement in the central thiadiazole, resulted in an excellent n-type characteristic of OFET devices based on **5.23a**. However, the other isomer **5.23b** had higher LUMO levels and a less ordered arrangement compared to the other isomer, thus making it devoid of any character in OFETs.

The connectivity of azulene has a direct influence on the charge carrier mobility as also explained in azulene and indacenodithiophene-based isomers **5.24a–b** (Figure 5.18) [70]. The isomer **5.24a** involving connectivity of azulene *via* 2-position has its HOMO energy level (−4.70 eV), 0.31 eV higher than the isomer involving connectivity of azulene *via* 6-position that is **5.24b**. In these isomers, it was observed that by changing the connectivity pattern of azulene the device characteristic can also be altered.

5.23a **5.23b**

Figure 5.17 The chemical structures of derivatives **5.23a** and **5.23b**.

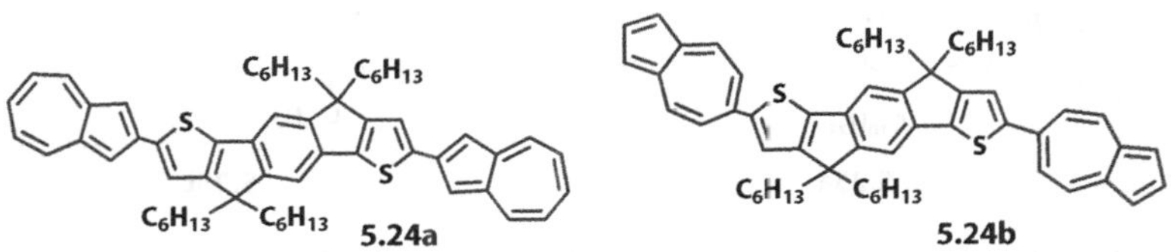

5.24a **5.24b**

Figure 5.18 The chemical structures of derivatives **5.24a** and **5.24b**.

For example, isomer **5.24a** which is a strong electron donor exhibited p-type characteristics with a hole mobility of around 4.14×10^{-3} $cm^2V^{-1}s^{-1}$. The isomer **5.24b** had a twisted geometry and this connectivity of azulene imparted the molecule with an electron mobility of 1.05×10^{-5} $cm^2V^{-1}s^{-1}$.

Another study emphasizing the influence of a connectivity pattern on the device characteristic was done by Gao *et al.* It involved azulene-naphthodithiophene diimide isomers (**5.25a** and **5.25b**) (Figure 5.19), involving connectivity of azulene *via* 2- and 6- positions, respectively [71]. The isomer **5.25a** had azulene connected to napthodithiophene diimide *via* the 2-position, whereas the isomer **5.25b** had the connectivity of the same *via* 6-position. Theoretical insights into the HOMO and LUMO of the two isomers proved to be significant in predicting the device characteristics. The LUMO density was delocalized over the entire molecule for the isomers, suggesting the potential for n-type transport behavior. The HOMO density of **5.25a** was delocalized over the entire molecule. For isomer **5.25b**, the same was confined at the two terminal azulene units, indicating the potential of p-type transport behavior in the former isomer. The theoretical results were consistent with experimental findings where devices based on **5.25a** exhibited electron mobilities and hole mobilities of 0.32 $cm^2V^{-1}s^{-1}$ and 0.03 $cm^2V^{-1}s^{-1}$, respectively. The devices based on isomer **5.25b** contrarily exhibited only electron mobility up to 0.13 $cm^2V^{-1}s^{-1}$.

The study conducted by Katagiri *et al.* also emphasized a similar observation by taking terazulene isomers. They reported the design and synthesis of three regioisomers of terazulene by varying the orientation of azulene in all the isomers (**5.26–5.28**) (Figure 5.20) [72]. In this case, theoretical

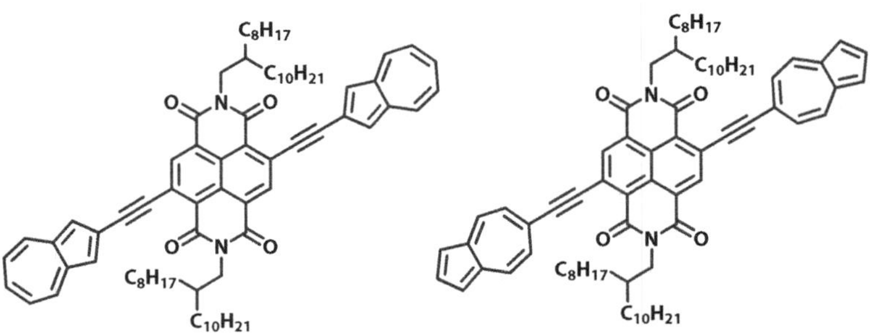

5.25a **5.25b**

Figure 5.19 The chemical structure of derivatives **5.25a** and **5.25b**.

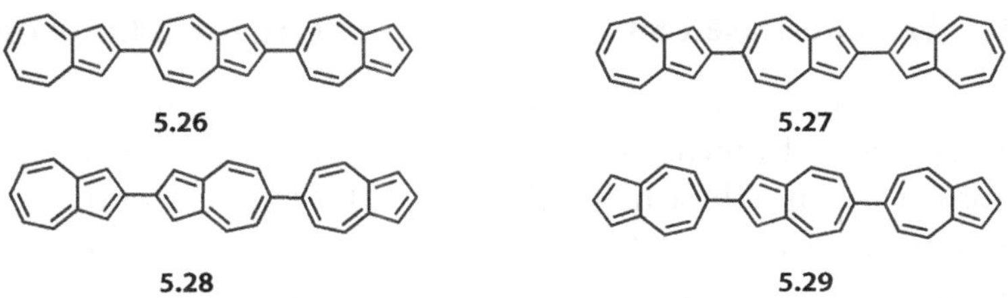

Figure 5.20 The chemical structure of derivatives **5.26–5.29**.

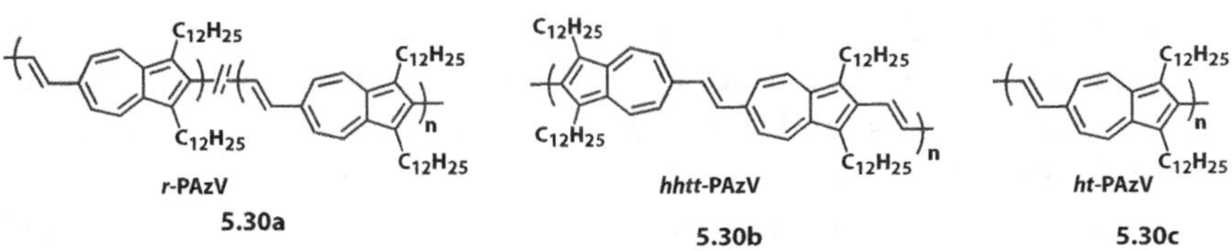

Figure 5.21 Chemical structures of *r*-PAzV, *hhtt*-PAzV and *ht*-PAzV (**5.30a–c**).

insights were also perfectly corroborated by the experimental findings, suggesting the importance of predicting the device characteristics in azulene-based systems prior to their experimental synthesis. The effect of connectivity variation in influencing the device performance was also done by Gao *et al.* in a series of poly(2,6-azulene vinylene)s that involved different the connectivity of azulene (**5.30a–c**) as shown in Figure 5.21 [73].

5.5 Application in Organic Photovoltaics

The unique photophysical and chemical properties of azulene have resulted in the potential application of azulene derivatives in NLO materials [74], NIR materials [75], organic light emitting diodes [76], organic solar cells [77], and OFETs [78]. Recent years have witnessed improvements in methods for the synthesis of azulene derivatives, further advancing their applications [79]. Azulene-based materials have also been found to be potential candidates for applications in OPVs. Some examples of such azulene derivatives are discussed below in brief.

5.5.1 Azulene-Based Small Molecules in Organic Photovoltaics

Xin *et al.* reported the design and synthesis of azulene diimide derivatives [80] (Figure 5.22) by incorporating different aromatic groups at 6- and 6'-positions of azulene. These molecules were further studied for their optical and electrochemical properties, as well as their semiconductor behavior. The study revealed that the aromatic substituents at 6- and 6'- positions of azulene can exert a significant influence on the optoelectronics properties of the molecules. The potential of these molecules, both in OFETs and OPVs, was explored. The power conversion efficiency (PCE) of these molecules were calculated using OPV devices. The data revealed that PCE of semiconductors **5.33** and **5.36** were around 1.38% and 1.36%, respectively.

Shoji *et al.* reported the design and synthesis of azulene-substituted-tetracyanobutadienes and heteroazulenyl tetracyanobutadienes derivatives (**5.38–5.43**) (Figure 5.23) [81]. Insight into absorption spectra revealed a significant bathochromic shift and high extinction coefficient with the enhancement of π-conjugation. Electrochemical insights revealed that azulene-tetracyanobutadiene derivatives can undergo electrochromism, making them potential candidates utilized in the development of new organic electronic materials.

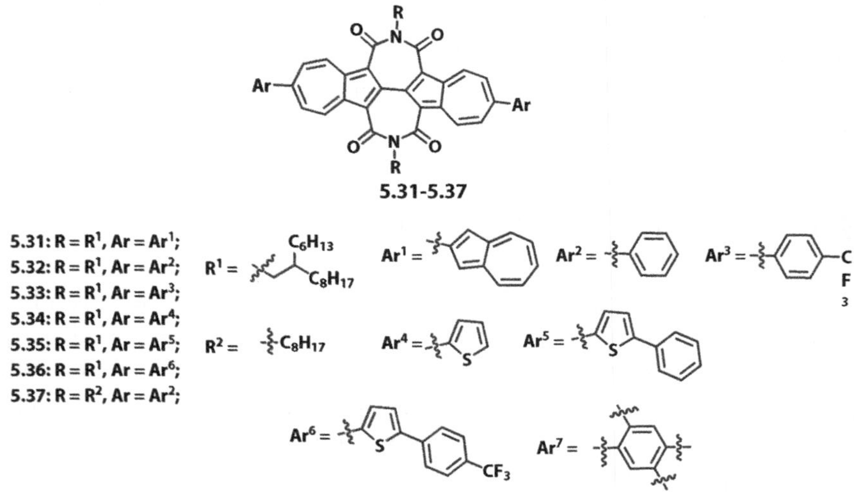

Figure 5.22 Structures of aryl-substituted biazulene diimide derivatives.

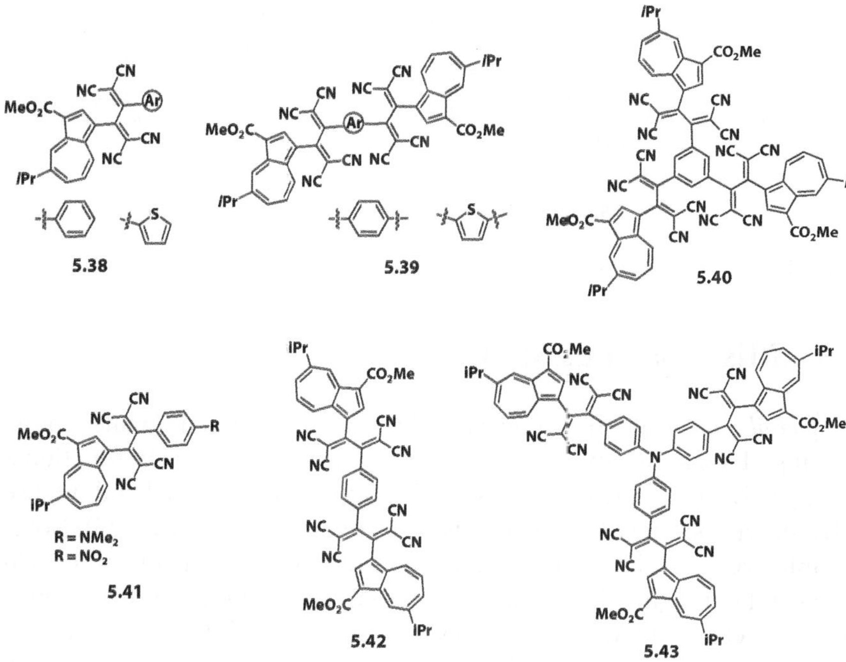

Figure 5.23 Structures of azulene-substituted-tetracyanobutadienes-based molecules.

5.5.2 Azulene-Based Small Polymers in Organic Photovoltaics

Conjugated polymers have been widely used in the development of organic optoelectronic devices such as solar cells and OFETs [82]. For example, Gao *et al.* reported the synthesis of azulene-diimides containing conjugated polymers **5.44a** and **5.44b**, as shown in Figure 5.24 [83]. Both the polymers exhibited excellent n-type characteristics in OFET devices with electron mobilities as high as 0.42 $cm^2V^{-1}s^{-1}$ and 0.24 $cm^2V^{-1}s^{-1}$ for **5.44a** and **5.44b**, respectively. Polymer **5.44a** was also used as an active material in an all-polymer solar cell (PSC) based on the PTB7-Th: **5.44a** blend. A PCE of approximately 1.82% was observed, making this polymer a potential candidate for application in polymer solar cells.

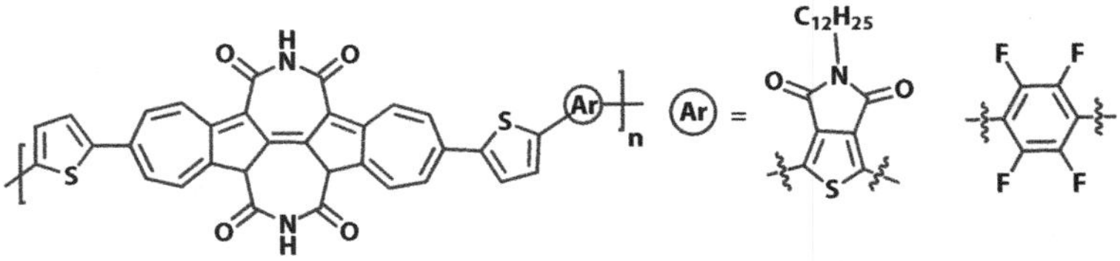

5.44a, 5.44b

Figure 5.24 Structures of azulene-based polymers.

5.6 Miscellaneous Studies

Jiang *et al.* performed theoretical insights on a series of tetra(azulene)porphyrins (TAzPs) derivatives by using time-dependent density functional theory (TD-DFT) calculations [84], which revealed the broad and intense NIR absorptions of H_2(TAzP) and its derivatives (**5.44c**, Figure 5.25). Novel organic dyes with absorption covering the entire solar spectrum (300–2500 nm) can be designed by including electron-donating units at the periphery of the core H_2(TAzP) unit. This study also revealed that low molecular symmetry and small HOMO/LUMO or HOMO/LUMO+1 gap are the key

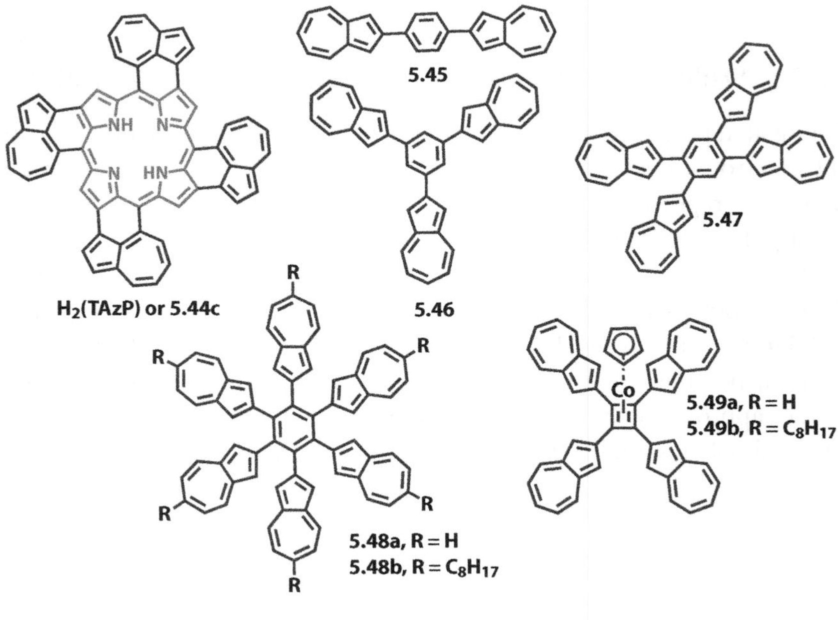

Figure 5.25 The chemical structure of derivatives **H2(TAzP)**, **5.44a–c**, **5.45a–b**, **5.46 a–b**, **5.47**, **5.48**.

factors to design novel organic dyes with a broad absorption profile. Such systems can be potential candidates for applications in an array of fields including organic solar antenna collectors and organic solar cells.

Korichi *et al.* carried out interesting theoretical calculations on a series of azulene complexes and provided a rational explanation of the bonding within this huge class of complexes by analyzing the molecular and electronic structure with respect to the electron count [85]. The DFT calculations revealed that the dissymmetry of azulene makes the coordination chemistry of azulene very rich. This is also attributed to the fact that azulene can adapt to the demand of the metal ion by adjusting its electronic and structural factors. The coordination compounds of azulene are also expected to exhibit rich fluxional behavior. Another study supporting the above study was also done by Wang *et al.* at the same time [86].

Ito *et al.* carried out extensive work on the synthesis and properties of azulene-based star systems featuring benzene rings at the core (**5.45–5.49**) as shown in Figure 5.25 [87]. Electrochemical insights reveal that these types of systems are capable of exhibiting multielectron redox behavior. The molecules **5.48a** and **5.48b** were synthesized by employing a cyclotrimerization reaction of di(2-azulenyl)acetylenes. The molecules of these types are also potential candidates for application as liquid crystalline materials.

Nishimura *et al.* [11] reported the design and synthesis of molecules containing azulene and oxygen-bridged triarylamine skeletons (**5.50–5.52**) (Figure 5.26). The tetrasubstituted derivative, which is **5.52**, was a potential hole-transporting material with an excellent PCE of 16.5% observed in perovskite solar cells. Truong *et al.* [88] further synthesized different derivatives of molecule **5.52** by varying the alkyl chains to study the influence of the alkyl chain on the device performance. The best PCE of around 18.9% in polymer solar cells was observed for the butyl chain substituted derivative.

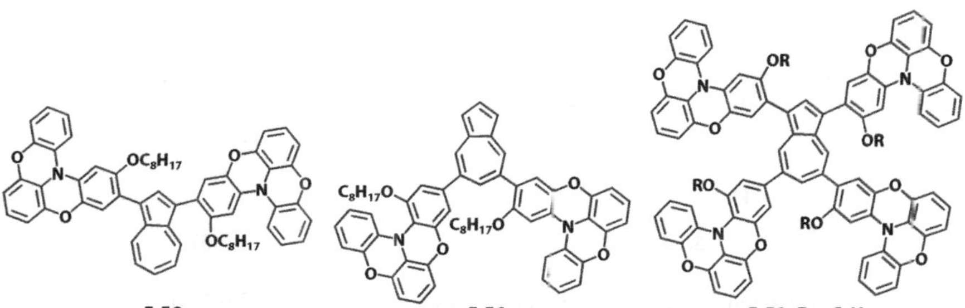

5.50 5.51 5.52: R = C_8H_{17}

Figure 5.26 The chemical structure of derivatives 5.50–5.52.

Replacing the four pyrrole units in porphyrin by azulene results in another class of interesting systems termed calix[4]azulene. Asao *et al.* [89] synthesized the first azulene analog of calixarene, which was named "azulenophane" (**5.53**). For the first time in 2002, Lash *et al.* [90] synthesized an all-hydrocarbon calix[4]azulene using azulene and paraformaldehyde as the starting material (**5.54a**) and further also reported the synthesis of tetra-6-*tert*-butylcalix[4]azulene (**5.54b**) [91]. Georghiou *et al.* [92] reported a phenyl-substituted calix[4]azulene system (**5.56**) that can encapsulate C_{60} or C_{70} fullerenes easily owing to the presence of electron donor 1- and 3-positions of azulene. Krug *et al.* reported the synthesis of an interesting hexamer macrocycle featuring C-Cu-C bonds by using 1,3-dibromo azulene as the starting material (**5.61**) (Figure 5.27) [93]. Such reactions generally lead to a competitive reaction between the formation of oligomer chains and macrocycles. However, control over the same can be achieved by using surface-assisted synthesis on a copper (111) surface.

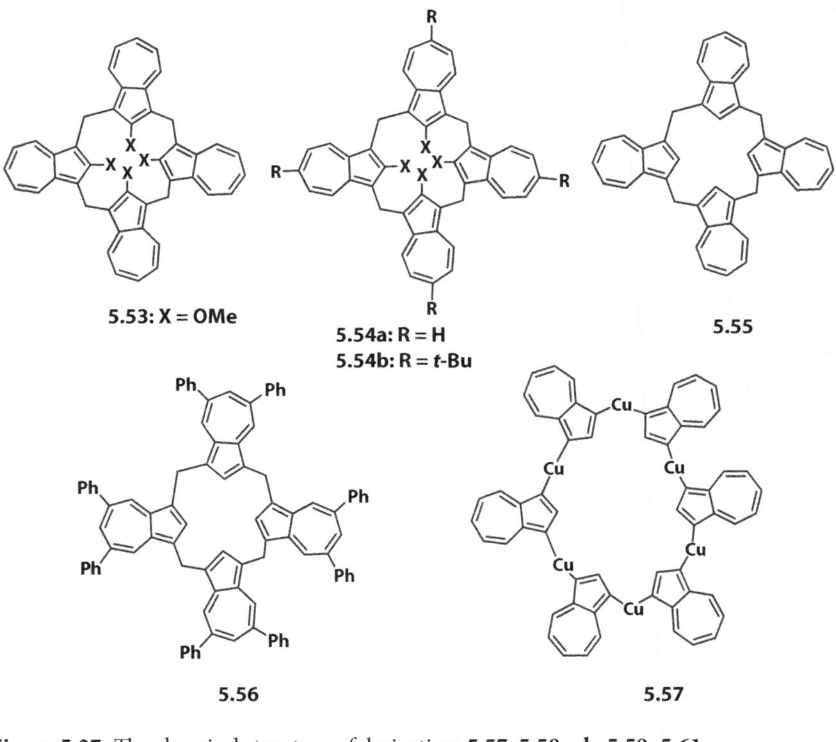

Figure 5.27 The chemical structure of derivatives **5.57**, **5.58a–b**, **5.59–5.61**.

Recently, Tian *et al.* [19] in their report discussed the importance of N-doped azulene-based nanographene systems as NLO materials. A molecular design strategy was adopted by swapping carbon atoms at the two ends of the azulene in the nanographene system with nitrogen atoms. This strategy was highly advantageous as not only was the stability of the nanographene system significantly enhanced but an improvement was also observed in the second- and third-order parameters of NLO properties. Theoretical insights revealed that replacing C atoms with N atoms enlarged HOMO-LUMO energy, which led to large second-order responses. Doping with heteroatoms proved to be significant in enhancing the NLO characteristics of carbon-based nanographene as potential multifunctional materials.

5.7 Conclusions

Azulene is an interesting molecule that has garnered the interest of researchers because of its unusually large dipole moment and a fluorescence that violates Kasha's rule ($S_2 \rightarrow S_0$). This is a result of the significant energy gap between the S_1 and S_2 states. The functionalization of azulene at 1-, 3- or 2-, 6- or 4-, 7- positions can alter not only its photophysical properties but can also change the molecular properties of resulting molecules. To shine a light on the different azulene derivatives in organic electronics, we have discussed the development of various small, or polymer-based azulene derivatives, with applications in OPVs, NLO, OFETs, and molecular devices. However, their device performance remains inferior to state-of-the-art organic optoelectronic materials. Developing and synthesizing high-performance azulene derivatives is a significantly challenging yet desirable task.

Further key points of consideration for developing desirable optoelectronic materials rely on factors such as:

1. Regioselective functionalization at desirable positions such as 2-, 6- or 1-, 3.
2. Modulating HOMO, LUMO, and LUMO+1 by wisely incorporating electron-donating and electron-rich groups.
3. Developing easy synthetic routes for azulene-based conjugated systems involving 2,6-connectivity to enable efficient utilization of the resonance structure of azulene, thus more efficient π–π stacking, which enhances the device performance.

Furthermore, the orientation of azulene units within the main conjugated system influences the distribution of molecular frontier orbital densities and intermolecular interactions, thereby affecting the device characteristics. Thus, we hope the compilation of various reports on azulene will help researchers revitalize this classic molecule with innovative chemistry and novel strategies to expand its applications in the development of organic optoelectronic materials.

AI disclosure statement: The author acknowledges the use of AI-based tools for minor language editing or grammatical correction. However, all intellectual contributions, analyses, and conclusions presented in this chapter are entirely the author's own.

References

1. Ruzicka, L. and Rudolph, E.A., Höhere terpenverbindungen XXVII. Zur kenntnis der azulene. *Helv. Chim. Acta*, 9, 131, 1926.
2. Gordon, M., The azulenes. *Chem. Rev.*, 50, 1, 127, 1952.
3. Pfau, A.S. and Plattner, P.A., Zur kenntnis der flüchtigen pflanzenstoffe VIII. Synthese des vetivazulens. *Helv. Chim. Acta*, 22, 202, 1939.
4. Ziegler, K. and Hafner, K., Eine rationelle synthese des azulens. *Angew. Chem.*, 67, 301, 1955.
5. Mathey, P., Fernandez, I., Morin, J.-F., Exploring C-C bond formation reactions for expanding azulene derivatives linked at the 2- and/or 6-positions. *New J. Chem.*, 48, 4801, 2024.
6. Elwahy, A.H.M., Abdelhamid, I.A., Shaaban, M.R., Recent advances in the functionalization of azulene through Pd-catalyzed cross-coupling reactions. *ChemistrySelect*, 6, 47, 13664, 2021.
7. Shoji, T., Ito, S., Yasunami, M., Synthesis of azulene derivatives from 2H-cyclohepta[*b*]furan-2-ones as starting materials: Their reactivity and properties. *Int. J. Mol. Sci.*, 22, 19, 10686, 2021.
8. Anderson Jr., A.G., Nelson, J.A., Tazuma, J.J., Azulene. III. Electrophilic substitution. *J. Am. Chem. Soc.*, 75, 20, 4980, 1953.
9. Dunlop, D., Ludvikova, L., Banerjee, A., Ottosson, H., Slanina, T., Excited-state (anti)aromaticity explains why azulene disobeys Kasha's rule. *J. Am. Chem. Soc.*, 145, 39, 21569, 2023.
10. Malpicci, D., Lucenti, E., Giannini, C., Forni, A., Botta, C., Cariati, E., Prompt and long-lived anti-Kasha emission from organic dyes. *Molecules*, 26, 22, 6999, 2021.
11. Nishimura, H., Ishida, N., Shimazaki, A., Wakamiya, A., Saeki, A., Scott, L.T., Murata, Y., Hole-transporting materials with a two-dimensionally expanded

π-system around an azulene core for efficient perovskite solar cells. *J. Am. Chem. Soc.*, 137, 15656, 2015.

12. Zhang, H., Li, J., Yang, C., Guo, X., Single-molecule functional chips: Unveiling the full potential of molecular electronics and optoelectronics. *Acc. Mater. Res.*, 5, 8, 971, 2024.
13. Zhang, C., Kong, Y., Xiang, J., Chen, S., Kornyshev, A.A., Ulstrup, J., Gao., X., Zhang, G., Li, Y., Li, J., Switchable modes of azulene-based single molecule-electrode coupling controlled by interfacial charge distribution. *Chem. Sci.*, 16, 1353, 2025.
14. Bektas, N., Dastan, A., Erdogan, M., Aydogan, A., A guaiazulene-based highly selective colorimetric sensor for cyanide ion detection in aqueous medium and food samples. *J. Mol. Liq.*, 423, 127038, 2025.
15. Bekir, M., Gurke, J., Reifarth, M., Photoswitchable surfactants-Are there alternatives to azobenzene-based systems. *ChemSystemsChem*, 6, 5, e202400026, 2024.
16. Baharfar, M., Hillier, A.C., Mao, G., Charge-transfer complexes: Fundamentals and advances in catalysis, sensing, and optoelectronic applications. *J. Org. Chem.*, 36, 42, 2406083, 2024.
17. Ariga, K., Responsive materials nanoarchitectonics at interfaces. *Responsive Mater.*, 2, 2, e20240011, 2024.
18. Gao, H., Yao, Y., Li, C., Zhang, J., Yu, H., Yang, X., Shen, J., Liu, Q., Xu, R., Gao, X., Ding, D., Fused azulenyl squaraine derivatives improve phototheranostics in the second near-infrared window by concentrating excited state energy on non-radiative decay pathways. *Angew. Chem. Int. Ed.*, 22, e202400372, 2024.
19. Zhang, Y.Q., Yang, C.-C., Ma, J.-Y., Tian, W.Q., The enhancement of nonlinear optical properties of azulene-based nanographene by N atoms: A finishing touch. *Chem. Sci.*, 15, 2100, 2023.
20. Hatakenaka, Y., Urabe, K., Ueno, S., Yamauchi, M., Mizuhata, Y., Yamada, H., Mikata, Y., Kamijo, S., Tani, F., Murafuji, Y., Doubly linked azulene dimar: A novel non-benzenoid isomer of perylene. *Chem. Eur. J.*, 31, e202404679, 2025.
21. Raheem, A.A. and Praveen, C., π-Distorted charge transfer chromophores and their materials chemistry in organic photovoltaics. *J. Mater. Chem. C*, 12, 8611, 2024.
22. Boschloo, G., Improving the performance of dye-sensitized solar cells. *Front. Chem.*, 7, 77, 2019.
23. Ghanavatkar, C.W., Mishra, V.R., Nagaiyan, S., Review of NLOphoric azo dyes–developments in hyperpolarizabilities in last two decades. *Dyes Pigm.*, 191, 109367, 2021.
24. (a) Zhou, L., Mao, J., Ren, Y., Han, S.T., Roy, V.A.L., Zhou, Y., Recent advances of flexible data storage devices based on organic nanoscaled materials. *Small*, 14, 1703126, 2018. (b) Ogawa, K., Two-photon absorbing molecules as potential materials for 3D optical memory. *Appl. Sci.*, 4, 1, 2014.

25. Nisic, F., Colombo, A., Dragonetti, C., Garoni, E., Marinotto, D., Righetto, S., De Angelis, F., Lobello, M.G., Salvatori, P., Biagini, P., Melchiorre, F., Functionalized Ruthenium Dialkynyl Complexes with High Second-Order Nonlinear Optical Properties and Good Potential as Dye Sensitizers for Solar Cells. *Organometallics*, 34, 94, 2015.
26. Shen, Y., Shuhendler, A.J., Ye, D., Jing, J., Chen, H., Two-photon excitation nanoparticles for photodynamic therapy. *Chem. Soc. Rev.*, 45, 6725, 2016.
27. (a) Bandara, H.M.D. and Burdette, S.C., Photoisomerization in different classes of azobenzene. *Chem. Soc. Rev.*, 41, 1809, 2012. (b) Grommet, A.B., Lee, L.M., Klajn, R., Molecular photoswitching in confined spaces. *Acc. Chem. Res.*, 53, 2600, 2020.
28. (a) Karamanis, P., Otero, N., Pouchan, C., Unleashing the quadratic nonlinear optical responses of graphene by confining white-graphene (h-BN) sections in its framework. *J. Am. Chem. Soc.*, 136, 7464, 2014. (b) Cox, J.D. and Garćıa de Abajo, F.J., Plasmon-enhanced nonlinear wave mixing in nanostructured graphene. *ACS Photonics*, 2, 306, 2015.
29. Li, H.P., Bi, Z.T., Xu, R.F., Han, K., Li, M.X., Shen, X.P., Wu, Y.X., Theoretical study on electronic polarizability and second hyperpolarizability of hexagonal graphene quantum dots: Effects of size, substituent, and frequency. *Carbon*, 122, 756, 2017.
30. Baumgartner, T. and Reáu, R., Organophosphorus π-Conjugated Materials. *Chem. Rev., 107*, 303, 2007.
31. Ito, S. and Morita, N., Creation of stabilized electrochromic materials by taking advantage of azulene skeletons. *Eur. J. Org. Chem.*, 2009, 4567, 2009.
32. Anderson, A.G. and Steckler, B.M., Azulene. VIII. A study of the visible absorption spectra and dipole moments of some 1- and 1,3-substituted azulenes. *J. Am. Chem. Soc.*, 81, 4941, 1959.
33. Michl, J. and Thulstrup, E.W., Why is azulene blue and anthracene white? a simple mo picture. *Tetrahedron*, 32, 205, 1976.
34. Herrmann, R., Pedersen, B., Wagner, G., Youn, J.H., Molecules with potential applications for non-linear optics: the combination of ferrocene and azulene. *J. Organomet. Chem.*, 571, 261, 1998.
35. Lacroix, P.G., Mulfant, I., Iftime, G., Razus, A.C., Nakatani, K., Delarie, J.A., Azo-azulene derivatives as second-order nonlinear optical chromophores. *Chem. Eur. J.*, 14, 6, 2000.
36. Yang, C.C., Ma, J.Y., Su, X., Zheng, X.L., Chen, J., He, Y.Y., Tian, W.Q., Li, W.Q., Yang, L., High performance nonlinear optical materials with simple aromatic hydrocarbons. *FlatChem*, 33, 100362, 2022.
37. Yang, C.C., Li, L., Tian, W.Q., Li, W.Q., Yang, L., Strong second order nonlinear optical properties of azulene-based porphyrin derivatives. *Phys. Chem. Chem. Phys.*, 24, 13275, 2022.
38. Tang, C., Ayinla, R.T., Wang, K., Beyond electrical conductance: Progress and prospects in single-molecule junctions. *J. Mater. Chem. C*, 10, 13717, 2022.

39. Yokota, K., Taniguchi, M., Tsutsui, M., Kawai, T., Molecule electrode bonding design for high single-molecule conductance. *J. Am. Chem. Soc.*, 132, 17364, 2010.
40. Xin, N., Hu, C., Al Sabea, H., Zhang, M., Zhou, C., Meng, L., Jia, C., Gong, Y., Li, Y., Ke, G., He, X., Selvanathan, P., Norel, L., Ratner, M.A., Liu, Z., Xiao, S., Rigaut, S., Guo, H., Guo, X., Tunable symmetry-breaking-induced dual functions in stable and photoswitched single-molecule junctions. *J. Am. Chem. Soc.*, 143, 20811, 2021.
41. Elbing, M., Ochs, R., Koentopp, M., Fischer, M., von Hänisch, C., Weigend, F., Evers, F., Weber, H.B., Mayor, M., A single-molecule diode. *Proc. Natl. Acad. Sci. U.S.A.*, 102, 8815, 2005.
42. Capozzi, B., Xia, J., Adak, O., Dell, E.J., Liu, Z.-F., Taylor, J.C., Neaton, J.B., Campos, L.M., Venkataraman, L., Single-molecule diodes with high rectification ratios through environmental control. *Nat. Nanotechnol.*, 10, 522, 2015.
43. Kan, M., Yang, C., Wang, Q., Zhang, Q., Yan, Y., Liu, K., Guan, A., Zheng, G., Defect-assisted electron tunneling for photoelectrochemical CO_2 reduction to ethanol at low over potentials. *Adv. Energy Mater.*, 12, 2201134, 2022.
44. Su, B.-W., Zhang, X.-L., Yao, B.-W., Guo, H.-W., Li, D.-K., Chen, X.-D., Liu, Z.-B., Tian, J.-G., Laser writable multifunctional van der Waals heterostructures. *Small*, 16, 2003593, 2020.
45. An, J., Sun, T., Wang, B., Xu, J., Li, S., Efficient graphene in-plane homogeneous p-n-p junction based infrared photodetectors with low dark current. *Sci. China Inf. Sci.*, 64, 140403, 2021.
46. Xinzhe, S., Arpan, S., Jean-François, S., Henri, D., Metal-catalyzed C–H bond activation of 5-membered carbocyclic rings: A powerful access to azulene, acenaphthylene and fulvene derivatives. *Chem. Asian J.*, 13, 143, 2018.
47. Li, M., Fu, H., Wang, B., Cheng, J., Hu, W., Yin, B., Peng, P., Zhou, S., Gao, X., Jia, C., Guo, X., Dipole-modulated charge transport through PNP-type single-molecule junctions. *J. Am. Chem. Soc.*, 144, 20797, 2022.
48. El-Nahas, A.M., Staykov, A., Yoshizawa, K., Electrical conductivity and diode-like behavior of substituted azulene electrical conductivity and diode-like behavior of substituted azulene. *J. Phys. Chem. C*, 121, 2504, 2017.
49. Xia, J., Capozzi, B., Wei, S., Strange, M., Batra, A., Moreno, J.R., Amir, R.J., Amir, E., Solomon, G.C., Venkataraman, L., Campos, L.M., Breakdown of interference rules in azulene, a nonalternant hydrocarbon. *Nano Lett.*, 14, 294, 2014.
50. Sun, S., Zhuang, X., Wang, L., Zhang, B., Ding, J., Zhang, F., Chen, Y., Azulene-bridged coordinated framework based quasi-molecular rectifier. *J. Mater. Chem. C*, 5, 2223, 2017.
51. Hou, B., Zhou, Z., Yu, C., Xue, X.-S., Zhang, J., Yang, X., Li, J., Ge, C., Wang, J., Gao, X., 2,6-Azulene-based homopolymers: Design, synthesis, and application in proton exchange membrane fuel cells. *ACS Macro Lett.*, 11, 680, 2022.

52. Dong, J.X. and Zhang, H.-L., Azulene-based organic functional molecules for optoelectronics. *Chin. Chem. Lett.*, 27, 1097, 2016.
53. Mateo-Alonso, A., π-Conjugated materials: Here, there, and everywhere. *Chem. Mater.*, *35*, 1467, 2023.
54. Liu, H., Zhong, J., Lee, C., Lee, S.W., Lin, L., A comprehensive review on piezoelectric energy harvesting technology: Materials, mechanisms, and applications. *Appl. Phys. Rev.*, *5*, 041306, 2018.
55. Lin, L., Wang, C., Deng, Y., Geng, Y., Isomerically pure oxindole-terminated quinoids for n-type organic thin-film transistors enabled by the chlorination of quinoidal core. *Chem. A Eur. J.*, 29, e202203336, 2022.
56. Wang, F., Lai, Y.-H., Han, M.-Y., Stimuli-responsive conjugated copolymers having electro-active azulene and bithiophene units in the polymer skeleton: Effect of protonation and p-doping on conducting properties. *Macromolecules*, 37, 3222, 2004.
57. Nozoe, T., Seto, S., Matsumura, S., Tropone and its derivatives. VIII on the reaction of halotropones. *Proc. Jpn. Acad.*, 28, 483, 1952.
58. Nozoe, T., Seto, S., Matsumura, S., Synthesis of 2-substituted azulenes by nucleophilic substitution reactions of 2-haloazulene derivatives. *Bull. Chem. Soc. Jpn.*, 35, 1990, 1962.
59. Xin, H., Hou, B., Gao, X., Azulene-based π-functional materials: Design, synthesis, and applications. *Acc. Chem. Res.*, 54, 1737, 2021.
60. Kurotobi, K., Miyauchi, M., Takakura, K., Murafuji, T., Sugihara, Y., Direct introduction of a boryl substituent into the 2-position of azulene: Application of the miyaura and smith methods to azulene. *Eur. J. Org. Chem.*, 2003, 3663, 2003.
61. Xin, H., Ge, C., Jiao, X., Yang, X., Rundel, K., McNeill, C.R., Gao, X., Impact of 2,6-connectivity in azulene: optical properties and stimuli responsive behavior. *J. Mater. Chem. C*, *1*, 7400, 2013.
62. Koch, M., Blacque, O., Venkatesan, K., Incorporation of 2,6-connected azulene units into the backbone of conjugated polymers: Toward high-performance organic optoelectronic materials. *Angew. Chem. Int. Ed.*, 57, 1322, 2018.
63. Yao, J., Cai, Z., Liu, Z., Yu, C., Luo, H., Yang, Y., Yang, S., Zhang, G., Zhang, D., Tuning the semiconducting behaviors of new alternating dithienyldiketopyrrolopyrrole-azulene conjugated polymers by varying the linking positions of azulene. *Macromolecules*, *48*, 2039, 2015.
64. Hou, B., Li, J., Zhou, Z., Tan, W.L., Yang, X., Zhang, J., McNeill, C.R., Ge, C., Wang, J., Gao, X., Incorporation of electron-rich indacenodithiophene units into the backbone of 2,6-azulene-based conjugated polymers for proton-responsive materials and p-type polymeric semiconductors. *ACS Mater. Lett.*, 4, 392, 2022.
65. Wang, Y., Tan, W.L., Xiang, J., Ge, C., McNeill, C.R., Gao, X., Tuning dipole orientation of 2,6-azulene units in conjugated copolymers by C-H activation

strategy toward high-performance organic semiconductor. *ACS Macro Lett.*, 12, 487, 2023.

66. Xiang, J., Tan, W.L., Zhang, J., Wang, Y., Duan, C., McNeill, C.R., Yang, X., Ge, C., Gao, X., Poly(2,6-azuleneethynylene)s: Design, synthesis, and property studies. *Macromolecules*, 55, 8074, 2022.
67. Puodziukynaite, E., Wang, H.W., Lawrence, J., Wise, A.J., Russell, T.P., Barnes, M.D., Emrick, T., Azulene methacrylate polymers: synthesis, electronic properties, and solar cell fabrication. *J. Am. Chem. Soc.*, 136, 11043, 2014.
68. Murakami, A., Kobayashi, T., Goldberg, A., Nakamura, S., Murakami, A., Kobayashi, T., Goldberg, A., Nakamura, S., CASSCF and CASPT2 Studies on the structures, transition energies, and dipole moments of ground and excited states for azulene. *J. Chem. Phys.*, 120, 1245, 2004.
69. Shibuya, Y., Matsunaga, A., Kumaki, D., Tokito, S., Katagiri, H., Azulene end-capped 1,3,4-thiadiazole as an n-type organic semiconductor with a herringbone–brickwork cooperative 2D layered structure. *Cryst. Growth Des.*, 22, 6554, 2022.
70. Peng, P., Li, J., Hou, B., Xin, H., Cheng, T., Gao, X., Design, synthesis and properties of indacenodithiophene derivatives end capped with azulene. *Chin. J. Org. Chem.*, 40, 3916, 2022.
71. Xin, H., Ge, C., Fu, L., Yang, X., Gao, X., Naphthalene diimides endcapped with ethynyl azulene: molecular design, synthesis and properties. *Chin. J. Org. Chem.*, 37, 711, 2017.
72. Yamaguchi, Y., Takubo, M., Ogawa, K., Nakayama, K.I., Koganezawa, T., Katagiri, H., Terazulene isomers: polarity change of OFETs through molecular orbital distribution contrast. *J. Am. Chem. Soc.*, 138, 11335, 2016.
73. Xiang, J., Vu, D., He, M., Duan, C., Ge, C., McNeill, C.R., Gao, X., Poly(2,6-azulene vinylene)s: Azulene Orientation Control and Property Studies. *Macromolecules*, 56, 9475, 2023.
74. Cristian, L., Sasaki, I., Lacroix, P.G., Donnadieu, B., Asselberghs, I., Clays, K., Razus, A.C., Donating strength of azulene in various azulen-1-yl-substituted cationic dyes: application in nonlinear optics. *Chem. Mater.*, 16, 3543, 2004.
75. Muranaka, A., Yonehara, M., Uchiyama, M., Azulenocyanine: a new family of phthalocyanines with intense near-IR absorption. *J. Am. Chem. Soc.*, 132, 7844, 2010.
76. Yamaguchi, Y., Ogawa, K., Nakayama, K., Ohba, Y., Katagiri, H., Terazulene: a high-performance n-type organic field-effect transistor based on molecular orbital distribution control. *J. Am. Chem. Soc.*, 135, 19095, 2013.
77. Wang, Z., Sun, Y., Liu, X., Zhai, M., Xia, Z., Tian, Y., Cheng, M., Li, H., Rational design of 1,3,6-trisubstituted azulene based hole transport materials for efficient and stable planar perovskite solar cells. *Dyes Pigm.*, 240, 112856, 2025.

78. Liu, S., Díaz-Fernández, M., Zhang, M., Huang, F., Chen, Y., Yang, Y., Marín-Beloqui, J. M., Lan, J., You, J., Casado, J., Zhang, C., Azuperylene: The non-alternant isomer of perylene. *Angew. Chem.*, 137, 29, e202505897, 2025.
79. Xin, H. and Gao, X., Application of azulene in constructing organic optoelectronic materials: New tricks for an old dog. *ChemPlusChem*, 82, 945, 2017.
80. Xin, H., Li, J., Ge, C., Yang, X., Xue, T., Gao, X., 6, 6′-Diaryl-substituted biazulene diimides for solution-processable high-performance n-type organic semiconductors. *Mater. Chem. Front.*, 2, 975, 2018.
81. Shoji, T. and Ito, S., Azulene-based donor–acceptor systems: synthesis, optical, and Electrochemical Properties. *Chem. Eur. J.*, 23, 16696, 2017.
82. Chamelot, G., Idir, M., Leclerc, M., Morin, J.F., Synthesis and properties of 2,6-azulene-based conjugated polymers and their applications in dispersing single-walled carbon nanotubes. *Polym. Chem.*, 14, 1206, 2023.
83. Xin, H., Ge, C., Jiao, X., Yang, X., Rundel, K., McNeill, C.R., Gao, X., Incorporation of 2,6-connected azulene units into the backbone of conjugated polymers: Towards high-performance organic optoelectronic materials. *Angew. Chem. Int. Ed.*, 57, 1322, 2018.
84. Qi, D., Zhang, L., Jiang, J., Toward panchromatic organic functional molecules: Density functional theory study on the nature of the broad UV–Vis–NIR spectra of substituted tetra(azulene)porphyrins. *J. Mol. Graphics Modell.*, 38, 304, 2012.
85. Korichi, H., Zouchoune, F., Zendaoui, S.M., Zouchoune, B., Saillard, J.Y., The coordination chemistry of azulene: A comprehensive DFT investigation. *Organometallics*, 29, 1693, 2010.
86. Wang, H., Sun, Z., Xie, Y., King, R.B., Schaefer, H.F., Unsaturation and variable hapticity in binuclear azulene iron carbonyl complexes. *Organometallics*, 29, 630, 2010.
87. (a) Ito, S., Terazono, T., Kubo, T., Okujima, T., Morita, N., Murafuji, T., Sugihara, Y., Fujimori, K., Kawakami, J., Tajiri, A., Efficient preparation of 2-azulenylboronate and Miyaura-Suzuki cross-coupling reaction with aryl bromides for easy access to poly (2-azulenyl) benzenes. *Tetrahedron*, 60, 5357, 2004. (b) Ito, S., Ando, M., Nomura, A., Morita, N., Kabuto, C., Mukai, H., Ohta, K., Kawakami, J., Yoshizawa, A., Tajiri, A., Synthesis and properties of hexakis(6-octyl-2-azulenyl)benzene as a multielectron redox system with liquid crystalline behaviour. *J. Org. Chem.*, 70, 10, 3939, 2005.
88. Truong, M.A., Lee, J., Nakamura, T., Seo, J.Y., Jung, M., Ozaki, M., Shimazaki, A., Shioya, N., Hasegawa, T., Murata, Y., Zakeeruddin, S.M., Grätzel, M., Murdey, R., Wakamiya, A., Influence of alkoxy chain length on the properties of two-dimensionally expanded azulene-core-based hole-transporting materials for efficient perovskite solar cells. *Chem. Eur. J.*, 25, 6741, 2019.
89. Asao, T., Ito, S., Morita, N., Synthesis of [1.1.1.1] (1,3)-2-methoxyazulenophane: azulene analogue of calixarenes. *Tetrahedron Lett.*, 29, 2839, 1988.
90. Colby, D.A. and Lash, T.D., Calix[4]azulene. *J. Org. Chem.*, 67, 1031, 2002.

91. Lash, T.D., El-Beck, J.A., Colby, D.A., Synthesis of a tetraazulene porphodimethene analogue. *J. Org. Chem.*, 74, 8830, 2009.
92. Georghiou, P.E., Rahman, S., Alodhayb, A., Nishimura, H., Lee, J., Wakamiya, A., Scott, L.T., Calixazulenes: Azulene-based calixarene analogues–an overview and recent supramolecular complexation studies. *Beilstein J. Org. Chem.*, 14, 2488, 2018.
93. (a) Krug, C.K., Fan, Q., Fillsack, F., Glowatzki, J., Trebel, N., Heuplick, L.J., Koehler, T., Gottfried, J.M., Organometallic ring vs. chain formation beyond kinetic control: steering their equilibrium in two-dimensional confinement. *Chem. Commun.*, 54, 9741, 2018. (b) Krug, C.K., Nieckarz, D., Fan, Q., Szabelski, P., Gottfried, M., The macrocycle versus chain competition in on-surface polymerization: Insights from reactions of 1, 3-dibromoazulene on Cu (111). *Chem. Eur. J.*, 26, 7647, 2020.

6

Azulene-Based Systems as Molecular Switches

Sariful Molla and Sheelbhadra Chatterjee*

Department of Chemical Sciences, IISER Kolkata, West Bengal, India

Abstract
Researchers have been particularly interested in photochromic molecules, owing to their unique ability to switch between multiple states in response to different external stimuli. This extraordinary property can utilize these materials in diverse areas such as optoelectronics, memory devices, and self-healing materials. Visible light photoswitches have recently gained attention to minimize the use of UV light used in photoisomerization. Azulene, a C_7H_8 hydrocarbon, shows an intense absorption in the visible region between 380 nm–480 nm. As a result, azulene-containing photoswitches might function in the visible range of the spectrum. Other unique features of azulene, such as polarity, luminescence, and proton abstraction, make the system a more promising candidate in the field of smart materials. This chapter discusses a verse series of azulene derivates functionalized with different photochromic compounds and their potential application in different fields.

Keywords: Photochromism, visible light photoswitch, photochromic material, azulene, smart materials

6.1 Introduction

Organic molecular switches are versatile systems that respond to specific external stimuli, producing one or more metastable (less stable) isomeric states. The isomerization process is accompanied by geometric and electronic changes in the molecule, leading to alterations in its physical and electronic properties. These make the molecular switches potential candidates

**Corresponding author*: sheelbhadra1994@gmail.com

Neha Rani Kumar and Abhijeet R. Agrawal (eds.) Azulene-Based Materials: Design, Synthesis, and Applications, (151–178)

or valuable tools for a wide range of applications, including catalysis, photopharmacology, and optoelectronics [1–4]. Stimuli that drive molecular switches include chemical agents [5–7], mechanical force [8], electrical impulses [9], heat [10], and light [11, 12]. As the utmost concern of scientists in the present era is sustainability, light-driven modulation of different properties of the system has gained attention. Light stands out as a sustainable and non-invasive stimulus. It is non-toxic, allows spatiotemporal control, and often produces minimal or no byproducts during light-driven reactions. This has made light-driven modulation of molecular properties, particularly through photoswitching, an area of intense research interest.

Photoswitches, or photochromic molecules, undergo reversible isomerization between two states depending on the wavelength of light applied. They are broadly classified into two categories: positive photochromic systems and negative photochromic systems. If the absorption maxima (λ_{max}) of the metastable state generated upon photoirradiation has bathochromically shifted (red shifted) i.e., λ_{max}(metastable)> λ_{max}(stable), it falls under the category of positive photochromism. It changes from a colorless state to a colored state, with azobenzene and dithienylethene as some examples [13]. On the other hand, in negative photochromism, the absorption maxima are hypsochromically shifted (blue shift) i.e., λ_{max}(metastable) < λ_{max}(stable). Often in negative photochromism, discoloration or bleaching takes place, for example dimethyldihydropyrene or imidazole dimeric photoswitches [14, 15]. Depending on the thermal activation energy barrier, molecular photoswitches are often classified into two types: P-type

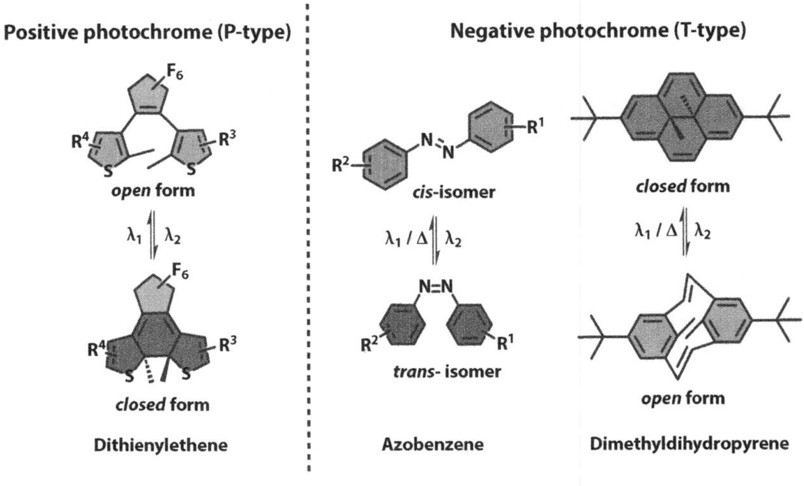

Scheme 6.1 Examples of different types of photoswitches.

and T-type photoswitches. P-type photoswitches rely exclusively on light exposure for isomerization, and their high activation energy barrier prevents thermal reversal in the dark. Dithienylethene is a common example of a P-type photoswitch [16]. In contrast, T-type photoswitches revert to their stable isomeric form upon heating due to their low activation energy barriers. Examples include azobenzene, spiropyran, and dimethyldihydropyrene [17–19]. This reversible isomerization allows a photo-controlled modulation of properties, such as color or π-electronic delocalization, without requiring external agents. Consequently, photochromic systems have found applications in stimuli-responsive materials, optoelectronics, and sensing. However, most photoswitches rely on ultraviolet (UV) light for isomerization, therefore significantly limiting their scope of application. Prolonged exposure to UV light can lead to photodegradation, and its low tissue-penetrating ability restricts its applications in biomedical fields [20]. To overcome these challenges, research efforts are focused on developing visible-light-responsive photoswitches. Visible light, being less energetic than UV, reduces the risk of photodegradation and improves compatibility with biological systems. Various strategies have been employed to achieve visible-light-mediated photoisomerization. These include the conjugation of benzimidazole units [21], ortho-substitution of azobenzene with functional groups [22, 23], incorporation of formyl groups into stiff-stilbene [24], and other molecular modifications. Structural engineering, such as the insertion of heteroatoms or functional group modifications, remains critical for improving photoswitch performance. This chapter explores recent advancements in azulene-coupled photochromic systems, highlighting their photophysical properties and potential applications.

6.2 Azulene-Fused Molecular Switches

To serve the purpose discussed earlier, different azulene (*Az*)-based molecular switches have been designed because of some unique combination of optical, luminescence, and stimuli-responsive properties. Some distinct characteristics of azulene include the absorption in the visible region, which causes a deep blue color and therefore allows for the harvesting of energy in the visible region in the spectrum. The combination of an electron-deficient, seven-membered ring, and an electron-rich, five-membered ring produces a dipole moment of 1.08 D that helps in controlling the electronic and optical properties of azulene-based functional materials (Figure 6.1a) [25]. Another surprising feature of azulene is the basicity at position 1 to give a stable Az-H^+ cation (Figure 6.1b). Intense absorption bands in the

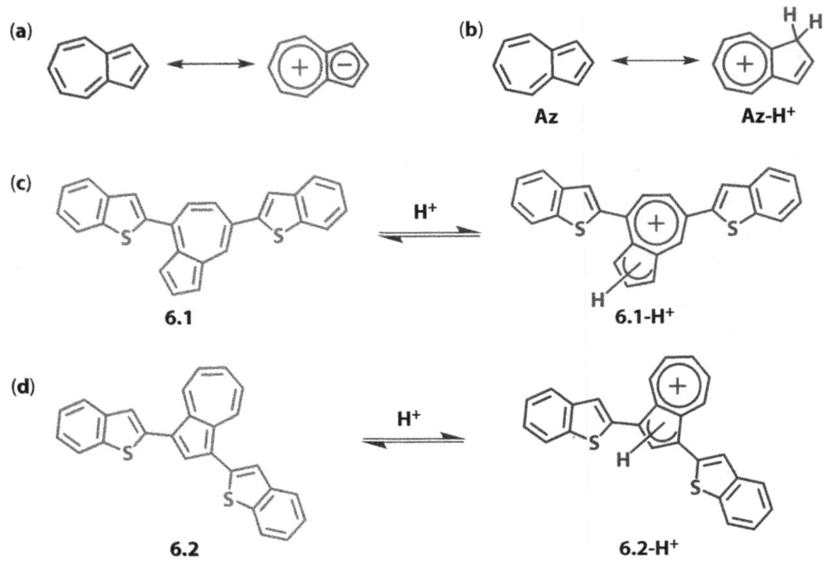

Figure 6.1 (a) Resonance in azulene; (b) protonation at 1 position in azulene; (c & d) previously reported azulene-based derivatives.

visible range are produced upon protonation of azulene (Az-H$^+$), which subsequently enhances the sensitivity to visible light. Hawker's group has demonstrated elegantly the UV-vis spectral change upon protonation in different azulene derivatives (Figures 6.1c, d) [26]. Azulene has been incorporated into several photochromic units such as azobenzene, stilbene, and dithienylethene (DTE) to develop a new generation of photoswitchable devices, mainly visible light-driven photoswitches.

6.2.1 Azulene-Functionalized Stilbene Systems

Tokumaru *et al.* reported a couple of azulene-containing photoswitches for the first time by incorporating different azulene units into stilbene (Figure 6.2) [27]. Explicitly *cis*-to-*trans* one-way isomerization has been accomplished with these systems. The *Cis* and *trans* isomer of the photoswitch **6.3** exhibited a low energy $S_0 \leftarrow S_1$ transition at 610 and 630nm, respectively, along with an intense absorbance at 390nm corresponding to high energy $S_0 \leftarrow S_2$ transition. Therefore, direct irradiation of **6.3-*cis*** with 366nm light (populating the S_2 state) or with 600nm light (populating the S_1 state) resulted in *cis*-to-*trans* isomerization with a quantum yield $\Phi^{direct}_{cis \rightarrow trans} = 0.03$. Surprisingly, when the system was irradiated with

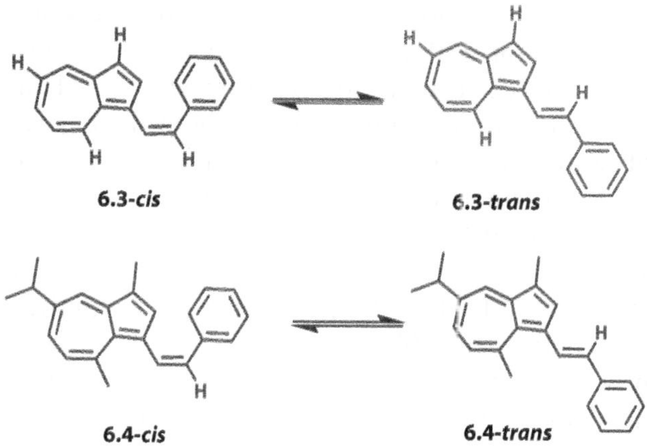

Figure 6.2 Azulene-containing stilbene-based photoswitchable systems.

a 450nm laser in the presence of a triplet sensitizer, biacetyl in this case, in benzene (purged with N_2) a remarkable enhancement in the *cis*-to-*trans* isomerization ($\Phi^{sensitizer}_{cis \to trans}$) was observed with an isomeric distribution of 98% trans and 2% cis in the photostationary state. The isomerization pathway involved a triplet energy transfer from the biacetyl to the 6.3-*cis*, resulting in the formation of a triplet *cis* state ($^3C^*$). The species further isomerized to a triplet trans state ($^3T^*$) through the C-C bond rotation, which ultimately led to the **6.3-*trans***. The triplet state energy of the sensitizer (E_T for biacetyl = 56 kcal mol^{-1}) was estimated to be higher compared to that of the *cis* and *trans* isomers (**6.3-*cis*** and **6.3-*trans***). Therefore, an efficient energy transfer was expected from the sensitizer to the *cis* form leading to successful isomerization to the *trans*. The following mechanistic approach has been proposed for this unique one-way isomerization. The variable $^3p^*$ is the perpendicularly twisted conformation in the triplet state. Surprisingly, it was observed that the $\Phi^{sensitizer}_{cis \to trans}$ increased with the concentration of the *cis* isomer. Therefore, the process is called by authors as the "quantum chain process." The other derivative (**6.4**) also showed similar results. Thus, the triplet energy level was achieved due to the azulene present in the system.

$$S \to {}^1S^* \to {}^3S^*$$
$${}^3S^* + c \to S + {}^3c^*$$
$${}^3c^* \to {}^3p^* \to {}^3t^*$$
$${}^3t^* \to t$$
$${}^3c^* + t \to c + {}^3t^*$$

This one-way isomerization in azulene functionalized stilbene was further studied by Fujimori with multiple azulene units present in the system (Figure 6.3) [28]. The *cis*-to-*trans* isomerization was studied for mono- and bis-azulene-substituted compounds through both direct irradiation and triplet sensitization. Benzil was chosen as the triplet sensitizer. The quantum yield of direct irradiation $\Phi^{direct}_{cis \to trans}$ when irradiated at 656nm light (populating the S_1 state) was found to be 0.04 and 0.0014 for compounds **6.5** and **6.6** at a millimolar concentration (1.0×10^{-3} M and 1.1×10^{-3} M), respectively. Adding a triplet sensitizer displayed a significant enhancement in the quantum yield ($\Phi^{sensitizer}_{cis \to trans} > \Phi^{direct}_{cis \to trans}$). The reason behind this observation was the absorption of most visible light by azulene, leading to less efficient isomerization. Similar to the previously discussed experiments here, $\Phi^{direct}_{cis \to trans}$ and $\Phi^{sensitizer}_{cis \to trans}$ increased with the increasing concentration of the *cis* isomer, thereby following the "quantum chain process." A mechanistic pathway was also proposed by the authors for both direct and sensitized isomerization.

To achieve reversible isomerization along with improved photoswitching performance in azulene-based stilbene systems, cyanostilbene was selected as the photochromic unit reported by Zhu *et al.* (Figure 6.4a) [29]. The fused systems **6.7** and **6.8** did not exhibit direct photoisomerization under UV or visible light exposure as no spectral change was recorded during the UV-vis experiment. Polyethylene glycol was incorporated in **6.8** for better water solubility and biocompatibility. Nevertheless, the systems displayed reversible isomerization between *Z* and *E* (**Z-6.7**→ ***E*-6.7** and vice-versa) in acidic

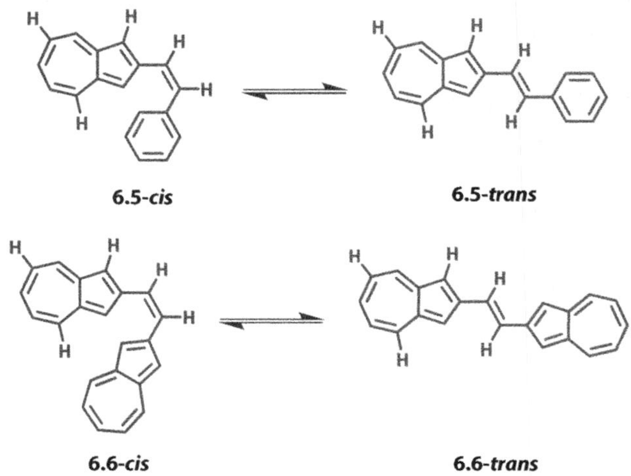

Figure 6.3 Azulene-containing stilbene-based photoswitchable systems.

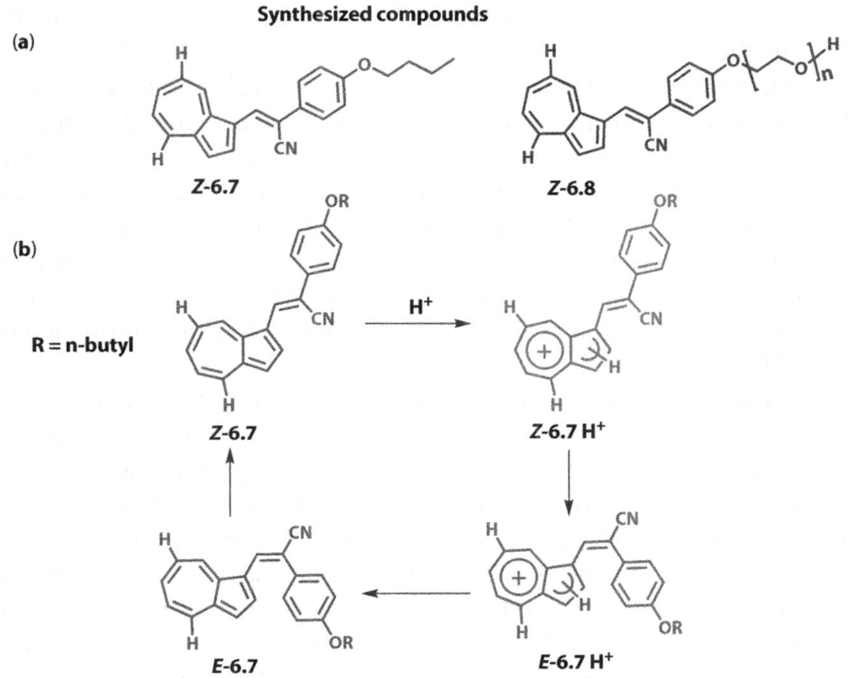

Figure 6.4 (a) Azulene-fused cyanostilbene derivatives; (b) proton-assisted photoisomeriztion scheme for compound **6.7**.

conditions (Figure 6.4b). Further upon thermal heating, the system reverted to the initial ***Z*-6.7** after deprotonation. A stable tropylium cation was formed upon the protonation of the electron-rich five-membered ring as evident from proton nuclear magnetic resonance ($_1$H-NMR) spectral changes. The exposure of 365 nm light to a photostationary state was attained in nearly 2 hrs of irradiation. The isomerization was monitored through both nuclear magnetic resonance (NMR), as well as the UV-vis spectral changes. While there was no emission in the initial ***Z*-6.7**, a substantial increase in emission was observed on isomerization to **E-6.7** (λ_{max} = 480 nm). The emission spectral analysis also revealed a long wavelength emission was intensified in the near-infrared (NIR) region for **E-6.7** (λ_{max} = 810 nm). The spectral pattern for each pair of isomers i.e., ***E*-6.7/ *E*-6.7 H⁺** and ***Z*-6.7/*Z*-6.7 H⁺** had remained unchanged (observed under UV-vis or emission studies). Thus, the configuration of the photochromic unit rather than the protonation was the key factor to control the spectral behavior. Thus, proton-gated photoisomerization, leading to a dual emission output, was demonstrated with these photoswitchable systems. Similar photo-responsive behaviour was observed with **6.8.** Being emissive in the NIR region, these compounds have the potential for bio-imaging.

The visible light-directed photoresponsivity of cyanostyryl azulene has been further employed in acidic sensing. The pH sensing technique was found to be effective in diagnosing several biomedical conditions [30, 31]. To serve this purpose a series of azulene substituted compounds were synthesized where the cyanostyryl functionality was attached to the seven-membered ring unlike the reported compound **6.7** (Figure 6.5a) [32]. Protonation in the azulene moiety led to a significant shift in the band gap to the higher wavelength region as observed during the UV-vis study. An intense absorption band at 520 nm appeared with the simultaneous diminishment of the absorption band at 400 nm upon protonation (formation of **Z-6.9 H⁺**).

The initial **Z-6.9** did not exhibit any photoisomerization under UV or visible light. However, the system showed photo-responsive behavior in acidic conditions along with an intense color change from pale green to red (Figures 6.5 b, c). Irradiation with green light (520 nm) resulted in the formation of ***E*-6.9 H⁺**. Further deprotonation led to the generation of the initial **Z-6.9**. The whole isomerization scheme was monitored through ^{1}H-NMR spectral changes and can be performed over multiple cycles. The protonation was completely reversible, and the addition of triethylamine completely restored the spectral characteristics. While no emission was

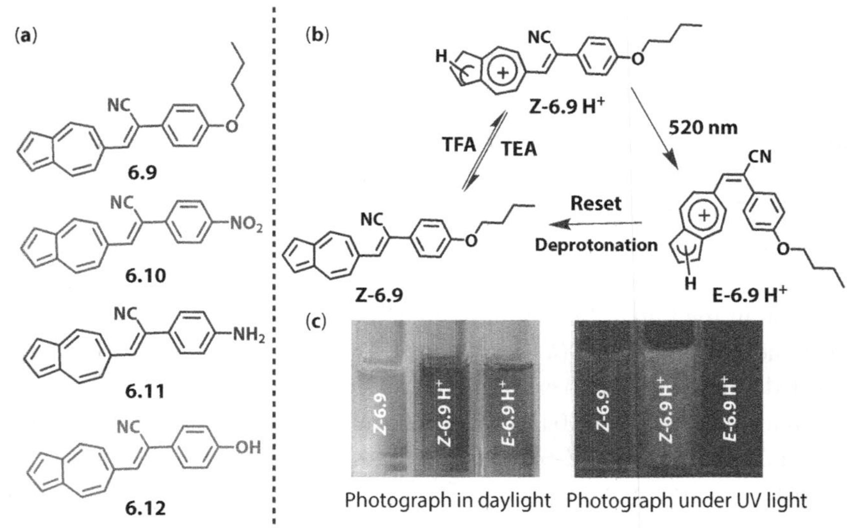

Figure 6.5 (a) Azulene-fused cyanostilbene compounds with different end groups; (b) Proton-assisted photoisomerization scheme for compound. **Z-6.9**; (c) Real image of the isomers of **6.9** in different conditions. Reproduced with permission [32]. Copyright 2018, American Chemical Society.

displayed by **Z-6.9**, a strong emission signal at 630 nm was produced from **Z-6.9 H**$^+$ because of the tropylium cation formation. Further, photoisomerization to *E* form led to the quenching of the emission. Compounds **6.10** and **6.11** did not form the corresponding tropylium cation.

The presence of electron-withdrawing $-NO_2$ group in **6.10** destabilized the cation formation and **6.11** formed a precipitate due to the protonation over the $-NH_2$ group. Although compound **6.12** was able to form the cation, no emission peak was detected. The –OH group easily formed intermolecular hydrogen bonding that quenched the emission. Therefore, the end-group substitution was crucial in enabling the proton-assisted photoisomerization accompanied by the emission enhancement. In addition, **6.9** was successfully employed to develop analytical tools for detecting gastric acid.

An azulene-based photochromic dyad **6.13** has been developed by Zhu *et al.* to modulate the fluorescence output from the system in a single-molecule level (Figure 6.6a) [33]. The system was synthesized by attaching the azulene-functionalized cyanostyryl and a piperazinyl naphthalimide through an alkyl spacer. In the initial state i.e., **6.13**, a green emission was observed from the system due to the excitation of the naphthalimide core. In the first stage of protonation, after adding trifluoracetic acid (TFA), the piperazine ring was protonated, which led to the formation of **6.13**$^+$, and the emission was shifted to blue. In this state, the photoinduced electron transfer from piperazine was suppressed. When the acid concentration was further increased, the protonation took place in the azulene ring resulting in the formation of a tropylium cation (**6.13**$^{2+}$) along with the change in emission color from blue to yellow. The protonation sites in different stages were confirmed from

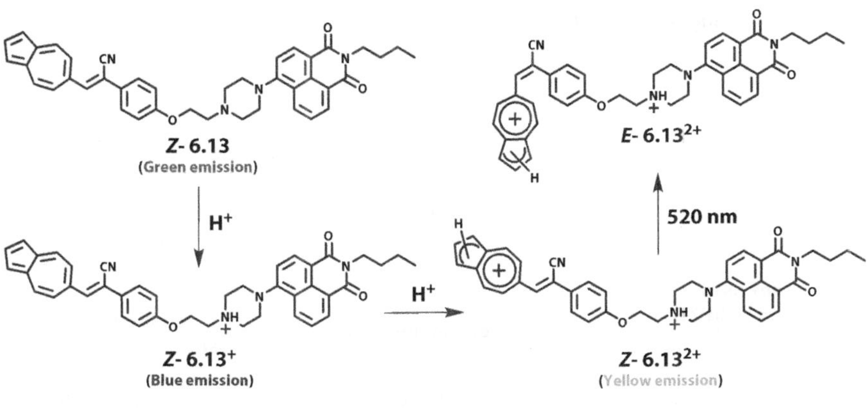

Figure 6.6 (a) Schematic representation of the protonation of **6.13** in different stages followed by isomerization in a visible light.

^{1}H-NMR spectral shifts. An emission enhancement in the long-wavelength region over 600 nm was also observed after the tropylium cation (**6.13^{2+}**) formation. All the emission spectral behavior can be efficaciously reversed by adding a base (triethylamine). Like other cyanostyryl systems discussed previously, this system also displayed proton-assisted photoisomerization from ***Z*-6.13^{2+}** to ***E*-6.13^{2+}** under visible light exposure. The isomerization eventually led to the restoration of the green emission. Therefore, a dual method of emission modulation has been demonstrated with this system by changing the pH of the medium or by irradiation with light.

Researchers used a different technique to control the emission output from the cyanostyryl-modified azulene systems [34]. Figure 6.7 shows the structure of hexathiobenzene-azulene functionalized photochromic compound **6.14** with the individual units **6.15** and **6.16** and the second part of Figure 6.7 is the schematic representation of the emission modulation by aggregation and photoisomerization. The compound **6.14** was developed by incorporating multiple cyanostibenes into a central hexathiobenzene

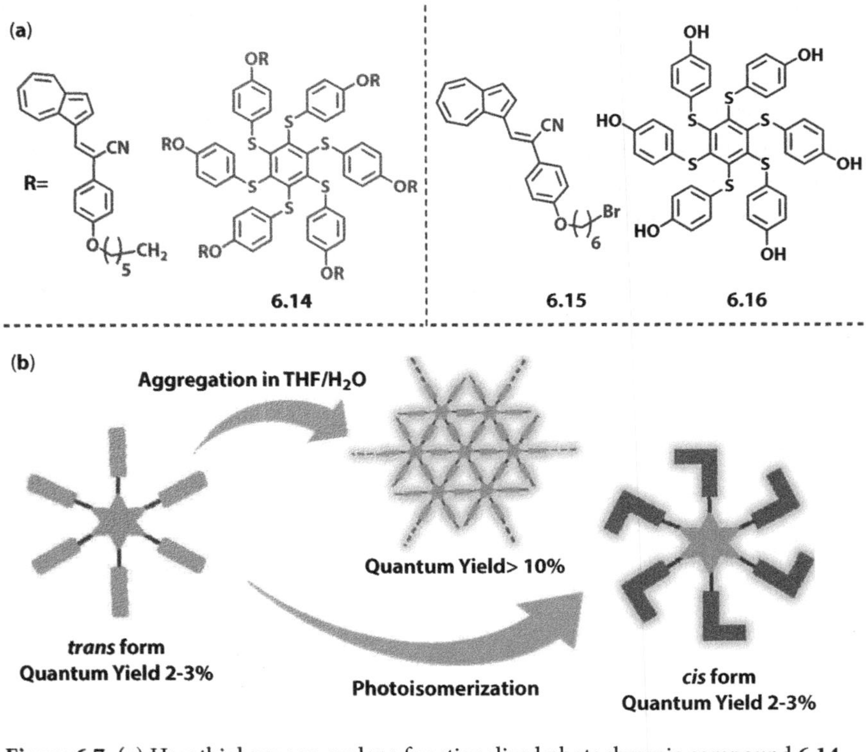

Figure 6.7 (a) Hexathiobenzene-azulene functionalized photochromic compound **6.14** with the individual units **6.15** and **6.16**; (b) Schematic representation of the emission modulation by aggregation and photoisomerization.

ring *via* a covalent bond. The persulfate arene ring was chosen to utilize the aggregation-induced emission property in an appropriate solvent condition [35, 36]. Initially, **6.14** displayed an intense cyan emission at 480 nm (λ_{ex}= 365 nm). Excitation spectra unveiled the presence of multiple electronic excitation (260nm, 310 nm, and 365nm) corresponding to different electronic states. The individual photochromic unit **6.15** did not show any significant fluorescence. The UV-vis studies revealed the overlap of emission spectra of a separate hexathiobenzene ring with the UV-vis spectra of compound **6.15**. Therefore, a resonance energy transfer from the core to the periphery resulted in high emission. For the same reason, the emission output corresponding to the hexathiobenzene was not observed, although the control experiment with compound **6.16** produced a yellow emission. An aggregation-induced enhanced emission was also observed due to the restricted intramolecular motion when 90% of water was introduced in the tetrahydrofuran solution of **6.14**. A 10% increase in quantum yield was detected upon aggregation. Control experiments with **6.16** also showed changes in emission property, demonstrating that the aggregation-induced emission behavior of hexathiobenzene was retained in **6.14**. The aggregate formation in different solvent systems was confirmed by electron microscope images and dynamic light scattering studies. Like other systems discussed previously, proton-assisted *Z*→*E* photoisomerization was also investigated with **6.14**. Upon isomerization, the emission intensity further strengthened because it became more steric as it was introduced into the system. The fluorescence quantum yield of **6.14** was almost four times higher than **6.15**, suggesting a favorable energy transfer from its sulfur-rich core to the azulene units. These feature a dual-mode approach demonstrated by the authors to tune the emission signal of a system.

6.2.2 Azulene-Functionalized Diarylethene Systems

Another type of photochromic system that can play a major role in controlling diverse molecular properties is the diarylethene (DAE) system [37]. The thermodynamically stable open form undergoes a ring closure isomerization under exposure to UV light while reverting to the initial isomeric state under irradiation of visible light. The significant stability of both states makes the system a potential candidate for applications in optoelectronics [38], sensing [39], molecular magnetism [40], and enzyme mimicry [41]. As a result, numerous azulene-linked DAE molecular switches have been developed by attaching azulene to the ethene moiety or the thiophene rings to use them in the aforementioned areas. In this section, a series of azulene-containing DAE photochromic systems are discussed in detail.

The photochromism in an azulene-linked DAE was demonstrated by Uchida *et al.* in **6.16o** and **6.17o** (Figure 6.8b) [42]. The DAE derivative **6.16o** displayed two absorption bands at 594 nm and 290 nm corresponding to $S_1 \leftarrow S_0$ and $S_2 \leftarrow S_0$ transitions, respectively. Surprisingly, no UV-vis spectral changes were detected with this compound when irradiated with either UV or visible light suggesting the absence of any photoisomerization to the closed form. On the other hand, **6.17o** showed the absorbances at 288 nm and 594 nm. Upon exposure to UV light (313 nm), two new absorption bands at 495 nm and 700 nm appeared along with the diminishment of the 288 nm band indicating clear photoisomerization (**6.17o→6.17c**). Again, irradiation with visible light (>500 nm) caused the regeneration of the initial spectra as a result of reverse isomerization (**6.17c→6.17o**). Therefore, the system demonstrated a reversible photoisomerization with a distinct color change. The density functional theory study revealed that DAE could acquire different conformational states where the anti-parallel geometry led to photocyclization. The steric repulsion between the hydrogens in thiophene rings and azulene moiety prevented conformational changes in isomerization. The intramolecular hydrogen bonding in **6.17o** stabilized the antiparallel geometry, which resulted in a more efficient

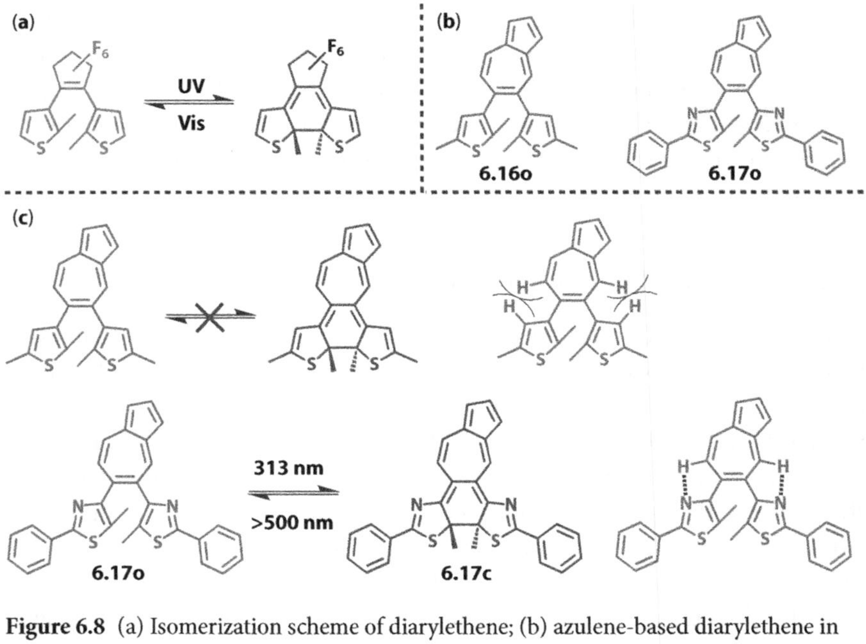

Figure 6.8 (a) Isomerization scheme of diarylethene; (b) azulene-based diarylethene in the photochromic system **6.16o** and **6.17o**; (c) isomerization scheme of **6.16o** and **6.17o**, respectively, with probable reason.

photoisomerization (Figure 6.8c). Energy optimization for both compounds revealed that **6.17o** had lower energy of the anti-parallel geometry than **6.16o**, which explained the nonexistence of the photoisomerization behavior in 6.16o.

The presence of donor-accepter conjugation can modulate the electronic and optical properties of DAE [43]. To examine the influence of the donor-acceptor pair on the photoisomerization of azulene-fused DAE, a series of derivatives have been synthesized by incorporating different accepting and donating groups onto the azulene ring (Figure 6.9) [44]. Both **6.18o** and **6.19o** consist only of donor moieties, while **6.20o** and **6.21o** consist of acceptor moieties. Only **6.18o** showed photoisomerization from open to closed under 313 nm light and closed to open in >486 nm light in a reversible manner like **6.17o**. A detailed theoretical investigation was performed to derive the reason for this surprising behavior of those compounds. Theoretical studies also revealed that the **6.18o** photo cyclized upon HOMO-1 to LUMO+2 excitation corresponding to the S_{11} electronic state. The highest bond order between the two reactive carbon centers was observed in the S_2 electronic state of **6.17o**. Therefore, photocyclization occurs from this state. Thus, the photoisomerization of these compounds came from higher-energy excited states.

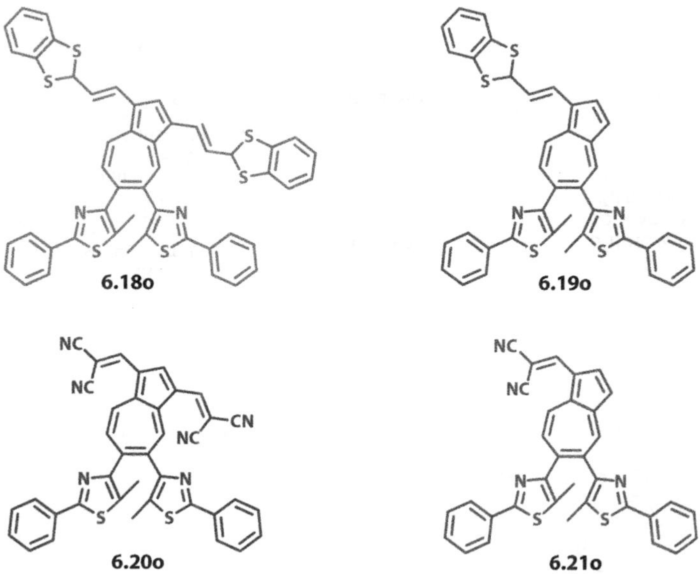

Figure 6.9 Structures of azulene-containing molecular switches **6.18o**, **6.19o**, **6.20o** and **6.21o**.

A different approach to attaching azulene to the DAE was taken by choosing the five-membered ring as the attachment site. Lyov *et al.* reported 1,2-dithienyl- and 1,2,3-trithienyl substituted azulene (Figure 6.10a) [45]. Authors proposed a convenient synthetic approach to prepare for different azulene-fused DAE. Like previous compounds, both **6.22o** and **6.23o** did not exhibit any photoisomerization under UV and visible light irradiation. The geometric orientation of the azulenyl group and the thiophene rings anticipated the reason for the photo-inactive nature. Upon protonation with TFA, **6.22o** generates a stable cation **6.22o-H**$^+$ with a distinct color change from colorless to red. The cation is still inactive to any photoirradiation. Despite the lack of photoisomerization, the cation **6.22o-H**$^+$ displayed an intense color change from red to blue upon treatment with aromatic aldehydes along with a drastic change in absorption spectra. The ^{1}H-NMR spectral changes revealed the formation of the desired product through condensation (Figure 6.10c). On the other hand, upon gradual

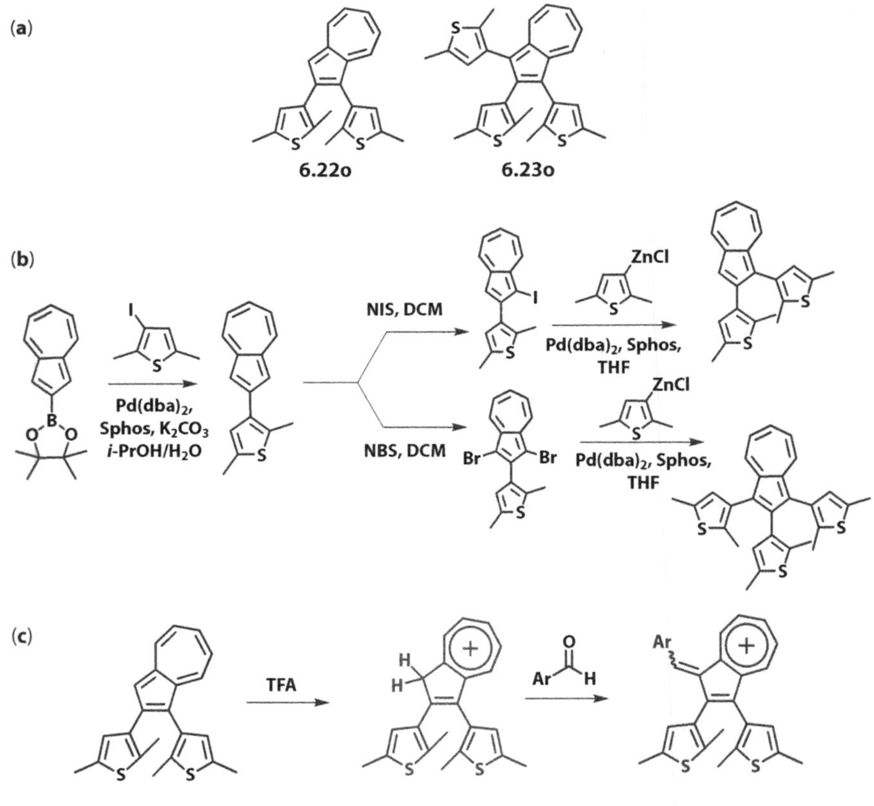

Figure 6.10 (a) Azulene-fused DAE **6.22o** and **6.23o**; (b) synthetic pathway; (c) acid-mediated condensation of **6.22o** with aromatic aldehyde.

addition of TFA to the solution of **6.23o**, no suitable NMR spectral change was observed, implying the decomposition of the compound.

Hecht *et al.* reported a more promising photoisomerization where the substituted azulene was chosen as one of the aryl groups for the DAE system (Figure 6.11) [46]. In the neutral form, the minor change in UV-vis spectra under 365nm light irradiation for 5 min suggests an insufficient photocyclization (**6.24o** →**6.24c**) due to charge transfer from the five-membered ring to the seven-membered ring. Upon addition of acid, **6.24o** forms **6.24o-H**$^+$ with an increase in UV-vis spectra between 400–600 nm. Under exposure of 546nm light, **6.24o-H**$^+$ isomerizes to **6.24c-H**$^+$ in less than 2 mins showing an enhanced photoswitching performance. Being electron-deficient in the cationic form, a charge transfer in azulene is impossible, and photoirradiation increases the electron density in hexatriene linkage, resulting in improved switching. A hypsochromic shift of almost 100nm on photocyclization of **6.24o-H**$^+$→**6.24c-H**$^+$ set an example of proton-gated negative T-type photochromism in DAE systems, although most of the DAE systems exhibit positive photochromism. The **6.24c-H**$^+$ exhibits thermal ring opening when kept in the dark with a half-life of 7 min. The reverse isomerization of **6.24c-H**$^+$→**6.24o-H**$^+$ does not proceed efficiently. *In situ* deprotonation followed by irradiation with visible light restores the initial isomeric state of **6.24o**.

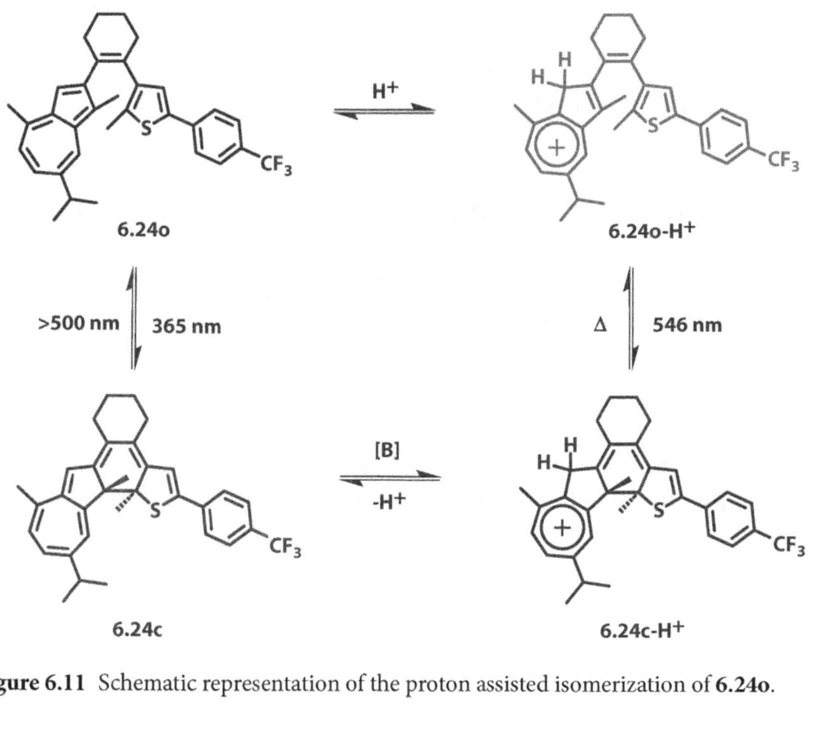

Figure 6.11 Schematic representation of the proton assisted isomerization of **6.24o**.

6.24; R= H
6.25; R= Br
6.26; R= CHO

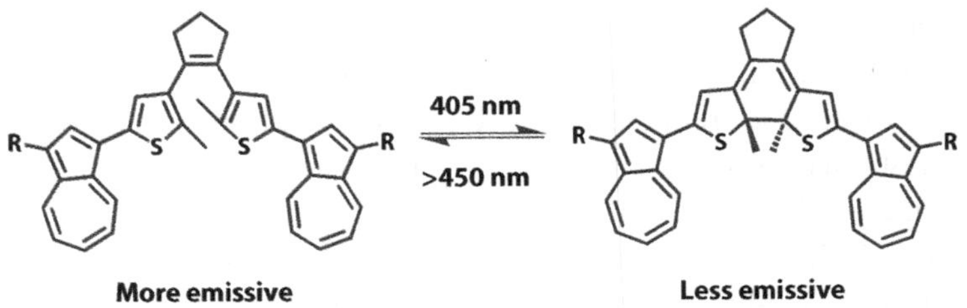

Figure 6.12 Schematic representation of the bidirectional visible light switching.

Most of the DAE systems endure ring closing under exposure to UV light. Bidirectional visible light photo switch with DAEs is still a topic of great interest. Azulene can serve as a potential candidate to address this problem. Therefore, derivatives have been synthesized by Nica *et al.* where substituted azulene groups are attached to the peripheral thiophene rings (Figure 6.12) [47]. The systems undergo ring-closing isomerization under 405nm light and revert to the original open isomer when exposed to >500nm light. Fluorescence experiment studies prove that all open isomers are more emissive. A drastic reduction in emission intensity in the closed form upon isomerization (*open* → *closed*) provides a tool to perform photo-controlled fluorescence switching. The extent of emission spectral changes is different for all derivatives depending on the substituent on the azulene ring. Cyclic voltammetry with these compounds provides an oxidative ring closure above 1.7 V. All these substances may be utilized in the future to prepare all visible light photoswitches.

6.3 Azulene-Functionalized Azobenzene Switches

Azobenzenes are an important class of organic photochromic molecules, having a nitrogen-nitrogen double bond (N=N) connecting two aryl groups. In particular, its photo-responsive property allows it to undergo reversible isomerization between the *trans* and the *cis* forms upon exposure to specific wavelengths of light [48, 49] Often, this isomerization accompanies significant changes in its physical and chemical properties, for example, molecular geometry, dipole moment, and chemical reactivity, thus making azobenzene highly valuable across a range of applications [50, 51].

Over the last few decades, scientists have been working to exploit this feature to build molecular machines and data storage devices in light-responsive smart materials and phytopharmacology [52]. In this context, several groups have also explored the azulene-functionalized azobenzene derivatives because azulene has a unique electronic structure that allows the absorption of light in the visible range, often resulting in blue or green coloration [53]. When combined with azobenzene, these derivatives can exhibit a broader and tunable absorption spectrum, making them more effective in applications like optical data storage and light-harvesting materials. Lacroix *et al.* explored the nonlinear optical (NLO) properties of an azulene-coupled azobenzene system [54]. Nonlinear optical materials are particularly interesting for their applications in optical information processing and telecommunication. Azulene with its polar resonance structures and the electronic drift from the seven-member ring to the five-member ring has a dipole moment of about 0.8 D. The seven-member ring carries a positive charge, and the five-member ring has a negative charge. Therefore, azulene can act as an electron donor (through the five-member ring) and acceptor (through the seven-member ring). Lacroix *et al.* have coupled the azulene through a five-member donor ring with the phenyl diazonium salts to get the azo-azulene derivatives **6.27a** and **6.27b**. The donor azulene and acceptor azo provided an expanded π-conjugation system. Hence, the push-pull chromophoric system results in low-lying charge transfer excited states, which are appealing for NLO materials, particularly **6.27b**. The dependence of electronic structure and NLO properties has been discussed using different spectroscopic and electric field-induced second harmonic techniques. Among the investigated azo-azulene derivatives **6.27a–c**, the dipole moment of **6.27b** is almost comparable to **6.28**, which supports the possible resonance structures of **6.27b** (Figure 6.13b) and the push-pull nature of the system. The high thermal stability and the non-centrosymmetric environment result in 420 times higher second-harmonic generation efficiency compared to urea. These features make azo-azulene derivatives a suitable candidate for NLO applications.

In 2009, Razus *et al.* reported the synthesis and the photoelectronic behavior of an azo dye with one azulene unit and two azo units coupled sequentially [55]. The synthesis includes a simple condensation with nitrosobenzene and coupling with diazonium salts, giving final derivatives **6.29a–c**. The changes in electronic spectra are obvious, owing to the donor character of the azulene five-member ring. Upon incorporating one phenyl-azo unit through a 3-position, as in compound **6.29a**, a 5–7

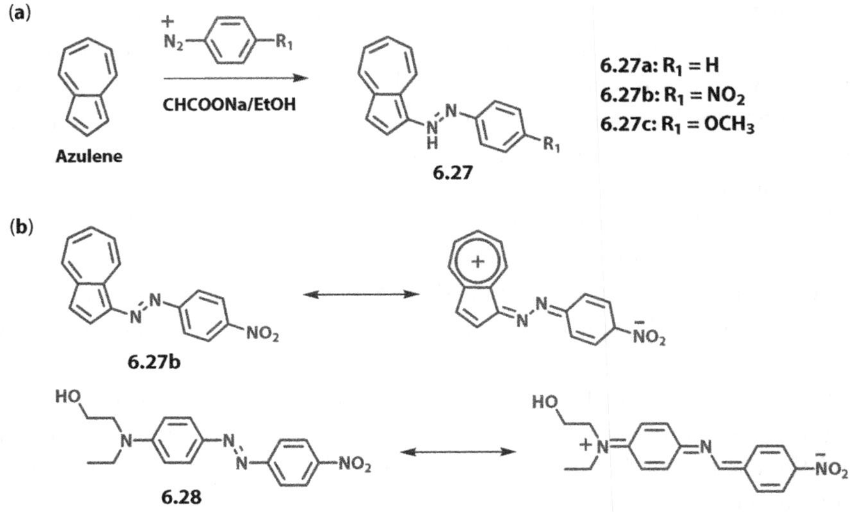

Figure 6.13 (a) Synthetic scheme for the azo-azulene derivatives; (b) Resonating structure.

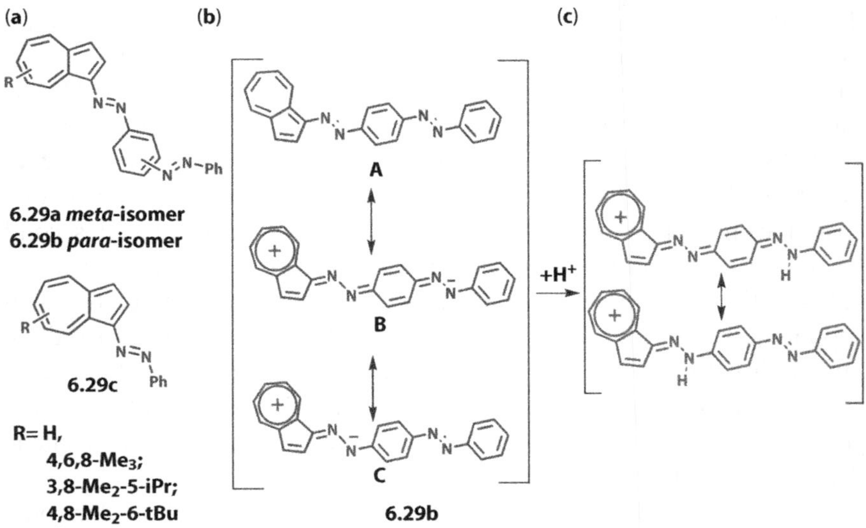

Figure 6.14 (a) Synthesized azo-azulene derivatives; (b) resonating structure of **6.29b**; (c) protonation of **6.29b**.

nm bathochromic shift has been observed because of the donor-acceptor nature of the system. Also, the bathochromic shift of 35 nm in **6.29b** is justified because of the increased dipolar structure, as well as the stable tropylium cation formation of the seven-member ring (Figure 6.14b).

Protonation also resulted in a bathochromic shift of the λ_{max} for all three derivatives (Figure 6.14c). Inspired by the results, the authors also investigated the acid-base equilibrium to determine the pK_a values the conjugated acids.

The pK_a values of **6.29a** and **6.29c** are similar, and the reason can be attributed to the practically decoupled nature of the azo substituent in **6.29a**. However, the conjugate acid of the **6.29b** is much weaker, and this is because of the extensive conjugation that increases electron density over the nitrogen atoms, thereby, increasing the basicity. In all the derivatives, alkylation on the seven-membered ring increases the pK_a values, and it is expected due to the stabilization of the tropylium cation by inductive effect. In 2010, the same group synthesized a series of azulenyl diazenes using benzothiazole-2-yl moieties [56]. The hyperpolarizability of the molecules because of the largely separated charges and stability of the tropylium cation further reduces the gap between frontier molecular orbitals HOMO and LUMO. Also, the presence of NO_2 in the 6th carbon position further increases the electron-accepting ability (Figure 6.15a).

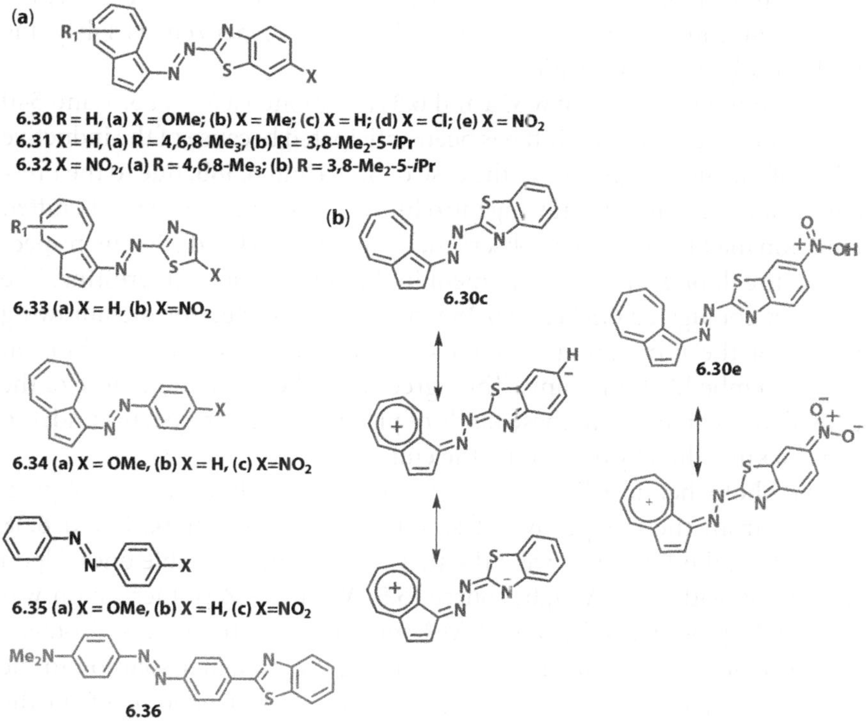

Figure 6.15 (a) Synthesized azo-azulene derivatives; (b) resonating structure of **6.30c**; (c) resonating structure of **6.30e** showing extended conjugation because of the electron.

The correlation between the pH and electronic spectra was employed to determine the pKa values of the derivatives. The electronic spectra for both electron-withdrawing and donating substituents resulted in a bathochromic shift as the conjugation length increased. However, the pK_a values of the conjugated acids have shown a linear correlation with Hammett parameters. Therefore, with the increase in the electron-withdrawing ability, basicity has decreased.

The substituent presence of the azulene and benzothiazolyl moiety deeply influences these compounds' oxidation and reduction potential. Cyclic voltammetric studies revealed that all the azulenyl derivatives **6.30–6.34** are more oxidizable compared to others. Later, the same group synthesized mono- and bis-azo azulenylpyridines and discussed the electronic behaviors by cyclic voltammetry. The pKa values were investigated by correlating electronic spectra and pH [57]. In 2014, Nica *et al.* reported the photoisomerization of the azo functionality in an azulenylpyridine system for the first time [58]. Azulene and its derivatives are known to exhibit fluorescence in solution owing to its $S_2 \rightarrow S_0$ (high quantum yield) and $S_1 \rightarrow S_0$ (low quantum yield) electronic transitions [59]. Azobenzene molecules are usually non-fluorescent owing to their E/Z isomerization. Therefore, the fluorescence in the azulenylpyridine-substituted azobenzene is solely due to the $S_2 \rightarrow S_0$ azulene transition.

The emission spectra of **6.37a** and **6.37b** exhibited a λ_{max} at 531 and 540 nm, respectively. The redshift has been anticipated because of the inductive effect of the methyl groups. In the case of **6.38a** and **6.38b**, due to the presence of an electron-withdrawing nitro functional group, further red-shifted emission maxima have been observed, λ_{max} around 554 and 556 nm, respectively. The fluorescence measurement in thin films is quite interesting. The highly π-conjugated **6.38a** and **6.38b**, because of the electron-withdrawing nature of the nitro group, exhibited significant emission-quenching in PMMA-embedded thin films. This agrees with the fact that because of the extended π-conjugation, these kinds of planar molecular systems undergo π-π stacking, thereby quenching the emission.

The photochemical *E*/*Z* isomerization of the azo bond can be of great interest from the perspective of smart applications. For **6.37** and **6.38** derivatives, the *E* isomers were the thermodynamically stable ones. Upon irradiating with the UV light source (5.5 W), an E→Z isomerization was observed by both UV-Vis and NMR spectroscopy. The reverse isomerization (*Z*→*E*) has been achieved by exposing the *Z*-isomer solution to the sunlight (Figure 6.16b). Although the authors do not comment on the changes in emission bands of the derivatives associated with isomerization,

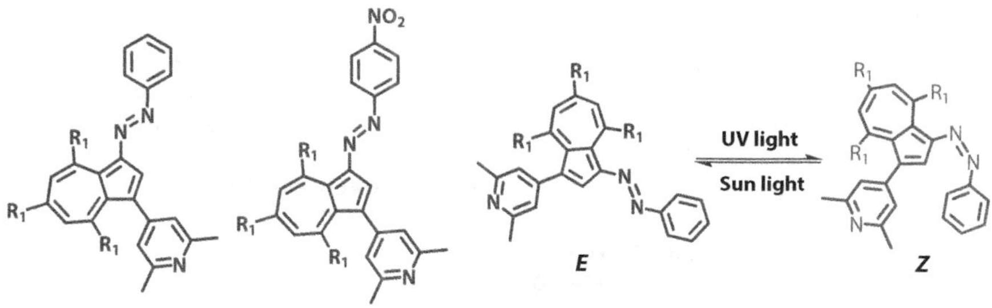

6.37a R = H
6.37b R = CH_3

6.38a R = H
6.38b R = CH_3

Figure 6.16 (a) Studied arylazoazulene derivatives; (b) schematic for the cis-trans isomerization.

it would be interesting to investigate. Shoji *et al.* have reported that arylazoazulenes can be obtained by simple aromatic nucleophilic substitution also as shown in Figure 6.17a [60]. The 6-arylazulenes (**6.39a–c**) exhibit strong absorption bands around 390 nm and 470–490 nm. In comparison with 2-arylazoazulene (**6.42a–c**), this is red-shifted, suggesting an extended conjugation of the 6-arylazoazulene systems. Investigations of the redox properties of **6.39a–c** by cyclic voltammetry shows quasi-reversible processes. The electrochemical reduction produces **6.39a**$^{\bullet-}$- **6.39c**$^{\bullet-}$ radical anions, which is expected to be delocalized over the azulene ring. Because of this stabilization, **6.39a–c** exhibit better electrochemical reversibility compared to 2-arylazoazulenes. The electrochemical energy gap (E_1^{ox}-E_1^{red}) of **6.39a–c** is smaller in comparison with 2-arylazoazulene **6.42a–c**, and this has been explained based on the difference in crystal structures of **6.39a–c** and **6.42a–c**. The mechanism of the electrochromic changes upon reduction in 6-arylazoazulenes is shown in Figure 6.17b.

X-ray structure analysis revealed a twisted structure for the 2-arylazoazulene derivative, whereas 6-arylazoazulenes are highly planar. The effective π-conjugation increases with the increasing planarity of the system, leading to the reduced HOMO-LUMO gap. However, both 6-arylazoazulkenes (**6.39a–c**) and 2-arylazoazulene (**6.42a–c**) have failed to show photoswitching properties (Figure 6.17c). Exposure to ultra-violet light of 254nm resulted in the decomposition of the compound and with 365nm light, no isomerization has been observed. The spectral overlap between azo moieties and azulene could be the possible reason for this observation. However, the ease of synthesis and the electrochemical properties of the derivatives would contribute to the development of new organic electronic materials involving the azulene skeleton.

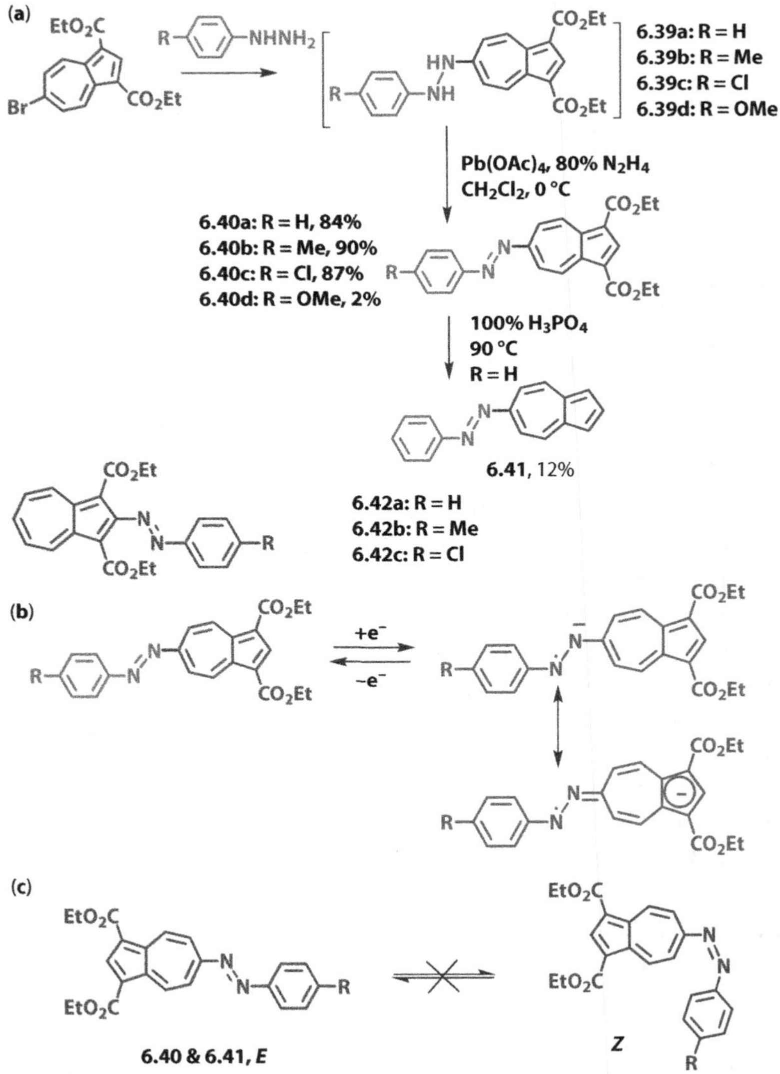

Figure 6.17 (a) Scheme for the synthesis of the studied azoazulenyl derivatives; (b) mechanism of the electrochromic changes upon reduction; (c) expected possible photoisomerization mechanism.

6.4 Conclusion and Future Prospects

Azulene proved to be a multifunctional motif that can enhance optical and photo-responsive properties, as well as stimuli-responsive properties significantly of a system. Due to its high absorbance in the visible region, incorporating this unit into photoswitches eventually leads to achieving more promising results including bidirectional visible light photoswitches. Several studies have revealed that azulene can significantly affect the switching behavior of the system, sometimes leading to enhancement in the photoswitching rate or the opposite. Different fused photochromic systems e.g., stilbene, DAE, and azobenzene have been discussed throughout the chapter along with their photochemical behavior. Some systems have found their potential applications in pH sensing, fluorescence switching, or biomedical detection. Acid-mediated negative photochromism has been demonstrated for the first time in a diarylethene-azulene conjugated system. Therefore, using azulene in photoswitches expands the possibilities for developing advanced stimuli-responsive smart materials.

AI disclosure statement: The author confirms that no artificial intelligence (AI)-generated content or AI-assisted writing tools were used in the preparation of this chapter (including summaries and the abstract). All text, analysis, and conclusions are the result of the author's original work.

References

1. Soto, F., Karshalev, E., Zhang, B., de Avila, B.E.F., Nourhani, A., Wang, J., Smart materials for microrobots. *Chem. Rev.*, 122, 5365, 2022.
2. She, P., Qin, Y., Wang, X., Zhang, Q., Recent progress in external-stimulus-responsive 2D covalent organic frameworks. *Adv. Mater.*, 34, 2101175, 2022.
3. Yang, P., Zhu, F., Zhang, Z., Cheng, Y., Wang, Z., Li, Y., Stimuli-responsive polydopamine-based smart materials. *Chem. Soc. Rev.*, 50, 8319, 2021.
4. Chatterjee, S., Molla, S., Ahmed, J., Bandyopadhyay, S., Light-driven modulation of electrical conductance with photochromic switches: Bridging photochemistry with optoelectronics. *Chem. Commun.*, 59, 12685, 2023.
5. Zhu, R., Lubben, J., Dittrich, B., Clever, G.H. Stepwise halide-triggered double and triple catenation of self-assembled coordination cages. *Angew. Chem. Int. Ed.*, 54, 2796, 2015.
6. Maity, M.L., Mahato, S., Bandyopadhyay, Visible-light-switchable chalcone-flavylium photochromic systems in aqueous media. *Angew. Chem. Int. Ed.*, 62, e202311551, 2023.

7. Kar, S., Chatterjee, S., Bandyopadhyay, S., A multi-stimuli responsive spiropyran–tetraphenylethene conjugate for the control of near-infrared emission. *Synthesis*, 57, 01, 189, 2025.
8. Zhang, Y., Ray, S.S., Yarin, A.L., Mechanoresponsive polymer nanoparticles, nanofibers and coatings as drug carriers and components of microfluidic devices. *J. Mater. Chem.*, 21, 8269, 2011.
9. Nie, J.H., Chen, X.Y., Wang, Z.L., Electrically responsive materials and devices directly driven by the high voltage of triboelectric nanogenerators. *Adv. Funct. Mater.*, 29, 1806351, 2019.
10. Vancaeyzeele, C., Olivier, F., Petroffe, G., Peralta, S., Vidal, F., Nanostructured thermal responsive materials synthesized by soft templating. *ACS Appl. Mater. Interfaces*, 9, 12706, 2017.
11. Saha, M., Chatterjee, S., Hossain, M.S., Ghude, A., Bandyopadhyay, S., Modulation of electronic mobility of a one-dimensional coordination polymeric molecular wire with light. *Chem. Asian J.*, 14, 4659, 2019.
12. Pooler, D.R.S., Lubbe, A.S., Crespi, S., Feringa, B.L., Designing light-driven rotary molecular motors. *Chem. Sci.*, 12, 14964, 2021.
13. Bouas-Laurent, H. and Dürr, H., Organic photochromism (IUPAC Technical Report). *Pure Appl. Chem.*, 73, 639, 2001.
14. Molla, S. and Bandyopadhyay, S., Visible light mediated efficient photoswitching of dimethyldihydropyrenes in thin films for all-photonic logic gate applications and dynamic encryption/decryption capabilities. *J. Mater. Chem. C*, 12, 17511, 2024.
15. Hatano, S., Horino, T., Tokita, A., Oshima, T., Abe, J., Unusual negative photochromism via a short-lived imidazolyl radical of 1,1'-binaphthyl-bridged imidazole dimer. *J. Am. Chem. Soc.*, 135, 3164, 2013.
16. Majee, D. and Presolski, S., Dithienylethene-based photoswitchable catalysts: State of the art and future perspectives. *ACS Catal.*, 11, 2244, 2021.
17. Hossain, M.S., Chatterjee, S., Bandyopadhyay, S., Azobenzene-based photochromic delivery vehicles for ions and small molecules. *Chem. Eur. J.*, 28, e202201902, 2022.
18. Chatterjee, S., Molla, S., Hossain, M.S., Bandyopadhyay, S., Modulation of conductance using light-induced morphological change of a layered π-structure. *J. Mater. Chem. C*, 10, 14993, 2022.
19. Molla, S., Chatterjee, S., Adak, S., Bandyopadhyay, S., A dimethyldihydropyrene π-switch for light-mediated dynamic control of charge transportation. *ACS Appl. Opt. Mater.*, 2, 1488, 2024.
20. Ash, C., Dubec, M., Donne, K., Bashford, T., Effect of wavelength and beam width on penetration in light-tissue interaction using computational methods. *Lasers Med. Sci.*, 32, 1909, 2017.
21. Steinmüller, S.A.M., Odaybat, M., Galli, G., Prischich, D., Fuchter, M.J., Decker, M., Arylazobenzimidazoles: versatile visible-light photoswitches with tuneable *Z*-isomer stability. *Chem. Sci.*, 15, 5360, 2024.

22. Beharry, A.A., Sadovski, O., Woolley, G.A., Azobenzene photoswitching without ultraviolet light. *J. Am. Chem. Soc.*, 133, 19684, 2011.
23. Samanta, S., McCormick, T.M., Schmidt, S.K., Seferos, D.S., Woolley, G.A., Robust visible light photoswitching with ortho-thiol substituted azobenzenes. *Chem. Commun.*, 49, 10314, 2013.
24. Xu, F., Sheng, J., Stindt, C.N., Crespi, S., Danowski, W., Hilbers, M.F., Buma, W.J., Feringa, B.L., All-visible-light-driven stiff-stilbene photoswitches. *Chem. Sci.*, 15, 6763, 2024.
25. Lvov, A.G. and Bredihhin, A., Azulene as an ingredient for visible-light- and stimuli-responsive photoswitches. *Org. Biomol. Chem.*, 19, 4460, 2021.
26. Amir, E., Murai, M., Amir, R.J., Cowart, J.S., Chabinyc, M.L., Hawker, C.J., Conjugated oligomers incorporating azulene building blocks-Seven- vs. five-membered ring connectivity. *Chem. Sci.*, 5, 4483, 2014.
27. Karatsu, T., Kitamura, A., Arai, T., Sakuragi, H., Tokumaru, K., Photochemical cis to trans one-way isomerization of styrylazulenes on their triplet excited state. Examination of ethylenes with non-benzenoid substituents of low triplet energies. *Bull. Chem. Soc. Jpn.*, 67, 1674, 1994.
28. Arai, T., Hozumi, Y., Takahashi, O., Fujimori, K., One-way cis → trans photoisomerization of azulenylethenes. *J. Photochem. Photobiol. A Chem.*, 104, 85, 1997.
29. Zhou, Y., Zhuang, Y., Li, X., Ågren, H., Yu, L., Ding, J., Zhu, L., Selective dual-channel imaging on cyanostyryl-modified azulene systems with unimolecularly tunable visible–Near infrared luminescence. *Chem. Eur. J.*, 23, 7642, 2017.
30. Shiraishi, Y., Tokitoh, Y., Nishimura, G., Hirai, T., Solvent-driven multiply configurable on/off fluorescent indicator of the pH window: A diethylenetriamine bearing two end pyrene fragments. *J. Phys. Chem. B*, 111, 5090, 2007.
31. de Silva, A.P., Gunaratne, H.Q.N., Gunnlaugsson, T., Huxley, A.J.M., McCoy, C.P., Rademacher, J.T., Rice, T.E., Signaling recognition events with fluorescent sensors and switches. *Chem. Rev.*, 97, 1515, 1997.
32. Zhou, Y., Zou, Q., Qiu, J., Wang, L., Zhu, L., Rational design of a green-light-mediated unimolecular platform for fast switchable acidic sensing. *J. Phys. Chem. Lett.*, 9, 550, 2018.
33. Zhou, Y. and Zhu, L., Involving synergy of green light and acidic responses in control of unimolecular multicolor luminescence. *Chem. Eur. J.*, 24, 10306, 2018.
34. Gong, Y., Zhou, Y., Yue, B., Wu, B., Sun, R., Qu, S., Zhu, L., Multiwavelength anti-Kasha's rule emission on self-assembly of azulene-functionalized persulfurated arene. *J. Phys. Chem. C*, 128, 22511, 2019.
35. Zhu, C., Kwok, R.T.K., Lam, J.W.Y., Tang, B.Z., Aggregation-induced emission: A trailblazing journey to the field of biomedicine. *ACS Appl. Bio Mater.*, 1, 1768, 2018.
36. Hong, Y., Lam, J.W.Y., Tang, B.Z., Aggregation-induced emission: Phenomenon, mechanism and applications. *Chem. Commun.*, 29, 4332, 2009.

37. Matsuda, K. and Irie, M., Diarylethene as a photoswitching unit. *J. Photochem. Photobiol. C Photochem. Rev.*, 5, 169, 2004.
38. Rashid, U., Chatir, E., Medrano Sandonas, L., Sreelakshmi, P., Dianat, A., Gutierrez, R., Cuniberti, G., Cobo, S., Kaliginedi, V., Dithienylethene-based single molecular photothermal linear actuator. *Angew. Chem. Int. Ed.*, 62, e202218767, 2023.
39. Huang, S., Li, Z., Li, S., Yin, J., Liu, S., Imidazole-based dithienylethenes as a selective chemosensors for iron(III) ions. *Dyes Pigm.*, 92, 961, 2012.
40. Matsuda, K. and Irie, M., Photoswitching of intramolecular magnetic interaction using a diarylethene dimer. *J. Am. Chem. Soc.*, 123, 9896, 2001.
41. Wilson, D. and Branda, N.R., Turning "on" and "off" a pyridoxal 5'-phosphate mimic using light. *Angew. Chem. Int. Ed.*, 51, 5431, 2012.
42. Kitai, J.I., Kobayashi, T., Uchida, W., Hatakeyama, M., Yokojima, S., Nakamura, S., Uchida, K., Photochromism of a diarylethene having an azulene ring. *J. Org. Chem.*, 77, 3270, 2012.
43. Gilat, S.L., Kawai, S.H., Lehn, J.-M., Light-triggered molecular devices: Photochemical switching of optical and electrochemical properties in molecular wire type diarylethene species. *Chem. Eur. J.*, 1, 275, 1995.
44. Hattori, Y., Maejima, T., Sawae, Y., Kitai, J.I., Morimoto, M., Toyoda, R., Nishihara, H., Yokojima, S., Nakamura, S., Uchida, K., Cyclization from higher excited states of diarylethenes having a substituted azulene ring. *Chem. Eur. J.*, 26, 11441, 2020.
45. Lvov, A.G., Bredihhin, A., Peet, J., Yadykov, A.V., Dmitrienko, A.O., Shirinian, V.Z., 1,2-Bis- and 1,2,3-tris(2,5-dimethylthiophen-3-yl)azulenes: Synthesis, structure and properties. *Dyes Pigm.*, 172, 107843, 2020.
46. Hou, I.C.Y., Berger, F., Narita, A., Müllen, K., Hecht, S., Proton-gated ring-closure of a negative photochromic azulene-based diarylethene. *Angew. Chem. Int. Ed.*, 59, 18532, 2020.
47. Dragu, E.A., Ion, A.E., Shova, S., Bala, D., Mihailciuc, C., Voicescu, M., Ionescu, S., Nica, S., Visible-light triggered photoswitching systems based on fluorescent azulenyl-substituted dithienylcyclopentenes. *RSC Adv.*, 5, 63282, 2015.
48. Jerca, F.A., Jerca, V.V., Hoogenboom, R., Advances and opportunities in the exciting world of azobenzenes. *Nat. Rev. Chem.*, 6, 51, 2022.
49. Bandara, H.M.D. and Burdette, S.C., Photoisomerization in different classes of azobenzene. *Chem. Soc. Rev.*, 41, 1809, 2012.
50. Fedele, C., Ruoko, T.P., Kuntze, K., Virkki, M., Priimagi, A., New tricks and emerging applications from contemporary azobenzene research. *Photochem. Photobiol. Sci.*, 21, 1719, 2022.
51. Saha, M., Hossain, M.S., Bandyopadhyay, S., A photoregulated racemase mimic. *Angew. Chem. Int. Ed.*, 60, 5220, 2021.
52. Bozovic, O., Jankovic, B., Hamm, P., Using azobenzene photocontrol to set proteins in motion. *Nat. Rev. Chem.*, 6, 112, 2022.

53. Murfin, L.C. and Lewis, S.E., Azulene—a bright core for sensing and imaging. *Molecules*, 26, 353, 2021.
54. Lacroix, P.G., Malfant, I., Iftime, G., Razus, A.C., Nakatani, K., Delaire, J.A., Azo-azulene derivatives as second-order nonlinear optical chromophores. *Chem. Eur. J.*, 6, 2599, 2000.
55. Razus, A.C., Birzan, L., Cristea, M., Tecuceanu, V., Blanariu, L., Enache, C., Novel mono- and bis-azo dyes containing the azulen-1-yl moiety: Synthesis, characterization, electronic spectra and basicity. *Dyes Pigm.*, 80, 337, 2009.
56. Razus, A.C., Birzan, L., Cristea, M., Ungureanu, E.M., Cretu, M.S., Investigation on the correlation between the structure of azulen-1-yl-benzothiazol-2-yl diazenes and their properties. Acidity and electrochemical redox potentials. *Dyes Pigm.*, 86, 1, 2010.
57. Dragu, E.A., Hanganu, A., Rau, I., Kajzar, F., Tane, A., Razus, A.C., Nica, S., Photochemistry of fluorescent azobenzenes substituted with azulenylpyridine moiety. *Mol. Cryst. Liq. Cryst.*, 604, 4, 2014.
58. Abou-Zied, O.K., Sinha, H.K., Steer, R.P., S_2-S_0 Spectroscopy of azulene and its 2-chloro, 2-methyl, and 1,3-dimethyl derivatives. *J. Mol. Spectrosc.*, 18, 42, 1997.
59. Yeh, H.C., Wu, W.C., Wen, Y.S., Dai, D.C., Wang, J.K., Chen, C.T., Derivative of α,β-dicyanostilbene: Convenient precursor for the synthesis of diphenylmaleimide compounds, E-Z isomerization, crystal structure, and solid-state fluorescence. *J. Org. Chem.*, 69, 6455, 2004.
60. Shoji, T., Sugiyama, S., Takeuchi, M., Ohta, A., Sekiguchi, R., Ito, S., Yatsu, T., Okujima, T., Yasunami, M., Synthesis of 6-Amino- and 6-Arylazoazulenes via nucleophilic aromatic substitution and their reactivity and properties. *J. Org. Chem.*, 84, 1257, 2019.

7

Azulene-Based π-Functional Materials and Their Application as Sensors

Bagmita Bhattacharyya* **and Kangkana Deka**

Department of Chemistry, Tihu College, Tihu, Assam, India

Abstract

Azulene, a non-alternant aromatic hydrocarbon and structural isomer of naphthalene, is a molecule of utmost importance on account of its strong chromophoric nature, bright blue color, and fluorescence properties owing to narrow energy gap between the highest occupied molecular orbital (HOMO) and lowest unoccupied molecular orbital (LUMO). The azulene core functions as a fluorescent probe that aids in analyte identification due to the anti-Kasha fluorescence feature from the $S_2 \rightarrow S_0$ state. They work well in situations where a colorimetric response is inappropriate, such as in a biological matrix or other highly colored environment. Spontaneous electron drift occurs between the two rings in azulene, which facilitates it to functionalize electrodes. Because of its distinct electronic and structural characteristics, including its dipolar structure with a dipole moment of 1.08 D, high electron affinity, low ionisation energy, and unusual $S_2 \rightarrow S_0$ fluorescence, azulene and its derivatives make excellent anions sensors, heavy metal sensors, optical sensors, polymer sensors, and many other types.

Keywords**:** Azulene, sensors, fluorescence, chromophore

7.1 Introduction

Azulene is an interesting compound having a unique structure consisting of a five-membered, electron-rich ring connected with a seven-membered, electron-deficient ring (Figure 7.1). This distinctive electronic property of the aromatic azulene compounds with a highly polarized electronic system

**Corresponding author*: bagmitabhattacharyya13@gmail.com

Neha Rani Kumar and Abhijeet R. Agrawal (eds.) Azulene-Based Materials: Design, Synthesis, and Applications, (179–198) © 2026 Scrivener Publishing LLC

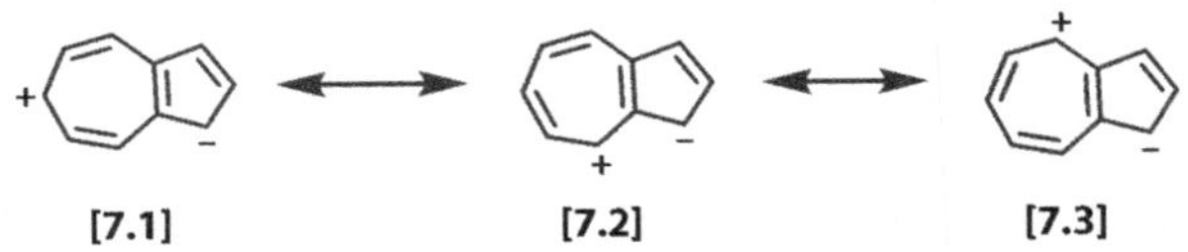

Figure 7.1 Resonance structure of azulene.

helps interact with other systems with polarized electronic environment. Azulene, an inherently push-pull system, exhibits low ionization energy and high electron affinity [1]. This makes it an attractive building block for advanced materials synthesis. Polyazulene, synthesized through the oxidative polymerization of azulene, is particularly promising due to its wide range of potential applications, including electronic materials, sensors, and switching devices. Azulene and its derivatives are of interest in recent times owing to its distinctive bright blue color, easy interaction with polar moiety, and high dipole moment, which adds to its significant electrochemical behavior, such as irreversible electrooxidation and quasi-reversible reduction [2]. The probable sites of functionalization in azulene are shown in Figure 7.2.

Because of its unique photophysical characteristics and subsequent impacts on the absorption spectra, azulene and its derivatives have sparked a lot of research as a building block for the development of multi-responsive functional materials [3]. Color-changing reactions can be utilized as crucial indicators in chelatometry, acid-base titrations, and other processes that involve the identification and quantitative study of compounds. Azulene derivatives have diverse applications such as sensors in heavy metal detection, anion detection, and optoelectronics. Herein, we will discuss the diverse applications that can be achieved by azulene and its derivative compounds. This chapter will shine a light into the use of azulene and its derivatives in sensing applications.

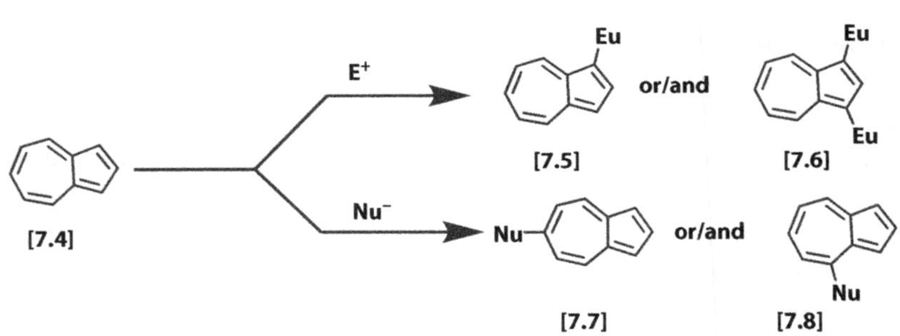

Figure 7.2 Reactivity of azulene in presence of electrophile and nucleophiles.

7.2 Application of Azulene and Its Derivatives for Sensing of Heavy Metals

Azulene and its derivatives are extensively used in heavy metal detection. Efficient identification of poisonous metals such as Pd, Hg, Cd, and Cr, which impart adverse effects to human health and the environment, is of utmost importance. The classical methods for detecting heavy metal ions include inductively coupled mass spectrometry, atomic absorption, and emission spectroscopy, and cold vapor atomic fluorescence spectrometry. While these techniques offer high sensitivity, they often require intricate sample preparation, precise experimental conditions, and costly equipment. Electrochemical detection methods that include anodic stripping voltammetry are highly sensitive, cost-effective, and compatible with turbid and colored samples. Among the electrodes employed in stripping techniques, chemically modified electrodes are particularly noteworthy due to their ability to enhance the sensitivity and selectivity of electroanalytical methods and their ease of fabrication. These electrodes are typically prepared by coating the electrode with polymeric films, and the film thickness can be easily controlled, ensuring reliability and reproducibility. Among the various monomers used for electrode functionalization, azulene stands out for its ease of polymerization due to its polar nature. Several methods have been adopted for the synthesis of azulene derivatives. Knoevenagel condensation is one of the most common methods adopted for the preparation of azulene derivatives. Many such azulene derivatives have been reported to have very high selectivity with a diverse high range of detection limits.

In 2017, Ungureanu *et al.* reported (Z)-2-thioxo-5-((4,6,8- trimethylazulen-1-yl)methylene) imidazolidin-4-one (**7.9**) [4], a compound prepared by the Knoevenagel condensation method and characterized by electrochemical methods such as cyclic voltammetry, differential pulse voltammetry, and rotating disk electrode methods. For the sensing purpose, the metal electrodes were modified by electropolymerization of the ligand in the positive electrode. This electrode can be used in the detection of Pd and Cu ions in water when the concentration is higher than 10^{-6} M. Another compound 2,6-bis((E)-2-(furan-2-yl)vinyl)-4-(5-isopropyl-3,8-dimethylazulen-1-yl)pyridine (**7.10**), reported by the same group in 2017, was investigated using cyclic voltammetry, differential pulse voltammetry, and rotating disk electrode methods [5]. The study focused on developing modified electrodes through potential cycling or controlled potential electrolysis for detecting heavy metal ions such as mercury, cadmium, copper, and lead. The experiments were conducted in a 0.1 M buffer acetate solution at pH 5.5 and 25°C under argon atmosphere. The modified electrodes, specifically those coated with polymerized **7.10**, were characterized and tested for their ability to detect heavy metals, with the best results observed for lead ions (Pb) at a detection limit of 10^{-7} M (Figure 7.3)

Similar compound 4-azulen-1-yl-2,6-bis(2-thienyl)pyridine (**7.11**) was used to create polymer-coated electrodes for metal ion complexation, specifically targeting lead and cadmium ions [6]. These modified electrodes were synthesized *via* oxidative electropolymerization of the compound in acetonitrile with 0.1 M tetrabutylammonium perchlorate as the supporting electrolyte. The resulting polymer films were characterized by similar methods such as cyclic voltammetry, differential pulse voltammetry, and scanning electron microscopy. The complexation properties of these polymer-coated electrodes were evaluated using a chemical preconcentration-anodic stripping method. The study revealed that the electrodes effectively complexed lead and cadmium ions, demonstrating their potential for use in electrochemical sensors. The electrochemical properties indicated that higher oxidation potential led to less conserved azulene structures, resulting in more amorphous and insulating materials. Whereas chelating modified electrodes with 4-azulen-1-yl-2,6-bis(2-thienyl)pyridine (**7.11**) [7] exhibited notable selectivity for lead ions, even when other metal cations were present in large excess. The detection limits achieved were 0.7 nM for lead(II) and 10 nM for cadmium(II) ions. These modified electrodes were also successfully applied for detecting lead in both natural and tap water. The compound 2,2'-(ethane-1,2-diylbis((2-(azulen-2-ylamino)-2-oxoethyl)azanediyl)) diacetic acid (**7.12**) [8] has been reported to exhibit selective colorimetric properties in response to Hg(II) ions, changing from purple to yellow (Figure 7.4). Compound **7.12** and Hg(II) ions can form a

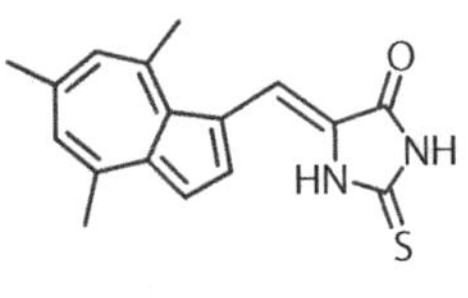

(Z) -2-thioxo-5-((4,6,8- trimethylazulen-1-yl) methylene) imidazolidin-4-one [7.9]

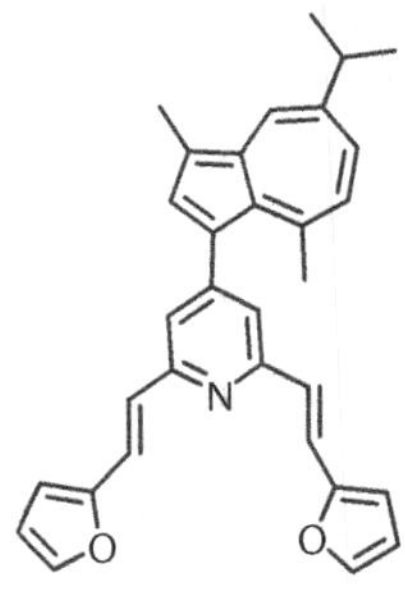

2,6-bis((E)-2-(furan-2-yl)vinyl)-4-(5-isopropyl -3,8-dimethylazulen-1-yl)pyridine [7.10]

Figure 7.3 Structure of compounds **7.9** and **7.10**.

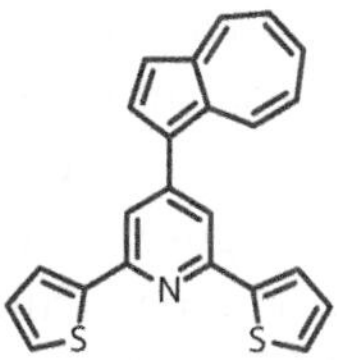

4-azulen-1-yl-2,6-bis(2-thienyl)pyridine

[7.11]

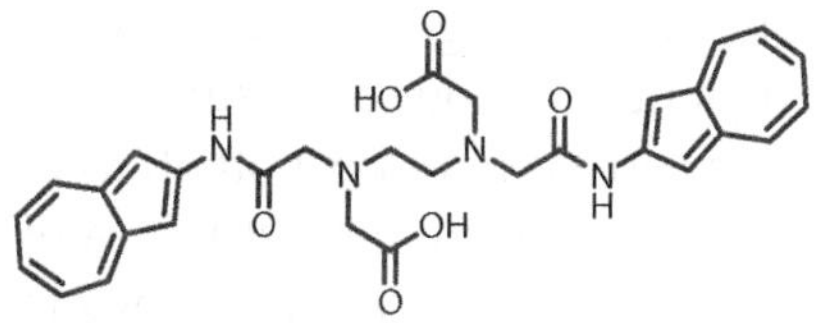

2,2'-(ethane-1,2-diylbis((2-(azulen-2-ylamino)-2-oxoethyl)azanediyl))diacetic acid

[7.12]

Figure 7.4 Structures of **7.11** and **7.12.**

1:1 complex, and the formation of these complexes can be easily confirmed with the help of UV-Vis spectral titration and electrospray ionization-mass spectrometry.

Another azulene derivative (E)-5-(azulen-1-yldiazenyl)-1H-tetrazole (**7.13**) [9] has a tetrazole unit that acts as a complexing agent that can be attached to the electrode surface, and it can easily be used to detect metal ions such as lead ions. The azulene-based electrode demonstrated high sensitivity with a detection limit of 5×10^{-8} M and a linear detection range between 5×10^{-8} M and 10^{-6} M. Besides this, the approach offers the possibility of detecting a wide range of heavy metal ions such as Hg(II), Cd(II), and Pb(II). The modified electrodes demonstrated an effective complexation and detection of the target metal ions, suggesting potential applications in environmental monitoring for heavy metals. The study indicates that these modified electrodes are suitable for analyzing water samples containing Pb and Cu ions at concentrations above 10^{-6} M. Similarly 4-(azulen-1-yl)-2,6-bis((E)-2-(thiophen-3-yl)vinyl)pyridine (**7.14**) [10] can detect lead and copper in water when concentration is higher than 10^{-6} M (Figure 7.5).

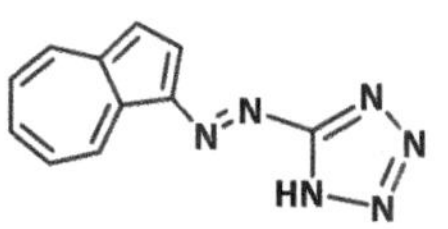

(E)-5-(azulen-1-yldiazenyl)-1H-tetrazole

[7.13]

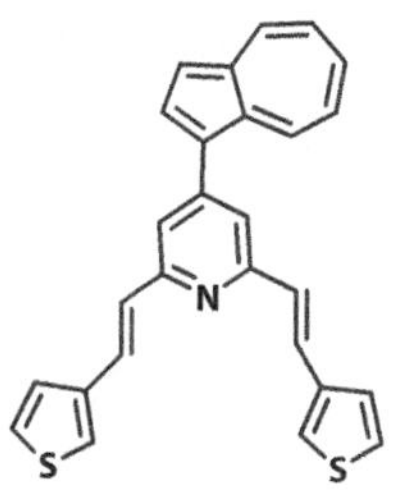

4-(azulen-1-yl)-2,6-bis((E)-2-(thiophen-3-yl)vinyl)pyridine

[7.14]

Figure 7.5 Structures of **7.13** and **7.14.**

The Knoevenagel condensation of azulene-1-carbaldehyde with active methylene compounds, resulting in various derivatives, can show significant variations in redox potentials depending on the substituents attached. A series of azulene derivatives were synthesized by Ungureanu *et al.* and tested for their ability to act as ligands for metal ions, with particular attention to their interaction with heavy metals like lead (Pb), cadmium (Cd), copper (Cu), samarium (Sm) and mercury (Hg) (Scheme 7.1).

The study revealed that some of these azulene derivatives [11], especially those with corresponding auxiliary groups, could effectively be complex with metal ions, indicating their potential use as metal sensors. The electrochemical properties of these compounds, such as oxidation and reduction potentials, are significantly influenced by the nature of the substituents. Another set of 1-vinylazulene was reported by the same group in the year 2016 [12] using the Knoevenagel condensation method using piperidine as a catalyst (Figure 7.6). The oxidation potential varied significantly based on the nature of the substituents attached to the azulene moiety. Strong electron-withdrawing groups (EWGs) such as nitro and trifluoroacetyl increased the oxidation potential, while EDGs decreased it. The study highlighted the ability of the synthesized ligands to interact with metal ions, particularly samarium (Sm), magnesium (Mg), and cerium (Ce). These interactions were observed through shifts in UV-Vis absorption spectra, indicating the formation of metal-ligand complexes. The complexes were stable and showed potential for use in metal ion detection.

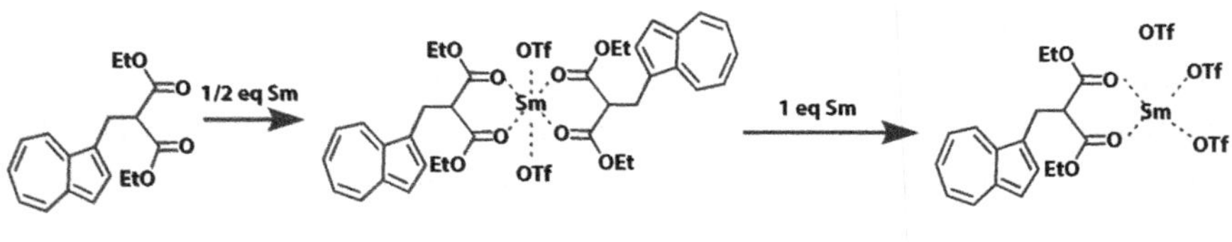

[7.15] [7.16] [7.17]

Scheme 7.1 Interaction of an azulene ligand **7.15** with metallic ion.

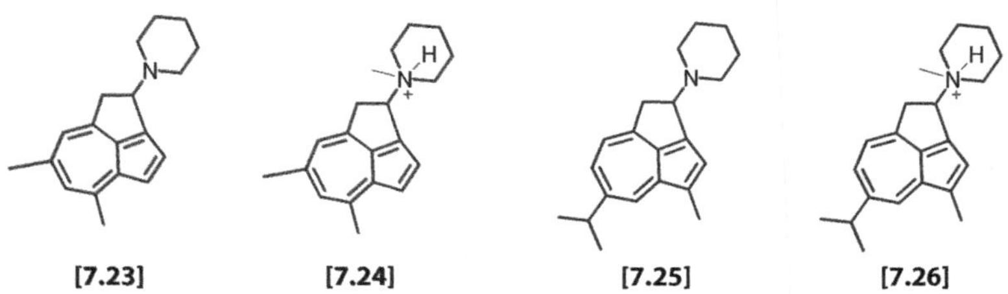

[7.23] [7.24] [7.25] [7.26]

Figure 7.6 Condensation products of azulene-1-carbaldehyde in the presence of weak activated methylene compounds.

A study conducted by Buica *et al.* [13] explores the interactions between lead Pb(II) and cadmium Cd(II) ions and complexing polymer films (**7.18**), specifically those based on azulene derivatives, which are used in metal detection sensors (Scheme 7.2). The polymer films were created *via* controlled potential electrolysis, and their affinities for the metal ions were evaluated using sorption isotherms at different temperatures. Specifically for the azulene polymer poly(4-azulen-1-yl-2,6-bis(2-thienyl)pyridine), the Langmuir isotherm model provided the best fit, indicating that polyazulene derivative films have a higher affinity for lead ions compared to cadmium, with sorption capacities of approximately 42 mg/g for Pb(II) and 17 mg/g for Cd(II). Insights into thermodynamic parameters revealed that the sorption occurs spontaneously for both metal ions. The Pb(II) sorption is predominantly a chemisorption process, while Cd(II) involves a blend of chemisorption and physisorption. The study emphasizes the importance of thermodynamic analysis in understanding the metal ion sorption process within polymer films and their potential application in electrochemical metal ion detection.

The ability of an azulene derivative to sense heavy metals is greatly influenced by its surrounding environment and the presence of EWGs or EDGs. Environmental factors such as solvent choice, pH, electrode surface, electropolymerization conditions, and ionic strength can impact the solubility, binding affinity, and sensitivity of the azulene derivative. Additionally, attaching EWGs increases selectivity and stability but may reduce sensitivity, while EDGs enhance sensitivity but may lower selectivity and risk overoxidation. Fine-tuning these factors optimizes the azulene-based sensors for specific heavy metals.

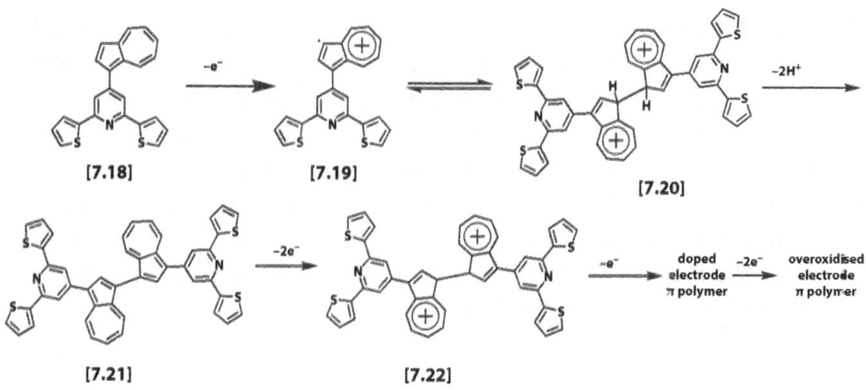

Scheme 7.2 Polymerization of **7.18**.

7.3 Azulene-Based Sensors for Detection of Anions

7.3.1 Sensors for Phosphate

The determination of phosphate is important for the clinical diagnosis of various conditions. Measurement of phosphate concentration in the body fluids can help in the diagnosis of several disorders such as hyperparathyroidism, hypertension, vitamin D deficiency, and mineral and bone disorder. Additionally, unusually high levels of phosphate in the body fluid can have adverse effects, mostly leading to the deposition of calcium phosphate in the blood, which can cause kidney damage. Apart from biological issues, phosphate is also a well-known pollutant of ground and surface water. An increase in the level of phosphate concentration in a water body causes rapid growth of plankton, which makes such water unsuitable for drinking.

For all these reasons, monitoring phosphate levels both in water bodies and body fluids of humans is crucial. Different kinds of methods have been employed for measuring phosphate concentrations in various systems. The harmful effects of highly elevated levels of phosphate in blood such as calcium phosphate deposition can lead to kidney damage. The detection of phosphate levels in body fluids provides useful information about diseases like kidney failure [14].

Azulene is an interesting subunit for the construction of anion receptors due to few of its unique characteristics. It has an uncommon geometry of a seven-membered ring, and it is a strong chromophore with a large dipole moment of 0.8 D.J. Jurczak *et al.* reported in 2008 about azulene-1,3- dicarboxylic acid bisamides and azulene-5,7-dicarboxylic acid bisamides as building blocks for phosphate anion receptors. The five-membered ring dicarboxylic acid derivatives were prepared by the direct acylation of azulene with trifluoroacetic acid anhydride, followed by hydrolysis to afford the acid. The availability of the acid dichloride is of great advantage for the five-membered derivatives, as it offers a convenient synthetic route to more efficient ligands. After preparing the model ligands, the anion-binding properties were studied. It was fascinating to see the change in optical properties of the azulene-based bisamides upon complex formation with anions as all azulene derivatives are colorful. Azulene-1,3-dicarboxyamide derivatives showed the potential to sense dihydrogen phosphate anions. However, the low anion affinity of these receptors prevented efficient interactions under competitive conditions [15].

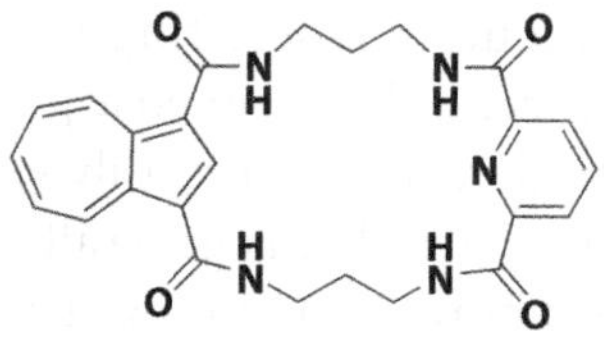

[7.27]

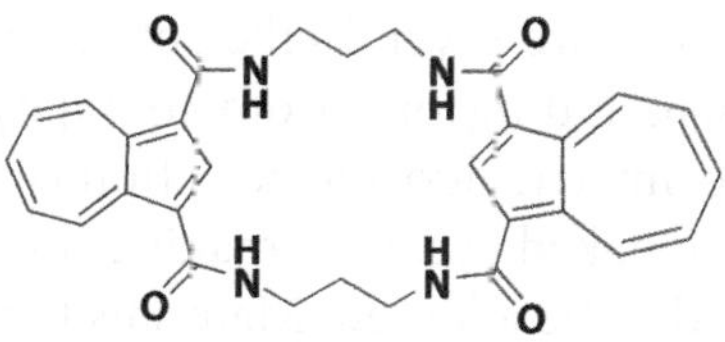

[7.28]

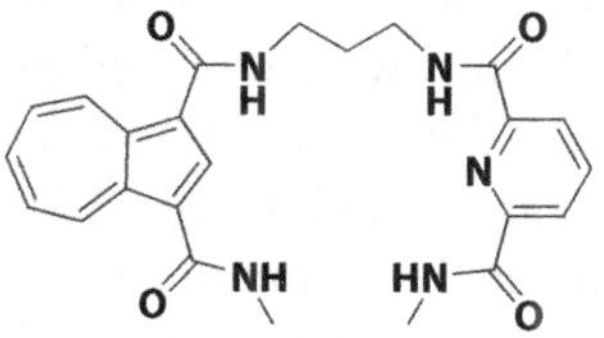

[7.29]

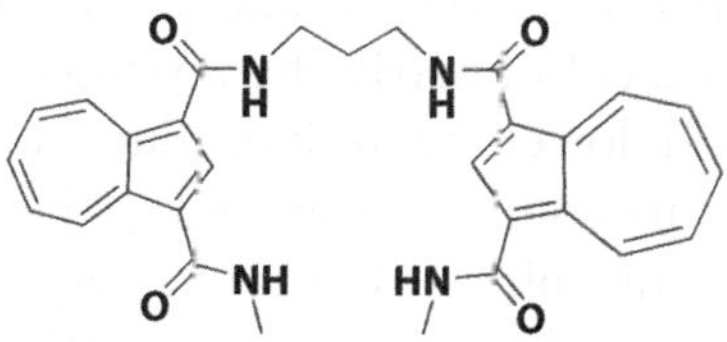

[7.30]

Figure 7.7 Azulene receptors **7.27–7.30**.

In 2016, J. Jurczak *et al.* reported a pair of macrocyclic receptors that possess both the properties of the azulene-1,3-dicarboxyamide unit and the benefit of a macrocyclic topology of receptors, which allows proficient phosphate anion binding and sensing under highly competitive circumstances (**7.27–7.30**). They also prepared and evaluated two linear analogs. The macrocyclic receptors fused with chromophoric azulene moiety become easily detectable to the naked eye providing the potential to serve as phosphate anion sensors, particularly the phosphate anions $H_2PO_4^-$ and $HP_2O_7^{3-}$ that show the change in color in solution (Figure 7.7) [16].

7.3.2 Sensors for Fluoride

The determination of fluoride is important in drinking water, as well as in biological systems, due to health and environmental protection. Just like any other trace elements, fluoride also has its standard acceptable daily intake, which must be followed. The World Health Organization recommends a human daily intake below 10 mg. An overexposure to fluoride can cause harmful effects like nephrotoxic changes, skeletal fluorosis, acute gastric, urolithiasis, and kidney failure, and it can induce metabolic disorders in the immune system. Higher levels of fluoride can even exert adverse effects on the human brain, causing lower IQ and even cancer at very high concentrations [17]. Therefore, quantitative detection of fluoride is necessary including different analytical methods.

There are several efficient analytical techniques for detecting fluoride ions, including ion chromatography, fluorescence, and colorimetric probes. Colorimetric fluoride detection is simple to understand and readily visible to the naked eye, thus making it the easier approach that is generally preferred to be adopted. Materials that combine a boron-containing receptor with an azulene chromophore in one molecule may be useful for fluoride ion detection. The production of a dimeric azulene boronate ester and its use as a colorimetric chemosensor for fluoride ions were reported by Tao *et al.* To lower the limit for fluoride detection, a dimeric azulene boronate structure containing sp^2 and sp^3 boron was created (**7.32**, Figure 7.8) [18]. Here, the color of the probe responds immediately, changing from light green to yellow and demonstrating a lower limit of detection, excellent selectivity, and stability. The detection limit in this case was found to be 0.6 $mg{\cdot}L^{-1}$ which is lower than the WHO limit.

Xia *et al.* developed an efficient [5+2] annulation strategy to rapidly construct an uncommon metal-containing [5-5-7] scaffold skeleton, termed as metalla-dual-azulene (MDA). This is a cross between an MDA and an organic azulene, both of which share a seven-membered tropylium motif (**7.36**, Figure 7.9).

The dipole moment of the molecule can be significantly affected by replacing carbon with a transition metal in the [5-5-7] skeleton, resulting in completely different reactivities of the two azulene fragments. The seven-membered ring can undergo a facile nucleophilic addition, and this facilitates the inclusion of a wide variety of substituents in the ring systems and also provides a way to finely tune the electronic properties of the compound. An MDA of type **7.40**, as shown in Figure 7.10, is capable of excellently detecting fluoride anions even in the presence of 13 other anions in the mixture. These types of systems pave the way for the construction of new enlarged π-conjugated polycyclic systems with intriguing properties [19].

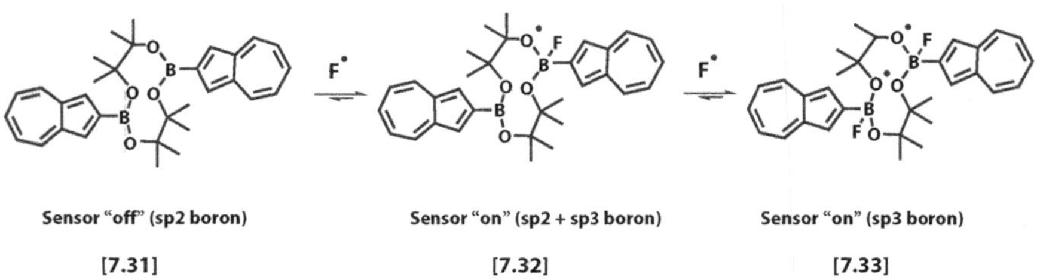

Figure 7.8 The chemical structure of a dimeric azulene boronate ester and a binding mechanism between host and guest molecules.

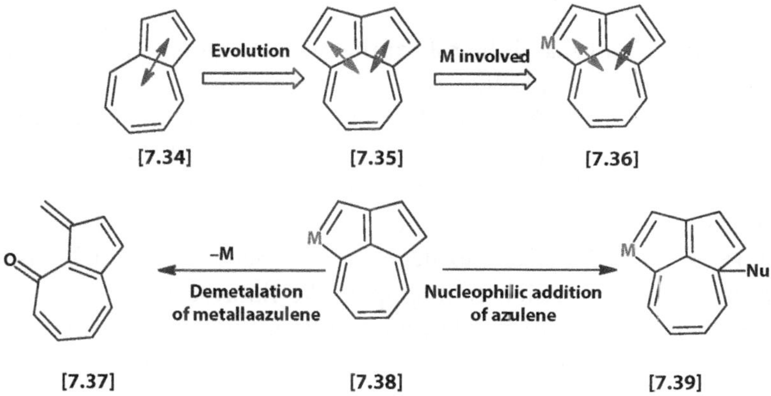

Figure 7.9 Structural skeletons of azulene, cyclopenta azulene, MDA, and the reactivities and demetallation of MDA.

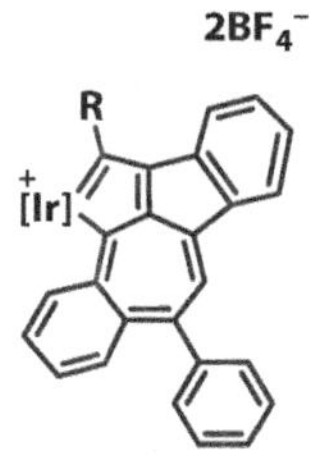

[7.40]

Figure 7.10 Fused-polycyclic metallacycles.

In 2017, López-Alled *et al.* reported an azulene-based colorimetric sensor Az-1-Bpin **7.41** for fluoride anion with significant selectivity for fluoride in the tetrahydrofuran (THF) solution. A color change from purple to yellow was observed.

In 2020, Murfin *et al.* reported an azulene-based colorimetric fluoride sensor NAz-6-Bpin **7.42** that can selectively sense fluoride much more easily and effectively than other halides and common anions. The supremacy of the approach lies in the fact that this azulene-based sensor works effectively in a water-ethanol system and does not necessitate the use of a surfactant. However, the sensor becomes ineffective in a solvent system comprising a mixture of water and an organic solvent. Another azulene-containing sensor Az-1-Bpin is also effective in detecting fluoride in water when surfactants like cetyltrimethylammonium bromide and tetra-*n*-butylammonium hydrogen sulfate are used. In this case, a change in color from purple to purple-blue takes place. Here, azulene plays the role of the reporter motif, and the role of the receptor motif is played by pinacolborane (Figure 7.11) [20].

Eichen *et al.* in 2015 reported an azulene-based chemosensor, which could selectively detect fluoride ions with the help of fluorescence. The exceptionally short-lived emission of azulene from S_2 makes it an excellent candidate to be used as luminophore. An example of this class of luminophore is 1,3-di(2-pyrrolyl)azulene (**7.47**, Scheme 7.3), which has been used as positron emission tomography-like fluorescent anion receptors in dimethyl sulfoxide for the detection of fluoride anion. The detection is accompanied by a change in color or turning "on" its fluorescence emission [21].

Son *et al.* in 2014 reported a chemosensor based on guaiazulene and indole derivatives. It showed high sensitivity to target anions. The chemosensor **7.50** involves a planar form containing large conjugation, which can provide it with spectral changing properties by better electron migration between the donor-acceptor unit. The chemosensor could detect CN^- and F^- selectively by undergoing optical and spectral changes (Scheme 7.4) [22].

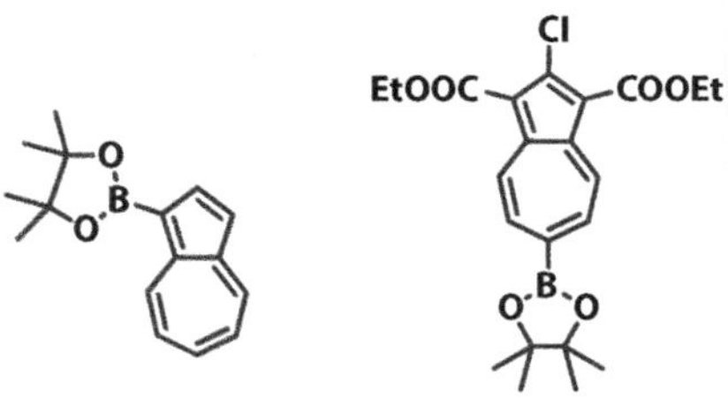

Az-1-Bpin [7.41] NAz-6-Bpin [7.42]

Figure 7.11 A comparison between Az-1-Bpin and NAz-6-Bpin.

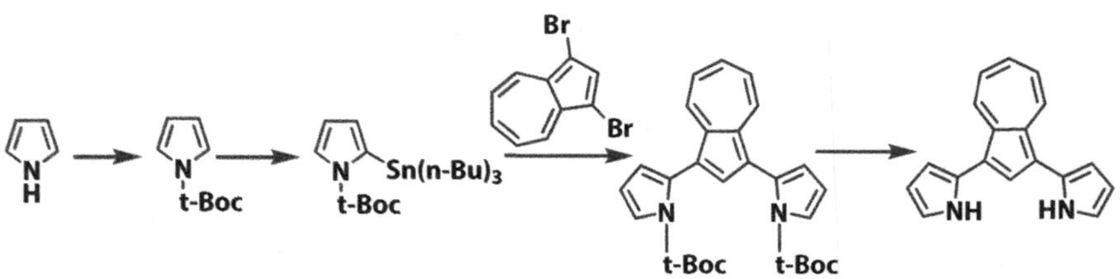

[7.43] [7.44] [7.45] [7.46] [7.47]

Scheme 7.3 Synthesis of 1,3-Di(2-pyrrolyl)azulene.

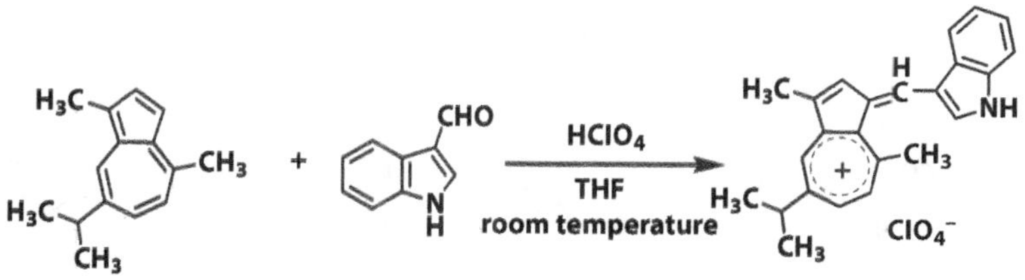

[7.48] [7.49] [7.50]

Scheme 7.4 Synthetic scheme of chemosensor **7.50.**

7.3.3 Sensors for Nitrite

Cured meat products, such as dry-cured ham, can be preserved for a long time by using salts of nitrite such as potassium nitrite (KNO_2) and sodium nitrite ($NaNO_2$) as preservatives. These salts are generally used as a part of a salting mixture along with NaCl and corresponding nitrate salts to reduce microbial spoilage and increase the shelf life of meat products. The nitrite salts not only alter the color of the meat products but also positively impact the organoleptic properties of these meat products. However, the side effects of nitrite consumption cannot be ignored as they are carcinogenic. In the acidic environment of the body, nitrite can get converted to N-nitrosamines and N-nitrosamides *via* amine and amide intermediates. These intermediates are known to be potentially carcinogenic. The limited intake of 0-0.06 mg/kg of body weight per day has been specified by the World Health Organization for nitrite ions. Besides that, nitrite ion is a common pollutant with potential negative health impacts that are often found in water, thus making its quantitative analysis significant.

The concentration of nitrite ions in solutions is generally determined by employing the Griess reaction in which nitrite forms azo derivatives with probes like arylamine, gold nanoparticles, silver nanoparticles, aza-BODIPY, pyrrole, organometallic complexes, and many more. Sawicki *et al.* reported that, under an acidic medium, azulene and nitrite ions together formed a species that exhibited an absorption maxima at 617nm leading to an easy detection of nitrite [23]. In an independent report, Garcia *et al.* reported the detection of nitrite by first diazotizing the same with para-nitroaniline followed by the reaction with azulene to form an azulene-containing azo dye [24].

L. C. Murfin *et al.* reported the selective sensing of nitrite anion by employing a colorimetric chemodosimeter. This colorimetric chemodosimeter was based on a 6-bromo-2-amino azulene derivative (**7.51**) and involves the formation of diazonium salt and then a diazoquinone in the

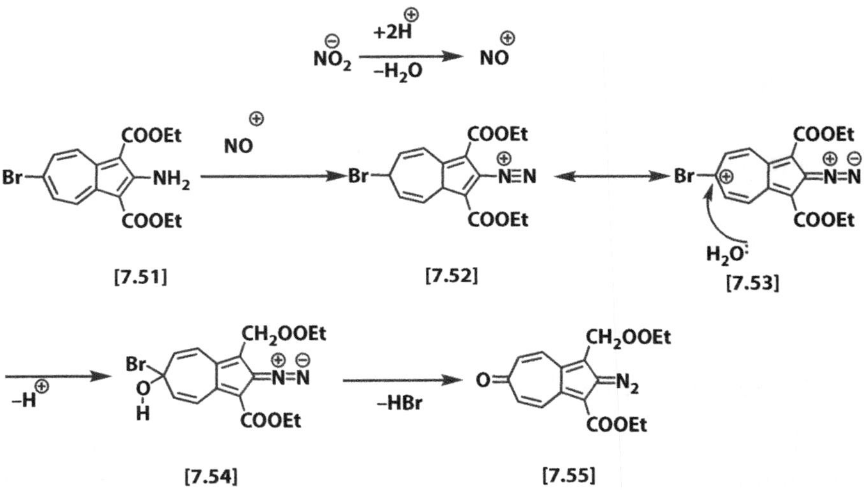

Scheme 7.5 Reaction of an azulene compound with NO_2-.

detection process (Scheme 7.5). In one minute, a color change in the solution from yellow to grey was observed, and if kept for a prolonged time (approximately 10 min), a change from grey to orange was encountered. The mechanistic insights were validated by UV-Vis spectroscopy.

The detection limit for the azulene sensor was also calculated. The limit of the detection for spectrophotometric detection was found to be 0.23 mg/L nitrite. The detection limit for the naked eye i.e., clear visibility response of the sensor for nitrite was found to be at the concentration of ≥50 mg/L. The sensor was then used to detect nitrite content in meat samples, and the obtained values were comparable to those of the British Standard [25].

7.4 Bioimaging

Contemporary biology is dedicated to unraveling how individual components come together to build and sustain complex biological systems. A key focus is understanding the organization of cellular structures, the interactions between various biomolecules, and the mechanisms governing their formation and removal throughout different stages of the life cycle. Additionally, studying the dynamic movement of these components within the cellular environment is essential for grasping how cells function as integrated systems.

Advancements in "bioimaging" technologies now make it possible to observe these intricate processes with remarkable precision across multiple levels of resolution. This emerging and rapidly evolving field enables scientists to visualize cellular and molecular events in living organisms in real-time, providing deeper insights into the fundamental workings of life. Through bioimaging, researchers can capture the complexity of biological systems in action, opening new frontiers in understanding health, disease, and the intricate machinery of life at the cellular level [26].

7.4.1 Use of Azulene in Bioimaging

Probes that excite or emit in the red to near-infrared (NIR) spectrum (650–900 nm) are considered advantageous for a wide array of biological applications. These advantages include reduced background absorption from water and hemoglobin, as well as non-invasive deep-tissue penetration that ranges from micrometers to centimeters. They also lessen the possibility of photobleaching and sample damage, which are frequently linked problems to higher-energy UV light sources. It is also important to avoid autofluorescence from endogenous molecules, which typically occur at wavelengths below 600 nm, to ensure clearer and more accurate imaging.

Zhou *et al.* in 2017 developed a cyanostyryl-modified azulene system with tunable photophysical properties, and *in situ* stimuli responsiveness has been developed. This system enables NIR emission and introduces a dual-pathway luminescent strategy (visible and NIR) with adjustable intensity, driven by the S_0 to S_1 transition regulation and molecular conformation control. Intracellular studies demonstrate its use for selective dual-channel imaging [27].

Azulenylsquaraines have emerged as valuable tools in biological applications, particularly NIR fluorescence resonance energy transfer quenchers in peptide-based systems. In 2002, Tung *et al.* designed and developed NIRQ750/**7.56** (Figure 7.12), a molecule specifically engineered to leverage the benefits of NIR absorption. The researchers proposed that the molecule's extended π-conjugated system would facilitate strong absorbance in the 650–900 nm range. This spectral range is especially advantageous for biological applications due to its ability to penetrate deeper into tissues with minimal interference from water and hemoglobin, while also reducing background noise from autofluorescence [28].

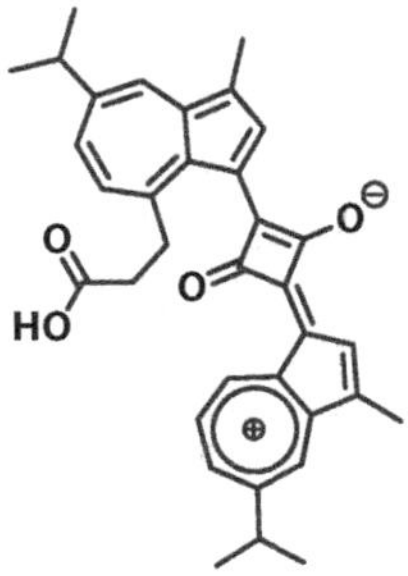

NIRQ750 [7.56]

Figure 7.12 Near-infrared fluorescent quenchers in the peptide.

Pham *et al.* in 2013, demonstrated the potential of an ^{18}F-labeled azulene derivative for cancer imaging. By targeting COX-2 as a biomarker, they achieved selective retention of the probe in breast cancer tumors in mice, enabling tumor visualization through positron emission tomography [29].

In 2004, Pispisa *et al.* investigated the fluorescent properties of **7.57** in hexapeptides by varying the distances between the fluorescent probe and the quencher 2,2′,6,6′-tetramethyl-1-oxyl-4-amino-4-piperidine carboxylic acid. The azulenyl residue exhibits spectral and photophysical properties that complement those of tryptophan. This innovative fluorescent amino acid (**7.57**, Figure 7.13), which is nearly isosteric with its natural counterpart, offers a promising alternative probe for fluorescence studies of synthetic proteins. Importantly, it can likely be used without disrupting the functional activity of the native protein structure [30].

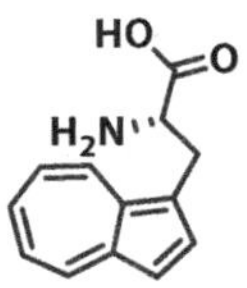

L-Aa1 [7.57]

Figure 7.13 Non-canonical amino acid AzA1a/7.57.

7.5 Conclusion

Azulene and its derivatives have exceptional chemical and physical properties that make them ideal for sensing applications. The application of azulene derivatives in the detection of heavy metals and different anions including fluoride, phosphonate, and nitrates are owing to its high sensitivity and selectivity as a sensor. The non-benzenoid aromatic structure, which has a distinct electron distribution and an intrinsic dipole moment, allows the development of extremely precise and responsive sensing materials, in addition to the possibility of fine-tuning the electrical characteristics of azulene derivatives. Azulene-based sensors are particularly well-suited for colorimetric and fluorometric detection techniques due to their frequent interactions with target analytes, which often result in notable changes in optical properties, such as bathochromic shifts in absorption or fluorescence spectra. The ongoing research and development of these sensors hold significant interest in environmental monitoring and public health, where the accurate detection of hazardous substances like heavy metals and anions is crucial.

AI disclosure statement: The author acknowledges the use of AI-based tools for minor language editing or grammatical correction. However, all intellectual contributions, analyses, and conclusions presented in this chapter are entirely the author's own.

References

1. Shoji, T., Ito, S., Yasunami, M., Synthesis of azulene derivatives from 2 H-cyclohepta [b] furan-2-ones as starting materials: Their reactivity and properties. *Int. J. Mol. Sci.*, 22, 19, 10686, 2021.
2. Razus, A.C., Azulene moiety as electron reservoir in positively charged systems; A short survey. *Symmetry*, 13, 4, 526, 2021.
3. Liu, R.S. and Asato, A.E., Tuning the color and excited state properties of the azulenic chromophore: NIR absorbing pigments and materials. *J. Photochem. Photobiol., C*, 4, 3, 179, 2003.
4. Arnold, G.L., Lazar, I.G., Buica, G.O., Ungureanu, E.M., Birzan, L., New azulene based modified electrodes for heavy metals sensing. *U.P.B. Sci. Bull. Ser. B*, 79, 2, 89, 2017.
5. Pop, M.D., Anastasoaie, V., Popescu, M.A., Oprisanu, A., Ungureanu, E.M., Cristea, M., New azulene modified electrodes for heavy metal sensing. *Rev. Chim.*, 68, 2172, 2017.

6. Buica, G.O., Ungureanu, E.M., Birzan, L., Razus, A.C., Bujduveanu, M.R., Films of poly (4-azulen-1-yl-2, 6-bis (2-thienyl) pyridine) for heavy metal ions complexation. *Electrochim. Acta*, 56, 14, 5028, 2011.
7. Buica, G.O., Ungureanu, E.M., Birzan, L., Razus, A.C., Mandoc, L.R., Voltammetric sensing of lead and cadmium using poly (4-azulen-1-yl-2, 6-bis (2-thienyl) pyridine) complexing films. *J. Electroanal. Chem.*, 693, 67, 2013.
8. Buica, G.O., Lazar, I.G., Birzan, L., Lete, C., Prodana, M., Enachescu, M., Tecuceanu, V., Stoian, A.B., Ungureanu, E.M., Azulene-ethylenediamine tetraacetic acid: A versatile molecule for colorimetric and electrochemical sensors for metal ions. *Electrochim. Acta*, 263, 382, 2018.
9. Enache, L.B., Anăstăsoaie, V., Birzan, L., Ungureanu, E.M., Diao, P., Enachescu, M., Polyazulene-based materials for heavy metal ion detection. 2.(E)-5-(azulen-1-yldiazenyl)-1H-tetrazole-modified electrodes for heavy metal sensing. *Coatings*, 10, 9, 869, 2020.
10. Oprisanu, A., Ungureanu, E.M., Isopescu, R., Birzan, L., Mihai, M., Vasiliu, C., Polyazulene based materials for heavy metal ions detection. *IOP Conf. Ser.: Mater. Sci. Eng.*, 213, 1, 012050, 2017.
11. Birzan, L., Cristea, M., Tecuceanu, V., Ungureanu, E.M., Razus, A.C., Syntheses of ligands based on azulene derivatives used as metal sensors. *Rev. Roum. Chim.*, 62, 6-7, 489, 2017.
12. Birzan, L., Cristea, M., Draghici, C.C., Tecuceanu, V., Maganu, M., Hanganu, A., Arnold, G.L., Ungureanu, E.M., Razus, A.C., 1-vinylazulenes–potential host molecules in ligands for metal ion detectors. *Tetrahedron*, 72, 18, 2316, 2016.
13. Buica, G.O., Birzan, L., Popescu, L.R., Ivanov, A.A., Ungureanu, E.M., Thermodynamics of interactions between lead (II) and cadmium (II) ions and azulene-based complexing polymer films. *J. Solid State Electr.*, 20, 401, 2016.
14. Law, AT. and Adeloju, S.B., Progress and recent advances in phosphate sensors: A review. *Talanta*, 114, 191, 2013.
15. Zielinski, T., Kędziorek, M., Jurczak, J., Bisamides derived from azulene-1,3- and -5,7-dicarboxylic acids as new building blocks for anion receptors. *Chem. Eur. J.*, 14, 838, 2008.
16. Lichosyt, D., Dydio, P., Jurczak, J., Azulene-based macrocyclic receptors for recognition and sensing of phosphate anions. *Chem. Eur. J.*, 22, 1, 2016.
17. Dhiman, S., Ahmad, M., Singla, N., Kumar, G., Singh, P., Luxami, V., Kaur, N., Kumar, S., Chemodosimeters for optical detection of fluoride anion. *Coord. Chem. Rev.*, 405, 213, 2020.
18. Fang, H., Gan, Y., Wang, S., Tao, T., A selective and colorimetric chemosensor for fluoride based on dimeric azulene boronate ester. *Inorg. Chem. Commun.*, 95, 17, 2018.

19. Liu, H., Ruan, K., Ma, K., Fei, J., Lin, Y., Xia, H., Synthesis of metalla-dual-azulenes with fluoride ion recognition properties. *Nat. Commun.*, 14, 5583, 2023.
20. Murfin, L.C., Chiang, K., Williams, G.T., Lyall, C.L., Jenkins, A.T.A., Wenk, J., James, T.D., Lewis, S.E., A colorimetric chemosensor based on a Nozoe azulene that detects fluoride in aqueous/alcoholic media. *Front. Chem.*, 8, 10, 2020.
21. Salman, H., Abraham, Y., Tal, S., Meltzman, S., Kapon, M., Tessler, N., Speiser, S., Eichen, Y., 1,3-Di(2-pyrrolyl)azulene: An efficient luminescent probe for fluoride. *Eur. J. Org. Chem.*, 11, 2207, 2005.
22. Son, Y.A., Gwon, S., Kim, S., Characteristics of guajazulene based chemosensor toward CN^- and F– anions. *Mol. Cryst. Liq. Cryst.*, 600, 89, 2014.
23. Sawicki, E., Stanley, T.W., Pfaff, J., D'Amico, A., Comparison of fifty-two spectrophotometric methods for the determination of nitrite. *Talanta*, 10, 641, 1963.
24. Garcia, E.E., Determination of nitrite ion using the reaction with p-nitroaniline and azulene. *Anal. Chem.*, 39, 1605, 1967.
25. Murfin, L.C., López-Alled, C.M., Sedgwick, A.C., Wenk, J., James, T.D., Lewis, S.E., A simple, azulene-based colorimetric probe for the detection of nitrite in water. *Front. Chem. Sci. Eng.*, 14, 90, 2020.
26. Kim, J.H., Park, K., Nam, H.Y., Lee, S., Kim, K., Kwon, I.C., Polymers for bioimaging. *Prog. Polym. Sci.*, 32, 8, 1031, 2007.
27. Zhou, Y., Zhuang, Y., Li, X., Ågren, H., Yu, L., Ding, J., Zhu, L., Selective dual-channel imaging on cyanostyryl-modified azulene systems with unimolecularly tunable visible-near infrared luminescence. *Chem. Eur. J.*, 23, 32, 7642, 2017.
28. Pham, W., Weissleder, R., Tung, C.H., An azulene dimer as a near-infrared quencher. *Angew. Chem. Int. Ed.*, 41, 3659, 2002.
29. Nolting, D.D., Nickels, M., Tantawy, M.N., Yu, J.Y.H., Xie, J., Peterson, T.E., Crews, B.C., Marnetta, L., Gore, J.C., Pham, W., Convergent synthesis and evaluation of ^{18}F-labeled azulenic COX2 probes for cancer imaging. *Front. Oncol.*, 2, 1, 2013.
30. Venanzi, M., Valeri, A., Palleschi, A., Stella, L., Moroder, L., Formaggio, F., Toniolo, C., Pispisa, B., Structural properties and photophysical Behavior of conformationally constrained hexapeptides functionalized with a new fluorescent analog of tryptophan and a nitroxide radical quencher. *Biopolymers*, 75, 128, 2004.

8

Exploring the Synthesis and Characteristics of Nanographenes and PAHs with Azulene Embedment

Dileep Kumar Singh[1] and Bijoy P. Mathew[2]*

[1]*Department of Chemistry, Bipin Bihari College, Jhansi, Uttar Pradesh, India*
[2]*Department of Chemistry, Vimala College (Autonomous), Thrissur, Kerala, India*

Abstract

Azulene, characterized as a non-alternant aromatic hydrocarbon, is an isomer of naphthalene, featuring a 10-π-electron system and a notable high dipole moment. An intriguing aspect is the electron-rich nature of the five-membered ring in azulene, in contrast to the electron-poor seven-membered ring. The unique donor-acceptor structure makes azulene well-suited for intermolecular charge transfer properties, contributing to its utility in the development of advanced optoelectronic materials. Moreover, when contiguous pentagonal and heptagonal rings are present, they induce structural defects within molecular systems, leading to various alterations such as twisting, helical arrangements, and wrapping phenomena. These structural perturbations significantly impact the electronic, magnetic, and overall structural properties of the fused molecules, introducing complexities in their behavior and functionality. Researchers have fused or connected azulene with various chromophoric moieties to explore their photophysical and biological attributes. Notably, a diverse range of large-π-conjugated ring systems, such as nanographenes, porphyrins, phthalocyanines, corroles, calixarenes, naphthalene and perylene diimides, polymers, BODIPY's, polycyclic aromatic hydrocarbons (PAHs), have been integrated with azulene for comprehensive property investigations. Applications of azulene-based molecules span a wide spectrum, encompassing IR absorption materials, organic solar cells, organic semiconductors, nonlinear optical materials, conducting polymers, and organic field-effect transistors. This chapter endeavors to offer a comprehensive overview encompassing the synthesis methodologies and diverse array of properties exhibited by expansive structures

**Corresponding author*: mathew.bijoyp@gmail.com

Neha Rani Kumar and Abhijeet R. Agrawal (eds.) Azulene-Based Materials: Design, Synthesis, and Applications, (199–244) © 2026 Scrivener Publishing LLC

integrating azulene such as nanographenes and PAHs. An investigation of these systems will provide a deeper understanding of their structural properties and functional capabilities, thereby shedding light on the significance of these systems in materials science and beyond.

***Keywords*:** Synthesis, azulene, nanographenes, polycyclic aromatic hydrocarbons (PAHs), wrapped, helical, twisted

8.1 Introduction

Over the past century, chemists and material scientists alike have been fascinated by the unique structural and photophysical properties of azulene, a captivating blue pigment named because of its color [1]. As a product of the dehydrogenation of sesquiterpenoids, it is characterized by a unique bicyclic ring system, a fused five- and seven-membered ring that places it at the forefront of non-benzenoid aromatic chemistry [2, 3]. Azulene is known for its remarkable aromatic stability and dipole moment of 1.08 D [4] despite being an isomer of naphthalene since Pfau and Plattner [5] discovered its fundamental structure in 1936 (Figure 8.1), revealing an unsaturated bicyclic ring system with fused five- and seven-membered rings.

Azulene differs from conventional fused benzenoids due to its electron-rich five-membered and electron-poor seven-membered rings. As a result of these unique properties, azulene has made a significant contribution to the development of advanced optoelectronic materials. The derivatives of azulene have been utilized to create chemical switches [6], anion-based sensors [7], liquid crystals [8, 9], electrochromic materials [10, 11] organic and polymeric conductors [12], conducting charge-transfer complexes [13],

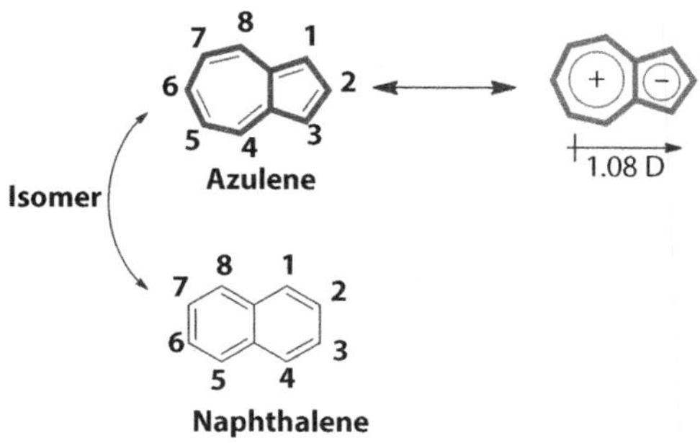

Figure 8.1 Azulene and naphthalene isomers and the polarized resonating structure of azulene.

as well as materials with near-infrared (NIR) resonance [14–16]. These diverse applications stem from azulene's ability to violate Kasha's rule with its robust $S_0 \rightarrow S_2$ absorption, attracting significant attention for its anti-Kasha-rule emissions [17, 18].

Despite azulene's promising properties, the synthesis and design of compounds based on it are historically challenging, particularly when it comes to achieving specific substitution patterns or incorporating the compound into larger molecular frameworks. Nevertheless, recent developments in synthetic strategies have led to the creation of more complex azulene derivatives, which have expanded the use of these materials in high-performance applications. Theoretical calculations, such as those from the Pariser–Parr–Pople and Heisenberg models, indicate that azulene-fused systems are capable of exhibiting intriguing multiferroic properties at ambient conditions [19], further enhancing their appeal [20].

One of the most exciting developments in azulene chemistry is the integration of azulene into polycyclic aromatic hydrocarbons (PAHs) and nanographenes [21, 22]. These azulene-embedded structures feature non-alternating topologies that contrast with the alternating topologies of traditional PAHs, leading to unique electronic characteristics. Additionally, when non-hexagonal polygons like pentagons and heptagons are incorporated into nanographene, geometries such as bowl [23], saddle [24], and tub [25] shapes can be formed, each with a distinct electronic structure. These structures have been extensively investigated for their potential in developing new functional materials.

The synthesis of azulene-embedded PAHs and nanographenes opens new avenues for creating materials with tailored properties for specific applications. These materials hold promise for use in high-entropy species of non-six-membered carbon rings-based aromatic molecules [26, 27], polymers, frameworks, and functional carbon materials. However, the development of these advanced materials is still constrained by synthetic challenges, particularly in achieving precise control over molecular design and functionality.

This chapter explores the synthesis and characteristics of azulene-embedded PAHs and nanographenes, highlighting the innovative strategies employed to overcome synthetic difficulties and the potential applications of these novel materials. By delving into the unique structural, electronic, and photophysical properties of azulene-embedded frameworks, this chapter aims to shed light on the future directions and possibilities in advanced organic materials.

8.2 Azulene-Based Polycyclic Aromatic Hydrocarbon

As a nonalternant hydrocarbon with distinct physicochemical properties, azulene acts as a foundation for creating large π-conjugated polycyclic organic materials. This section discusses the synthesis and characteristics of various azulene-based PAHs. The azulene group has been combined with different π-conjugated groups, including acenes, indacenes, helicenes, naphthalene, phenanthrene, and other heterocyclic rings, to form an extensive π-conjugated framework. The synthesis, characterization, and properties of these compounds are covered.

Chiral organic π-conjugated molecules have been extensively studied in recent years to create chiral electronic materials and devices. Chaolumen *et al.* [28] reported the synthesis, as well as the properties of axially chiral 1,1′-biazulene **8.4** and its π-extended analogue **8.6**. Compound **8.3** was obtained in approximately 57% yield by Suzuki–Miyaura cross-coupling reaction between azulen-2-ylboronic acid pinacol ester and 2,6-diphenyl-1-iodobenzene (Scheme 8.1). To prepare 1,1′-biazulenes **8.4**, azulene **8.3** was treated with $FeCl_3$ for oxidative dehydrogenation and induced intermolecular C-H/-C-H coupling, resulting in 1,1′-biazulene **8.4** in 88% yield and cyclodehydrogenated product **8.5** in 8%. When compound **8.4** was treated again under the same oxidative conditions, another π-extended 1,1′-biazulene derivative **8.6** was obtained in 33% yield *via* intramolecular cyclodehydrogenation. All attempts to produce

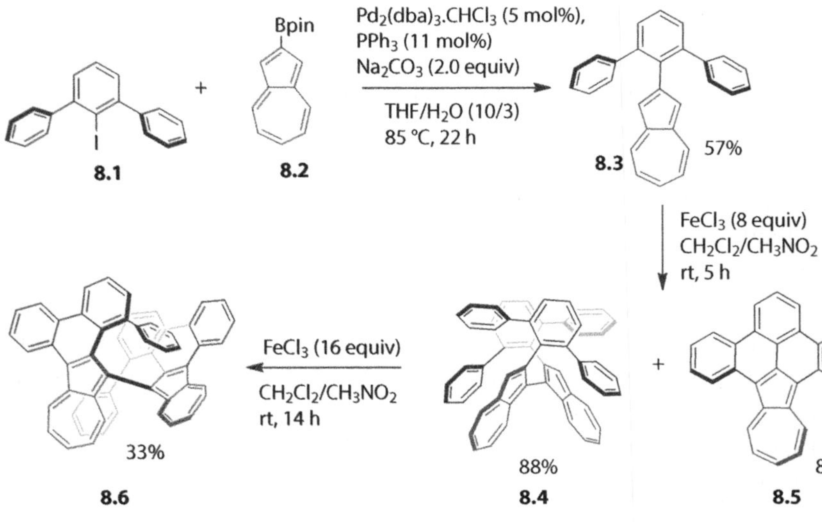

Scheme 8.1 Synthesis of 1,1′-biazulene **8.4** and its analog **8.6**.

compound **8.6** from **8.3** resulted in only trace amounts of **8.6**. This is most likely because intermediate **8.4** is rapidly protonated by *in situ*-formed HCl under oxidative conditions, preventing the subsequent oxidation of **8.4** from undergoing intramolecular cyclization reaction. X-ray diffraction studies confirmed the structures of compounds **8.4** and **8.6**, as well as the presence of two enantiomers in their crystal structures. Chiral HPLC accomplished the separation of these two enantiomers. They also displayed mirror-image CD spectra up to 800 nm and 900 nm, respectively. Furthermore, cyclic voltammetry (CV) analysis revealed that the compound **8.6** exhibited a high-lying highest occupied molecular orbital (HOMO) by the electron-rich five-membered ring of the azulene group, which shows the longest absorption band at 683 nm, tailing up to 900 nm. Thus, all the photophysical and electronic properties of compound **8.6** suggest that it can be a potential candidate in optoelectronic devices.

Tsuchiya *et al.* [29] described the synthesis and properties of 1,2-connected azulene oligomers and azulene-embedded polycyclic hydrocarbons (PHs) (Scheme 8.2). To synthesize twisted terazulene **8.8** with a 61% yield, 1,2'-biazulene **8.7** was first treated with one equivalent of N-bromosuccinimide (NBS) to brominate the 1-position of azulene, which was then coupled with 2-azulenylboronate (**8.2**) *via* a Pd-catalyzed Suzuki–Miyaura cross-coupling reaction. A similar reaction was repeated with terazulene **8.8** to produce quaterazulene **8.9** in 8% yield, alongside the formation of ring-closed trimer **8.10** in 36% yield. The formation of fused

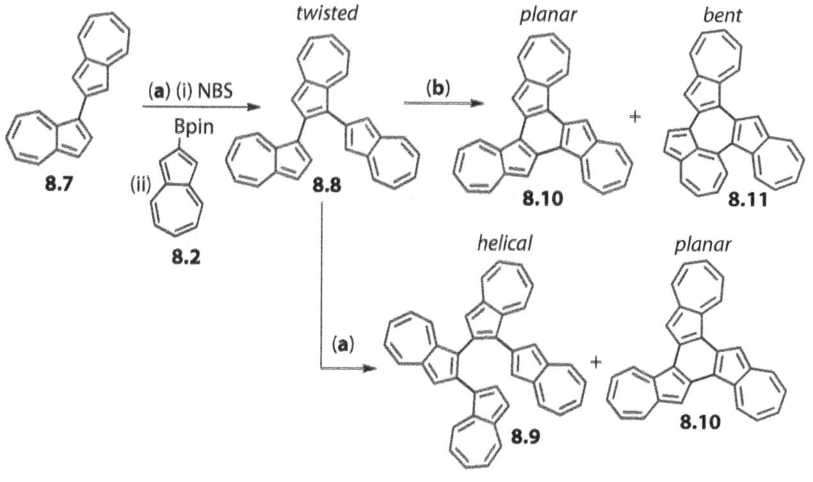

Scheme 8.2 Synthesis of 1,2-position-connected azulene oligomers **8.8–8.11**; Reagent and conditions; (a) (i) NBS, CH_2Cl_2, -78°C, 3 h, (ii) $Pd(OAc)_2$, SPhos, K_3PO_4, DME/H_2O (10:1), reflux; (b) (i) NBS, CH_2Cl_2, -78°C, 3 h, (ii) $Pd(dba)_2$, ligand, Cs_2CO_3, toluene, reflux.

terazulene **8.10** was likely due to the intramolecular C-H/C-Br arylation facilitated by the Pd-catalyst, and the yield of this product depended on the phosphine ligand employed (Table 8.1). Terazulene **8.10** was obtained in 69% yield using 2-dicyclohexylphosphino-2',6'-dimethoxybiphenyl (SPhos) as the ligand, whereas using 1,1'-bis(diphenylphosphino)ferrocene (DPPF) or PPh_3 resulted in the production of compound **8.11** in 18% and 12% yields, respectively. In addition to nuclear magnetic resonance (NMR) and other spectroscopic techniques, compounds **8.8, 8.10,** and **8.11** were characterized using single-crystal X-ray crystallography. The X-ray structure analysis revealed that compound **8.8** has a twisted structure, while 1,2"-closed terazulene **8.10** adopts a planar structure. In contrast, 1,8"-closed terazulene **8.11**, analyzed in a C_{60} co-crystal, exhibits a curved structure. Furthermore, nucleus-independent chemical shift (NICS) calculations for compound **8.11** yielded a positive value, indicating anti-aromatic properties. This characteristic, along with its smaller HOMO–LUMO gap, also known as its highest occupied molecular orbital and lowest unoccupied molecular orbital gap, and higher solubility compared to compound **8.10**, positions compound **8.11** as a promising candidate for designing and developing optical devices, catalysts, and functional electronic materials.

Table 8.1 Synthesis of compounds **8.10** and **8.11** using different phosphine ligands.

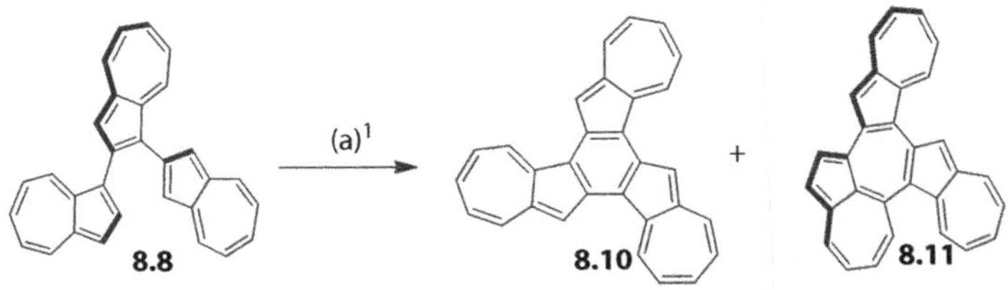

Ligands	Yield (%) Compound 8.10	Compound 8.11
DPPF	28	18
PPh_3	36	12
SPhos	69	-
XPhos	22	-
^{t}Bu-XPhos	16	-

[1]Reagent and conditions: (a) (i) NBS, CH_2Cl_2, −78°C, 3 h, (ii) $Pd(dba)_2$, ligand, Cs_2CO_3, toluene, reflux, 24 h

In recent years, non-benzenoid non-alternant PAHs have become an important topic of research owing to their intriguing optoelectronic properties. Liu *et al.* [30] reported the synthesis and properties of 18 examples of azulene-embedded PAHs of the type **8.15** using an efficient two-step synthetic protocol. These PAHs include furan, thiophene, and benzene condensed compounds, Z- and butterfly-shaped large compounds, and double [5]helicene structures containing two azulene units. In the first step, a series of structurally different building blocks **8.14** with various Ar^1 and Ar^2 substituents have been synthesized in 57–93% yields by using Pd-catalyzed Suzuki coupling reaction of diverse indene-based aryl boronic ester with the commercially available aryl halides containing an aldehyde group. In the second step, compound **8.14** underwent a Knoevenagel-type condensation in the presence of ^{t}BuOK under THF, from zero to room temperature, to produce compound **8.15** with yields up to 96% (Scheme 8.3). After synthesizing and characterizing the products, they were examined for aromaticity, structural topology, and photophysical properties using NMR, UV/Vis absorption spectroscopy, X-ray crystallographic analysis, and density functional theory (DFT) calculations. The UV/Vis

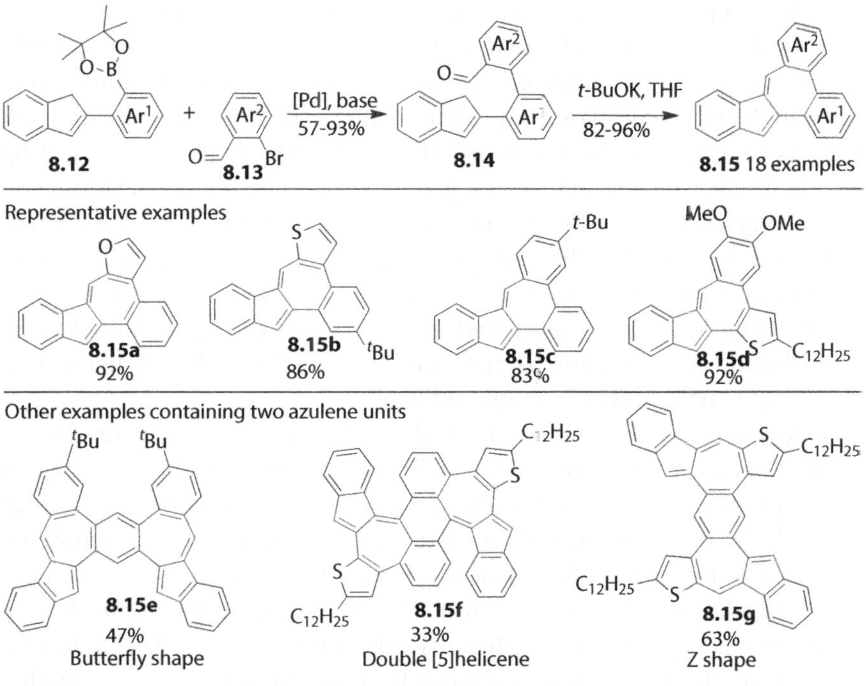

Scheme 8.3 Synthesis of non-alternant PAHs (**8.15**) containing one and two azulene units.

absorption spectra revealed that when the Ar^1 and Ar^2 parts are benzene rings, an intense absorption occurs between 300–410 nm with a broad λ_{max} at 449 nm. In contrast, when the Ar^1 and Ar^2 parts are furan or thiophene rings, the absorption maximum is red-shifted and found approximately at 500 nm, due to the electron-rich nature of the furan and thiophene rings. Anisotropy of the induced current density and NICS calculations suggested that the seven-membered rings in most compounds are anti-aromatic, while the five-membered rings are aromatic. It was concluded that when a thiophene ring is fused with an azulene core, it impacts the reactivity of the condensation protocol and enhances the π-conjugation and stability of large non-alternant PAHs.

Uehara *et al.* [31] synthesized a 1,8-naphthylene-bridged spiro-azulene dimer and a helicene-like structure using the Suzuki–Miyaura cross-coupling reaction between 1,8-dibromonaphthalene (**8.16**) and **8.2** in a dioxane-water mixture. However, instead of obtaining the desired product **8.17**, the reaction formed **8.18a** and **8.18b** with yields of 4% and 25%, respectively, due to an unusual rearrangement and oxidative fusion reaction. The structures of all these products were confirmed through X-ray crystallography. The UV-vis-NIR absorption spectra of both dimers were examined in dichloromethane (DCM). The spiro-azulene dimer **8.18a** showed a very broad absorption ranging from 750–500 nm, with intense peaks at 435, 414, and 342 nm. In contrast, the helical azulene dimer **8.18b**, due to a strong π-conjugation, displayed NIR absorption extending up to 900 nm. Notably, compared to pristine azulene, the molar absorption coefficients of dimers **8.18a** and **8.18b** are significantly increased, indicating that their electronic structures are greatly affected by the conjugation. The resonating structure of dimer **8.18b** is illustrated in Scheme 8.4. Resonating structure **A** highlights the cyclopentadienyl anion character of the five-membered ring, which bears five π-electrons on the C1-C2 and C1'-C2' bonds. Consequently, the observed antiaromaticity can be attributed to the contribution of the 8π-antiaromatic circuit (depicted as **B** in Scheme 8.4), which is formally a cycloheptatrienyl anion.

Due to their narrow energy gaps and potential applications in developing various optical and electronic materials, diverse antiaromatic π-conjugated PHs have been investigated recently. Among these, Yamamoto *et al.* [32] utilized the olefin metathesis reaction to synthesize a seven-membered ring containing polycyclic hydrocarbon **8.24**, which features a fused dicyclohepta[*cd,ij*]-s-indacene framework. Synthesizing compound **8.24** from **8.19** was accomplished in five steps, as illustrated in Scheme 8.5. In the first step, the di-allylated product **8.20** was formed with a 65% yield through a Pd-catalyzed Suzuki coupling reaction of the starting compound **8.19** with

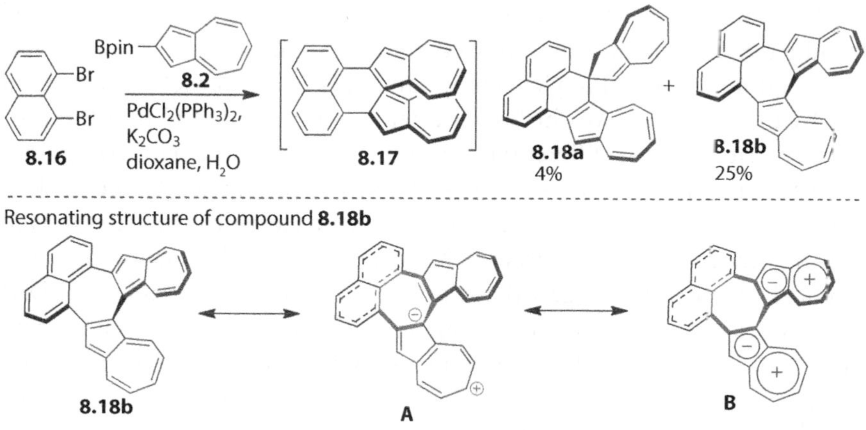

Scheme 8.4 Synthesis of 1,8-naphthalene-bridged azulene dimers **8.18a-b** and the resonating structure of **8.18b**.

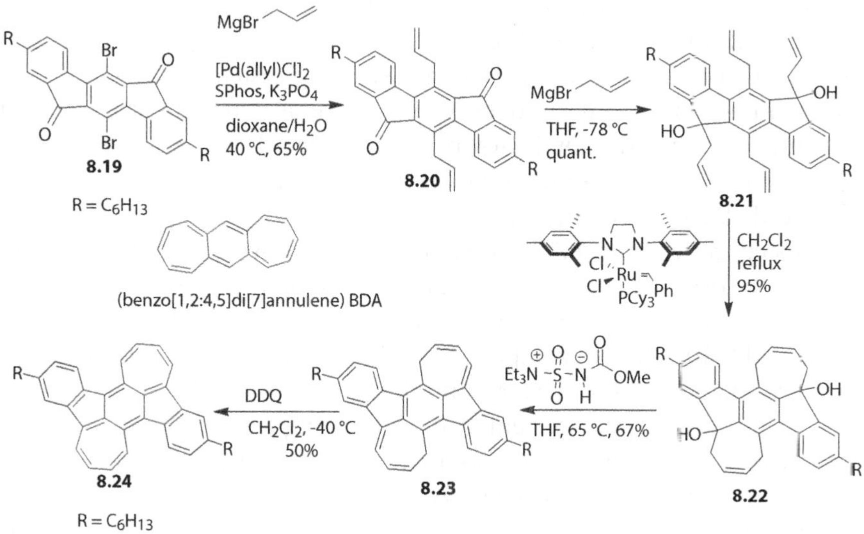

Scheme 8.5 Synthesis of dicyclohepta[*cd,ij*]-*s*-indacene PAH **8.24**.

allylboronic acid pinacol ester. In the subsequent step, **8.20** was treated with allylmagnesium bromide, resulting in the formation of the *tetra*-allylated compound **8.21** with an almost quantitative yield. The key intermediate **8.22**, which contains seven-membered rings, was then obtained *via* an olefin metathesis reaction of compound **8.21** using the second-generation Grubbs' catalyst, giving a yield of 95%. Finally, the treatment of intermediate **8.22** with Burgess reagent produced the dehydrated product **8.23**, which was oxidized with 2,3-Dichloro-5,6-dicyano-1,4-benzoquinone (DDQ) to yield compound **8.24** in a 50% yield. Compound **8.24** exhibited a 5% weight loss at a temperature of 443°C in a thermogravimetric analysis. Differential scanning calorimetry data for compound **8.24** revealed an endothermic peak at 234°C and an exothermic peak at 250°C during the first heating process, while no clear peaks were observed during the second heating process. X-ray crystallography analysis and theoretical calculations suggest that the Benzo[1,2:4,5]di[7]annulene (BDA) framework within the molecule significantly contributes to its antiaromatic character. Given that compound **8.24** demonstrated hole-transporting properties, improvements in molecular design could further aid in the development of high-performance electronic materials.

Among the various benzenoid PHs, acenes stand out as significant and promising candidates for semiconducting materials in organic electronic devices. Jiang *et al.* [33] reported the synthesis and the photophysical properties of four derivatives of the class of two azulene-fused s-indacene isomers (**8.31a-b, 8.32a-b**). These compounds feature a 28π-electron system in their backbone, comprising a fused ring system of 7-5-5-6-5-5-7. The synthesis of these derivatives involved several steps. Initially, a Suzuki coupling reaction was conducted between azulen-2-yl-boronic acid pinacol ester (**8.2**) and either dibromo compound **8.25** or **8.26**. This reaction yielded dialdehyde intermediates **8.27** and **8.28**, respectively. In the subsequent step, compounds **8.27** and **8.28** were treated with an anthryl lithium reagent to form the corresponding diols. These diols underwent $BF_3 \cdot OEt_2$-catalyzed Friedel–Crafts alkylation reactions to produce dihydro-products **8.29a** and **8.30a**. Finally, compounds **8.29a** and **8.30a** underwent oxidative dehydrogenation to give the target products **8.31a** and **8.32a** with 54% and 57% yields, respectively (Scheme 8.6). Interestingly, these products exhibited excellent photostability at room temperature, but proved to be highly sensitive to silica gel under light. To address this issue, compounds **8.29a** and **8.30a** were converted to compounds **8.29b** and **8.30b**, respectively, by treatment with trifluoroacetic anhydride. Subsequently, compounds **8.29b** and **8.30b** were dehydrogenated to yield compounds **8.31b** and **8.32b**. An X-ray crystallographic analysis of compounds **8.31b** and **8.32b** confirmed their nearly planar backbones, with the anthryl groups positioned almost perpendicular to the backbone. Importantly, in both synthesized products (**8.31a-b, 8.32a-b**), the aromatic character of the azulene unit and the anti-aromatic character of the s-indacene unit were preserved. These properties also imparted the compounds with open-shell diradical character, and their magnetic properties depended on the fusion mode and substituents.

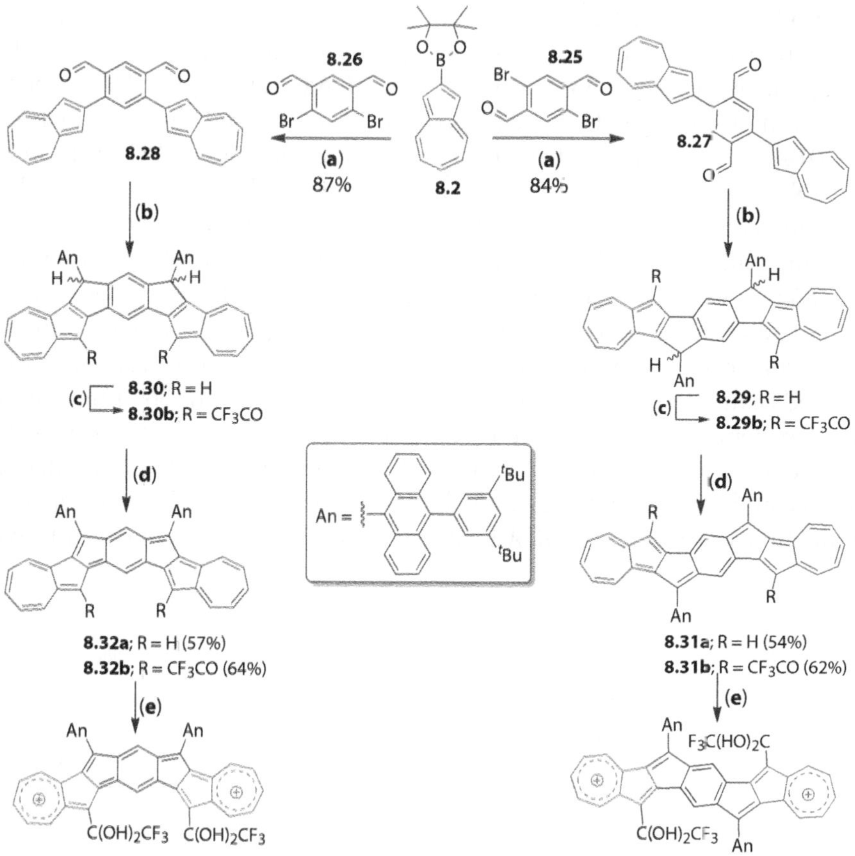

Scheme 8.6 Synthesis of azulene-fused s-indacenes **8.31a-b** and **8.32a-b**; Reagents and conditions: (a) $Pd(PPh_3)_4$, K_2CO_3, THF/H_2O (5:1), reflux; (b) (i) Anthryl lithium reagent, dry THF, rt (ii) $BF_3{\cdot}OEt_2$, dry DCM, rt; (c) trifluoroacetic anhydride, rt.; (d) DDQ, dry toluene, rt.; (e) $NO{\cdot}SbF_6$.

Acenes containing PAHs have shown significant potential in organic electronics in recent years due to their excellent photoelectronic properties. Wang *et al.* [34] reported the efficient synthesis and photophysical properties of four nonalternant isomers of oligoacenes fused with multiple azulene moieties using a two-step [4 +3] annulation protocol. These four products include one nonalternant isomer of anthracene, dibenzo[e,g]azulene (**DBA**), two pentacene, benzo[1,2-f:5,4-f']diazulene (**BDA1**), benzo[1,2-f:4,5-f']diazulene (**BDA2**), and one hexacene, naphtho[2,3-f:6,7-f']diazulene (**NDA**). In the first step of the synthesis, an easily available substrate 1,2,3,4-tetraarylcyclopentadienes (**8.33**) was treated with *o*-haloaryl aldehydes **8.13, 8.25, 8.26,** and **8.34** in the presence of NaOMe under methanol at reflux conditions to produce products **8.35–8.38** in 20–49% yields after the dehydration condensation process. In the second step, Pd-catalyzed intramolecular C-H coupling of products **8.35–8.38** gave **DBA**, **BDA1**, **BDA2**, and **NDA**, respectively, in 40–70% yields (Scheme 8.7). The X-ray crystallographic analysis of **DBA** reveals a saddle-shaped conformation. Compounds **BDA1**, **BDA2**, and **NDA** displayed slightly S-shaped conformations due to their fused bulky 7-membered rings. The UV/Vis absorption spectrum of **DBA** showed λ_{max} at 400 nm with a broad shoulder at 523 nm, while λ_{max} of **BDA1** was observed at 483 nm, which is even more red-shifted than **NDA** (472 nm). Benzo[1,2-f:4,5-f']diazulene has a λ_{max} of 405 nm despite being highly conjugated with two azulene units. This is likely due to its twisted conformation. All four compounds exhibited negligible emission when irradiated with 365 nm. When electrochemical properties are investigated, the CV curve of **DBA** reveals only one reversible oxidation and one reduction wave. On the other hand, **BDA1** and **NDA** show two reversible oxidation waves and one reversible reduction wave, whereas **BDA2** exhibits two reversible oxidation and two

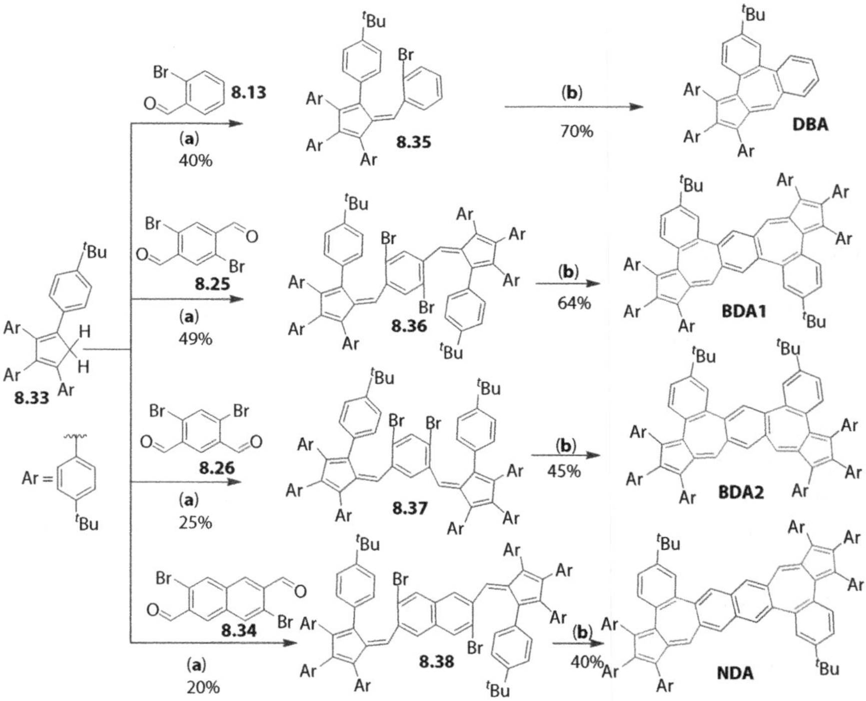

Scheme 8.7 Syntheses of azulene-fused PAHs – **DBA**, **BDA1**, **BDA2**, and **NDA**; Reagents and conditions: (a) NaOMe, MeOH, reflux; (b) $Pd(PPh_3)_4$, Cs_2CO_3, toluene, reflux.

reduction waves. Similar to azulene, these four nonalternant isomers had a narrow HOMO–LUMO band gap. It was discovered that the middle portion of PAHs can be significantly stabilized by efficient electron density dispersal caused by the fusion of electron-poor seven-membered rings. These synthetic strategies can produce a diverse range of multiazulene-fused PAHs.

Ninomiya *et al.* [35] prepared 1,1'-bi(2-pyridylazulene) (**8.41**), as well as its protonated structure by employing an aryl-aryl coupling strategy. The synthesis of compound **8.41** beginning from azulene is depicted in Scheme 8.8. To obtain the protonated structure of compound **8.41**, a toluene solution of **8.41** was treated with an aqueous solution of HBF_4, resulting in blue single crystals after recrystallization from a DCM/hexane (2:3) mixture. After successful synthesis, an X-ray diffraction analysis revealed the presence of an intramolecular hydrogen bond between the two pyridyl groups in compound **8.41**, which exists as a racemic mixture of axial chiral species. Compound **8.41** was further investigated for its acid-base characteristics, revealing a two-step protonation process of its pyridyl moieties. During protonation, two energy shifts related to intramolecular transitions were observed. The first protonation induced a blue shift in the absorption maximum, while the second protonation caused a red shift. The unusual blue shift was attributed to the stabilization of the HOMO due to an extended electronic structure between the two azulene skeletons.

Murai *et al.* [36] synthesized azulene-fused PAHs **8.43–8.45** from commercially available azulene in four steps (Scheme 8.9). At first, Ir-catalyzed borylation of azulene was done to produce 2-boryazulene **8.2** in 52% yields. The Suzuki–Miyaura cross-coupling reaction of borylazulene **8.2** with 2-bromobenzaldehyde in the presence of $Pd(PPh_3)_4$ and Na_2CO_3 and subsequent Wittig reaction gives methyl vinyl ether **8.42** in 85% yields

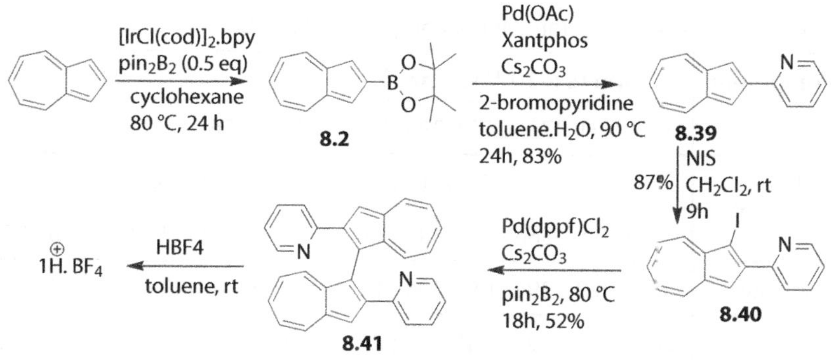

Scheme 8.8 Synthesis of 1,1'-Bi(2-pyridylazulene) (**8.41**).

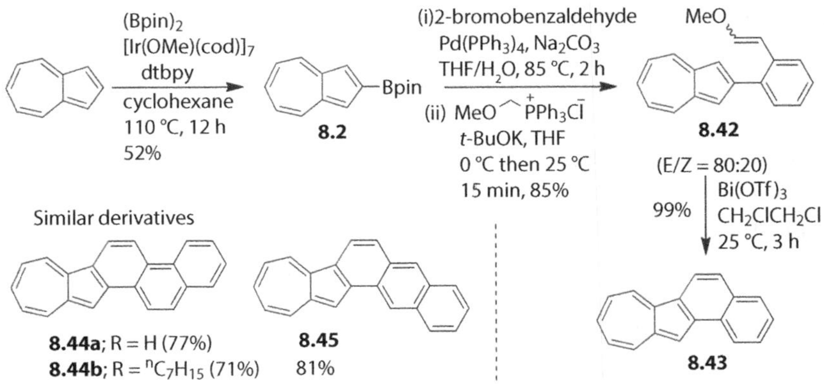

Scheme 8.9 Synthesis of azulene-fused PAHs **8.43–8.45**.

(E/Z = 80:20). Finally, naphtha[2,1-a]azulene **8.43** was synthesized in 99% yields by bismuth-catalyzed cyclization and aromatization of compound **8.42**. By following this procedure, other derivatives **8.44a, 8.44b**, and **8.45** were also prepared in 71–81% yields. After successfully synthesizing these azulene-based PAHs, their photophysical properties were thoroughly evaluated. The UV/Vis absorption spectra of compounds **8.43–8.45**, along with similar 18π-conjugated systems like chrysene ([4]phenacene) and tetracene ([4]acene), are summarized in Table 8.2. Compound **8.43** exhibited three main absorption peaks at 326 nm, 375 nm, and 396 nm, which show a bathochromic shift when compared to the parent azulene (269 nm, 321 nm, and 343 nm). The $S_0 \rightarrow S_1$ transition also resulted in a weak absorption at 627 nm and contributed to its blue color. These synthesized azulene conjugates demonstrate a narrow HOMO–LUMO bandgap, high air stability, excellent photooxidative resistance, and improved solubility.

Table 8.2 UV/Vis absorption data of compounds **8.43-8.45**.

Compound	λ_{abs}/nm ($\varepsilon \times 10^{-4}$/$M^{-1}$ cm^{-1})[a]
8.43	326 (6.3), 375 (0.75), 396 (0.76), 627 (0.036)
8.44a	335 (8.2), 379 (1.3), 401 (1.2), 418 (0.49), 628 (0.049)
8.45	341 (10.8), 366 (4.7), 387 (4.6), 402 (2.6), 424 (1.6), 646 (0.055)
tetracene	276 (18.9), 294 (1.9), 396 (0.16), 417 (0.34), 443 (0.65), 473 (0.69)
chrysene	261 (6.7), 269 (11.4), 297 (1.2), 306 (1.3), 321 (1.2)

[a] In DCM (1×10^{-4} M)

In a similar approach, Murai *et al.* [37] extended their previous report and described the synthesis of azulene-fused linearly π-conjugated PAHs (**8.51**) (Scheme 8.10). To synthesize the target molecule, azulene **8.46** was used as a building block. To increase the solubility of the final product, a *n*-heptyl group was introduced at the 6-position of starting compound **8.46**. Initially, azulene **8.46** was exposed to an iridium-catalyzed borylation reaction through its C-H bond and produced 2-borylazulene (**8.47**) with 54% yield and 1-borylazulene in 29% yield. 2-Borylazulene was then coupled with 2-Bromoazulene *via* Suzuki–Miyaura cross-coupling reaction to give the compound in 95% yield, which on subsequent formylation under Vilsmeier reaction results in 1,1'-diformyl-2,2'-biazulene (**8.49**) in 75% yield. Finally, reacting compound **8.49** with *p*-toluenesulfonyl hydrazine, subsequent Rh-catalyzed cyclization, and N_2 elimination produced target compound **8.51** in 76% yield. In this synthesis, after the fusion of azulene, the HOMO–LUMO gap decreased, making compound **8.51** highly stable in air and light. The absorption profile of compound **8.51** revealed the absorption maxima to be at 375 nm with a broad band between 400–450 nm, which are red-shifted compared to previously synthesized compounds **8.43–8.45** (Scheme 8.9). Among all these π-extended molecules, phenanthrene-fused derivative **8.44a** demonstrated the highest hole mobility 5.3 $\times$ 10^{-2} $cm^2V^{-1}s^{-1}$ in the thin film, which is enough to show a probable candidate for p-type semiconductors. Based on these findings, it is possible that adding a functional handle to carbo- and heterocycles containing azulene would result in more desirable electronic and photophysical characteristics.

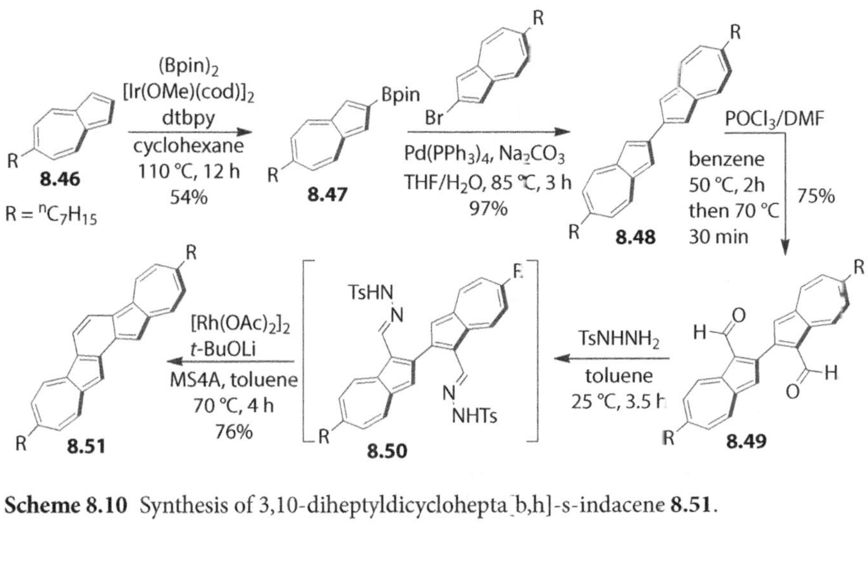

Scheme 8.10 Synthesis of 3,10-diheptyldicyclohepta[b,h]-s-indacene **8.51**.

Konishi *et al.* [38] described a straightforward synthesis of azulene-embedded PAH **8.54** using silver(I) cation and explored its optoelectronic properties. The process involved a three-fold transannulation reaction between acetylene and aromatic units of **8.52** catalyzed by $AgBF_4$ in chlorobenzene, yielding azulenium cation **8.53** with a 78% yield, which on subsequent treatment with Et_3N produced compound **8.54** with a 77% yield. A plausible mechanism for the transannulation of compound **8.52** to compound **8.53** is illustrated in Scheme 8.11. Initially, a silver(I) ion facilitates the single-electron oxidation of the *ortho*-phenylene ethynylene units in compound **8.52**, generating cation radical **A**. The electron transfer from the acetylene unit to the silver (I) ion is assisted by the weak coordination of the Ag (I) in **A**. Subsequent cyclization occurs in tandem, progressing through intermediates **B** to **E**, ultimately yielding the neutral azulene-based PAH **8.54.** The series of reactions proceed by electrophilic and radical pathways. The *in situ*-generated HBF_4 leads to the protonation of **8.54** to produce the isolable azulenium cation **8.53**. Electrochemical insights into compound **8.54** revealed its amphoteric redox behavior in the solution phase and weak absorption profile in higher wavelength regions. These are the characteristics of non-alternant hydrocarbons originating from azulene moiety, and thus, compound **8.54** can be used for application in optoelectronic devices.

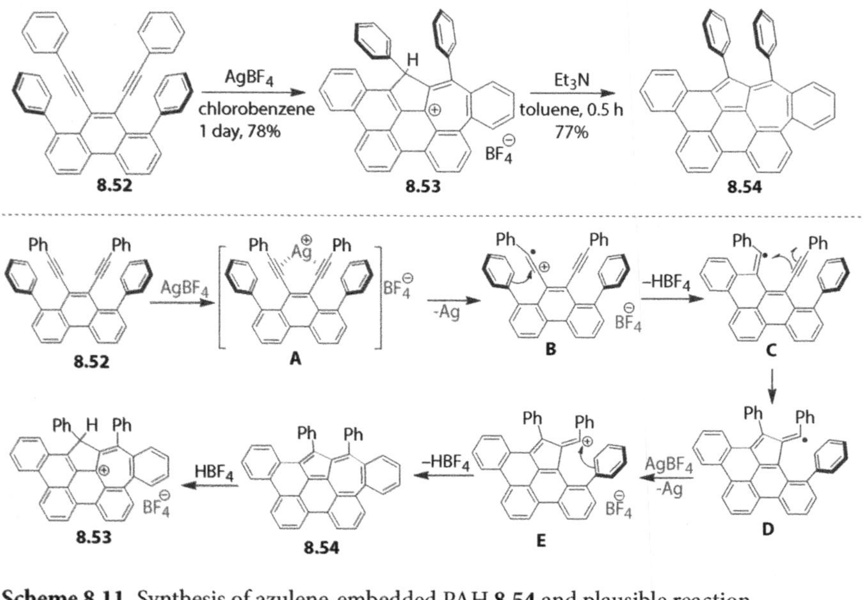

Scheme 8.11 Synthesis of azulene-embedded PAH **8.54** and plausible reaction mechanism for the conversion of **8.52** to **8.54** and **8.53**.

Matsuda *et al.* [39] synthesized various azulenophenanthrene derivatives **8.56a–e** through a platinum (II)-catalyzed skeletal rearrangement of 2,2'-di(arylethynyl)biphenyls *via* C–C bond cleavage of the benzene ring (Scheme 8.12). To synthesize compound **8.56a**, first, 2,2'-bis[(4-chlorophenyl)ethynyl]biphenyl (**8.55**) was heated in *p*-xylene at 120°C in the presence of platinum (II) catalyst generated *in situ* from [PtCl(C_6F_5)(cod)] and P(OCH_2CF_3)$_3$. The reaction yielded **8.56a** in 55% yield. Another polycyclic aromatic compound, **8.57**, formed as a minor product, yielded 20%. Compound **8.57** (benzo[*f*]tetraphene) was identified as an intramolecular dehydro-Diels–Alder product. Similarly, other azulenophenanthrenes **8.56b–e** were produced as major products in 45–78% yields.

The proposed mechanism for the formation of compounds **8.56a** and **8.57** from 2,2'-bis[(4-chlorophenyl)ethynyl]biphenyl (**8.55**) is illustrated in Scheme 8.13. Initially, the alkyne moiety in compound **8.55** serves as a π-base and coordinates with the π-acidic Pt(II) metal center, initiating a 6-exo-dig cyclization reaction with the adjacent alkyne group. This process forms a vinylic cation intermediate **A**. Subsequently, intermediate **A** undergoes intramolecular attack by the 4-chlorophenyl ring at two possible positions: at the ipso-carbon atom, forming a spirocyclic intermediate **B** that leads to the major product **8.56a**; or at the *ortho*-position, resulting in intermediate **E**, which ultimately provides the minor product **8.57**. The cyclopentadienyl group in intermediate **B** reacts with the cyclohexadienyl cation to form intermediate **C**, a norcaradiene. Subsequently, the

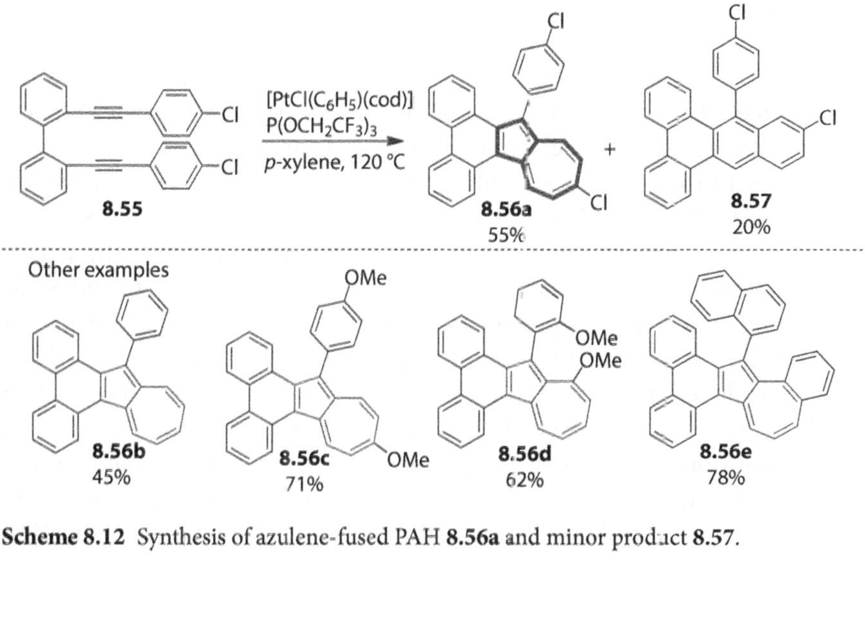

Scheme 8.12 Synthesis of azulene-fused PAH **8.56a** and minor product **8.57**.

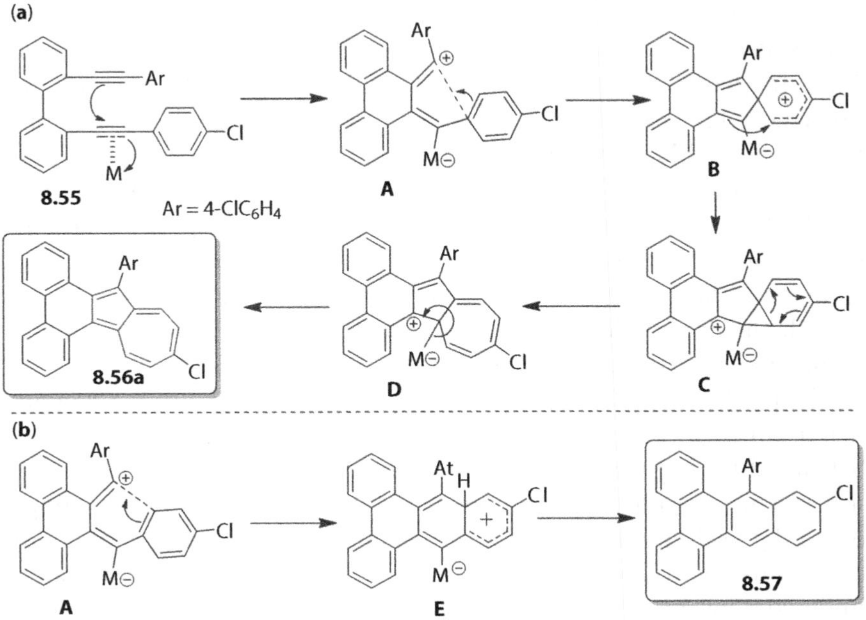

Scheme 8.13 Plausible mechanism for the formation of **8.56a** and **8.57**.

norcaradiene undergoes a Cope rearrangement, opening its six-membered ring to yield cycloheptatriene **D**. The demetalation of **D** introduces aromaticity, forming the azulene moiety and producing compound **8.56a** (Scheme 8.13a). In contrast, intermediate **E** produces the minor product, dehydro-Diels–Alder adduct **8.57**, upon the rearomatization and demetalation (Scheme 8.13b).

Wu *et al.* [40] synthesized three open-shell non-alternant PHs (compounds **8.63**, **8.67**, and **8.71**) based on the cyclohepta-[*def*]fluorine skeleton using a strategy that efficiently stabilizes them through the fusion of benzene rings to their periphery. Initially, they prepared three key dialdehyde intermediates (**8.61**, **8.65**, and **8.69**) through various Pd-catalyzed Suzuki–Miyaura couplings using different halides (**8.58**, **8.64**, and **8.68**). These dialdehydes were then converted into dihydroprecursors (**8.62**, **8.66**, and **8.70**) containing a pentagon-heptagon pair. Finally, oxidative dehydrogenation of these dihydroprecursors using DDQ resulted in the formation of the desired azulene-based products (**8.63**, **8.67**, and **8.71**) in nearly quantitative yields (Scheme 8.14). The synthesized products were characterized using a high-resolution matrix-assisted laser desorption/ionization time-of-flight mass spectrometry analysis and UV/Vis-NIR spectroscopy. The electronic absorption spectra exhibited a red-shifted absorption pattern, indicating

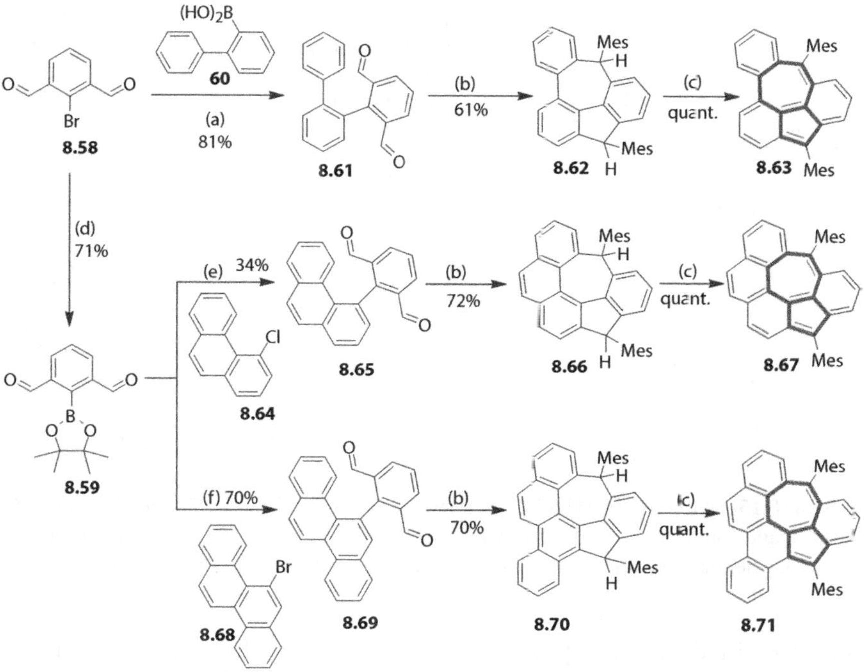

Scheme 8.14 Synthesis of non-alternant PAH **8.63**, **8.67**, and **8.71**; Reagent and conditions: (a) $Pd(PPh_3)_4$, K_2CO_3, dioxane/H_2O, 95°C, 36 h (b) (i) 2-mesitylmagnesium bromide, THF, rt, overnight; (ii) $BF_3{\cdot}OEt_2$, CH_2Cl_2, rt, 40 min; (c) DDQ, chlorobenzene for **8.63** and toluene for **8.67** and **8.71**, rt, 30 min (d) B_2Pin_2, $Pd(PPh_3)_2Cl_2$, KOAc, 1,4-dioxane, 90°C, overnight, (e) $Pd(OAc)_2$, SPhos, K_3PO_4, toluene/EtOH/H_2O, 80°C, 40 h, (f) $Pd_2(dba)_3$, SPhos, K_3PO_4, toluene/THF, 100°C, 36 h.

an extremely narrow optical energy gap of 0.54 eV, 0.52 eV, and 0.69 eV for compounds **8.63**, **8.67**, and **8.71**, respectively. Furthermore, variable temperature electron paramagnetic resonance (VT-EPR) results suggested that compounds **8.63**, **8.67**, and **8.71** possess a singlet ground state with an exceptionally small singlet-triplet energy gap. Incorporating nonalternant units into polycyclic hydrocarbons (PHs) is a potential way to construct open-shell PHs with remarkably low singlet-triplet energy gaps.

Zhu *et al.* [41] described the synthesis of an azulene-embedded PAH **8.74**, which incorporates two fused heptagon and pentagon units and their charged derivatives. This compound **8.74** was produced with a 15% isolated yield through a two-step Pd-catalyzed [5+2] annulation process involving 3,9-diboraperylene **8.72** and 1,2-dibromoacenaphthylene **8.73** (Scheme 8.15). The resulting compound **8.74** exhibited excellent solubility in a range of organic solvents, demonstrated good solid-state stability,

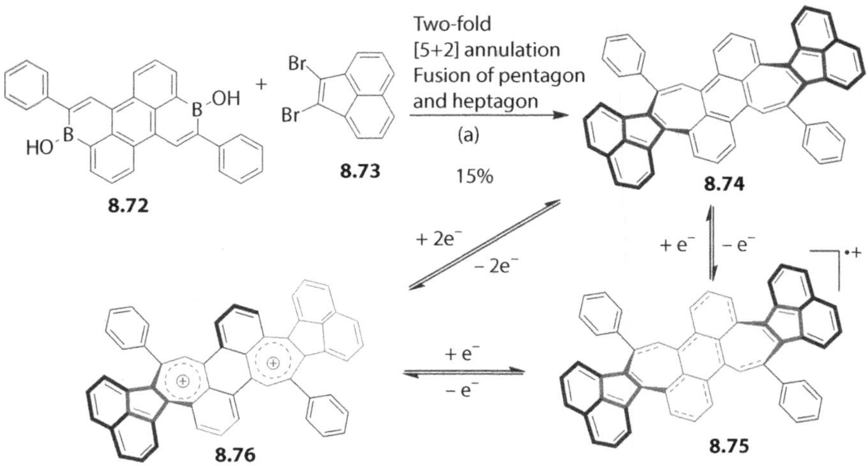

Scheme 8.15 Synthesis of PAH **8.74** and formation of radical cation **8.75** and dication **8.76**; Reagent and conditions: (a) $Pd_2(dba)_3{\cdot}CHCl_3$, tBu_3PHBF_4, Cs_2CO_3, H_2O, tAmOH, 100°C, N_2, 48 h. Chemical oxidation conditions: $NO{\bullet}SbF_6$, dichloromethane, N_2, r.t., 5 min.

and remained undecomposed for several weeks. The stepwise chemical oxidation of compound **8.74** with $NO{\bullet}SbF_6$ in DCM under nitrogen at room temperature resulted in the formation of the radical cation **8.75** and the dication **8.76**. X-ray crystallographic studies revealed changes in the conformation and aromatic ring current of the saddle-shaped PAH as its oxidation state increased. The radical cation **8.75** was observed to adopt a π-dimer type cisoid conformation, whereas the dication **8.76** exhibited a transoid conformation. Besides that a slip-stacked π-overlap was observed with its nearest neighbours.

Mastalerz *et al.* [42] synthesized and investigated the photophysical properties of PAHs **8.80–8.86**, which feature negative curvatures and contain two azulene moieties. Initially, bisnaphthyl diareno compound **8.78** was synthesized with a 67% yield through a Pd-catalyzed Suzuki–Miyaura cross-coupling reaction between the corresponding bromide **8.77** and 1-naphthylboronic acid (Scheme 8.16). To prepare azulene-embedded PAHs, the oxidative cyclization of compound **8.78** was attempted using $FeCl_3$ in a CH_3NO_2/DCM system. However, this reaction selectively formed a five-membered ring due to two-fold chlorination, resulting in the double-helicene product **8.79** with a 94% yield. Even with the addition of excess oxidant (20 equivalents), the product with seven-membered rings was not obtained. In contrast to using $FeCl_3$, when 8 equivalents of DDQ under trifluoromethanesulfonic acid (TfOH)/DCM were employed for the oxidative cyclization of compound **8.78**, both cycloheptannulation and

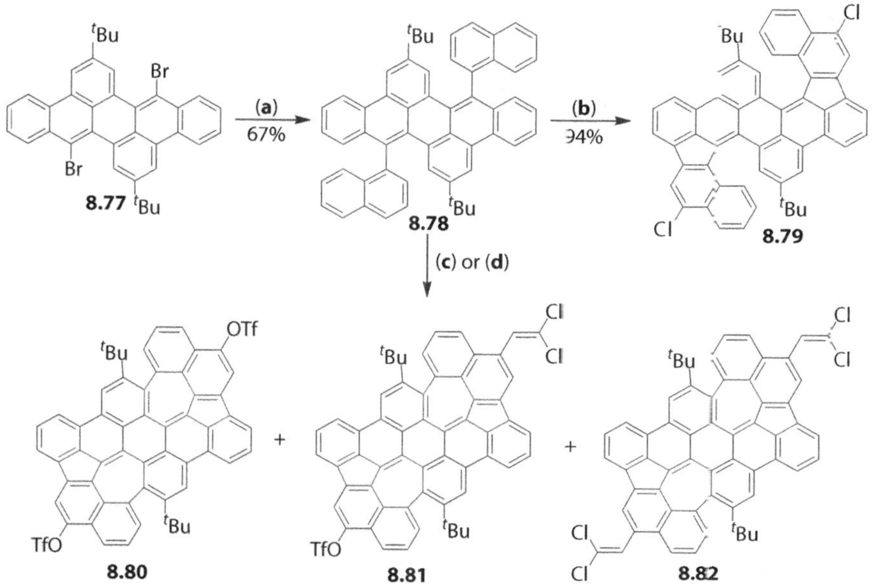

Scheme 8.16 Synthesis of azulene-based PAHs **8.80–8.82**; Reagent and conditions: (a) 1-naphthylboronic acid, $Pd_2(dba)_3$, tBu_3PHBF_4, THF, 2N K_2CO_3(aq), 80°C, 16 h; (b) $FeCl_3$, CH_3NO_2/CH_2Cl_2, 0°C to reflux, 1 h; (c) DDQ (8 equiv.), $TfOH/CH_2Cl_2$, rt, 1 h, **8.80** (14%), **8.81** (30%), **8.82** (9%); (d) DDQ (10 equiv.), $TfOH/CH_2Cl_2$, rt, 1 h, **8.80** (25%), **8.81** (11%), **8.82** (2%).

cyclopentannulation occurred, resulting in a mixture of azulene-containing PAHs **8.80, 8.81**, and **8.82** in yields of 14%, 30%, and 9%, respectively. Alongside the cyclization, triflate and dichlorovinylidene (DCV) groups were also attached to the aromatic ring, with triflyloxylation competing with the attachment of the DCV unit. The authors suggested that DCV originates from DCM solvent impurities, specifically tetrachloroethane. To minimize the competing substitution by DCV in place of trifluoromethanesulfonate (OTf) groups, the DCM amount was reduced, the concentration of compound **8.78** was doubled, and the DDQ amount was increased from 8 to 10 equivalents. Consequently, the yield of bistriflate **8.80** increased from 14% to 25%, while the combined yields of **8.81** and **8.82** decreased from 39% to 13%.

This procedure represents a novel method for introducing the DCV group onto the aromatic ring during the oxidative cyclization of PAHs. The triflate group in PAH **8.80** is particularly significant because it can participate in Pd-catalyzed cross-coupling reactions with boronic acids or esters. As illustrated in Scheme 8.17, compound **8.80** undergoes coupling reactions with

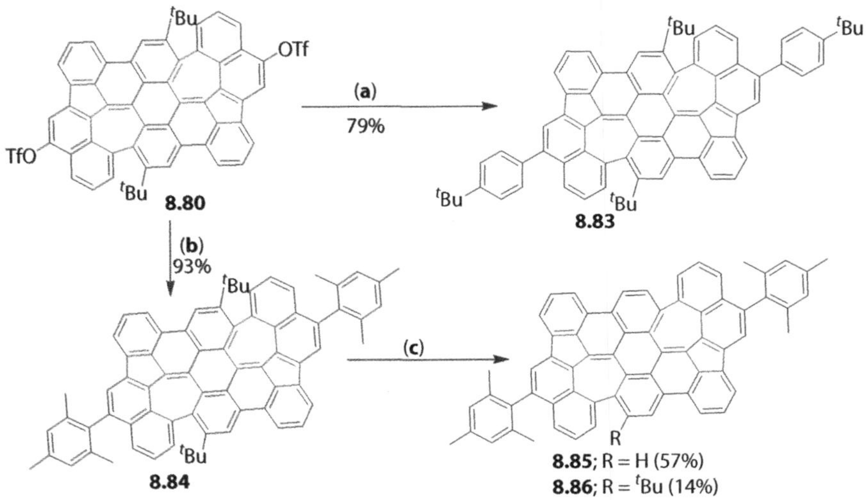

Scheme 8.17 Synthesis of azulene-fused compound **8.83-8.86**; Reagent and conditions: (a) 4-*tert*-Butylphenyl boronic acid, $Pd(PPh_3)_4$, toluene/EtOH, K_2CO_3, 100°C, 16 h (b) mesityl boronic acid, $Pd(PPh_3)_4$, toluene/EtOH, 2N K_2CO_3(aq), 100°C (c) $AlCl_3$, toluene, 70°C, 48 h.

p-tert-butyl phenylboronic acid or mesityl boronic acid, yielding azulene-containing PAHs **8.83** and **8.84** in 79% and 93% yields, respectively. Additionally, PAHs **8.85** and **8.86** were obtained in 57% and 14% yields, respectively, following treatment of compound **8.84** with $AlCl_3$ in toluene at 70°C.

After successfully synthesizing and characterizing the compounds, all of them were studied using UV/Vis absorption and emission spectroscopy. Additionally, frontier molecular orbitals (FMOs) and their energies were calculated using DFT methods. In the first set of compounds, PAHs **8.80–8.82**, compound **8.80**, containing two triflate groups, exhibited an absorption peak in the UV/Vis spectrum at 628 nm. In contrast, compound **8.81**, which includes one triflate and one DCV group, displayed a red-shifted absorption band at 636 nm, and compound **8.80** showed an even further red-shifted peak at 644 nm. This shift towards longer wavelengths (bathochromic shift) is attributed to the enlargement of the π-system through the attachment of DCV units to the aromatic ring. Similarly, the emission spectra also showed a red shift: from 648 nm for PAH **8.81**, through 662 nm for PAH **8.81**, to 671 nm for PAH **8.82**, with fluorescence quantum efficiencies ranging between ϕ = 20% and 27%. Comparing PAHs **8.84–8.86** in the second series is particularly interesting because they share the same π-system but differ in curvature, offering insights into how changes in curvature affect aromatic properties. PAH **8.84**, with two *tert*-butyl groups, exhibited the longest wavelength absorption at 633 nm. This absorption peak shifted to 630 nm in PAH **8.85** (one *tert*-butyl group) and further to 618 nm in PAH **8.86** (no *tert*-butyl group). Notably, all these compounds emit light within the near-infrared (NIR) window, making them suitable for biological applications due to their ability to penetrate biological tissues. Bistriflate **8.80** is particularly noteworthy in this context because it can undergo further functionalization with various groups that can selectively interact with specific sites on biological probes, similar to graphene quantum dots, making it potentially valuable for imaging applications.

The introduction of five- and seven-membered rings in the form of azulene in PAHs was also reported by Kirschbaum *et al.* [43]. The synthesis and characterization of azulene-embedded PAHs benzotrisazulene, is based on reducing and eliminating its trioxo derivative (Scheme 8.18). At first, tris(formylnaphthyl)truxene (**8.88**) was synthesized by the Pd-catalyzed Suzuki–Miyaura cross-coupling of triborylated truxene **8.87**

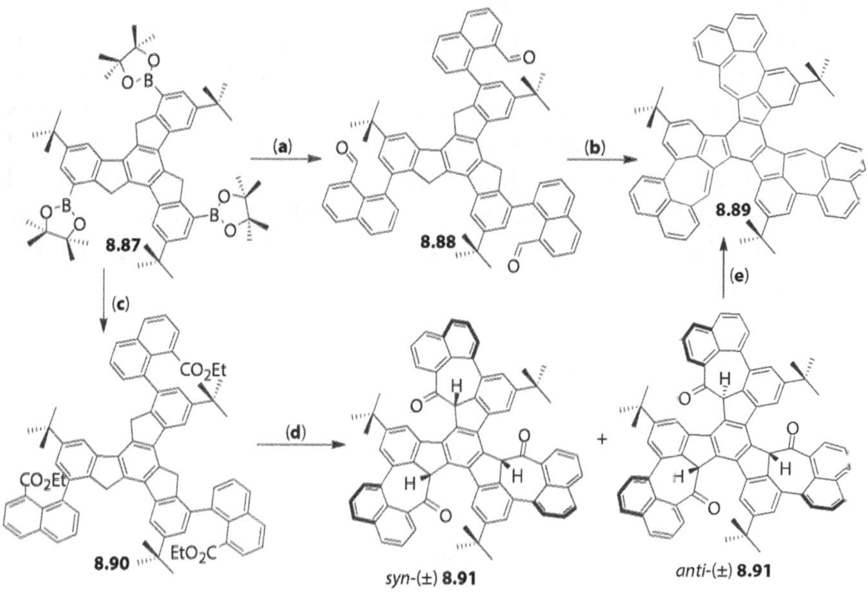

Scheme 8.18 Synthesis of benzotrisazulene **8.89**; Reagents and conditions: (a) 8-bromo-2-naphthaldehyde, $Pd(PPh_3)_4$, K_2CO_3, DME, 90 °C, 24 h; (b) KOH, THF, 60°C, 24 h, 9% over two steps; (c) ethyl 8-bromo-2-naphthoate, Pd_2dba_3, tBu_3PHBF_4, K_2CO_3, THF, 80°C, 48 h, 85%; (d) KO^{t}Bu, THF, 75 °C, 72 h, 92% (anti/syn: 5:1); (e) (i) $LiAlH_4$, THF, 0°C, r.t., 18 h; (ii) Et_3N, then MsCl, DCM, 0°C, 80°C, 96 h, 29%.

with 8-bromo-2-naphthaldehyde under dimethoxyethane (DME) in the presence of aqueous K_2CO_3. This was followed by a ring-closing reaction using KOH powder as a base, yielding benzotrisazulene **8.89** with a total yield of 9%. An alternative synthetic route was also developed to enhance the overall yield of compound **8.89**. First, compound **8.90** was synthesized in 85% yield as a mixture of non-separable rotamers by the reaction between **8.87** and ethyl 8-bromo-2-naphthoate, which on cyclization gives a mixture of *syn*(±)**8.91** and *anti*(±)**8.91** in 92% yield. These two pairs of enantiomers of trioxobenzotrisazulene **8.91** were separated by column chromatography, with the reaction favoring the anti-(±)-isomer at a ratio of approximately 5:1. Notably, the syn(±) isomer exhibited very poor solubility in chlorinated solvents like chloroform and DCM, as well as in tetrahydrofuran, but was soluble in aromatic solvents. In contrast, the anti(±) isomer was easily soluble in most organic solvents. Therefore, due to its better solubility and higher proportion, anti(±)-**8.91** was reduced with lithium aluminum hydride, and subsequent *in situ* elimination produced trisazulene **8.89** in a 29% yield.

All the synthesized products were characterized by various spectroscopic techniques in addition to X-ray crystallography. Moreover, DFT calculations in combination with optoelectronic characterization revealed that the azulene unit in compound **8.89** is better described as isolated heptagons and pentagons, as it appears at a shorter wavelength (474 nm) in comparison to azulene (609 nm). To enlarge the PAHs **8.89**, the authors tried various functionalizations of the periphery. Interestingly, compound **8.89** was borylated by an iridium catalyst to produce selectively sixfold compound **8.92** in 92% yield (Scheme 8.19). This borylation opens the possibility of constructing various significant molecules using Suzuki–Miyaura cross-coupling reactions.

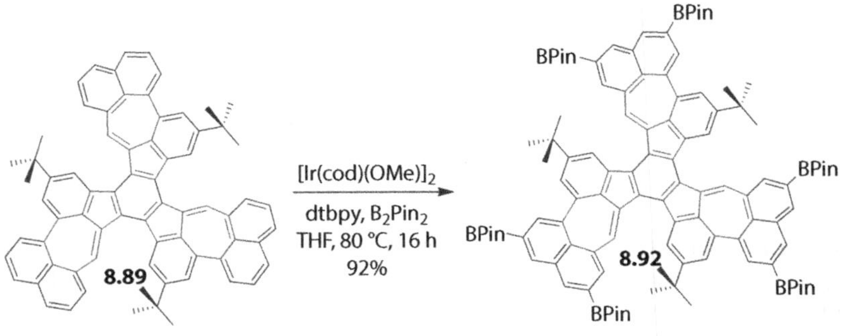

Scheme 8.19 Synthesis of compound **8.92** by borylation of benzotrisazulene **8.89**.

Liang *et al.* [44] reported the synthesis and photophysical properties of benzo[1,2-a:3,4-a':5,6-a"]triazulene (BTA) (**8.96**) and its nitro analog BTA-NO_2 (**8.97**) (Scheme 8.20). Initially, compound **8.94** was prepared with a 64% yield by treating 6-*tert*-butyl-indanone (**8.93**) with *p*-toluenesulfonic acid under reflux in propionic acid. The triborylated truxene **8.95** was then synthesized with a 90% yield through an iridium-catalyzed borylation reaction. Interestingly, BTA (**8.96**) was synthesized with a 26% yield *via* a two-step [3+4] annulation, combining Suzuki coupling and Knoevenagel condensation in one pot under optimized Suzuki coupling conditions. Finally, to oxidize BTA **8.96** with NO•BF_4 under THF at room temperature, a trinitrated product BTA-NO_2 (**8.97**) was obtained as the major product in 40% yield. The structures of compounds **8.96** and **8.97** were confirmed using NMR, HRMS, and single-crystal X-ray crystallography. Additionally, both BTA (**8.96**) and BTA-NO_2 (**8.97**) were found to be quite stable at room temperature and exhibited good solubility in common organic solvents. Single-crystal X-ray crystallography insights revealed the structure of BTA to be comprised of a dimer formed by two enantiomeric helicene conformers, and that of BTA-NO_2 comprises a π-tetramer made

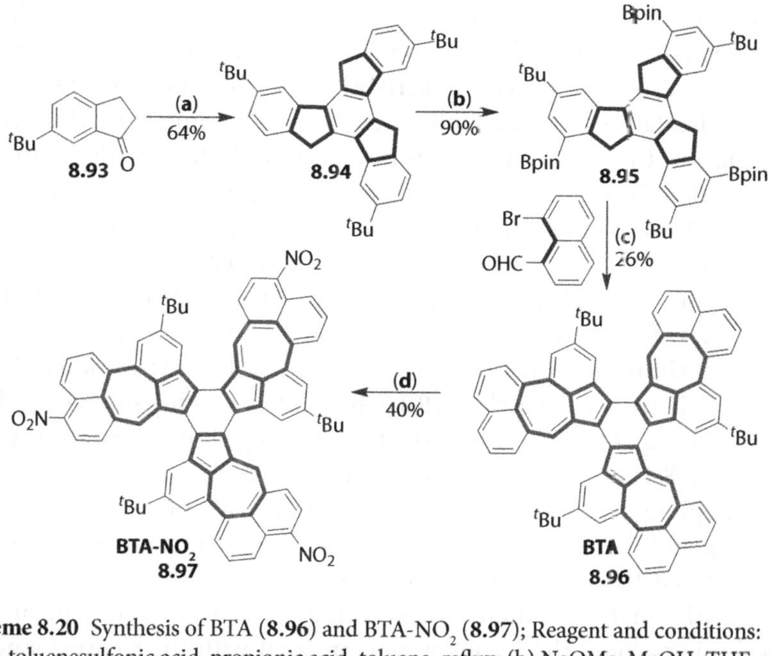

Scheme 8.20 Synthesis of BTA (**8.96**) and BTA-NO_2 (**8.97**); Reagent and conditions: (a) *p*-toluenesulfonic acid, propionic acid, toluene, reflux; (b) NaOMe, MeOH, THF, reflux (c) $[Ir(OMe)COD]_2$, B_2pin_2, dtbpy, THF, reflux (d) 8-bromo-1-naphthaldehyde, $Pd(PPh_3)_4$, Cs_2CO_3, *N,N*-Dimethylformamide (DMF), reflux; (e) NO•BF_4, THF, rt.

up of two enantiomeric dimers, resulting in four distinct helicene conformers. Both compounds displayed excellent fluorescence with large Stokes shifts of up to 3300 cm^{-1} for BTA and 5100 cm^{-1} for BTA-NO_2. Additionally, BTA-NO_2 demonstrated a solvatochromic effect in various solvents and exhibited hydrogen-bonding-induced transfer of fluorescence emission. These findings will aid in synthesizing diverse azulene-fused PAHs, stabilizing their structure-performance relationships, and exploring different applications.

Nakagawa *et al.* [45] have reported the synthesis and liquid crystal properties of azulene-embedded PAHs containing 6-hexadecyl substituents on each azulene moiety. Moreover, the synthesis of hexakis(6-hexadecyl-2-azulenyl)benzene (**8.105b**) was achieved by $Co_2(CO)_8$-catalyzed cyclotrimerization reaction of bis(6-hexadecyl-2-azulenyl) acetylene (**8.104b**), as similar to the synthesis of **8.105a** from dioctyl derivative **8.104a** as previously reported by Ito *et al.* [46]. The synthesis of **8.105b** was accomplished in a multistep synthesis using 6-bromo-2-methoxyazulene-1,3-dicarboxylate as a starting material, which gives compound **8.98** in three steps, i.e. Pd-catalyzed cross-coupling reaction with 1-hexadecyne, catalytic hydrogenation using Pd/C, and ester hydrolysis. Subsequently, treating compound **8.98** with 100% H_3PO_4 produced a mixture of compounds **8.99** and **8.100**, which contained hydroxyl and methoxy groups in yields of 71% and 10%, respectively. The hydroxyl product (**8.99**) was then reacted with PBr_3 to form the 2-bromo derivative (**8.101a**). This 2-bromo derivative underwent a Pd-catalyzed cross-coupling reaction with trimethylsilylacetylene, resulting in the formation of 6-hexadecynyl-2-(trimethylsilylethynyl) azulene (**8.102**) with an 89% yield. Furthermore, 2-ethynyl-6-octylazulene (**8.103**) was synthesized with a 95% yield by reacting compound **8.102** with potassium fluoride in DMF. The Pd-catalyzed cross-coupling reaction of compound **8.103** with the 2-bromo derivative produced **8.104b** in 77% yield, while the reaction with the 2-iodo derivative yielded **8.104b** in 86% yield. Ultimately, the $Co_2(CO)_8$-catalyzed cyclotrimerization of bis(6-hexadecyl-2-azulenyl)acetylene (**8.104b**) yielded a highly soluble product, **8.105b**, with a 78% yield. Additionally, for comparison of phase-transfer properties, **8.106b** was synthesized *via* a Pd-catalyzed Suzuki–Miyaura cross-coupling reaction of compound **8.101a** with phenylboronic acid, achieving an 85% yield (Scheme 8.21). The liquid crystalline behavior of compounds **8.105b**, **8.104b**, and **8.106b** was examined using X-ray crystallographic analysis, differential scanning calorimetry, and polarizing optical microscopy. The results were compared with their corresponding 6-octyl derivatives, **8.105a**, **8.104a**, and **8.106a**. Due to the increased carbon atoms in the side chains of compounds **8.105b**, **8.104b**, and **8.106b**,

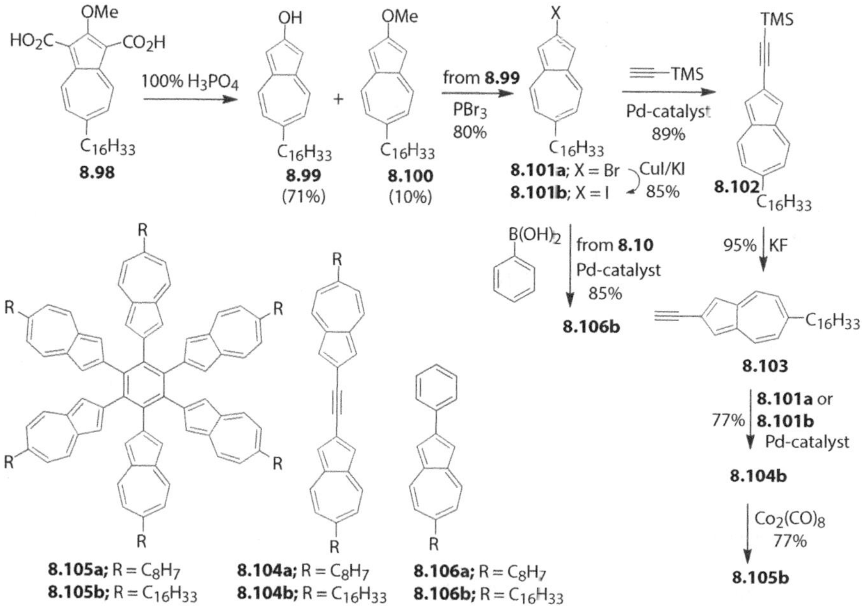

Scheme 8.21 Synthesis of hexakis(6-hexadecyl-2-azulenyl)benzene (**8.105b**).

their isotropization temperatures decreased, measuring 56.9°C, 33°C, and 23.6°C, respectively. Additionally, the phase-transition behavior changes, and the crystalline-mesophase transition temperatures decrease with the increased carbon chain length. Interestingly, when compound **8.105b** is cooled from the isotropic liquid phase, it readily self-organizes into columns perpendicular to the substrate, as indicated by the Col_{hd} mesophase data. Owing to the amphoteric redox characteristics of the azulene derivative, compound **8.105b** might exhibit superior charge carrier transport properties, making it a promising candidate for future semiconductor applications.

Sasaki *et al.* [47] reported on the synthesis and properties of azulene-fused azacoronene **8.110** (AzAC), a large molecule formed by circularly fusing both azulene and pyrrole groups. Azacoronene **8.110** was synthesized in three steps (Scheme 8.22). The first step involved Pd-catalyzed Suzuki–Miyaura cross-coupling of Bpin-azulene **8.107** with bromopentafluorobenzene, which yielded coupling adduct **8.108** in 85%. Next, compound **8.108** undergoes nucleophilic aromatic substitution reaction with 3,4-diethylpyrrole to give compound **8.109** in 35% yield. Finally, Scholl cyclization of **8.109** in the presence of $FeCl_3$ produced **8.110** in a moderate yield (46%). In addition, dications of compound **8.110** were also

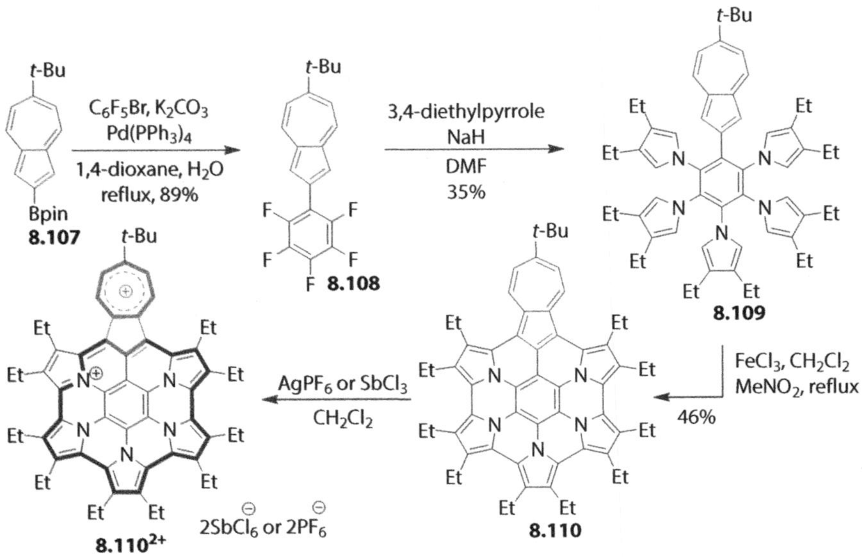

Scheme 8.22 Synthesis of AzAC **8.110** and its dication **8.110**$^{2+}$.

prepared by treatment of **8.110** with $AgPF_6$ or $SbCl_3$ in DCM. The UV-vis-NIR absorption spectroscopy revealed that compounds **8.108** and **8.109** displayed a weak and broad absorption from 720 to 440 nm with more intense peak at 369 and 389 nm, respectively. In contrast, due to effective π-conjugation, compound **8.110** exhibited a significantly perturbed spectrum at up to 950 nm, which indicates the characteristic properties of the azulene moiety. Additionally, compound **8.110** underwent investigations using CV and differential pulse voltammetry. The results show that in a diluted solution (0.09 mM), compound **8.110**'s CV displayed three oxidation waves, and a significant difference of around 0.66V was present between the second and third oxidation potentials. The X-ray diffraction analysis, anisotropy of the induced current density calculations, and NICS studies confirmed the formation of a 22π-electron conjugated system and a 6π-electron conjugated tropolium cation in the diacationic state.

8.3 Azulene-Embedded Nanographenes

In the last 20 years, the scientific community has shown increasing interest in extended PAH, also known as nanographenes [48–51]. These molecules, resembling finite segments of graphene with sp^2-conjugated carbon atoms, have garnered attention owing to their intriguing optoelectronic properties and potential applications in organic electronics [52, 53]. The synthesis of structurally well-defined nanographenes through bottom-up approaches [54, 55] has been pivotal for establishing reliable structure-property relationships. The field of nanographenes has gained momentum after the pioneering work by Scholl and Clar in the early 20th century that laid the groundwork [56–58]. Subsequent advances in analytical techniques facilitated remarkable synthetic breakthroughs, enabling the selective production of nanographenes under milder conditions [59]. By employing the bottom-up approach, researchers have synthesized nanographenes of varied shapes, sizes, symmetries, and edge topologies, including structurally defined graphene nanoribbons, promising semiconductors of the future [60]. However, most reported nanographenes have consisted solely of benzenoid rings, limiting our understanding of the impact of defects on graphene's electronic structure. Analogous to small-molecule aromatic compounds like naphthalene and azulene, where minor structural changes drastically alter electronic properties, perturbations in graphene's structure may similarly influence its electronic characteristics. Despite recent progress, current graphene fabrication methods lack rational control over nonbenzenoid topologies, motivating the pursuit of atomically precise synthesis and thorough characterizations. In parallel, there is a growing interest in open-shell nanographenes containing unpaired electrons [61], potentially useful for all-organic spintronic devices.

While numerous all-benzenoid open-shell structures have been achieved, nonbenzenoid counterparts have remained elusive until recently, when Liu *et al.* reported on a novel family of nonbenzenoid, nonalternant, open-shell nanographenes synthesized through a combined in-solution and on-surface approach, featuring azulene-type units (**8.112a**, Scheme 8.23) [62]. The synthetic pathway commencing with precursor **8.111a** leads to the formation of the azulene-incorporated PAH **8.112a** upon oxidation in solution, as delineated in Scheme 8.23. The HOMO–LUMO gap in compound **8.112a** was found to be 1.13 eV, and besides that, it possessed an open-shell singlet ground state. In extending this synthetic methodology to a surface context, a modified precursor, **8.111b**, was also prepared. Upon the deposition of **8.111b** onto an Au(111) substrate and subsequent annealing at 250°C under ultra-high vacuum conditions, a diverse array of products ensued. This on-surface synthesis led to the successful formation of a cyclopenta-fused azulene-embedded PAH denoted as **8.116** in Scheme 8.23. With the help of scanning tunneling spectroscopy measurements, the HOMO–LUMO gap of 0.27 eV for **8.116** on Au(111) was discerned.

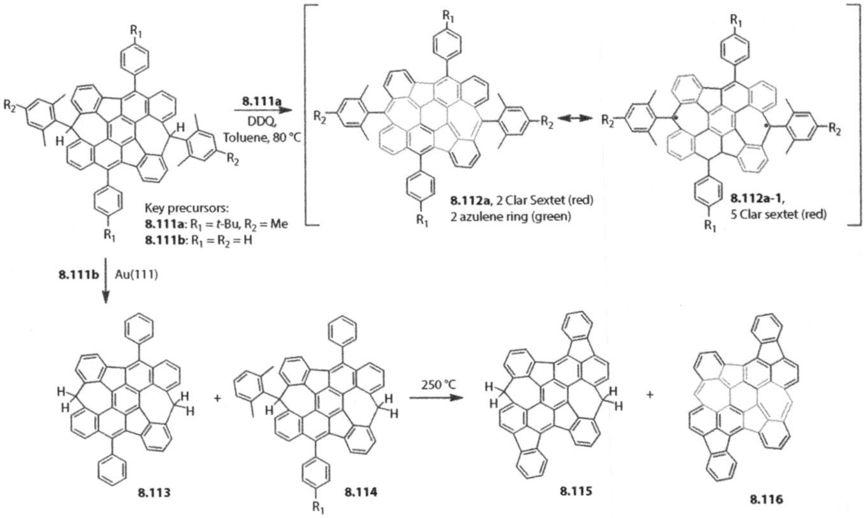

Scheme 8.23 Synthetic pathways leading to novel azulene-embedded nanographenes (**8.112**, **8.116**), centered around precursor **8.111**. Azulene rings are denoted by green, while red represents Clar sextets.

Hieulle *et al.* [63] devised an innovative method for crafting large, insoluble planar systems with unparalleled atomic precision, a feat often challenging through conventional solution-based chemistry techniques. Their approach hinges on the strategic use of solution chemistry to prepare fundamental building blocks (**8.117**), capable of facile deposition onto a surface, laying the groundwork for the final assembly of the desired nanomaterial. Initially, a minute amount of precursor material (**8.117**) (Scheme 8.24) is sublimated under ultra-high vacuum conditions onto a pristine Au(111) substrate, maintaining the substrate at 200°C to promote preferential Ullmann-like C-C coupling polymerization upon deposition. This process, monitored through scanning tunneling microscopy, reveals the transformation of most polymers into a graphenoid carbon structure *via* substrate-induced cyclodehydrogenation (CDH), with a few unreacted remnants discernible. Notably, a subset of these polymers adopts a trimer-like configuration, resembling the anticipated [18]annulene core (**8.119**) post-CDH, albeit with subtle deviations. Further examination uncovers trimers exhibiting a faint helical shape, akin to structure **8.120**, or even asymmetrical bending, denoted as structure **8.121**, indicative of an alternative novel cyclization pathway involving conjoined cove regions leading

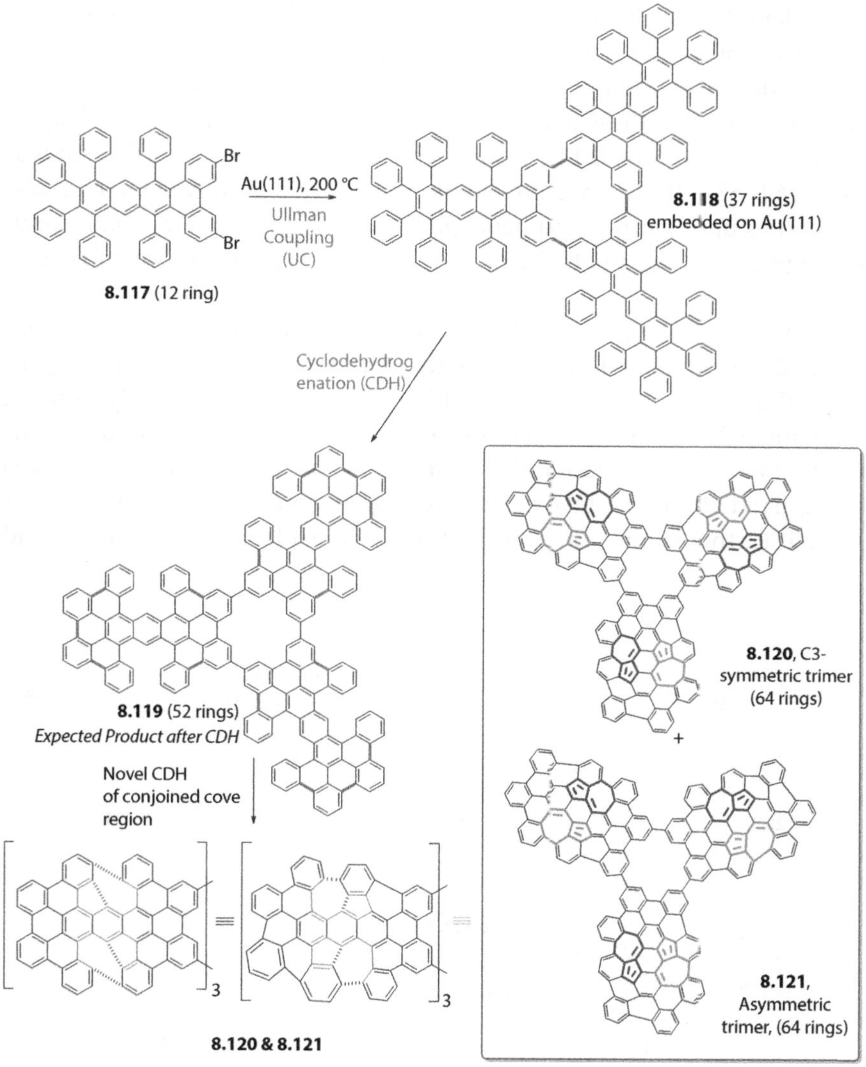

Scheme 8.24 Fabrication of large carbon nanostructures using a bottom-up approach. Molecular precursor **8.117**, synthesized *via* aryne cycloadditions (not shown), leads to nanographene **8.118** after Ullmann-like coupling of three precursor molecules, forming an [18]annulene core. Structure **8.119** results from the CDH of peripheral phenyl groups. New bonds are highlighted: pink Ullmann-like coupling and red for CDH. The scheme also shows the formation of two azulene moieties from conjoined cove regions in **8.119**'s blades, leading to symmetric (**8.120**) and asymmetric (**8.121**) planar nanographenes, with new bonds indicated by dashed lines.

to azulene fragment formation. This unexpected transformation ensures planarization of the nanostructure, a departure from the predicted outcome. Moreover, the presence of distinctive electronic states localized at the central pore of these trimers at elevated energies, reminiscent of super-atom molecular orbital-like resonances, was elucidated *via* DFT simulations [64, 65]. These novel image-like states are poised to confer unique properties upon porous graphene, potentially amplifying electronic bands or augmenting the photochemical reactivity of encapsulated molecular species.

Ogawa *et al.* [66] developed an innovative method for synthesizing helical azulene-embedded nanographenes (**8.127**) featuring contiguous pentagonal and heptagonal rings, resulting in a narrowed cove edge variant (Scheme 8.25). This synthesis was achieved through photoinduced oxidative cyclization of fluoranthene (**8.124**) to a key intermediate (**8.125**) using irradiation and DDQ. The cyclization potential of boronate ester (**8.124**) further facilitated the formation of the azulene ring *via* a C-B bond

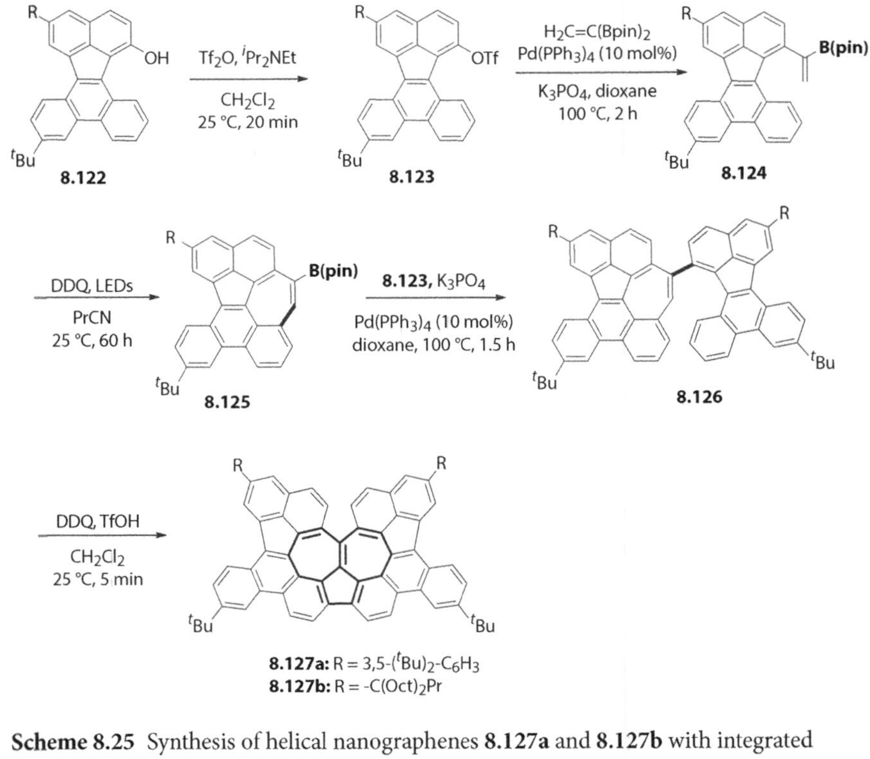

Scheme 8.25 Synthesis of helical nanographenes **8.127a** and **8.127b** with integrated contiguous azulene units and constricted cove-type edge.

between **8.125** and **8.123**, derived from the triflylation of hydroxyfluoranthene (**8.122**), yielding **8.126** in moderate to good yields. Subsequently, the Scholl reaction of **8.126** with DDQ and TfOH formed two C–C bonds, producing the desired nanographene **8.127a**. Attempts to generate only the seven-membered ring from **8.126** using DDQ/LEDs were unsuccessful. Interestingly, compound **8.127a** exhibited π-π stacking in the solid state due to self-association in solution, reflected in the ^{1}H NMR spectra by the upfield shifts of phenanthrene proton peaks due to face-to-face aggregations. The chiral variant **8.127b** was synthesized and separated using chiral HPLC, displaying clear mirror-symmetric CD spectra. Theoretical insights revealed a significant racemization barrier for the enantiomers of **8.127b**, attributed to the large interior angle of the seven-membered ring that caused a significant distortion in the transition state. The absorption profile of **8.127a** exhibited two weak bands at 660 and 600 nm and two strong bands at 477 and 404 nm in the visible region. Additionally, CV analysis confirmed that **8.127b** is stable under both electrochemical oxidation and reduction. This study marks the first reported synthesis of azulene-based narrowed cove edge nanographenes, paving the way for further experimental investigations.

For the first time, Han *et al.* [67] developed an innovative method to achieve a theoretical Stones–Wales rearrangement, transforming benzenoid PAH (naphthalene) into a non-benzenoid hydrocarbon (azulene). They synthesized nanographene (**8.131**) featuring a 5/7/7/5 ring-fused substructure with two formal azulene units, although only one exhibited azulene-like electronic structure (blue ring, Scheme 8.26). The precursor

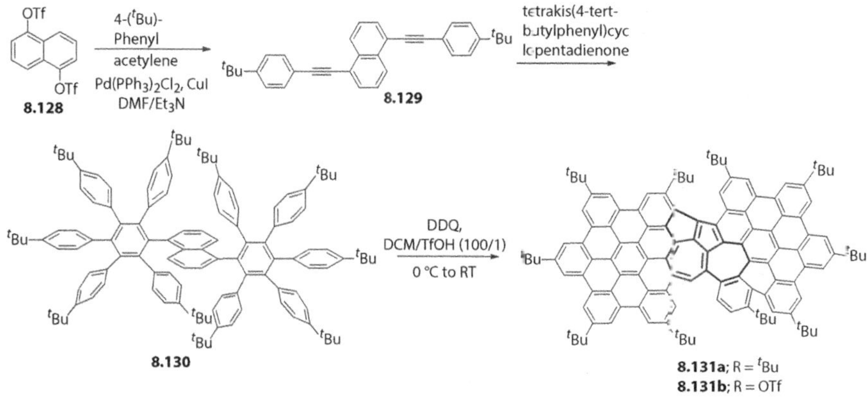

Scheme 8.26 Synthesis of unconventional nanographenes (**8.131a**, **8.131b**) *via* an unprecedented naphthalene to azulene rearrangement.

8.130 was obtained *via* Sonogashira coupling of reactants (**8.128**), followed by a Diels–Alder reaction with tetrakis(4-tert-butylphenyl)cyclopentadienone. Under the optimized reaction conditions, two major products were obtained. These products were identified as the azulene-embedded nanographene **8.131a** and its triflyloxylated product **8.131b**, isolated by silica gel column chromatography.

Mechanistic insights into the observed tert-butyl shift and the rearrangement from naphthalene to azulene were gained through DFT-D3 calculations by using B3LYP/6-31G(d) level and the IEPCM model for dichloromethane solvation. By comparing the mechanism to the Scholl reaction of hexaphenylbenzene, it was assumed that the Scholl reaction first yields the partially cyclized product **A** (Figure 8.2a) [68–70]. The shift of the tert-butyl group was attributed to either radical cation or arenium ion-mediated phenyl shift [71]. However, the lower energy barriers of the threanium ion pathway make it the more probable pathway (Figures 8.2a, 8.2b). In this pathway, the first step is the protonation step (of **A**) by using triflic acid. Protonation can occur either at position 4 of ring 'e' or position 2 of ring 'f'. The formation of arenium ion **B** necessitates overcoming an energy barrier of around 17.30 kcal/mol if the protonation is at position 2 in ring 'f' and an energy barrier of around 64.98 kcal/mol if the protonation is at position 4 in ring 'e'. This is followed by the formation of a spirocyclic intermediate **C** (Figure 8.2b). The formation of **D** from **C** is guided

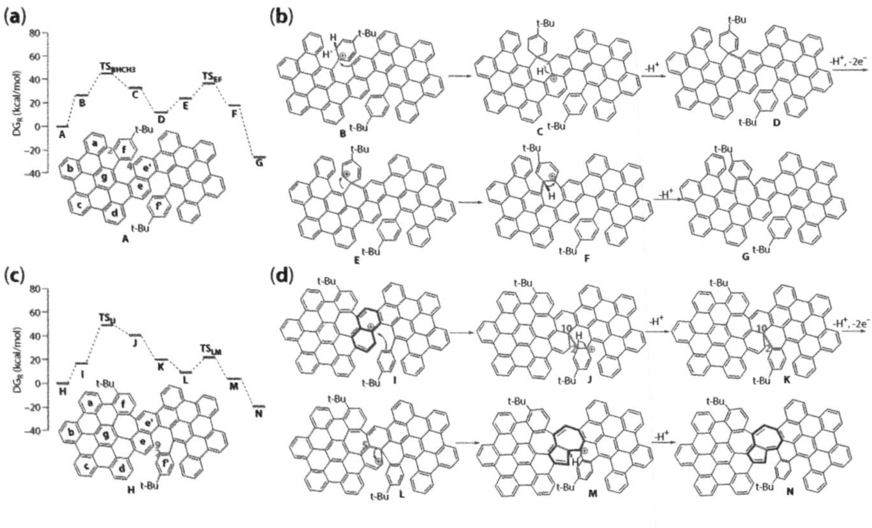

Figure 8.2 The proposed reaction mechanisms and Gibbs free energy diagrams for (a, b) the phenyl shift and (c, d) the rearrangement from naphthalene to azulene are illustrated.

by the aromaticity-induced proton elimination, followed by the formation of cation **E** from **D** by the removal of two electrons and one proton. The 1,2-phenyl migration in **E** leads to the formation of **F**, which is also accompanied by the formation of a heptagon and the shift of the tert-butyl group. Intermediate **F** further undergoes a loss of one proton to form partially fused **G**. The ortho-site of the tert-butyl group and the neighboring benzene ring undergo oxidative C–C bond formation to produce the cyclized product **H** (Figure 8.2c). Subsequently, the central naphthalene ring rearranges to azulene, with both radical cation and arenium ion pathways considered, though the arenium pathway again has a lower energy barrier (Figures 8.2c, 8.2d). The protonation of intermediate **H** to form the arenium ion **I** is facilitated by the more electronegativity of position 9 of ring 'e' than position 2 in ring 'f' (Figure 8.2d). This is followed by the formation of intermediate **J** (formed *via* electrophilic attack), and further, its deprotonation leads to the formation of **K**. A two-electron oxidation and deprotonation of intermediate '**K**' leads to the formation of cationic species **L**. A rearrangement of the ring skeleton from naphthalene to azulene occurs *via* a bond shift from position 10 to 9 to form **M**. This rearrangement involves an energy barrier of only 11.39 kcal/mol, making it a more favorable pathway than the previously reported high-energy barrier-free radical pathway [72, 73]. The formation of **131a** takes place from **M** in two steps involving deprotonation of **M** to form aromatic azulene-bridged intermediate **N**, and intramolecular oxidative C–C bond formation in **N**. The final cyclization step can again occur *via* radical cation or arenium mechanisms and theoretical insights reveal that both processes are equally favorable.

Ma *et al.* reported the synthesis of azulene-embedded helical nanographenes (compounds **8.137**, **8.142**, and **8.143**) starting from spatially crowded biaryl precursors and by employing a Scholl-type oxidative cyclization strategy [74]. The synthesis of compound **8.137** began as outlined in Scheme 8.27a. Initially, triisopropyl((2-(phenanthren-3-yl)phenyl)ethynyl) silane (**8.134**) was produced *via* a Suzuki coupling reaction of **8.133** and **8.132**. The compound **8.136** [(*S*)-8,8'-diiodo-7,7'-bibenzo[c]chrysene], which was the key precursor, was obtained from compound **8.134** in three steps, which involved deprotection using tetra-*n*-butylammonium fluoride (TBAF), followed by CuCl-catalyzed Glaser self-coupling and finally selective ICl-induced cyclization of **8.135**. Instead of producing a heptalene-embedded nanographene, treatment of **8.136** with DDQ and methanesulfonic acid yielded the unexpected azulene-embedded compound **8.137**. Similarly, larger and flexible precursors (**8.141**) were synthesized, leading to azulene-embedded superhelicenes (**8.142** and **8.143**). Precursor **8.141** was achieved *via* ICl-mediated benzannulation of diacetylene **8.140** in an 85% yield, which itself was prepared *via* the Diels–Alder cycloaddition of the butadiyne derivative (**8.138**) and cyclopentadienone derivative **8.139**. Treatment of **8.141** with DDQ/TfOH at 0°C produced the azulene-embedded π-expanded helical nanographenes **8.142** and **8.143** (Scheme 8.27b).

Compound **8.137** showed UV/Vis absorption at 474 nm, while compound **8.142** displayed a pronounced, red-shifted absorption band at 536 nm, a weak absorption at 629 nm, and a long tail up to 675 nm, attributable to the expanded π-system. The absorption spectrum of compound **8.143** revealed an absorption maximum at 573 nm and shoulder peaks at 535 nm and 507 nm. Additionally, compound **8.137** exhibited yellow luminescence in CH_2Cl_2 solution under UV light, with an emission spectrum peaking at 550 nm and shoulders at 508 nm and 587 nm. In contrast, the π-extended nanographenes **8.142** and **8.143**, in their red CH_2Cl_2 solutions, did not show detectable fluorescence upon excitation at their absorption maxima. Raman spectroscopy of compounds **8.137, 8.142,** and **8.143** revealed typical D and G bands, consistent with reported graphene molecules [75, 76]. Single-crystal X-ray diffraction analysis confirmed the structures of the azulene units in **8.137, 8.142,** and **8.143**, revealing their helically twisted geometry. Because of accumulated repulsions at the helical inner rim, these compounds have an inner non-planar and slightly antiaromatic azulene core. These π-extended superhelicenes also show effective HOMO–LUMO energy separation, amphoteric redox characteristics, and low optical energy band gaps.

Wu *et al.* developed an on-surface synthesis method to create unique cyclohepta[def]fluorene-based non-benzenoid nanographenes (NGs) [77]. This process combined solution-phase and on-surface synthesis, as shown in Scheme 8.28. While attempting to synthesize novel open-shell NGs (**T**), they unexpectedly produced unusual azulene-embedded non-benzenoid NGs (**8.150–8.152, 8.150', 8.152'**). The retrosynthesis of **T** originates from the precursor structure **P**, which was synthesized using solution chemistry. The compound **8.145** was synthesized using the Sonogashira reaction on the commercially available precursor 1-bromo-2-iodobenzene (**8.144**). This was followed by a Suzuki coupling to yield the trimethyl silane derivative (**8.146**). Further compound **8.146** was desilylated to give 2-chloro-2'-ethynyl-1,1'-biphenyl (**8.147**), and a subsequent $PtCl_2$-catalyzed cyclization produced 4-chlorophenanthrene (**8.148**). Compound **8.148** was subjected to a Suzuki coupling reaction with 2-(4,4,5,5-tetramethyl-1,3,2-dioxaborolan-2-yl)isophthalaldehyde to generate the key intermediate compound 2-(phenanthren-4-yl)isophthalaldehyde (**8.149**) [40]. Compound **8.149** was then reacted with (10-(2,6-dimethylphenyl)anthracen-9-yl)lithium to form a diol, which on being subjected to Friedel–Crafts alkylation led to the

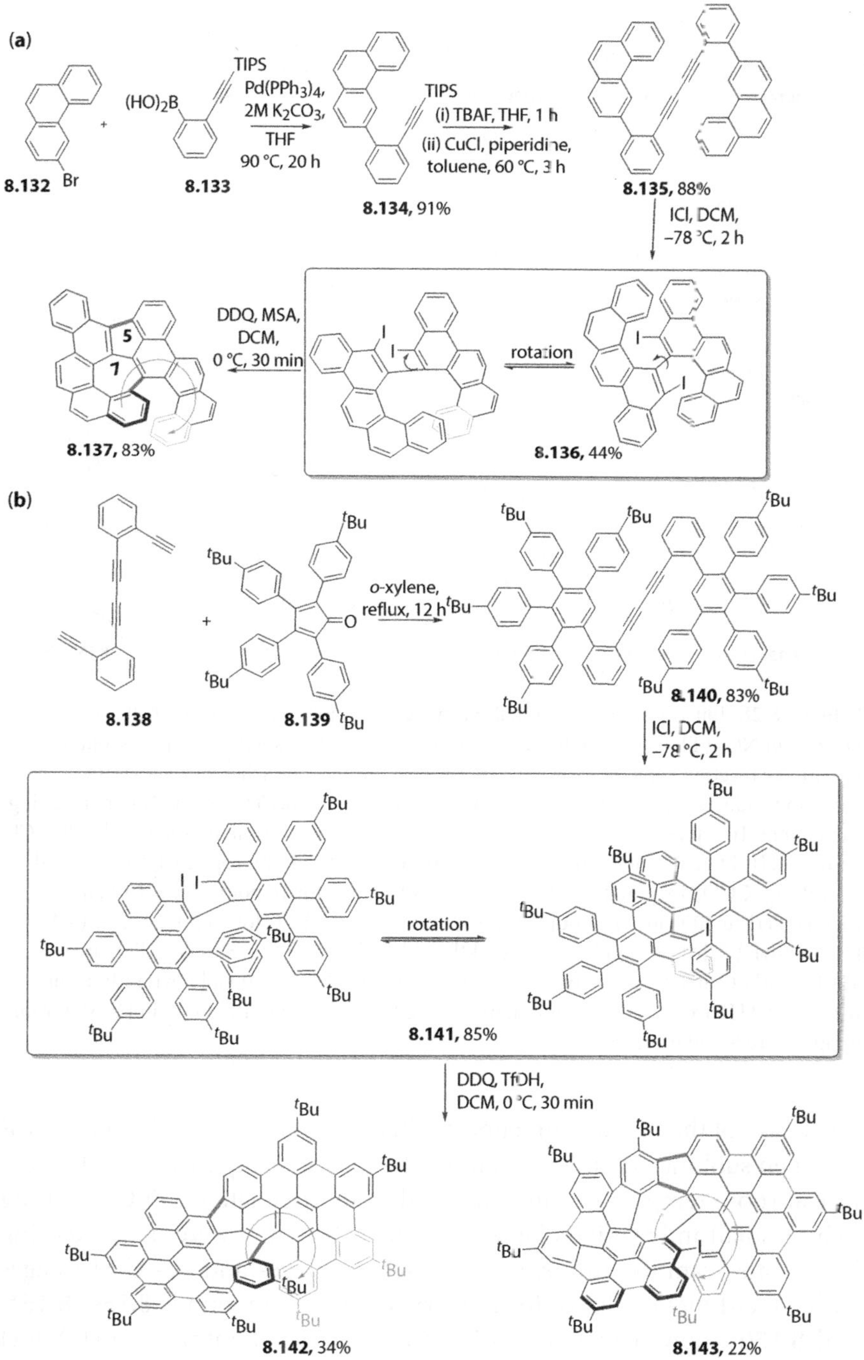

Scheme 8.27 Synthesis of helical nanographenes containing an azulene unit; (a) compounds **8.137** and (b) **8.142** and **8.143**.

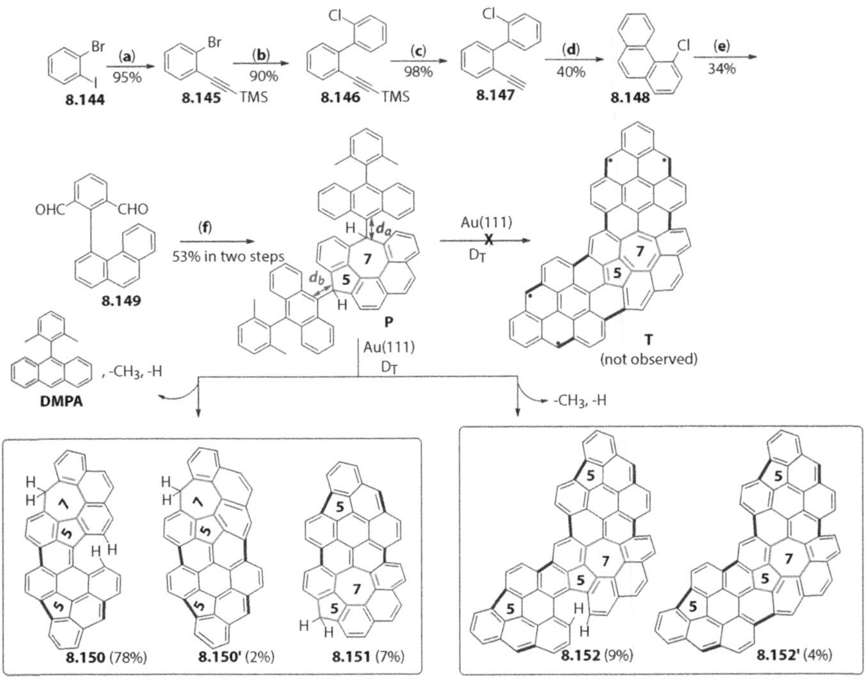

Scheme 8.28 Illustrates the conceptual synthetic routes for creating open-shell non-benzenoid NGs (**T**). It involves the synthesis of precursor **P** in solution and subsequent on-surface reactions of precursor **P** on an Au(111) surface. The schematic shows the reaction products **8.150–8.152** and their variants **8.150'** and **8.152'** obtained after annealing the sample. The steps and conditions are as follows: (a) Ethynyltrimethylsilane, $Pd(PPh_3)_2Cl_2$, CuI, in THF/Et_3N at room temperature for three hours; (b) (2-chlorophenyl)boronic acid, $Pd(PPh_3)_4$, K_2CO_3, in a mixture of Tol/EtOH/H_2O at 85°C for 16 hours; (c) K_2CO_3, in MeOH/THF at 40°C for one hour; (d) $PtCl_2$, in toluene at 110°C for 24 hours; (e) 2-(4,4,5,5-tetramethyl-1,3,2-dioxaborolan-2-yl)isophthalaldehyde, $Pd(OAc)_2$, SPhos, K_3PO_4, in Tol/EtOH/H_2O at 80°C for 40 hours; (f) i) 9-bromo-10-(2,6-dimethylphenyl)anthracene, n-BuLi, in THF from -78°C to room temperature for 12 hours; ii) $BF_3.OEt_2$, in DCM at room temperature for 40 minutes.

formation of the dihydro-precursor **P** in an overall 53% yield. Precursor **P** was then sublimated under vacuum onto an atomically clean Au(111) surface at room temperature and annealed at approximately 230°C, resulting in both small and relatively large NGs. Their structures were analyzed using large-scale scanning tunneling microscopy images and nc-AFM images with a CO-functionalized tip. The structures of compounds **8.150, 8.151**, and **8.150'** showed that the 9-(2,6-dimethylphenyl)anthracene (DMPA) segment often detaches from **P**, typically from the heptagon. Theoretical studies of the bonds (d_a and d_b) connecting DMPA with the seven- and five-membered rings in **P** indicated that d_b has a more double-bond character and d_a has more single-bond character, facilitating DMPA cleavage. The cleavage of the d_a bond results in the planarization of the central part of the molecule, which enhances molecule-surface interaction and reduces vertical strain. A small proportion of large NG compounds with the DMPA segment intact, having a "croissant-like" shape (**8.152** and **8.152'**), was also formed.

8.4 Conclusion

The synthesis of azulene-based compounds has remained a prominent research area due to their unique chemical structure and diverse properties. These molecules are used in creating π-functional materials like organic optoelectronic materials, organic solar cells, and organic field-effect transistors. Initially, research on azulene derivatives aimed to enhance π-conjugation to explore their anti-aromatic nature. Recently, azulene has been combined with various PAHs and nanographenes to study its optical and electronic properties for different biological and material applications. Moreover, with the implementation of effective synthetic strategies, including advancements in surface synthesis, various theoretical studies, and advanced spectroscopic and X-ray characterization, there is renewed interest in azulene chemistry. This chapter discusses the synthesis, characterization, and properties of azulene-embedded nanographenes and PAHs. Azulene has been fused with several significant π-conjugated units such as naphthalene, pyridine, indacene, acene, phenanthrene, and azacoronene, forming a large polycyclic aromatic framework. The study covers the synthesis and properties of various monoazulene, biazulene, and triazulene-embedded PAHs and nanographenes. The first part of the chapter focuses on the synthesis of azulene-based PAHs, while the second part addresses the synthesis and properties of azulene-embedded nanographenes. Many of these molecules exhibit intriguing photophysical and electrochemical properties, with a narrow HOMO–LUMO gap, making them suitable for various material applications. We believe this chapter will serve as a valuable guide in designing and developing molecules with fascinating properties and practical applications.

Acknowledgments

DKS is grateful to the Department of Chemistry, Bipin Bihari College, Jhansi, Uttar Pradesh, India, 284001, for the resources. BPM acknowledges DST-FIST, DBT-STAR College Scheme, and CURIE-PG for the financial assistance.

AI disclosure statement: The author acknowledges the use of AI-based tools for minor language editing or grammatical correction. However, all intellectual contributions, analyses, and conclusions presented in this chapter are entirely the author's own.

References

1. Michl, J. and Thulstrup, E.W., Why is azulene blue and anthracene white? A simple MO picture. *Tetrahedron*, 32, 2, 205, 1976.
2. Piesse, S., On the colouring principle of volatile oils. *Chem. News J. Phys. Sci.*, 8, 8, 245, 1863.
3. Sherndal, A.E., On the blue hydrocarbon occurring in some essential oils. [Preliminary Note.]. *J. Am. Chem. Soc.*, 37, 1, 167, 2002.
4. Wheland, G.W. and Mann, D.E., The dipole moments of fulvene and azulene. *J. Chem. Phys.*, 17, 3, 264, 1949.
5. Pfau, A.S. and Plattner, P., Zur kenntnis der flüchtigen pflanzenstoffe. IV. Über die konstitution der azulene. *Helv. Chim. Acta*, 19, 1, 858, 2004.
6. Lvov, A.G. and Bredihhin, A., Azulene as an ingredient for visible-light- and stimuli-responsive photoswitches. *Org. Biomol. Chem.*, 19, 20, 4460, 2021.
7. Murfin, L.C. and Lewis, S.E., Azulene-A bright core for sensing and imaging. *Molecules*, 26, 2, 353, 2021.
8. Estdale, S.E., Brettle, R., Dunmur, D.A., Marson, C.M., The azulene ring as a structural element in liquid crystals. *J. Mater. Chem.*, 7, 3, 391, 1997.
9. Schulz, F., Wank, B., Nacke, P., Frey, W., Laschat, S., First azulene liquid crystal with de Vries behavior and a SmA re-entrant phase. *Mater. Adv.*, 4, 5, 1306, 2023.
10. Shoji, T. and Ito, S., Azulene-based donor-acceptor systems: Synthesis, optical, and electrochemical properties. *Chemistry*, 23, 66, 16696, 2017.
11. Ito, S. and Morita, N., Creation of stabilized electrochromic materials by taking advantage of azulene skeletons. *Eur. J. Org. Chem.*, 27, 4567, 2009.
12. Huang, J., Huang, S., Zhao, Y., Feng, B., Jiang, K., Sun, S., Ke, C., Kymakis, E., Zhuang, X., Azulene-based molecules, polymers, and frameworks for optoelectronic and energy applications. *Small Methods*, 4, 10, 2000628, 2020.

13. Lambert, C., Nöll, G., Zabel, M., Hampel, F., Schmälzlin, E., Bräuchle, C., Meerholz, K., Highly substituted azulene dyes as multifunctional NLO and electron-transfer compounds. *Chemistry*, 9, 17, 4232, 2003.
14. Wang, F., Lin, T.T., He, C., Chi, H., Tang, T., Lai, Y.-H., Azulene-containing organic chromophores with tunable near-IR absorption in the range of 0.6 to 1.7 μm. *J. Mater. Chem.*, 22, 10448, 2012.
15. Tang, T., Lin, T., Wang, F., He, C., Azulene-based conjugated polymers with tuneable near-IR absorption up to 2.5 μm. *Polym. Chem.*, 5, 2980, 2014.
16. Tang, T., Lin, T., Wang, F., He, C., The origin of near-infrared absorption for azulene-containing conjugated polymers upon protonation or oxidation. *J. Phys. Chem. B*, 119, 25, 8176, 2015.
17. Beer, M. and Longuet-Higgins, H.C., Anomalous light emission of azulene. *J. Chem. Phys.*, 23, 8, 1390, 1955.
18. Viswanath, G. and Kasha, M., Confirmation of the anomalous fluorescence of azulene. *J. Chem. Phys.*, 24, 3, 574, 1956.
19. Thomas, S., Ramasesha, S., Hallberg, K., Garcia, D., Fused azulenes as possible organic multiferroics. *Phys. Rev. B*, 86, 180403, 2012.
20. Karwowski, J. and Flocke, N., Relations between Pariser–Parr–Pople and Heisenberg models. *Int. J. Quantum Chem.*, 90, 3, 1091, 2002.
21. Ong, A., Tao, T., Jiang, Q., Han, Y., Ou, Y., Huang, K.W., Chi, C., Azulene-fused acenes. *Angew. Chem. Int. Ed.*, 61, 44, e202209286, 2022.
22. Duan, C., Zhang, J., Xiang, J., Yang, X., Gao, X., Azulene-embedded [n] Helicenes (n=5, 6 and 7). *Angew. Chem. Int. Ed.*, 61, 18, e202201494, 2022.
23. Krzeszewski, M., Dobrzycki, L., Sobolewski, A.L., Cyrański, M.K., Gryko, D.T., Bowl-shaped pentagon- and heptagon-embedded nanographene containing a central pyrrolo[3,2-b]pyrrole core. *Angew. Chem.*, 133, 27, 15125, 2021.
24. Cruz, C.M., Marquez, I.R., Castro-Fernandez, S., Cuerva, J.M., Macoas, E., Campana, A.G., A triskelion-shaped saddle-helix hybrid nanographene. *Angew. Chem. Int. Ed.*, 58, 24, 8068, 2019.
25. Pun, S.H. and Miao, Q., Toward negatively curved carbons. *Acc. Chem. Res.*, 51, 7, 1630, 2018.
26. Feng, B. and Zhuang, X., The philosophy of carbon: Meso-entropy materials. *Faraday Discuss.*, 227, 80, 2021.
27. Feng, B. and Zhuang, X., Carbon-enriched meso-entropy materials: From theory to cases. *Acta Chim. Sin.*, 78, 9, 833, 2020.
28. Chaolumen, Ito, H., Itami, K., An axially chiral 1,1'-biazulene and its pi-extended derivative: Synthesis, structures and properties. *Chem. Commun.*, 55, 65, 9606, 2019.
29. Tsuchiya, T., Higashibeppu, M., Mazaki, Y., Synthesis and properties of twisted and helical azulene oligomers and azulene-based polycyclic hydrocarbons. *ChemistryOpen*, 12, 11, e202100298, 2023.
30. Liu, R., Fu, Y., Wu, F., Liu, F., Zhang, J.J., Yang, L., Popov, A.A., Ma, J., Feng, X., Modular synthesis of structurally diverse azulene-embedded polycyclic

aromatic hydrocarbons by Knoevenagel-type condensation. *Angew. Chem. Int. Ed.*, 62, 21, e202219091, 2023.
31. Uehara, K., Mei, P., Murayama, T., Tani, F., Hayashi, H., Suzuki, M., Aratani, N., Yamada, H., An anomalous antiaromaticity that arises from the cycloheptatrienyl anion equivalent. *Eur. J. Org. Chem.*, 2018, 33, 4508, 2018.
32. Yamamoto, K., Ie, Y., Tohnai, N., Kakiuchi, F., Aso, Y., Antiaromatic character of cycloheptatriene-bis-annelated indenofluorene framework mainly originated from heptafulvene segment. *Sci. Rep.*, 8, 1, 17663, 2018.
33. Jiang, Q., Tao, T., Phan, H., Han, Y., Gopalakrishna, T.Y., Herng, T.S., Li, G., Yuan, L., Ding, J., Chi, C., Diazuleno-s-indacene diradicaloids: Syntheses, properties, and local (anti)aromaticity shift from neutral to dicationic state. *Angew. Chem. Int. Ed.*, 57, 51, 16737, 2018.
34. Wang, S., Tang, M., Wu, L., Bian, L., Jiang, L., Liu, J., Tang, Z.-B., Liang, Y., Liu, Z., Linear nonalternant isomers of acenes fusing multiple azulene units. *Angew. Chem. Int. Ed.*, 61, 33, e202205658, 2022.
35. Ninomiya, K., Harada, Y., Kanetou, T., Suenaga, Y., Murafuji, T., Tsunashima, R., Synthesis and acid–base properties of a proton-bridged biaryl compound based on pyridylazulene. *New J. Chem.*, 39, 12, 9079, 2015.
36. Murai, M., Iba, S., Ota, H., Takai, K., Azulene-fused linear polycyclic aromatic hydrocarbons with small bandgap, high stability, and reversible stimuli responsiveness. *Org. Lett.*, 19, 20, 5585, 2017.
37. Murai, M., Iba, S., Hamao, S., Kubozono, Y., Ota, H., Takai, K., Azulene-fused linearly π-extended polycyclic aromatic compounds: Synthesis, photophysical properties, and OFETs applications. *Bull. Chem. Soc. Jpn.*, 96, 9, 1077, 2023.
38. Konishi, A., Morinaga, A., Yasuda, M., Construction of polycyclic π-conjugated systems incorporating an azulene unit following the oxidation of 1,8-Diphenyl-9,10-bis(phenylethynyl)phenanthrene. *Chemistry*, 24, 34, 8548, 2018.
39. Matsuda, T., Goya, T., Liu, L., Sakurai, Y., Watanuki, S., Ishida, N., Murakami, M., Azulenophenanthrenes from 2,2'-di(arylethynyl)biphenyls through C-C bond cleavage of a benzene ring. *Angew. Chem. Int. Ed.*, 52, 25, 6492, 2013.
40. Wu, F., Ma, J., Lombardi, F., Fu, Y., Liu, F., Huang, Z., Liu, R., Komber, H., Alexandropoulos, D.I., Dmitrieva, E., Lohr, G.T., Israel, N., Popov, A.A., Liu, J., Bogani, L., Feng, X., Benzo-extended cyclohepta[def]fluorene derivatives with very low-lying triplet states. *Angew. Chem.*, 61, 23, e202202170, 2022.
41. Zhu, C., Shoyama, K., Wurthner, F., Conformation and aromaticity switching in a curved non-alternant sp(2) carbon scaffold. *Angew. Chem. Int. Ed.*, 59, 48, 21505, 2020.
42. Yang, X., Rominger, F., Mastalerz, M., Contorted polycyclic aromatic hydrocarbons with two embedded azulene units. *Angew. Chem. Int. Ed.*, 58, 49, 17577, 2019.
43. Kirschbaum, T., Rominger, F., Mastalerz, M., Synthesis of a benzotrisazulene *via* trioxobenzotrisazulene. *Chemistry*, 29, 42, e202301470, 2023.

44. Liang, Y., Wang, S., Tang, M., Wu, L., Bian, L., Jiang, L., Tang, Z.-B., Liu, J., Guan, A., Liu, Z., Cascade synthesis of benzotriazulene with three embedded azulene units and large Stokes shifts. *Angew. Chem. Int. Ed.*, 62, 25, e202218839, 2023.
45. Nakagawa, K., Yokoyama, T., Toyota, K., Morita, N., Ito, S., Tahata, S., Ueda, M., Kawakaimi, J., Yokoyama, M., Kanai, Y., Ohta, K., Synthesis and liquid crystalline behavior of azulene-based liquid crystals with 6-hexadecyl substituents on each azulene ring. *Tetrahedron*, 66, 42, 8304, 2010.
46. Ito, S., Ando, M., Nomura, A., Morita, N., Kabuto, C., Mukai, H., Kazuchika, O., Kawakami, J., Yoshizawa, A., Tajiri, A., Synthesis and properties of hexakis(6-octyl-2-azulenyl)benzene as a multielectron redox system with liquid crystalline behavior. *J. Org. Chem.*, 70, 10, 3939, 2005.
47. Sasaki, Y., Takase, M., Okujima, T., Mori, S., Uno, H., Synthesis and redox properties of pyrrole- and azulene-fused azacoronene. *Org. Lett.*, 21, 6, 1900, 2019.
48. Drummer, M.C., Singh, V., Gupta, N., Gesiorski, J.L., Weerasooriya, R.B., Glusac, K.D., Photophysics of nanographenes: from polycyclic aromatic hydrocarbons to graphene nanoribbons. *Photosynth. Res.*, 151, 2, 163, 2022.
49. Jassas, R.S., Mughal, E.U., Sadiq, A., Alsantali, R.I., Al-Rooqi, M.M., Naeem, N., Moussa, Z., Ahmed, S.A., Scholl reaction as a powerful tool for the synthesis of nanographenes: a systematic review. *RSC Adv.*, 11, 51, 32158, 2021.
50. Matsuoka, W., Ito, H., Sarlah, D., Itami, K., Diversity-oriented synthesis of nanographenes enabled by dearomative annulative π-extension. *Nat. Commun.*, 12, 1, 3940, 2021.
51. Borissov, A., Maurya, Y.K., Moshniaha, L., Wong, W.S., Zyla-Karwowska, M., Stepien, M., Recent advances in heterocyclic nanographenes and other polycyclic heteroaromatic compounds. *Chem. Rev.*, 122, 1, 565, 2022.
52. Pham, P.V., Bodepudi, S.C., Shehzad, K., Liu, Y., Xu, Y., Yu, B., Duan, X., 2D Heterostructures for ubiquitous electronics and optoelectronics: Principles, opportunities, and challenges. *Chem. Rev.*, 122, 6, 6514, 2022.
53. Patra, A., More, M.A., Late, D.J., Rout, C.S., Field emission applications of graphene-analogous two-dimensional materials: recent developments and future perspectives. *J. Mater. Chem. C*, 9, 34, 11059, 2021.
54. Jolly, A., Miao, D., Daigle, M., Morin, J.F., Emerging bottom-up strategies for the synthesis of graphene nanoribbons and related structures. *Angew. Chem.*, 132, 12, 4652, 2019.
55. Hao, Z., Zhang, H., Ruan, Z., Yan, C., Lu, J., Cai, J., Tuning the electronic properties of atomically precise graphene nanoribbons by bottom-up fabrication. *ChemNanoMat*, 6, 4, 493, 2020.
56. Clar, E. and John, F., Zur kenntnis mehrkerniger aromatischer kohlenwasserstoffe und ihrer abkömmlinge, V. Mitteil.: Naphtho-anthracene, ihre oxydationsprodukte und eine neue klasse tiefgefärbter kohlenwasserstoffe. *Ber. Dtsch. Chem. Ges. (A B Ser.)*, 62, 11, 3021, 2006.

57. Clar, E., Ironside, C.T., Zander, M., 28. The electronic interaction between benzenoid rings in condensed aromatic hydrocarbons. 1 : 12-2 : 3-4 : 5-6 : 7-8 : 9-10 : 11-hexabenzocoronene, 1 : 2-3 : 4-5 : 6-10 : 11-tetrabenzoanthanthrene, and 4 : 5-6 : 7-11 : 12-13 : 14-tetrabenzoperopyrene. *J. Chem. Soc.*, 142, 1959.
58. Scholl, R., Seer, C., Weitzenböck, R., Perylen, ein hoch kondensierter aromatischer Kohlenwasserstoff $C_{20}H_{12}$. *Ber. Dtsch. Chem. Ges.*, 43, 2, 2202, 2006.
59. Yoon, K.-Y. and Dong, G., Liquid-phase bottom-up synthesis of graphene nanoribbons. *Mater. Chem. Front.*, 4, 1, 29, 2020.
60. Liu, Z., Fu, S., Liu, X., Narita, A., Samori, P., Bonn, M., Wang, H.I., Small size, big impact: Recent progress in bottom-up synthesized nanographenes for optoelectronic and energy applications. *Adv. Sci.*, 9, 19, e2106055, 2022.
61. Zdetsis, A.D., Open-shell magnetic states in alternant and non-alternant nanographenes: Conceptions and misconceptions. *Carbon Trends*, 14, 100330, 2024.
62. Liu, J., Mishra, S., Pignedoli, C.A., Passerone, D., Urgel, J.I., Fabrizio, A., Lohr, T.G., Ma, J., Komber, H., Baumgarten, M., Corminboeuf, C., Berger, R., Ruffieux, P., Mullen, K., Fasel, R., Feng, X., Open-shell nonbenzenoid nanographenes containing two pairs of pentagonal and heptagonal rings. *J. Am. Chem. Soc.*, 141, 30, 12011, 2019.
63. Hieulle, J., Carbonell-Sanroma, E., Vilas-Varela, M., Garcia-Lekue, A., Guitian, E., Pena, D., Pascual, J.I., On-surface route for producing planar nanographenes with azulene moieties. *Nano Lett.*, 18, 1, 418, 2018.
64. Zhao, J., Feng, M., Yang, J., Petek, H., The superatom states of fullerenes and their hybridization into the nearly free electron bands of fullerites. *ACS Nano*, 3, 4, 853, 2009.
65. Feng, M., Zhao, J., Petek, H., Atomlike, hollow-core-bound molecular orbitals of C60. *Science*, 320, 5874, 359, 2008.
66. Ogawa, N., Yamaoka, Y., Takikawa, H., Yamada, K.I., Takasu, K., Helical nanographenes embedded with contiguous azulene units. *J. Am. Chem. Soc.*, 142, 31, 13322, 2020.
67. Han, Y., Xue, Z., Li, G., Gu, Y., Ni, Y., Dong, S., Chi, C., Formation of azulene-embedded nanographene: Naphthalene to azulene rearrangement during the Scholl reaction. *Angew. Chem. Int. Ed.*, 59, 23, 9026, 2020.
68. Kübel, C., Eckhardt, K., Enkelmann, V., Wegner, G., Müllen, K., Synthesis and crystal packing of large polycyclic aromatic hydrocarbons: hexabenzo[b-c,ef,hi,kl,no,qr]coronene and dibenzo[fg,ij]phenanthro[9,10,1,2,3-pqrst] pentaphene. *J. Mater. Chem.*, 10, 4, 879, 2000.
69. Rempala, P., Kroulik, J., King, B.T., A slippery slope: mechanistic analysis of the intramolecular Scholl reaction of hexaphenylbenzene. *J. Am. Chem. Soc.*, 126, 46, 15002, 2004.
70. Di Stefano, M., Negri, F., Carbone, P., Müllen, K., Oxidative cyclodehydrogenation reaction for the design of extended 2D and 3D carbon nanostructures: A theoretical study. *Chem. Phys.*, 314, 1–3, 85, 2005.

71. Ajaz, A., McLaughlin, E.C., Skraba, S.L., Thamatam, R., Johnson, R.P., Phenyl shifts in substituted arenes *via* ipso arenium ions. *J. Org. Chem.*, 77, 21, 9487, 2012.
72. Alder, R.W., East, S.P., Harvey, J.N., Oakley, M.T., The azulene-to-naphthalene rearrangement revisited: A DFT study of intramolecular and radical-promoted mechanisms. *J. Am. Chem. Soc.*, 125, 18, 5375, 2003.
73. Mirzaei, M.S., Taherpour, A.A., Wentrup, C., Thermal rearrangement of azulenes to naphthalenes: A deeper insight into the mechanisms. *J. Org. Chem.*, 87, 5, 3296, 2022.
74. Ma, J., Fu, Y., Dmitrieva, E., Liu, F., Komber, H., Hennersdorf, F., *et al.*, Helical nanographenes containing an azulene unit: Synthesis, crystal structures, and properties. *Angew. Chem. Int. Ed.*, 59, 14, 5637, 2020.
75. Evans, P.J., Ouyang, J., Favereau, L., Crassous, J., Fernandez, I., Perles, J., Martin, N., Synthesis of a helical bilayer nanographene. *Angew. Chem. Int. Ed.*, 57, 23, 6774, 2018.
76. Fernandez-Garcia, J.M., Evans, P.J., Rivero, S.M., Fernandez, I., Garcia-Fresnadillo, D., Perles, J., Casado, J., Martin, N., π-Extended corannulene-based nanographenes: Selective formation of negative curvature. *J. Am. Chem. Soc.*, 140, 49, 17188, 2018.
77. Wu, F., Barragan, A., Gallardo, A., Yang, L., Biswas, K., Ecija, D., Mendieta-Moreno, J.I., Urgel, J.I., Structural expansion of cyclohepta[def]fluorene towards azulene-embedded non-benzenoid nanographenes. *Chem. Eur. J.*, 29, 51, e202301739, 2023.

9

Azulene-Based Systems of Biological Significance

Sunita Singh[1]*, Pragati Kushwaha[2], Pankaj Kumar Chaurasia[3], Shashi Lata Bharati[4], Ayman Younes Fathy Allam[5], Dhananjay Kumar Singh[6] and Naveen Kumar[7]

[1]*Department of Chemistry, Navyug Kanya Mahavidyalaya, University of Lucknow, Lucknow, Uttar Pradesh, India*
[2]*Department of Chemistry, University of Lucknow, Lucknow, Uttar Pradesh, India*
[3]*P.G. Department of Chemistry, L.S. College, B.R.A. Bihar University, Muzaffarpur, Bihar, India*
[4]*Department of Chemistry, North Eastern Regional Institute of Science and Technology, Nirjuli, Arunachal Pradesh, India*
[5]*Department of Food Science and Technology, Faculty of Agriculture, Menoufia University, Shibin El Kom, Egypt*
[6]*Command Hospital (EC), Kolkata, West Bengal, India*
[7]*Department of Chemistry, Sanatan Dharm College, Muzaffarnagar, Uttar Pradesh, India*

Abstract
Azulene is a non-alternant hydrocarbon, an isomer of naphthalene with a fused ring system of five and seven members. This aromatic organic compound has a blue color, a unique molecular structure, and biological properties. Some well-known azulene derivatives, such as guaiazulene and chamazulene, are found naturally in various plants and fungi. Azulene and its derivatives are potential candidates in technology, especially in devices that use light and electricity due to their fluorescence properties. They also have medicinal properties and have been used for centuries as ingredients of plants that can treat allergies, infections, and inflammation. This chapter discusses the various biological properties of azulenes and their related derivatives, which include antimicrobial, anti-inflammatory, anti-ulcer, antidiabetic, and antiviral activities, along with the challenges and opportunities in the development of novel derivatives of azulenes of biological importance.

**Corresponding author*: oceans.singh@gmail.com

Neha Rani Kumar and Abhijeet R. Agrawal (eds.) Azulene-Based Materials: Design, Synthesis, and Applications, (245–274) © 2026 Scrivener Publishing LLC

***Keywords*:** Azulenes, antimicrobial, anti-inflammatory, anti-ulcer, antidiabetic, antiviral

9.1 Azulenes, Their Sources, and Therapeutic Innovations

Azulene (**9.1**), known for its unique blue hue, consists of a cycloheptatriene ring with seven carbons fused to a cyclopentadiene ring with five carbons (**9.2**). This structure contrasts with its colorless counterpart, naphthalene. Figures 9.1a and 9.1b demonstrate the structure of azulene, its electronic distribution, famous derivatives, and biological effects. Azulene is predominantly found in the natural essential oils derived from plants and micro-organisms [1–7] such as *Matricaria chamomilla* L., *Achillea millefolium* L., and *Guaiacum officinale* L., with the latter also containing chamazulene (**9.3**) and guaiazulene (GA) (**9.4**). Additional azulene

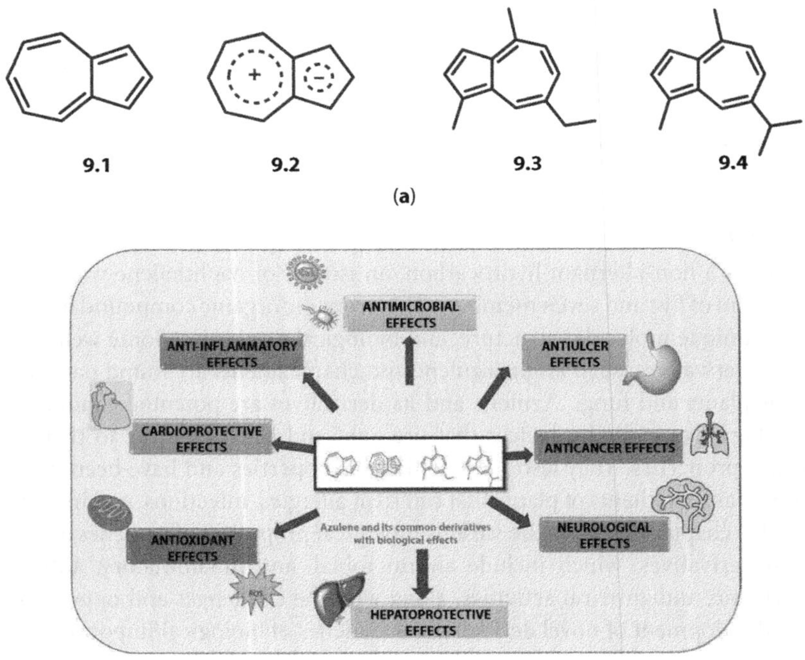

Figure 9.1 (a) Azulene and its electron distribution (**9.2**), chamazulene (**9.3**), and guaiazulene (**9.4**); (b) biological effects of azulene and its derivatives.

variants include the methyl (7-isopropenyl-4-methylazulen-1-yl) stearate from the *Lactarius indigo* fungus and 7-acetylo-1,4-dimethylazulene in the *Entoloma hochstetteri* fungus, as well as 2,3-dihydrolinderazulene and linderazulene found in *Acalycigorgia* sp. gorgonians. The scientific community has shown a growing interest in azulene and its derivatives, especially for treating skin conditions. The fundamental alterations to azulene involve replacing hydrogen atoms with suitable groups, commonly alkyl groups like methyl, ethyl, isopropyl, or alkoxyl groups.

Derivatives such as chamazulene, found in wormwood or achillea oil, are recognized for their antiallergic, anti-inflammatory, and antibacterial properties. They are found in cosmetic applications, particularly calming creams and ointments [8, 9]. An oleogel formulation derived from chamomile oil is effective against pain, nausea, and vomiting. It has also been successful in treating photophobia and phonophobia associated with migraines without aura [10]. Guaiazulene, used in a local pomade, has facilitated quick healing in newborns with stubborn diaper dermatitis, without adverse effects [11].

Azulene serves as an intriguing framework in medicinal chemistry, resembling the bicyclic aromatic structures found in numerous drugs and acting as a structural isomer to naphthalene. The modification of azulene and its derivatives typically involves the addition of substituents or their combination with other biologically active entities. Properties like solubility, absorption rate, and metabolic reactions can be enhanced by adding substituents to the terpene moieties. Combining the azulene moiety with other pharmacophores can yield more significant outcomes. The development of different molecular conjugates and complexes has accelerated drug discovery into a new horizon, which was analyzed by Pawełczyk *et al.* [8, 12].

Azulene derivatives are emerging as a significant class of molecules with potential uses in the pharmaceutical and organic materials sectors [13]. Yanagisawa *et al.* [55] synthesized various alkylazulene sodium sulfonates, discovered 1-ethyl-5-isopropylazulene (**9.3**), which demonstrated remarkable efficacy in preventing ulcers [13, 14]. This compound has since become a common treatment for gastric ulcers. In the realm of materials chemistry, azulene derivatives are also making strides. The organic field-effect transistor behavior with good carrier mobility was also reported in the literature [13–17].

9.2 Azulene and Its Derivatives with Biological Significance

The azulene compounds are particularly noteworthy for their multifaceted therapeutic roles, encompassing the treatment of inflammatory conditions and peptic ulcers, combating leukemia, managing diabetes, targeting HIV-1 in antiretroviral therapies, enhancing antimicrobial photodynamic therapy (PDT), and exerting antifungal effects. This section will discuss the antimicrobial, antidiabetic, anti-inflammatory, anticancer, and anti-ulcer effects of azulenes and their derivatives [5, 7–18].

9.2.1 Antidiabetic Action of Azulenes

The global incidence of diabetes has escalated alarmingly, soaring from 108 million individuals in 1980 to a staggering 422 million by 2014. This widespread metabolic disorder is a primary contributor to a host of severe health issues, such as vision loss, renal dysfunction, myocardial infarctions, cerebrovascular accidents, and amputations of the lower extremities. The criticality of diabetes management is underscored by the monitoring of glycemic levels, with unchecked diabetes precipitating extensive and enduring harm to various bodily systems.

Amid the ongoing search for efficacious therapeutic options, the spotlight has turned to azulene derivatives, which have surfaced as a promising avenue for intervention (Figure 9.2). These naturally occurring compounds, which are sourced from certain flora and fungi, have demonstrated notable antidiabetic potential [8].

The therapeutic potential of azulene in diabetes care is attributed to several mechanisms. A principal mechanism is the suppression of **cyclooxygenase-2 (COX-2)**, an enzyme integral to prostaglandin synthesis, which are pivotal agents in the inflammatory cascade. This suppression can diminish inflammation, a key element in the pathogenesis of insulin

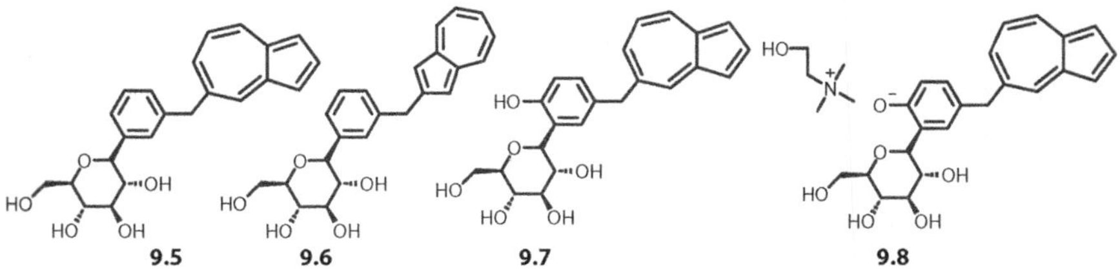

Figure 9.2 Chemical structures of **9.5–9.8**.

resistance and diabetes. Furthermore, azulene derivatives can modulate cytokine release, pivotal proteins in the immune signaling pathway. Such modulation can enhance insulin responsiveness and glucose processing, both of which are vital for diabetes control.

In addition, certain azulene derivatives possess antioxidant properties. Oxidative stress is implicated in the onset and progression of diabetes; antioxidants can counteract this stress, offering a defensive mechanism against the disease. It is imperative to recognize that these promising therapeutic pathways necessitate further scientific validation to ascertain the full spectrum of azulene derivatives' effectiveness and safety within a clinical context. Present insights are derived from a synthesis of laboratory-based research, animal studies, and initial human trials.

In one of the studies by Ikegai *et al.* [19], a range of C-glucosides containing azulene was researched within the aglycon part. They assessed the compounds' potential to inhibit human sodium-glucose transport proteins hSGLT1 and hSGLT2. They conducted *in vitro* tests using Chinese hamster ovary cells genetically modified to express these proteins. The effectiveness of these compounds was quantified by their half-maximal inhibitory concentration (IC50) values against hSGLT2 and their selectivity over hSGLT1. The key compounds demonstrated the following results: The synthesized compound 9.5 had an IC_{50} of 99 nM and was 140 times more selective for hSGLT2; synthesized compound 9.6 showed an IC_{50} of 22 nM with 590-fold selectivity; and compound **9.7** had an IC_{50} of 8.9 nM with 280-fold selectivity. These findings were consistent across at least two separate experiments.

Further evaluation involved preclinical trials using KK/Ay mice, a model for type 2 diabetes, to verify the antihyperglycemic effects of **9.7** in a living organism. A notable reduction in blood glucose levels—46% lower than the control—was observed following a single dose of 3 mg/kg of compound **9.5**. Subsequent refinement of **9.5** led to the creation of compound **6**, which emerged as the most promising candidate for further development. It was proposed for clinical trials in the form of a monocholine salt **9.8**, designated as YM543.

Pharmacokinetic analysis was performed on normal rats to understand how YM543 behaved within a biological system. The compound **9.8** showed rapid absorption when administered orally, with a peak concentration time (Tmax) of 0.5 hours, a bioavailability of 29%, a clearance rate of 2483 L/h/kg, and a distribution volume of 3360 L/kg. Notably, the renal concentration of the drug was found to be 23-fold higher than in plasma. Compound **9.7** led to a dose-dependent elevation in urinary glucose excretion, which was sustained for over 12 hours at the lowest dose of 0.3 mg/kg.

Additionally, compound **9.6** demonstrated significant antihyperglycemic effects in both streptozotocin-induced diabetic rats and KK/Ay mice without showing hypoglycemic effects.

The pivotal role of natural remedies in the diabetes management spectrum is increasingly acknowledged. Interventions such as physical activity, dietary modifications, and the strategic use of natural supplements have gained recognition for their efficacy in reducing blood glucose concentrations, thus offering a synergistic strategy alongside standard medical treatments. Studies with azulene derivatives' antidiabetic attributes further accentuate the integral part these natural solutions play in the comprehensive management of diabetes.

9.2.2 Antimicrobial Action of Azulene and Related Compounds

Derivatives of azulenes like guaiazulene and chamazulene have been recognized for their antimicrobial properties. They have been used in traditional medicine for their antiallergic, antibacterial, and anti-inflammatory effects. Natural origin alternatives for antimicrobial compounds are gaining attention due to the rise of antimicrobial resistance. Plants have historically been a rich source of antimicrobial agents, with only a fraction of Earth's plant species having been investigated for their medicinal properties. Azulene derivatives are fascinating compounds that are sourced from a variety of natural organisms [5, 8].

In the context of antimicrobial compounds of natural origin, azulene and its derivatives have shown potential in various medical applications, including antimicrobial PDT [5, 8]. However, there are considerations regarding the toxicity of these molecules, especially related to their phototoxic properties when exposed to light.

The exploration of azulene, its derivatives, and other natural compounds opens new avenues for developing antimicrobial agents (Figure 9.3) that could potentially address the challenges posed by antimicrobial resistance. Ongoing research and clinical studies are crucial to fully understand their mechanisms of action and to ensure their safety and efficacy in medical applications.

Nagai *et al.* have shown that azulenocyanine has a significant photodynamic antimicrobial impact on dentine plates infected with human pathogens [20]. In their research, human molar dentine plates were contaminated by soaking them in a *Streptococcus mutans* solution. The treatment involved using 0.01% azulenocyanine (0.1 ml) and, after a 5-minute

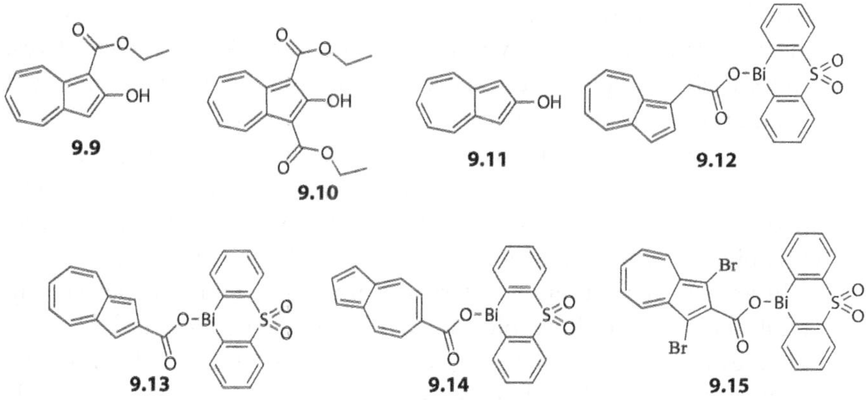

Figure 9.3 Chemical structures of azulene-related compounds with antimicrobial properties.

dark period, irradiating with a 1.5W laser at 940 nm, which proved to be the most effective in killing bacteria, as evidenced by a decrease in both colony-forming units (CFU/ml) and ATP levels. The lack of significant differences between treatment groups suggests that PDT could be a non-invasive method to decrease bacterial presence in periimplantitis.

The research conducted by Damrongrungruang *et al.* [21] reported that it can generate singlet oxygen in peripheral blood mononuclear cells (PBMCs) when exposed to light at 625 nm, with concentrations ranging from 5 to 500 μM. This process is thought to work through producing reactive oxygen species (ROS) that can damage cellular components, especially DNA. Azulene also acts as a photosensitizer in the PDT, effectively reducing the microbial colonies of *Streptococcus beta-haemolyticus*, *Prevotella* sp., and *Fusobacterium* sp., in dogs suffering from periimplantitis. Notably, its efficacy was comparable to the results of conventional treatments.

The studies also indicated that without light activation, azulene does not exhibit antibacterial properties, which supports the theory that its antibacterial effects are due to photodynamic action [5]. Furthermore, GA (**9.4**) was tested against various species of *Mycoplasma hominis* and compared with cinnamon oil, linoleic acid, carvacrol, and eugenol [22]. GA was found to be effective against the *M. hominis* strain when compared to natural alternatives such as cinnamon oil, carvacrol, and eugenol and exhibiting the lowest minimum inhibitory concentration (MIC) ranging from 62.5 to 4000 s μg/mL [22]. However, GA derived from *Euplexaura erecta* showed minimal antibacterial activity against *Pseudomonas aeruginosa* [23].

Beyond the photodynamic applications, azulene and its derivatives have shown promise in combating viruses and microbes. Azulene derivatives, known for their vivid blue hue and presence in essential oils, have emerged as compounds of interest in the fight against viral infections. Recent research has shed light on their potential as antiretroviral agents, particularly against the HIV-1 virus. These derivatives exhibit a unique mechanism of action, primarily inhibiting the reverse transcriptase enzyme, which is pivotal in the viral replication process. Exploring azulene derivatives in antiretroviral therapy offers a promising avenue for developing new treatments that could revolutionize the management of viral diseases. Peet *et al.* evaluated 18 azulene derivatives for their cytotoxicity and anti-HIV-1 properties [24]. They first assessed the toxicity of these compounds on U2OS cells, then tested the least toxic ones against HIV-1 using virus-like particles. The most promising results came from 2-hydroxyazulenes **9.9**, particularly compounds **9.10** and **9.11**, with compound **9.11** showing the highest efficacy with an IC_{50} value of 8.5 μM and a selectivity index of 7. These findings suggest that azulene derivatives could inhibit reverse transcriptase, although other antiviral actions were not ruled out, indicating their potential as antiretroviral drugs.

The antimicrobial capabilities of azulene were further explored against specific fungi and bacteria. Notably, Murafuji *et al.* synthesized a series of carboxylate derivatives of heterocyclic organobismuth (III) with azulene and investigated their antifungal effects on *Saccharomyces cerevisiae*, using DMSO as a control [25]. Compounds **9.12–9.15**, which are highly lipophilic, showed a notable ability to inhibit fungal growth. The bismuth-containing cationic heterocyclic structure was key to this inhibitory effect, while the carboxylate anion enhanced drug uptake and bioavailability.

9.2.3 Anticancer Effects of Azulene and Its Derivatives

Azulene and its derivatives are gaining attention as potential anticancer agents [8, 18]. The PDT with guaiazulene has shown promise in generating singlet oxygen and reducing inflammation in blood cells. Chamazulene (**9.3**), a potent antioxidant found in *Artemisia arborescens*, has demonstrated effectiveness against melanoma cells. Studies by Uehara *et al.* revealed that the cytotoxicity of alkylaminoguaiazulene derivatives against cancer cells increases with longer alkyl chains, while adding certain atoms reduces it [26]. One derivative exhibited high specificity and induced apoptosis in oral cancer cells. Alkoxyl guaiazulene-3-carboxylates are also being explored as anticancer agents. Research by Hong *et al.* on azulene-indoles showed that one compound was especially effective against lung and breast cancer cell lines, while others displayed selective activity against leukemia and other cancers [27]. The exact mechanisms of these compounds remain to be fully understood.

Numerous natural substances are being explored for their potential in cancer treatment [28]. Guaiazulene (**9.4**) and its derivatives are particularly promising, with chamazulene—derived from *Artemisia arborescent* leaves and comprising 63% GA—demonstrated effective pro-apoptotic actions on cancerous cells [29]. Chamazulene (**9.3**) also surpasses ascorbic acid and tocopherol in antioxidant capacity. GA has been observed to hinder the growth of neuronal and N2a neuroblastoma (N2a-NB) cells, altering levels of free radicals and antioxidants [30]. Although GA exhibits cytotoxic effects on N2a-NB and rat neuron cells *in vitro*, its limited absorption in N2a-NB cells and toxicity to healthy neurons pose challenges for its use in cancer chemotherapy [31]. Other sesquiterpenes like trans-farnesol, cis-nerolidol, and alpha-humulene have also shown *in vitro* anti-proliferative properties against Caco-2 cells.

Studies have revealed that GA can combat non-small cell lung cancer (NSCLC) by disrupting mitochondrial function, triggering oxidative stress, and depleting mitochondrial ATP, leading to anoikic cell death [32]. GA also promotes autophagy and hampers the Akt/mTOR pathway in NSCLC cells. When used alongside paclitaxel, GA's anticancer efficacy is enhanced. Additionally, GA treatment has been found to reduce integrin-3 levels and Akt phosphorylation in A549 and H1975 NSCLC cell lines [33]. Alkoxyl guaiazulene-3-carboxylates are being considered as a foundational compound for new cancer drugs due to their proven effectiveness against oral cancer cells [34]. Further research into GA's anticancer effects across various types of cancer could establish it as a viable anticancer agent.

Research by Nakamura and Yamamoto introduced azulene-modified carborane compounds that, despite their low toxicity, show a high uptake in melanoma cells, suggesting their utility as boron delivery agents for neutron capture therapy [35]. Similarly, azuleno[6,5-b]indole derivatives have shown promising anticancer properties against a range of cancer cell types [27].

The innovative development of azulene synthesis represents a significant advancement in medicinal chemistry, particularly due to its potential in cancer therapy. By employing 2H-cyclohepta[b]furan-2-ones as starting materials and reacting them with a variety of reactants such as olefins, active methylenes, enamines, and silyl enol ethers, researchers have been able to explore and document their reactivity and properties extensively [13].

This pioneering work contributes substantially to the field of azulene derivatives, which are increasingly recognized for their promising anticancer properties. AzuFluor® 483-Bpin, a fluorescent azulene derivative, reacts with ROS and nitrogen species within living cells, illuminating the presence of these molecules linked to various diseases, including cancer and cardiovascular conditions [36]. The anticancer potential of a specific azulene derivative, identified as 3,8-dimethyl-5-(propan-2-yl)-N-propylazulene-1-carboxamide **9.16** [7, 37]. This compound was tested for its cytotoxic effects on four human oral squamous cell carcinoma cell lines and three normal oral cell lines. The findings revealed that this derivative had a higher tumor specificity and partial selectivity strength compared to the standard chemotherapy drug doxorubicin. The derivative's tumor-targeting ability was closely associated with its hydrophobic nature and molecular structure, as determined by quantitative structure–activity relationship analysis. These promising results suggest that further research is warranted to confirm the anticancer efficacy of this guaiazulene derivative, which shows selective action against cancer cells and the ability to trigger cell death. Imanari *et al.* [38] evaluated the cytotoxic effects of 21 azulene amides on cancerous and normal cells. One derivative, **9.17**, stood out for its high cancer cell specificity, supported by TS and PSE metrics. This compound's action may involve non-apoptotic pathways, as suggested by apoptosis analysis. The study also found a link between anticancer properties and the molecular shape and lipophilicity. Interestingly, some azulene derivatives might promote normal cell growth, highlighting the need for further research to develop targeted azulene-based anticancer therapies.

Naitoh *et al.* in 2019 [18] presented a significant advancement in the search for effective anticancer agents. The researchers synthesized a series of 4,6,8-trimethylazulene amide (9.18) derivatives with 15 different substitutions and assessed their antitumor potential against oral malignant cells. Their methodical approach involved evaluating the anticancer activity of these compounds by their specificity to four human oral squamous cell carcinoma cell lines, compared to three normal oral cell types. Additionally, they assessed neurotoxicity by comparing cytotoxic effects on three neuronal cell lines against normal oral cells. The results were promising; among the 15 derivatives tested, compounds with substitution of $(CH_2)_3CH_3$, $(CH_2)_5CH_3$, $(CH_2)_2C_6H_5$ exhibited the highest anticancer activity. Notably, these compounds demonstrated lower neurotoxicity compared to several established chemotherapy agents. Furthermore, they induced a significant accumulation of a subG1 population, indicative of apoptosis, albeit with a comparatively lower level of caspase activation compared to actinomycin D. The study also delved into a quantitative structure–activity relationship analysis, which revealed a significant correlation between tumor specificity and the three-dimensional shape of the molecules. The ability of the compounds to target cancer cells with specificity and reduced neurotoxicity positions them as promising candidates in the ongoing battle against cancer. The study's insights into the molecular mechanisms of these compounds open new avenues for further research and development in the field of oncology.

A study concentrated on identifying effective treatments that target energy metabolism within cancer cells [39]. Given the resistance of cancer cells to targeted molecular therapies, the compound **9.19** garnered significant attention. Research on this compound's mechanism was conducted using cancer cell lines and the *C. elegans* animal model. Findings indicate that derivative **9.19** disrupts oxidative phosphorylation by inhibiting mitochondrial electron transport chain complex II, thereby impeding cellular energy production. This compound is a potential new anticancer agent with selective action against cancer cell mitochondria. Its disruptive impact on energy metabolism and its role in inducing apoptosis warrant further investigation for its application in cancer treatment. Researchers have developed new azulene derivatives that inhibit tyrosine kinase receptors, including FLT-3, which is crucial for leukemia development. Through structure-activity relationship (SAR) analysis and subsequent modifications, a particularly effective compound, **9.20**, was identified as an FLT-3 inhibitor. This breakthrough contributes significantly to the development of therapeutic tyrosine kinase inhibitors and opens new avenues for creating more effective anticancer drugs targeting FLT-3-activated cancers, highlighting azulene's importance in drug design against tyrosine kinase receptors.

The investigation of three azulene hydrazide-hydrazone derivatives (**9.21–9.23**) demonstrated a strong affinity for iron(II) ions and selective cytotoxicity against pancreatic cancer cell lines over healthy fibroblasts BJh-TERT. The anticancer effects are associated with increased NDRG1, HIF-1α, and TfR1 expression, and compound accumulation in mitochondria and lysosomes, indicating mitophagy induction as the action mechanism. These findings suggest azulene hydrazide-hydrazones as potential candidates for developing targeted anticancer therapies [40]. Figure 9.4 shows potential derivatives of azulenes with anticancerous properties.

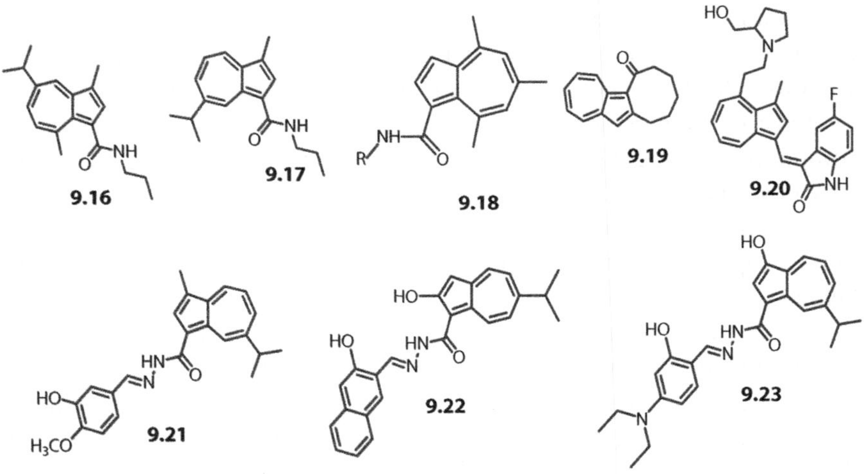

Figure 9.4 Structures of 3,8-dimethyl-5-(propan-2-yl)-*N*-propylazulene-1-carboxamide and derivatives of azulenes with anticancer properties (**9.16–9.23**).

9.2.4 Anti-Inflammatory Action of Azulene Derivatives

The body uses inflammation as a defensive mechanism to improve physiological adjustments that reduce tissue damage and pathogenic infection. These processes involve a complex cascade of cellular and modular reactions, such as fluids containing plasma proteins exuding from arterioles, venules, and capillaries, and dilatation of these anatomical structures with enhanced vascular permeability. However, under constant stimulation, the acute inflammatory response will turn into a chronic inflammation cascade if not controlled. Non-steroidal anti-inflammatory medicines (NSAIDs) are the most widely used anti-inflammatory medications [41]. The NSAIDs mainly attack cyclooxygenase (COX) biologically [42]. The essential cyclooxygenase enzyme is found in two isoforms: COX-2 is the second form, and COX-1 is the constitutive form. While COX-2 is expressed during inflammation, COX-1 is found in the majority of tissues. On the other hand, selective COX-2 inhibitors have been shown to have serious adverse effects on the cardiovascular system, whereas conventional NSAIDs block both cyclooxygenase enzymes, which causes gastrointestinal side effects [43]. Thus, the scientific community is searching to find out innovative molecules with improved profiles against chronic inflammation to address these disadvantages.

Owing to the extraordinary physicochemical properties of azulene derivatives, these derivatives can be used in various technical applications, e.g., molecular switching [44], sensors [45], different elements of optoelectronic devices [46], and solar cells [47]. However, azulene and its derivatives are known for their different biological properties. Including other bioactivities, the anti-inflammatory potential of azulene derivatives (Figure 9.5) has been reported in various studies, which are discussed below.

Yokota *et al.* reported the synthesis and anti-inflammatory activity of hydroxylated-isopropylazulenes sulphonamide derivatives [48]. These molecules were generated with 6-isopropylazulenes *via* regioselective oxidation and exhibited potential activity as TXA_2/PGH_2 receptor antagonists. The biological activities were carried out in isolated rat thoracic aorta preparations and compared with two non-prostanoid antagonists: sulotroban 28 and KT2-962. Most synthesized azulene derivatives have shown comparable or even better activities than the control used. Structure-activity relationships revealed that chlorine substitution at the 4-position of the aryl ring and position 3 of the azulene ring (compounds **9.24** and **9.25**) showed the most potent activities. The *in vivo* results displayed that pre-treatment with **9.24** and **9.25** in a 3.0 mg kg^{-1} p.o. dose stopped sudden death in mice, which was U-46619-induced [48].

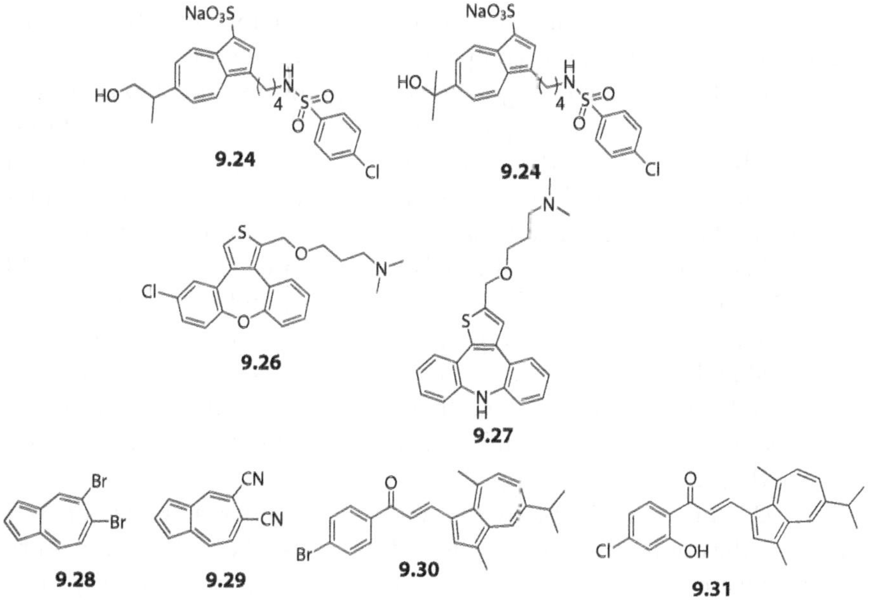

Figure 9.5 Structures of azulene derivatives with anti-inflammatory properties.

Landek *et al.* reported two series of dibenzo[*e,h*]azulenes *viz* 2,8-dithia-dibenzo[*e,h*]azulenes and 8*H*-1-Thia-8-aza-dibenzo[*e,h*]azulene analogues [49]. These two series of molecules were synthesized starting with the oxidation of 11*H*-dibenzo[*b,f*]thiepin-10-one and 5*H*-dibenzo[*b,f*]azepine, respectively. All the molecules in the series were assessed for *in vitro* anti-inflammatory effects. The synthesized first series was evaluated for its ability against TNF-α secretion in the LPS-activated human peripheral blood mononuclear cells assay. Among the series, the compounds (**9.26**), having an aminoalkoxy substituent at the thiophene ring, displayed efficiency to inhibit TNF-α production *in vitro* in a very low range (micromolar) with IC_{50} in the range of 1-3 μM. The second series of novel tetracyclic molecules was also evaluated against TNF-α secretion using the same assay. In this series, compound **9.27**, having an aminoalkoxy substituent at the C-2 position, displayed inhibitory effects towards TNF-α production *in vitro* at a very low range with IC_{50} values of 1–3 μM. Thus, both series had the same biological profile [49, 50].

In an interesting report by Ayaz *et al.*, the anti-inflammatory potential of cyano and bromo azulene derivatives (compounds **9.28** and **9.29**) was evaluated [51]. The bromo-substituted azulene exhibited the capability to inhibit TNFα and IL6 production, whereas the cyano-substituted molecule only suppressed IL6 production in LPS-stimulated macrophages and left the TNF-α production unaffected. These derivatives demonstrated their immunostimulatory potential, which was absent in unsubstituted azulene. Additionally, they displayed comparable or enhanced anti-inflammatory effects on activated macrophages compared to unsubstituted azulene and salicylic acid, commonly employed in anti-inflammatory medications, particularly in reducing IL6 production. The group further extended their studies and assessed the photodynamic anti-inflammatory activities of **9.28** and **9.29**. In the investigation, it was observed that at the time of mammalian macrophages, J774.2 cells were exposed to compounds **9.28** and **9.29** alongside LPS under dark conditions, and these compounds exhibited anti-inflammatory properties at their highest concentrations, as evidenced by ELISA analysis showing reduced levels of pro-inflammatory cytokines. Upon exposure to light, both derivatives demonstrated robust anti-inflammatory effects, significantly diminishing the production levels of TNF, IL6, GMCSF, and IL12p40 cytokines [52].

In 2023, Li *et al.* evaluated the anti-inflammatory activities of four newly synthesized series of guaiazulene derivatives. The anti-inflammatory activities were evaluated three days post-fertilization with a healthy macrophage fluorescent transgenic zebrafish model. During this experiment, the drug Indomethacin was used as a positive control. Most of the guaiazulene-chalcone hybrids exhibited excellent anti-inflammatory activities, with the inhibitory rate ranging from 9.23% to 34.29%. With compounds **9.30** and **9.31**, neutrophil accumulation in the damaged areas significantly decreased following treatment with these guaiazulene derivatives. Interestingly, the inhibitory rate of **9.31** was found to be 34.29%, which was even better than the standard Indomethacin (27.40%), and the migration rate of inflammatory cells was decreased potentially [53].

9.2.5 Anti-Ulcer Properties of Azulene Derivatives

Ulcers, particularly peptic ulcers, represent a prevalent gastrointestinal disorder affecting millions of individuals worldwide. Characterized by painful sores or lesions in the stomach lining (gastric ulcers) or the upper part of the small intestine (duodenal ulcers), these conditions can lead to significant discomfort, complications, and reduced quality of life if left untreated. Historically, the management of ulcers has been challenging, with conventional therapies focusing primarily on symptom relief rather than addressing the underlying causes [54]. There has been a growing emphasis on developing anti-ulcer molecules to alleviate symptoms, promote ulcer healing, and prevent recurrence. These molecules encompass a diverse range of compounds, including pharmaceutical drugs, natural products, and synthetic agents, each with unique mechanisms of action and therapeutic potential. Several azulene derivatives have been used to treat ulcers (Figure 9.6).

Yanagisawa *et al.* synthesized several sodium alkylazulene sulfonate derivatives and evaluated their anti-ulcer potential in Shay pylorus-ligated rats. These azulene derivatives were synthesized *via* multi-step reactions and obtained in good yields. Among the series compounds, **9.32** displayed an extremely potent inhibitory action in the used model. It was even showing better results compared to the guaiazulene sodium sulfonate. The values of log *P* were also studied for the SAR. The optimal value of log *P* was displayed by compound **9.32**, which was found to be -1.0 [14]. The same group has reported another series of alkylazulene-1-sodium sulfonates, which have an isopropyl group at position 6. These derivatives were also assessed for anti-ulcer behavior and were found effective. The most potent compound, **9.33**, exhibited a log *P* value of -0.46 (Figure 9.6) [55].

Yin *et al.* reported several derivatives of azulene sulfonamides, along with some Schiff bases and sulfonyl hydrazones variants of guaiazulene. They synthesized, characterized, and assessed the compounds for their efficacy in treating gastric ulcers. A significant portion of the tested compounds exhibited promising activity against gastric ulceration. Notably,

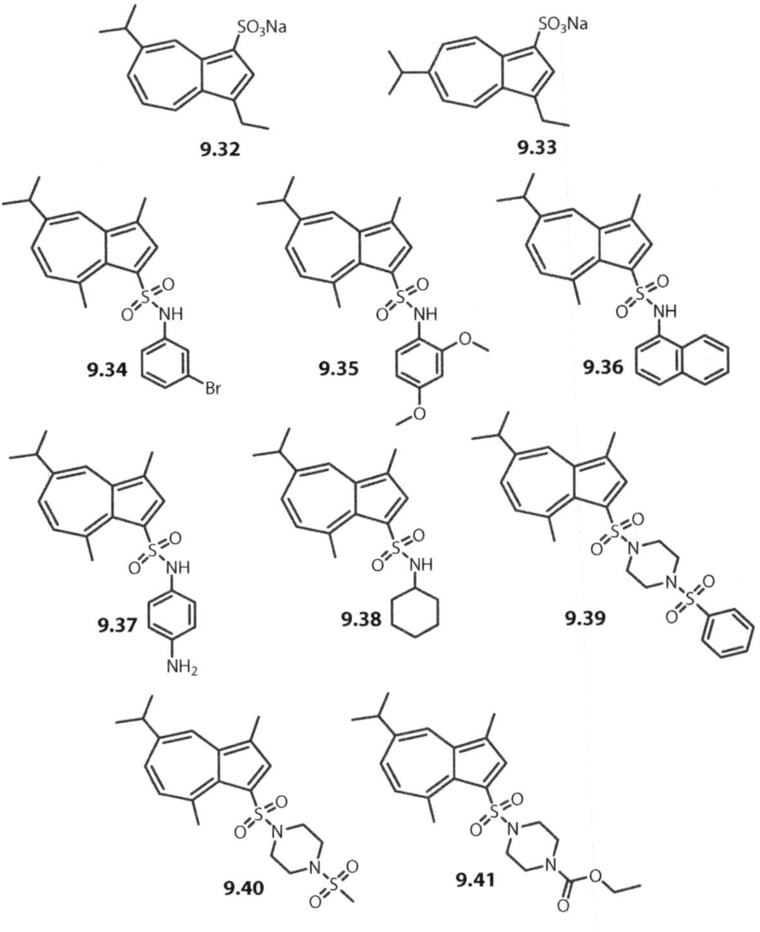

Figure 9.6 Structures of azulene derivatives with anti-ulcerative properties.

modifications at the C-1 position of guaiazulene, particularly the introduction of sulfanilamide, led to substantial inhibitory effects against gastric ulcers. Among these modifications, the lipophilic flat structure demonstrated notably potent antigastric ulcer activity, exemplified by compounds **9.34–9.38** [56].

Inspired by the research results that demonstrated the noteworthy antigastric ulcer effects of guaiazulene and sodium guaiazulene sulfonate, the same group synthesized two series of guaiazulene derivatives and subsequently assessed their efficacy in combating gastric ulcers. The discussion revolved around the data acquired from *in vivo* testing of these compounds utilizing a rodent ethanol-induced stomach injury model. Notably, among the compounds tested, **9.39–9.41** displayed superior antigastric ulcer activity compared to the positive control Omeprazole (2.005 ±1.011*), as evidenced by their ulcer index values (1.125 ±1.246**, 1.714 ±0.756*, 1.875 ±1.126*) [57].

9.2.6 Antioxidant Effects of Azulene and Its Derivatives

Studies suggest that azulene **9.1** and its derivatives possess antioxidant properties, which can help neutralize ROS generated during UV radiation exposure [7]. Azulene, abundant in *M.chamomile* with a concentration ranging from 1–15% in dried flowers, is known for its antioxidant effects [58]. Cadalene from Myrcia *sylvatica* leaves, with concentrations between 4.7–8.2%, shows antioxidant properties [59]. Guaiazulene, derived from *M. chamomile* with a dried flower concentration of 12.42%, exhibits notable antioxidant properties [60]. The dried flowers of *Euplexaura erecta*, containing 0.2% guaiazulene, display strong bioactive properties [61]. Similarly, *Pachyclavularia violacea* soft coral, which comprises 0.4% guaiazulene, also showcases these antioxidant properties [62].

These findings underscore the diverse antioxidant activities of these compounds, highlighting their potential therapeutic applications. These properties are attributed to their ability to scavenge free radicals and inhibit oxidative stress, thereby protecting cells and tissues from damage. During the 2-diphenyl-1-picrylhydrazyl antioxidant assay, 3-Vinylguaiazulene **9.42**, a derivative of guaiazulene, exhibits significant antioxidant activity [63]. The investigation performed in a rat liver model has shown that guaiazulene **9.4** effectively scavenges hydroxyl radicals and 2-diphenyl-1-picrylhydrazyl, inhibiting lipid peroxidation. The *in vitro* antioxidant assays result demonstrates that guaiazulene **9.4** inhibits lipid peroxidation in microsomal cell membranes, performing comparably well to traditional antioxidants like quercetin and propyl gallate, with a significant IC_{50} value [64].

Guaiazulene's mechanistic pathway involves scavenging peroxyl or alkoxyl intermediate radicals through the allylic hydrogen atom present at position 7 of the linked isopropyl group, thereby preventing allylic hydrogen abstraction from lipids [5, 65, 66]. The resultant guaiazulene radical is stabilized by resonance with the benzene ring moiety, making it less reactive. The high lipophilicity of GA, indicated by a conceptual LogP value of 5.74, enhances its accessibility, interaction, and retention within biomembranes undergoing oxidative stress [67, 68].

9.2.7 Dermatological Effects of Azulene and Its Derivatives

Recent research shows that azulene and its derivatives are promising for treating dermatological diseases like itching and atopic dermatitis [7, 69]. In 2022, a clinical study with 23 patients found that using 0.02% guaiazulene and 1% ceramides improved skin health and condition [70]. After four weeks, skin moisture increased by about 64%, roughness decreased by 4%, and itching reduced by nearly 70%. Additionally, sleep quality improved by 74% in atopic dermatitis patients, highlighting the potential benefits of this treatment.

Stresses, physiological reactions, and infections often cause nipple discomfort and fissures in women. Hydrogels, warm compresses, tea bags, mint sweat, collagenase ointment, honey, dexpanthenol, breast milk, and lanolin are some of the several therapies available. However, prolonged usage of these treatments may result in negative side effects and skin issues [71]. Applying guaiazulene **9.4** to the nipple and areola in non-breastfeeding women may increase sensitivity, suppleness, and protect against different discomfort. In a study with 20 patients, a topical pomade with guaiazulene (0.05 g/100 g) showed quick and positive effects on diaper dermatitis in newborns, with noticeable improvements by the 1st and 3rd days [72]. This pomade prevents skin dryness, reduces inflammation, strengthens the outer skin layer, and protects against irritants. Additionally, guaiazulene can treat canker sores and gastritis [73]. The outcomes showed that guaiazulene was a more successful alternative treatment than conventional techniques. These findings offer promising evidence for the potential of azulene and its derivatives in treating dermatological diseases, paving the path for advanced future research.

9.2.8 Cardioprotective Effects of Azulene and Its Derivatives

Cardiovascular diseases are the leading cause of death globally, affecting 1.7 billion people [74]. Azulene derivatives such as guaiazulene have shown positive effects on cardiovascular health [75]. Researchers from the Medical University of Yamanashi in Japan developed these compounds to be cardioprotective, using the antianginal drug nicorandil as a model [76]. They synthesized and compared the effectiveness of guaiazulene, 4,6,8-trimethylazulene, and azulene on coronary blood flow using perfused dog hearts. The effectiveness was ranked as follows: Guaiazulene > 4,6,8-trimethylazulene > azulene. In rabbit aorta strips, the azulene-1-carboxamidine **9.42** derivative HNS-321 exhibited potential vascular relaxant effects [74]. The SAR studies indicated that HNS-32's depressor action is due to the presence of pyridine ring, while its anti-arrhythmic and adverse chronotropic effects are linked to the methyl groups present at positions 3- and 8- on guaiazulene **9.4**.

9.2.9 Hepatoprotective Effects of Azulene and Its Derivatives

During the COVID-19 pandemic, liver damage has been linked to higher mortality and ICU rates, with a study showing 18% and 56% of non-severe and severe patients had high aspartasetransferase levels [5, 77, 78]. Treatments for COVID-19, including paracetamol and other drugs, can also cause liver damage [79]. As liver disease becomes more common, new therapies are needed to prevent hepatotoxicity from synthetic drugs. Herbal medications, including guaiazulene and related molecules, show promise for liver protection [80].

Guaiazulene neutralizes harmful radicals and prevents lipid peroxidation in the liver, providing potential antioxidant benefits [81]. It protects against paracetamol toxicity by inhibiting liver enzymes (CYP450s) and preventing glutathione (GSH) depletion [82]. Paracetamol is metabolized into N-acetyl-p-benzoquinone imine (NAPQI) by liver enzymes (CYP450s), which in turn depletes GSH in a dose-dependent manner. Glutathione is responsible for detoxifying NAPQI, but when GSH gets depleted, NAPQI attaches to sulfhydryl groups on proteins, forming the toxic adduct, leading to liver necrosis [83]. This suggests possible therapeutic uses of guaiazulene in treating various liver diseases. Guaiazulene, administered intraperitoneally at a single dose of 250 mg/kg in rats, reduces paracetamol-induced liver damage by inhibiting lipid peroxidation, preventing oxidative stress, and increasing glutathione levels, showing antioxidant and hepatoprotective effects [64]. The mechanism involves chain-breaking of the oxidant and protecting the liver from paracetamol-induced hepatotoxicity. These properties suggest potential therapeutic use in treating other hepatotoxic conditions [5].

9.2.10 Neuroprotective Effects of Azulene and Its Derivatives

Azulene and its related analogs were evaluated for their activities against various neurological disorders. In 2012, Lober *et al.* reported a novel pathway for the synthesis of azulene derivatives, expressing their activities for monoaminergic G protein-coupled receptors, which also included dopamine, serotonin, histamine, and α-adrenergic receptors, and performed their SAR analysis [84]. Among the series, molecule **9.43** showed a higher activity than apomorphine and revealed excellent D_4 receptor affinity and selectivity with EC_{50} at 0.41 nM. The study suggested that these derivatives can be beneficial for erectile dysfunction (Figure 9.7).

In 2017, Luo *et al.* [85] discovered an azulene derivative *ergosterol ganotheaecolum* **9.44**, which is present in the mushroom *Ganoderma theaecolum*. It was observed that this molecule, **9.44**, was used as a medicine in China for treating neurological disorders. The group has evaluated the neurite outgrowth-promoting activity in PC12 cells using nerve growth factors, which further confirmed that molecule **9.44** can stimulate cell differentiation in a dose-dependent manner and reach the maximum effect at 10μM.

In another study conducted by Leino *et al.*, a series of 70000 azulene-based compounds were screened, which target orexin receptors [86]. Among all the compounds, two compounds, **9.45** and **9.46**, showed the most promising binding affinity. These molecules were evaluated orthogonally in a functional assay, and it was found that they have K*i* values in the low micromolar range (3–9 μM). Furthermore, molecules **9.47** and **9.48** were identified as weak OX receptor agonists, including orexin receptor-mediated Ca^{+2} responses of 5% of the maximal response to orexin-A.

In 2019, Turku *et al.* developed a series of compounds **9.49** and **9.55** for synthesizing orexin (hypocetine) OX1 and OX2 receptor ligands by utilizing *in silico* software studies [87]. These receptors were involved in sleep-wake regulation and showed activity for treating insomnia, narcolepsy, and

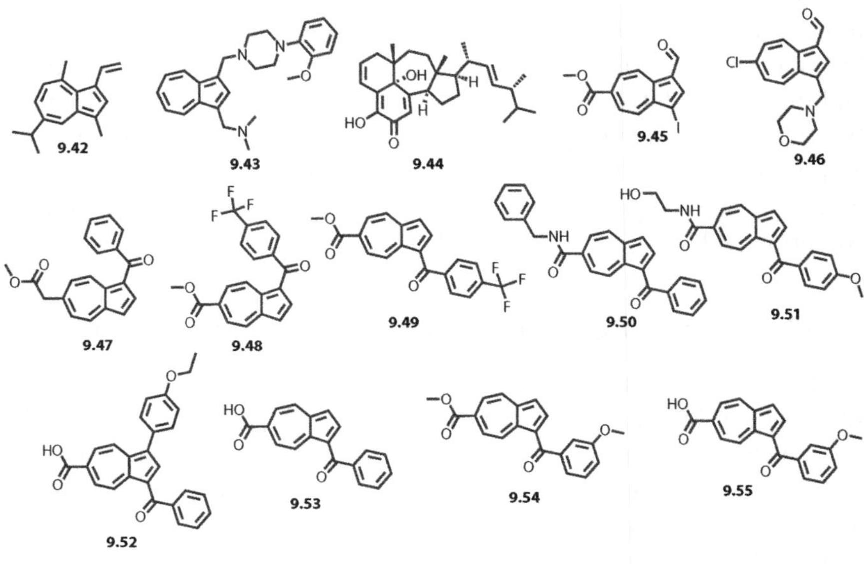

Figure 9.7 Structure of 9.50–9.56 having neurological effects.

certain cancers. The group has identified the seven potentiators and two weak dual orexin receptors at a concentration of 10μM. The study of these compounds gave high densities of human OX1 and OX2 receptors when monitored by Ca^{+2} response in Chinese hamster ovary (CHO)- K1 cells.

9.3 Challenges and Future Prospects

The synthesis and application of azulene derivatives present both challenges and opportunities. The **complexity of synthesis**, **stability issues**, **biocompatibility concerns**, and **regulatory hurdles** pose significant challenges. However, the **diverse biological activities** of azulene derivatives, their potential in **optoelectronic applications**, possibilities with **nanotechnology integration**, and the prospects for **personalized medicine** present exciting opportunities. In 2023, a team led by Maruoka K. isolated an azulene derivative from the fungus *L. indigo*, aiming to develop a blue food dye [88]. However, the use of Tween 80 as a solubilizer introduced a bitter taste and azulene's strong odor, which could affect food flavor, but these issues are not problematic for therapeutic or skincare applications. In 2024, García-Martínez *et al.* synthesized phenethylamine-azulene conjugates, focusing on the aziridine ring opening mechanism, highlighting the importance of the β-aryloethylamine scaffold in medicinal chemistry [89]. A.C. Razus investigated different synthesis techniques for benz[α]azulenes and related compounds, suggesting potential in cancer treatment and skin disease therapy [90]. Examples of new derivatives of azulene with significant potential are given in Figure 9.8.

The exploration of azulene and its derivatives is a multifaceted endeavor, aiming to harness their unique properties for various applications. The **design and synthesis** of new azulene-containing chalcones are at the forefront, with a focus on enhancing their optical and fluorescent characteristics. These improvements could render them invaluable as **chemical sensors and dosimeters**, changing color or fluorescence in response to

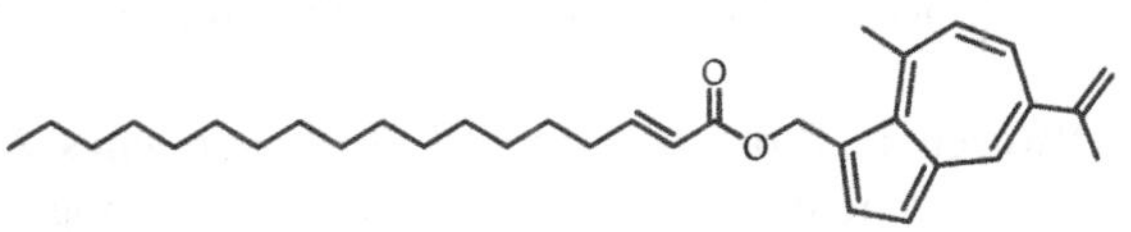

9.56

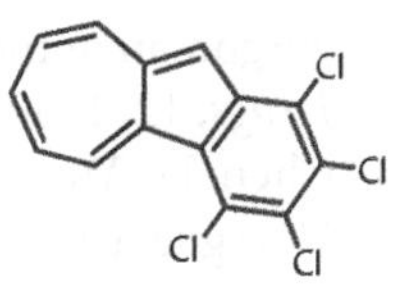

9.57

Figure 9.8 Structures of azulene derivatives with potential medicinal applications in future research.

specific stimuli, thus providing a visual or measurable indication of environmental changes. Simultaneously, there is a concerted effort to **investigate the pharmacological aspects** of azulene-based systems. This includes assessing their toxicity, stability, solubility, bioavailability, and selectivity within different biological environments. The goal is to optimize their pharmacokinetic and pharmacodynamic profiles, ensuring that these compounds are safe, stable, and effective when used in medical contexts. Furthermore, there is a keen interest in **understanding the molecular mechanisms** by which azulene-based systems exert their effects on various diseases. By exploring their role in conditions such as inflammation, cancer, diabetes, HIV, and fungal infections, researchers aim to uncover how these compounds interact with biological pathways and targets. This could lead to the development of more targeted and effective treatments, leveraging the specificity of azulene derivatives to combat disease. Lastly, the **development of novel azulene-based systems** for sensing and imaging represents a cutting-edge application. These systems are being engineered to possess enhanced specificity and sensitivity, making them suitable for use in diagnostics, therapeutics, and biosensors. Their application could revolutionize the detection and visualization of biological and chemical substances, offering precise and tailored approaches to medical diagnostics and treatment strategies. Combined, these research avenues underscore the potential of azulene derivatives to contribute significantly to the fields of medicine, pharmacy, and environmental monitoring.

AI disclosure statement: The author acknowledges the use of AI-based tools for minor language editing or grammatical correction. However, all intellectual contributions, analyses, and conclusions presented in this chapter are entirely the author's own.

References

1. Zheng, J.-J., Shao, C.-L., Chen, M., Gan, L.-S., Fang, Y.-C., Wang, X.-H., Wang, C.-Y., Ochracenoids A and B, Guaiazulene-based analogues from gorgonian anthogorgia ochracea collected from the South China Sea. *Mar. Drugs*, 12, 1569, 2014.
2. Chen, D., Yu, S., van Ofwegen, L., Proksch, P., Lin, W., Anthogorgienes, AO., New guaiazulene-derived terpenoids from a Chinese gorgonian anthogorgia species, and their antifouling and antibiotic activities. *J. Agric. Food. Chem.*, 60, 112, 2012.

3. Podkowa, A., Kryczyk-Poprawa, A., Opoka, W., Muszyńska, B., Culinary–medicinal mushrooms: A review of organic compounds and bioelements with antioxidant activity. *Eur. Food Res. Technol.*, 247, 513, 2021.
4. Nicholas, G.M., Australasian fungi: A natural product study, Ph.D. Thesis, University of Canterbury, Christchurch, New Zealand, 1998.
5. Akram, W., Tagde, P., Ahmed, S., Arora, S., Emran, T.B., Babalghith, A.O., Sweilam, S.H., Simal-Gandara, J., Guaiazulene and related compounds: A review of current perspective on biomedical applications. *Life Sci.*, 316, 121389, 2023.
6. Hoferl, M., Wanner, J., Tabanca, N., Ali, A., Gochev, V., Schmidt, E., Kaul, V.K., Singh, V., Jirovetz, L., Biological activity of matricaria chamomilla essential oils of various chemotypes. *Planta Med. Int. Open*, 7, e114, 2020.
7. Slon, E., Slon, B., Kowalczuk, D., Azulene and its derivatives as potential compounds in the therapy of dermatological and anticancer diseases: New perspectives against the backdrop of current research. *Molecules*, 29, 9, 2020, 2024.
8. Bakun, P., Czarczynska-Goslinska, B., Goslinski, T., Lijewski, S., *In vitro* and *in vivo* biological activities of azulene derivatives with potential applications in medicine. *Med. Chem. Res.*, 30, 834, 2021.
9. Andersen, F.A., Final report on the safety assessment of azulene. *Int. J. Toxicol.*, 18, 27, 1999.
10. Zargaran, A., Borhani-Haghighi, A., Salehi-Marzijarani, M., Faridi, P., Daneshamouz, S., Azadi, A., Sadeghpour, H., Sakhteman, A., Mohagheghzadeh, A., Evaluation of the effect of topical chamomile (Matricaria chamomilla L.) oleogel as pain relief in migraine without aura: a randomized, double-blind, placebo-controlled, crossover study. *Neurol. Sci.*, 39, 1345, 2018.
11. Gunes, T., Akin, M.A., Sarici, D., Hallac, K., Kurtoglu, S., Hashimoto, T., Guaiazulene: a new treatment option for recalcitrant diaper dermatitis in NICU patients. *J. Matern. Fetal Neonatal Med.*, 26, 197, 2013.
12. Pawełczyk, A., Sowa-Kasprzak, K., Olender, D., Zaprutko, L., Molecular consortia—various structural and synthetic concepts for more effective therapeutics synthesis. *Int. J. Mol. Sci.*, 19, 1104, 2018.
13. Shoji, T., Ito, S., Yasunami, M., Synthesis of azulene derivatives from 2H-cyclohepta[b]furan-2-ones as starting materials: Their reactivity and properties. *Int. J. Mol. Sci.*, 22, 10686, 2021.
14. Yanagisawa, T., Wakabayashi, S., Tomiyama, T., Yasunami, M., Takase, K., Synthesis and anti-ulcer activities of sodium alkylazulene sulfonates. *Chem. Pharm. Bull.*, 36, 641, 1988.
15. Yamaguchi, Y., Maruya, Y., Katagiri, H., Nakayama, K.-I., Ohba, Y., Synthesis, properties, and OFET characteristics of 5,5'-di(2-azulenyl)-2,20-bithiophene (DAzBT) and 2,5-Di(2-azulenyl)-thieno[3,2-b]thiophene (DAzTT). *Org. Lett.*, 14, 2316, 2012.
16. Yamaguchi, Y., Ogawa, K., Nakayama, K.-I., Ohba, Y., Katagiri, H., Terazulene: A high-performance n-type organic field-effect transistor based

on molecular orbital distribution control. *J. Am. Chem. Soc.*, 135, 19095, 2013.

17. Yamaguchi, Y., Takubo, M., Ogawa, K., Nakayama, K.-I., Koganezawa, T., Katagiri, H., Terazulene isomers: Polarity change of OFETs through molecular orbital distribution contrast. *J. Am. Chem. Soc.*, 138, 11335, 2016.
18. Naitoh, K., Orihara, Y., Sakagami, H., Miura, T., Satoh, K., Amano, S., Bandow, K., Iijima, Y., Kurosaki, K., Uesawa, Y., Hashimoto, M., Wakabayashi, H., Tumor-specificity, neurotoxicity, and possible involvement of the nuclear receptor response pathway of 4,6,8-trimethyl azulene amide derivatives. *Int. J. Mol. Sci.*, 23, 2601, 2022.
19. Ikegai, K., Imamura, M., Suzuki, T., Nakanishi, K., Murakami, T., Kurosaki, E., Noda, A., Kobayashi, Y., Yokota, M., Koide, T., Kosakai, K., Ohkura, Y., Takeuchi, M., Tomiyama, H., Ohta, M., Synthesis and biological evaluation of C-glucosides with azulene rings as selective SGLT2 inhibitors for the treatment of type 2 diabetes mellitus: Discovery of YM543. *Bioorg. Med. Chem.*, 21, 3934, 2013.
20. Nagai, Y., Suzuki, A., Katsuragi, H., Shinkai, K., Effect of antimicrobial photodynamic therapy (aPDT) on the sterilization of infected dentin *in vitro*. *Odontology*, 106, 154, 2018.
21. Damrongrungruang, T., Kitchindaopat, N., Thanasothon, P., Theeranut, K., Tippayawat, P., Ruangsuwan, C., Suwannee, B., Effects of photodynamic therapy with azulene on peripheral blood mononuclear cell viability and singlet oxygen formation. *Photodiagn. Photodyn. Ther.*, 24, 318, 2018.
22. Sleha, R., Mosio, P., Vydrzalova, M., Jantovska, A., Bostikova, V., Mazurova, J., *In vitro* antimicrobial activities of cinnamon bark oil, anethole, carvacrol, eugenol and guaiazulene against mycoplasma hominis clinical isolates. *Biomed. Pap. Med. Fac. Univ. Palacky Olomouc Czech Repub.*, 158, 2, 208, 2014.
23. Datta, D., Talapatra, S.N., Swarnakar, S., Bioactive compounds from marine invertebrates for potential medicines–An overview. *Int. Lett. Nat. Sci.*, 34, 42, 2015.
24. Peet, J., Selyutina, A., Bredihhin, A., Antiretroviral (HIV-1) activity of azulene derivatives. *Bioorg. Med. Chem.*, 24, 1653, 2016.
25. Murafuji, T., Kitagawa, K., Yoshimatsu, D., Kondo, K., Ishiguro, K., Tsunashima, R., Miyakawa, I., Mikata, Y., Heterocyclic bismuth carboxylates based on a diphenyl sulfone scaffold: synthesis and antifungal activity against Saccharomyces cerevisiae. *Eur. J. Med. Chem.*, 63, 531, 2013.
26. Uehara, M., Minemura, H., Ohno, T., *In vitro* antitumor activity of alkylaminoguaiazulenes. *In Vivo*, 32, 541, 2018.
27. Hong, B.-C., Jiang, Y.-F., Kumar, E.S., Microwave-assisted [6+4]-cycloaddition of fulvenes and α-pyrones to azulene–indoles: Facile syntheses of novel antineoplastic agents. *Bioorg. Med. Chem. Lett.*, 11, 1981, 2001.

28. Anwar, S., Malik, J.A., Ahmed, S., Kameshwar, V.A., Alanazi, J., Alamri, A., Ahemad, N., Can natural products targeting EMT serve as the future anticancer therapeutics? *Molecules*, 27, 22, 7668, 2022.
29. Russo, A., Bruno, M., Avola, R., Cardile, V., Rigano, D., Chamazulene-rich Artemisia arborescens essential oils affect the cell growth of human melanoma cells. *Plants*, 9, 8, 1000, 2020.
30. Tŏgar, B., Türkez, H., Tatar, A., Hacimuftuoglu, A., Geyikoglu, F., Cytotoxicity and genotoxicity of zingiberene on different neuron cell lines *in vitro*. *Cytotechnology*, 67, 6, 939, 2015.
31. Tŏgar, B., Çelik, K., Türkez, H., *In vitro* cytotoxic, genotoxic and antioxidant/oxidant effects of guaiazulene on human lymphocytes. *Braz. Arch. Biol. Technol.*, 58, 1, 61, 2015.
32. Chen, C.H., Lee, O., Yao, C.N., Chuang, M.Y., Chang, Y.L., Chang, M.H., Wen, Y.F., Yang, W.H., Ko, C.H., Chou, N.T., Lin, M.W., Lai, C.P., Sun, C.Y., Wang, L.M., Chen, Y.C., Hseu, T.H., Chang, C.N., Hsu, H.C., Lin, H.C., Chang, Y.L., Shih, Y.C., Chou, S.H., Hsu, Y.L., Tseng, H.W., Liu, C.P., Tu, C.M., Hu, T.L., Tsai, Y.J., Chen, T.S., Lin, C.L., Chiou, S.J., Liu, C.C., Hwang, C.S., Novel azulene-based derivatives as potent multireceptor tyrosine kinase inhibitors. *Bioorg. Med. Chem. Lett.*, 20, 20, 6129, 2010.
33. Ye, Q., Zhou, L., Jin, P., Li, L., Zheng, S., Huang, Z., Liu, J., Qin, S., Liu, H., Zou, B., Xie, K., Guaiazulene triggers ROS-induced apoptosis and protective autophagy in non-small cell lung cancer. *Front. Pharmacol.*, 12, 621181, 2021.
34. Teratani, M., Nakamura, S., Sakagami, H., Fujise, M., Hashimoto, M., Okudaira, N., Bandow, K., Iijima, Y., Nagai, J., Uesawa, Y., Wakabayashi, H., Antitumor effects and tumor-specificity of Guaiazulene-3-carboxylate derivatives against Oral squamous cell carcinoma *in vitro*. *Anticancer Res.*, 40, 9, 4885, 2020.
35. Nakamura, H., Sekido, M., Yamamoto, Y., Synthesis of carboranes containing an azulene framework and *in vitro* evaluation as boron carriers. *J. Med. Chem.*, 40, 2825, 1997.
36. Murfin, L.C., Weber, M., Park, S.-J., Kim, W.-T., Lopez-Alled, C.M., McMullin, C.L., Pradaux-Caggiano, F., Lyall, C.L., Kociok-Kohn, G., Wenk, J., Bull, S.D., Yoon, J., Kim, H.M., James, T.D., Lewis, S.E., Azulene-derived fluorescent probe for bioimaging: Detection of reactive oxygen and nitrogen species by two-photon microscopy. *J. Am. Chem. Soc.*, 141, 19389, 2019.
37. Wada, T., Maruyama, R., Irie, Y., Hashimoto, M., Wakabayashi, H., Okudaira, N., Uesawa, Y., Kagaya, H., Sakagami, H., *In Vitro* Anti-Tumor Activity of Azulene Amide Derivatives. *In Vivo Athens Greece*, 32, 479, 2018.
38. Imanari, K., Hashimoto, M., Wakabayashi, H., Okudaira, N., Bandow, K., Nagai, J., Tomomura, M., Tomomura, A., Uesawa, Y., Sakagami, H., Quantitative Structure–Cytotoxicity Relationship of Azulene Amide Derivatives. *Anticancer Res.*, 39, 3507, 2019.

39. Kasami, C., Yamaguchi, J.-I., Inoue, H., Guaiazulene derivative 1,2,3,4-tetra hydroazuleno[1,2-b] tropone reduces the production of ATP by inhibiting electron transfer complex II. *FEBS Open Bio*, 11, 2921, 2021.
40. Brogyányi, T., Kaplánek, R., Kejík, Z., Hosnedlová, B., Antonyová, V., Abramenko, N., Veselá, K., Martásek, P., Vokurka, M., Richardson, D.R., Jakubek, M., Azulene hydrazide-hydrazones for selective targeting of pancreatic cancer cells. *Biomed. Pharmacother.*, 155, 113736, 2022.
41. Lavi, Y., Kogan, N.M., Topping, L.M., Liu, C., McCann, F.E., Williams, R.O., Breuer, A., Yekhtin, Z., Ezra, A.F., Gallily, R., Feldmann, M., Mechoulam, R., Novel synthesis of C-methylated phytocannabinoids bearing antiinflammatory properties. *J. Med. Chem.*, 66, 5536, 2023.
42. Chahal, S., Rani, P., Kiran, Sindhu, J., Joshi, G., Ganesan, A., Kalyaanamoorthy, S., Mayank, Kumar, P., Singh, R., Negi, A., Design and development of COX-II inhibitors: Current scenario and future perspective. *ACS Omega*, 20, 17446, 2023.
43. Stiller, C.-O. and Hjemdahl, P., Lessons from 20 years with COX-2 inhibitors: Importance of dose–response considerations and fair play in comparative trials. *J. Intern. Med.*, 292, 557, 2022.
44. Achatz, J., Fischer, C., Salbeck, J., Daub, J., Functionalized photochromics for molecular switching: The multistabilities of a dihydroazulene–anthraquinone system. *J. Chem. Soc., Chem. Commun.*, 7, 504, 1991.
45. Buica, G.-O., Lazar, I.-G., Birzan, L., Lete, C., Prodana, M., Enachescu, M., Tecuceanu, V., Stoian, A.B., Ungureanu, E.-M., Azulene-ethylenediaminetetraacetic acid: A versatile molecule for colorimetric and electrochemical sensors for metal ions. *Electrochim. Acta*, 263, 382, 2018.
46. Xin, H., Ge, C., Jiao, X., Yang, X., Rundel, K., McNeill, C.R., Gao, X., Incorporation of 2,6-connected azulene units into the backbone of conjugated polymers: Towards high-performance organic optoelectronic materials. *Angew. Chem. Int. Ed.*, 57, 1322, 2018.
47. Cowper, P., Pockett, A., Kociok-Köhn, G., Cameron, P.J., Lewis, S.E., Azulene—thiophene—cyanoacrylic acid dyes with donor-π-acceptor structures: Synthesis, characterisation and evaluation in dye-sensitized solar cells. *Tetrahedron*, 74, 2775, 2018.
48. Yokota, M., Uchibori, S., Hayashi, H., Koyama, R., Kosakai, K., Wakabayashi, S., Tomiyama, T., Azulene derivatives as TXA2/PGH2 receptor antagonists—II: Synthesis and biological activity of 6-mono- and 6-dihydroxylated-isopropylazulenes. *Bioorg. Med. Chem.*, 4, 575, 1996.
49. Landek, I.O., Pešić, D., Merćep, M., Stanić, B., Mesić, M., Synthesis and antiinflammatory activity of 8H-1-thia-8-aza-dibenzo[e,h]azulenes. *J. Heterocycl. Chem.*, 48, 856, 2011.
50. Landek, I.O., Pešić, D., Novak, P., Stanić, B., Nujić, K., Merćep, M., Mesić, M., 2,8-Dithia-dibenzo[e,h] azulenes and their 8-oxa analogs: Synthesis and anti-inflammatory activity. *Heterocycles*, 78, 2489, 2009.

51. Ayaz, F., Yuzer, A., Ince, T., Ince, M., Anti-cancer and anti-inflammatory activities of bromo- and cyano-substituted azulene derivatives. *Inflammation*, 43, 1009, 2020.
52. Yetkin, D., Ince, T., Ayaz, F., Photodynamic anti-inflammatory activity of azulene derivatives on mammalian macrophages and their intracellular mechanism of action. *Photodiagn. Photodyn. Ther.*, 39, 102963, 2022.
53. Ma, Z., Han, X., Ren, J., Liu, K., Zhang, W., Li, G., Design, synthesis, and biological activity of guaiazulene derivatives. *Chem. Biodivers.*, 20, e202201174, 2023.
54. Prayoga, D.K., Aulifa, D.L., Budiman, A., Levita, J., Plants with anti-ulcer activity and mechanism: A review of preclinical and clinical studies. *Drug Des. Dev. Ther.*, 18, 193, 2024.
55. Yanagisawa, T., Kosakai, K., Tomiyama, T., Yasunami, M., Takase, K., Studies on anti-ulcer agents. II. Synthesis and anti-ulcer activities of 6-isopropylazulene-1-sodium sulfonate derivatives. *Chem. Pharm. Bull.*, 38, 3355, 1990.
56. Zhang, L.Y., Yang, F., Shi, W.Q., Zhang, P., Li, Y., Yin, S.F., Synthesis and anti-gastric ulcer activity of novel 5-isoproyl-3,8-dimethylazulene derivatives. *Bioorg. Med. Chem. Lett.*, 21, 5722, 2011.
57. Cao, T., Li, Y., Yang, Z., Yuan, M., Li, Y., Yang, H., Feng, Y., Yin, S., Synthesis and biological evaluation of 3,8-dimethyl-5-isopropylazulene derivatives as anti-gastric ulcer agent. *Chem. Biol. Drug Des.*, 88, 264, 2016.
58. Gupta, V., Mittal, P., Bansal, P., Khokra, S.L., Kaushik, D., Pharmacological potential of Matricaria recutita-a review. *Int. J. Pharm. Sci. Drug Res.*, 2, 1, 12, 2010.
59. de Moraes, A.A., Ferreira, O.O., da Costa, L.S., Almeida, L.Q., Varela, E.L., Cascaes, M.M., Franco, C., de, J.P., Percário, S., Nascimento, L.D., de Oliveira, M.S., Phytochemical profile, preliminary toxicity, and antioxidant capacity of the essential oils of *Myrciaria floribunda* (H. West ex Willd.) O. Berg. and Myrcia sylvatica (G. Mey) DC. (Myrtaceae). *Antioxidants*, 11, 10, 2076, 2022.
60. Romeilah, R.M., Anticancer and antioxidant activities of Matricaria chamomilla L. and Marjorana hortensis essential oils. *Res. J. Med. Med. Sci.*, 4, 2, 332, 2009.
61. Fusetani, N., Matsunaga, S., Konosu, S., Bioactive marine metabolites I. Isolation of guaiazulene from the gorgonian Euplexaura erecta. *Experientia*, 37, 7, 680, 1981.
62. Bowden, B.F., Coll, J.C., Tapiolas, D.M., Studies of Australian soft corals. XXX. A novel trisnorsesquiterpene from a cespitularpia species and the isolation of guaiazulene from a small blue alcyonium species. *Aust. J. Chem.*, 36, 211, 1983.
63. Pratsinis, H. and Haroutounian, S., A.Synthesis and antioxidant activity of 3-substituted guaiazulene derivatives. *Nat. Prod. Lett.*, 16, 3, 201, 2002.

64. Kourounakis, A.P., Rekka, E.A., Kourounakis, P.N., Antioxidant activity of guaiazulene and protection against paracetamol hepatotoxicity in rats. *J. Pharm. Pharmacol.*, 49, 9, 938, 1997.
65. Rekka, E. and Kourounakis, P.N., Effect of hydroxyethyl rutosides and related compounds on lipid peroxidation and free radical scavenging activity. Some structural aspects. *J. Pharm. Pharmacol.*, 43, 7, 486, 1991.
66. Flynn, G.L., Substituent constants for correlation analysis in chemistry and biology. *J. Pharm. Sci.*, 69, 9, 1109, 1980.
67. Müller, K., Leukel, P., Ziereis, K., Gawlik, I., Antipsoriatic anthrones with modulated redox properties. 2. Novel derivatives of chrysarobin and isochrysarobin-antiproliferative activity and 5-lipoxygenase inhibition. *J. Med. Chem.*, 37, 11, 1660, 2002.
68. Ayala, A., Munoz, M.F., Arguelles, S., Lipid peroxidation: production, metabolism, and signaling mechanisms of malondialdehyde and 4-hydroxy-2-nonenal. *Oxid. Med. Cell. Longev.*, 2014, 360438, 2014.
69. Furau, C., Grossmann, H., Furau, G., Vormann, J., Effect of a guaiazulene-containing ointment on nipple and areola area health of women. *J. Cosmet. Dermatol. Sci. Appl.*, 6, 4, 167, 2016.
70. Visscher, M., Odio, M., Taylor, T., White, T., Sargent, S., Sluder, L., Smith, L., Flower, T., Mason, B., Rider, M., Huebner, A., Skin care in the NICU patient: effects of wipes versus cloth and water on stratum corneum integrity. *Neonat.*, 96, 4, 226, 2009.
71. Niazi, A., Rahimi, V.B., Soheili-Far, S., Askari, N., Rahmanian-Devin, P., Sanei-Far, Z., Sahebkar, A., Rakhshandeh, H., Askari, V.R., A systematic review on prevention and treatment of nipple pain and fissure: are they curable? *J. Pharmacopuncture*, 21, 3, 139, 2018.
72. Kazaks, E.L. and Lane, A.T., Diaper dermatitis. *Pediatr. Clin. N. Am.*, 47, 909, 2000.
73. Nelson, S.D., Mechanisms of the formation and disposition of reactive metabolites that can cause acute liver injury. *Drug Metab. Rev.*, 27, 147, 1995.
74. Saglietto, A., Manfredi, R., Elia, E., D'Ascenzo, F., De Ferrari, G.M., Biondi-Zoccai, G., Munzel, T., Cardiovascular disease burden: Italian and global perspectives. *Minerva Cardiol. Angiol.*, 69, 231, 2021.
75. Tanaka, Y. and Shigenobu, K., A review of HNS-32: a novel azulene-1-carboxamidine derivative with multiple cardiovascular protective actions. *Cardiovasc. Drug Rev.*, 19, 297, 2001.
76. Markham, A., Plosker, G.L., Goa, K.L., Nicorandil. An updated review of its use in Ischaemic heart disease with emphasis on its cardioprotective effects. *Drugs*, 60, 955, 2000.
77. Prakash, C., Chhikara, S., Kumar, V., Mitochondrial dysfunction in arsenic-induced hepatotoxicity: pathogenic and therapeutic implications. *Biol. Trace Elem. Res.*, 200, 261, 2022.

78. Ahmad, A., Ishtiaq, S.M., Khan, J.A., Aslam, R., Ali, S., Arshad, M.I., COVID-19 and comorbidities of hepatic diseases in a global perspective. *World J. Gastroenterol.*, 27, 1296, 2021.
79. Assa, A., Benninga, M.A., Borrelli, O., Broekaert, I., de Carpi, J.M., Saccomani, M.D., Dolinsek, J., Mas, E., Miele, E., Thomson, M., Tzivinikos, E., Gastrointestinal committee of European society for paediatric gastroenterology, hepatology and nutrition. Gastrointestinal perspective of coronavirus disease 2019 in children-an updated review. *J. Pediatr. Gastroenterol. Nutr.*, 73, 299, 2021.
80. Pandey, B., Baral, R., Kaundinnyayana, A., Panta, S., Promising hepatoprotective agents from the natural sources: A study of scientific evidence. *Egypt. Liver J.*, 13, 14, 2023.
81. Wang, L., Yan, J., Fu, P.P., Parekh, K.A., Yu, H., Photomutagenicity of cosmetic ingredient chemicals azulene and guaiazulene. *Mutat. Res.*, 530, 19, 2003.
82. Banerjee, S.B., Melnyk, S.B., Krager, K.J., Aykin-Burns, N., McCullough, S.S., James, L.P., Hinson, J.A., Trifluoperazine inhibits acetaminophen-induced hepatotoxicity and hepatic reactive nitrogen formation in mice and in freshly isolated hepatocytes. *Toxicol. Rep.*, 4, 134, 2017.
83. Nakamichi, K., Nakano, T., Yasuura, H., Izumi, S., Kawashima, Y., Stabilization of sodium guaiazulene sulfonate in granules for tableting prepared using a twin-screw extruder. *Eur. J. Pharm. Biopharm.*, 56, 347, 2003.
84. Löber, S., Hübner, H., Buschauer, A., Sanna, F., Argiolas, A., Melis, M.R., Gmeiner, P., Novel azulene derivatives for the treatment of erectile dysfunction. *Bioorg. Med. Chem. Lett.*, 22, 7151, 2012.
85. Luo, Q., Yang, Z.L., Yan, Y.M., Cheng, Y.X., Ganotheaecolin, A., A neurotrophic conjugated ergosterol with a naphtho[1,8-ef]azulene scaffold from ganoderma theaecolum. *Org. Lett.*, 19, 718, 2017.
86. Leino, T.O., Turku, A., Yli-Kauhaluoma, J., Kukkonen, J.P., Xhaard, H., Wallén, E.A.A., Azulene-based compounds for targeting orexin receptors. *Eur. J. Med. Chem.*, 157, 88, 2018.
87. Turku, A., Leino, T.O., Karhu, L., Yli-Kauhaluoma, J., Kukkonen, J.P., Wallén, E.A.A., Xhaard, H., Structure-activity relationships of 1-benzoylazulenes at the OX1 and OX2 orexin receptors. *ChemMedChem*, 14, 965, 2019.
88. Maruoka, K., Suzuki, R., Kamishima, T., Koseki, Y., Ngoc Dao, A.T., Murafuji, T., Kasai, H., Total Synthesis of azulene derivative, A blue pigment isolated from Lactarius Indigo, and colorant application of its aqueous dispersion. *J. Agric. Food. Chem.*, 71, 11607, 2023.
89. García-Martínez, P., Bernardo, O., Borge, J., González, J., López, L.A., Synthesis of phenethylamine-azulene conjugates enabled by regioselective ring opening of Aziridines. *Adv. Synth. Catal.*, 366, 1007, 2024.
90. Razus, A.C., Syntheses of Azulene Embedded Polycyclic Compounds. *Symmetry*, 16, 382, 2024.

Part 3

AZULENE IN POROUS AND COORDINATION CHEMISTRY

10

Azulene-Based Porous Materials: Synthesis and Applications

Debajit Bora, Bikash Chandra Mushahary, Chayanika Goswami and Sanjeev Pran Mahanta*

Department of Chemical Sciences, Tezpur University, Napaam, Tezpur, Assam, India

Abstract

Azulene is a blue-colored, non-benzenoid, non-alternant, bicyclic aromatic hydrocarbon, having a considerably larger dipole moment (=1.08D) than its isomer naphthalene. Azulene possesses unique physicochemical properties such as a narrow band gap and anti-Kasha fluorescence from S_2 to S_0. Although azulene has been known for 100 years, the exploration for the potential use of azulene-based derivatives as structural building units for constructing advanced organic materials is less explored compared to its naphthalene analogs. This chapter primarily discusses how azulene-based novel building blocks are utilized to synthesize various π-functional porous materials for applications in gas storage, organic electronics, energy storage, catalysis, sensing, etc. Also, we describe the design concepts, the synthesis of azulene-based building blocks, and the corresponding porous materials such as metal-organic frameworks, covalent organic frameworks, covalent triazine frameworks, and other porous organic polymers, along with their applications. Finally, we outline key challenges and perspectives regarding azulene-based porous materials.

***Keywords*:** Porous organic polymer, azulene, anti-Kasha fluorescence, metal-organic frameworks, covalent organic frameworks, conjugated microporous polymers, hydrogen-bonded organic frameworks

**Corresponding author*: samahanta@gmail.com; spm@tezu.ernet.in

Neha Rani Kumar and Abhijeet R. Agrawal (eds.) Azulene-Based Materials: Design, Synthesis, and Applications, (277–310) © 2026 Scrivener Publishing LLC

10.1 Porous Materials

Porous materials have regions of space that can selectively adsorb guest molecules, which makes them valuable for various industrial and domestic uses, such as gas separation [1], energy storage [2], ion exchange [3], heterogeneous catalysis [4], and green chemistry [5]. The porous materials are often crystalline or ordered. Pores can be cavities, channels, and interstices present in the materials. Pores can be of different types, like open, closed, blind, and through space based on the accessibility of the pores. Pores in the materials can have different shapes, like cylindrical open, cylindrical blind, ink bottle-shaped, and funnel shapes [6]. The porous structure of a material is characterized by parameters like pore volume, pore size (or diameter), porosity (ratio of the total pore volume to the apparent volume of the material), and surface area. Based on the pore width, porous materials are further classified as microporous (pore width < 2 nm), mesoporous (2 nm $<$ pore width < 50 nm), and macroporous (pore width > 50 nm) 3 as shown in Figure 10.1 [7].

The classifications of porous materials are shown in Figure 10.1. The popularity of porous organic polymers (POPs) and organic-inorganic hybrid materials like metal-organic frameworks (MOFs) as porous materials stems from their exceptional versatility, tunability, and performance across a wide range of applications, driven by their unique structural and chemical properties. Porous organic polymers are an emerging class of two- or three-dimensional porous materials that are made up of light elements (e.g., C, H, N, O, etc.). Organic molecules of various geometries and topologies, being the building blocks of POPs, are linked through strong

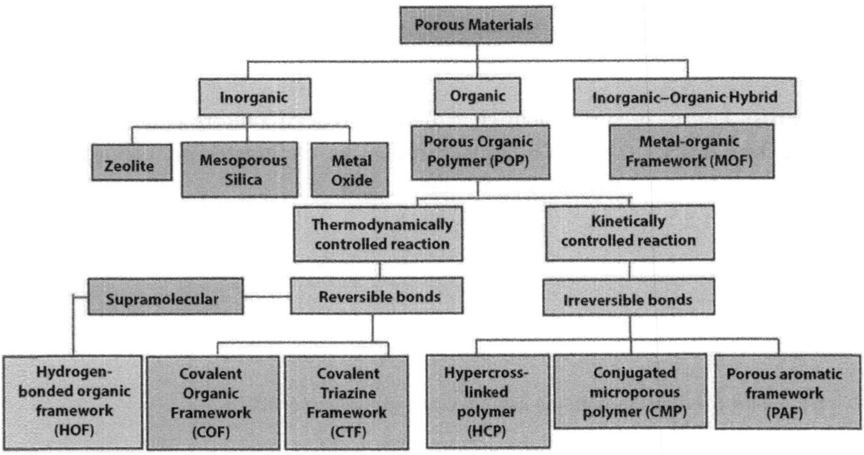

Figure 10.1 Classification of porous materials.

covalent bonds, resulting in a network structure [8]. They have received keen attention from researchers owing to their unique properties, including high surface area, high porosity, large pore volume, excellent chemical and thermal stability, low density, pre-designable structure, and convenient post-functionalization. These properties of POPs make them potential materials for applications in various fields namely gas adsorption [9], gas storage [10], energy storage [11], heterogeneous catalysis [12], water treatment [13], sensing [14], optoelectronic devices [15], biomedical [16] and electrochemical applications [17], etc. Also, POPs are categorized into two classes based on their degree of ordered structure of the skeleton, such as amorphous, having no order (e.g., HCPs, CMPs, PAFs, and PIMs), and crystalline, having long-range order (e.g., COFs and CTFs) [18].

10.2 Azulene and Its Electronic Properties

Azulene is a blue-colored, non-benzenoid, non-alternant, bicyclic aromatic hydrocarbon with a considerably higher dipole moment (=1.08D) compared to its isomer naphthalene. This unusual dipole moment arises due to the resonance delocalization (Figure 10.2a) of azulene, which makes it a donor-acceptor type molecule and differentiates it from the other conventional fused benzenoids with respect to reactivity, as well as optoelectronic properties, and makes them suitable for advanced optoelectronic applications [19]. Azulene shows unique physicochemical properties such as a narrow band gap and anti-Kasha fluorescence behavior ($S_2 \rightarrow S_0$ transition) (Figure 10.2b). In contrast to naphthalene where symmetrical HOMO–LUMO, or highest occupied molecular orbital and lowest unoccupied molecular orbital geometry, is present, azulene has unsymmetrical frontier molecular orbital (FMO) geometry with HOMO and LUMO density localized at odd and even positions, respectively (Figure 10.2c) [20]. Because of this characteristic of azulene FMOs, there is a lower energy difference between the ground state (S_0) and the first excited state (S_1), resulting in ($S_0 \rightarrow S_1$ transition) absorption seen in the visible range, which gives azulene its blue color [21]. In light of azulene's distinct molecular orbitals, the presence of an electron-donating substituent at an odd-numbered position and an electron-withdrawing substituent at an even position increases LUMO–LUMO+1 separation and lowers the HOMO–LUMO gap (Figure 10.2d).

In 1937, Plattner and Pfau reported their outstanding achievement in azulene synthesis [22]. However, due to the low yield of azulene, their work was not practical. In 1950, Ziegler and Hafner developed a highly effective method for the synthesis of a wide variety of azulene and its derivatives in

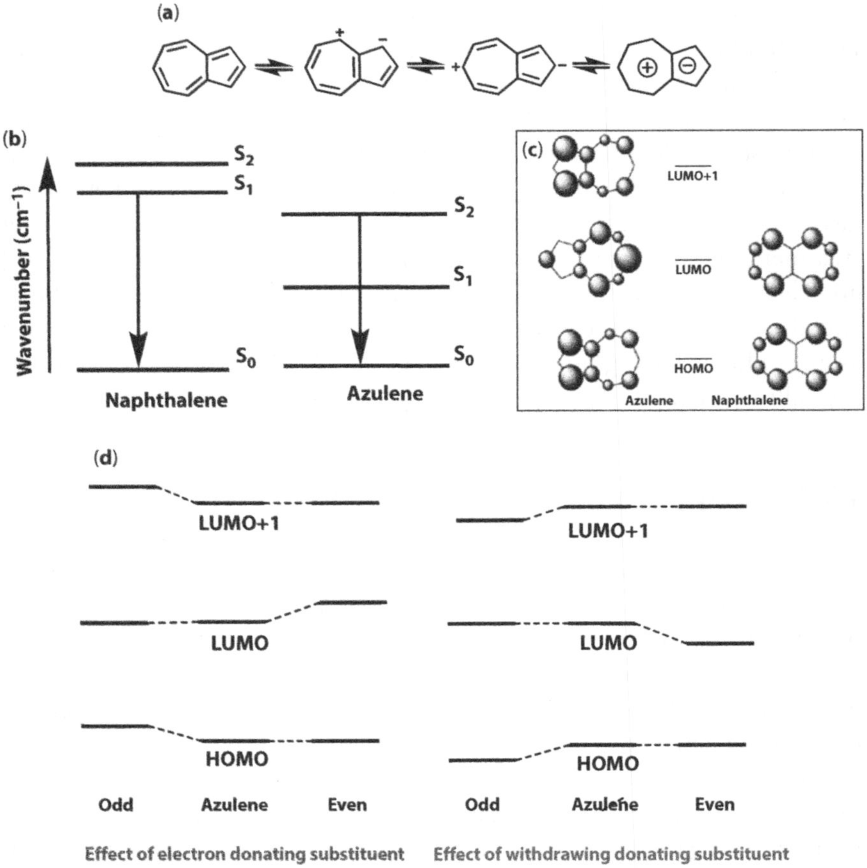

Figure 10.2 (a) Resonance structures of azulene; (b) comparison of energy spacing between the singlet states and radiative transitions of naphthalene and azulene; (c) probability of electrons in azulene's (left) and naphthalene's (right) HOMO, LUMO, and LUMO+1; (d) a qualitative description of the effect of different substituents on MO levels. Reproduced with permission [21]. Copyright 2003, American Chemical Society.

which pyridinium salts, along with cyclopentadienide ion, were used as a starting material [23]. Although azulene has been known for 100 years, due to the synthetic constraints, the potential use of azulene-based derivatives as structural building units in constructing advanced organic materials is less explored compared to its naphthalene analogs.

The intrinsic dipolar nature and the higher polarizability of azulene and its derivatives allow for the formation of covalent, coordinate, or hydrogen bonds with different types of molecules or metal ions to create two-dimensional and three-dimensional framework materials. Azulene has drawn much attention in recent years because of its distinctive optical and electronic properties. Therefore, azulene-based materials have emerged as an active field of study.

10.3 Functionalization of the Azulene Derivatives

Azulene derivatives are intriguing candidates due to their unique photophysical and redox characteristics, and their polarized π-electron density for synthesizing advanced materials. Among various types of azulene derivatives, aryl substitutions have many advantages as they substantially alter the electronic nature of the parent azulene by modifying the π-conjugation [24]. Aryl azulenes are synthesized *via* a multiple-step reaction, which generally requires the azulene derivatives containing the necessary functionalization as the starting material. The electron distribution in FMOs reveals the suitable positions for this purpose (Figure 10.3). The resonance delocalization of azulene showed that the odd positions 1-, 3-, 5-, and 7 of azulene are electron-rich and favor electrophilic reactions, whereas even positions 2-, 4-, 6-, and 8 are electron-deficient, thus favoring nucleophilic reactions [25]. However, the most reactive positions of azulene are 1- and 3-positions, and electrophilic substitution primarily occurs at these positions [26]. Recently, various types of azulene derivatives were reported, **10.1a–10.1k** (Figure 10.4). By following several organometallic C–C coupling reactions, such as the Stille or Suzuki–Miyaura reaction, the π-extension can be done at the 1- and 3-positions, resulting in compounds like **10.2–10.5** (Figure 10.4). Recently, Cowper *et al.* reported a series of azulenesulfonium salts **10.6** readily synthesized from azulene-based structures through an electrophilic aromatic substitution reaction with tetrahydrothiophene 1-oxide and trifluoroacetic anhydride [27]. The π-expanded azulenes **10.7** can be prepared by using these azulene sulfonium salts with various types of coupling partners through the Suzuki–Miyaura cross-coupling reactions. The substitution at the odd positions, such as 5- and 7-position, is often necessary to first protect the

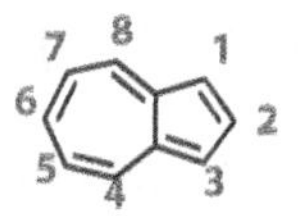

Figure 10.3 Reactivity of azulene at different positions.

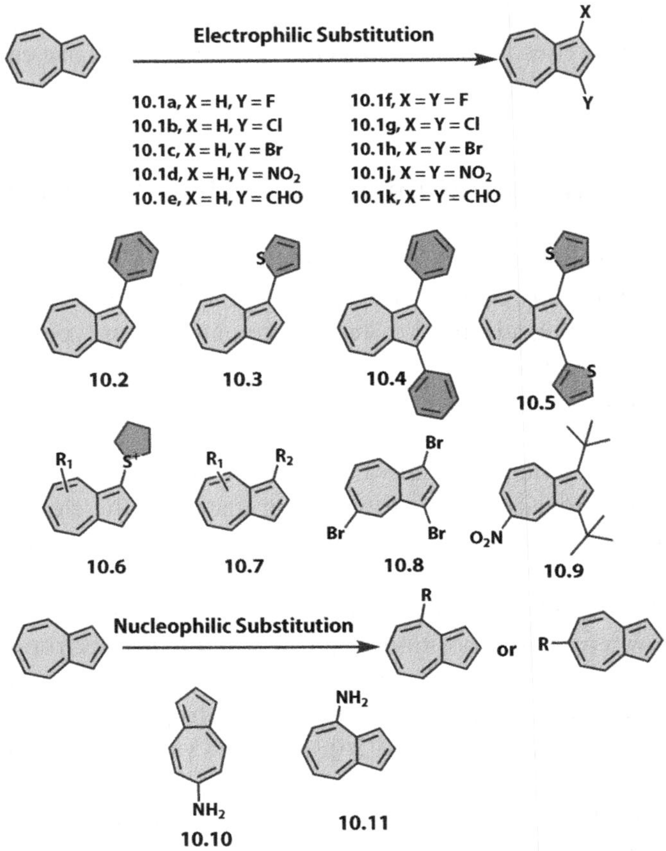

Figure 10.4 Azulene derivatives substituted at different positions.

1- and 3-positions. Examples of these types of compounds are compounds **10.8** and **10.9** (Figure 10.4). In general, nucleophilic addition to azulene gives a mixture of isomeric products. Sometimes, the nucleophilic addition of azulene at the 4- and 8-positions (**10.10** and **10.11**) generally gives an additional 6-substituted product (Figure 10.4) [28].

10.4 Synthetic Methods to Prepare Porous Materials

Several synthetic methods have been developed for the preparation of azulene-based porous materials. Conventionally, solvothermal and hydrothermal methods were used to synthesize COFs, MOFs, and other porous materials.

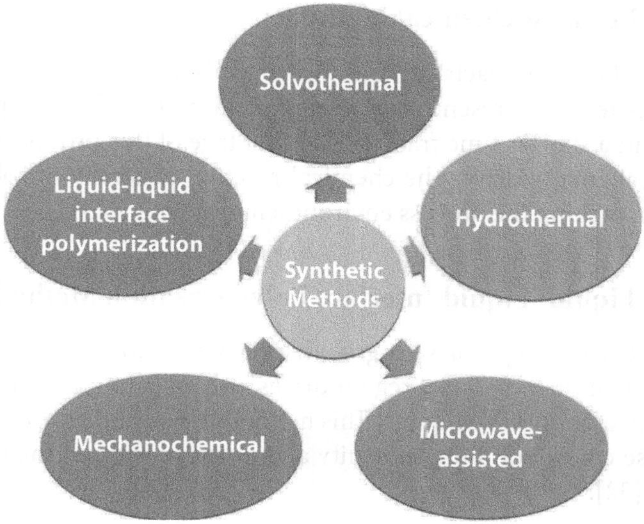

Figure 10.5 Representative diagram of different synthetic methods for porous materials.

Alternative techniques such as microwave-assisted synthesis and electrochemical deposition offer additional avenues to synthesize porous materials. The methods generally employed for the synthesis of porous polymers is shown in Figure 10.5.

10.4.1 Solvothermal Method

This method involves the condensation or polymerization of organic monomers in a high-boiling solvent within a Pyrex ampoule or autoclave, which is a sealed vessel that is heated to a high temperature (usually between 100 and 200°C) for a few hours to days [29]. The solvent transformed into a supercritical fluid under such conditions. On completion of the reaction, the autoclave is cooled to room temperature, and the desired product is recovered by removing solvents and impurities. When water is used as a solvent, the method is called hydrothermal synthesis.

10.4.2 Microwave-Assisted Method

In this technique, microwave radiation is the source of heat energy. Microwave-assisted synthesis is found to be advantageous over conventional hydrothermal methods in terms of giving relatively ordered porous materials in a shorter time scale [30].

10.4.3 Mechanochemical Method

In this method, the reactant molecules are mixed in a mortar and grind using a pestle in the presence of a catalyst (acid/base). Often, the product is obtained in a shorter time frame. The advantage of this process is that the mechanical energy drives the chemical reaction generally in solvent-free condition, making the process environmentally benign [31].

10.4.4 Liquid–Liquid Interface Polymerization Method

This process is a type of step-growth polymerization where the excess free energy of the interfacial region drives polymerization at the interface between two immiscible liquids. This method has advantages, especially in the precise control of dimensionality and porosity without the use of any template [32].

10.5 Different Types of Azulene-Based Porous Materials and Their Applications

10.5.1 Metal-Organic Frameworks

Metal-organic frameworks are crystalline porous materials formed from inorganic or organic moieties linked by strong coordination bonds. These materials, owing to their ultra-high porosity and enormous surface area (extending beyond 6000 m^2/g), along with the higher degree of variability for inorganic and organic moieties of their structures, serve as a potential material for sustainable development. For example, MOFs are widely investigated to be used in gas storage [33], such as hydrogen and methane gas, also as high-capacity adsorbents fulfilling various separation needs. Additionally, in recent years, MOFs have found various applications in fields such as membranes [34], thin film devices [35], various biomedical applications [36], and catalysis [37].

In 2010, Berke *et al.* prepared two azulene-based MOFs, MOF-645 and MOF-646 starting from 1,3-azulenedicarboxylic acid, and studied their H_2 adsorption potential [38]. The synthesis of 1, 3-azulene dicarboxylic acid (H_2L) is shown in Figures 10.6 & 10.7. Azulene contains two contiguous dipolar aromatic systems; therefore, inclusions of azulene moieties in MOF could help build up columbic fields essential for polarizing and binding to nonpolar molecules (like H_2) by dispersion forces, making them efficient adsorbents. Considerable stabilization of the polarized binding states of the H_2 molecule is achieved depending on

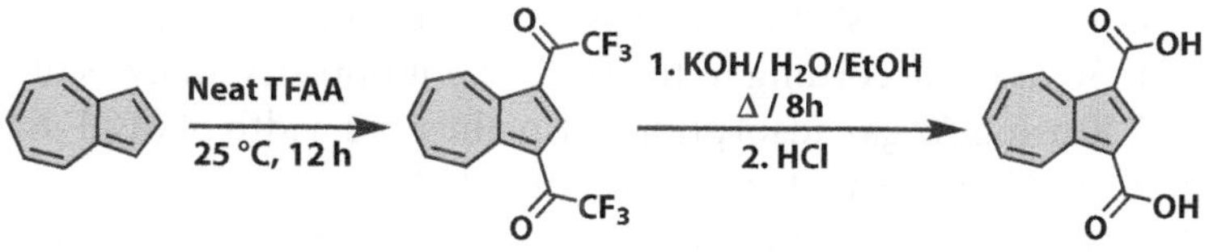

Figure 10.6 Synthesis of 1, 3-azulenedicarboxylic acid [41].

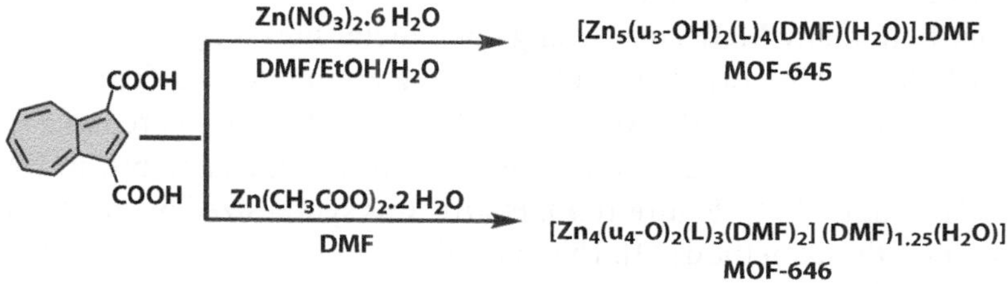

Figure 10.7 The synthesis of MOF-645 and MOF-646 [38].

the electric field size [39]. It was reported that due to the inherent bipolar nature of azulene, the interaction energy between H_2 and the azulene moiety is generally higher compared to benzene and naphthalene systems [40a, 40b]. Simply heating H_2L (L=1, 3-azulenedicarboxylic acid) and zinc nitrate in a DMF/EtOH/H_2O combination (1.0: 0.25: 0.25 mL) at 90°C produced the single crystals of MOF-645 (Figure 10.7). According to single-crystal structure studies, MOF-645 has a three-dimensional network structure with pentanuclear Zinc (II) clusters: $Zn_5(\mu_3\text{-}OH)_2(O_2C)_8(DMF)(H_2O)$. Secondary building units (SBU) are interconnected in the overall crystal structure, leading to produce a body-centered cubic lattice (bcu) net. On blocking this connectivity and removal of guests on coordination from the structure, an available void space was found to be around 41%. The guest-free form of MOF-645 exhibited structural displacement, as indicated by the PXRD signal, and could not adsorb N_2 at 77 K. $Zn(CH_3COO)_2$ and H_2L underwent a solvothermal reaction in DMF to synthesize MOF-646 (Figure 10.7).

The single-crystal structure of MOF-646 has a ligand L-connected Zn_4O unit. There is a small distortion in the tetrahedron structure of Zn_4O, and one of the Zn units is attached to two DMF molecules. The crystals were soaked in chloroform and evacuated at 45°C for 24 hours to extract the solvent guest molecules from the synthesized form of MOF-646. This condition can be resisted by the crystals of MOF-646, as confirmed by the PXRD analysis. Furthermore, eliminating the coordinated DMF validates the Zn_4O unit's open metal site. The activated material has a corresponding BET surface area of 925 m^2g^{-1} and a Langmuir surface area of 1130 m^2g^{-1}, according to the permanent porosity analysis of MOF-646 at 77 K using Ar gas adsorption. Also, the pore width of MOF-646 calculated from Non-Local Density Functional Theory (NLDFT) fit revealed that the pore width (<6 Å) is within the realm of an ultra-micropore. The H_2 adsorption analysis of MOF-646 showed that the MOF had a noticeably increased H_2 uptake, reaching up to 17.5 mg g^{-1} at 77 K and 1 bar. The estimated Q_{st} was found to be 7.8–6.8 $kJmol^{-1}$, which is considerably higher than many MOFs possessing open metal sites [42].

Kishida *et al.* reported an azulene having an interlocked framework of porous $\{[Zn(azdc)(bpy)].DMF\}_n$ (CID-13⊃DMF)(CID-13) prepared from $Zn(NO_3)_2.6H_2O$, 1,6-azulenedicarboxylic acid (H_2azdc), and bpy in DMF/MeOH solution [43] (Figure 10.8). H_2azdc was synthesized by the Ziegler–Hafner synthesis method [23]. They have also compared the gas adsorption of CID-13 with analogous isomeric porous framework CID-3 $[Zn(ndc)(bpy)]_n$ (ndc: 2,7-naphthalenedicarboxylate; bpy: 4,4′-bipyridyl) containing the naphthalene unit (Figure 10.8), which has flexible structural pores with highly selective sorption on CO_2 from $CO_2/O_2/N_2$ mixture. The constitutional isomers CID-13 and CID-3 both have similar structures, space groups, and cell parameters. The void volume of CID-3 is 25.8%, which is close to the void volume of CID-13; therefore, both have the identical 1D pore surrounded by polar azulene and nonpolar naphthalene for CID-13 and CID-3, respectively.

They investigated the gas adsorption behaviors of these compounds for polar H_2O and nonpolar CO_2 (0 D) to show how the polar nature of the azulene moiety affects the host-guest interactions compared to CID-3. Both compounds exhibit similar type-1 sorption behaviors for CO_2 at 195K with the same adsorption amount. However, the adsorption isotherms for polar H_2O reveal differences between the electric field potential of the isomeric pairs. The CID-3 exhibits type-V sorption behavior for H_2O, which shows weak host-guest interactions due to hydrophobic pore surfaces, but CID-13 adsorbs in the low-pressure region, which indicates the stronger host-guest interactions attributed to the dipolar nature of the azulene moiety.

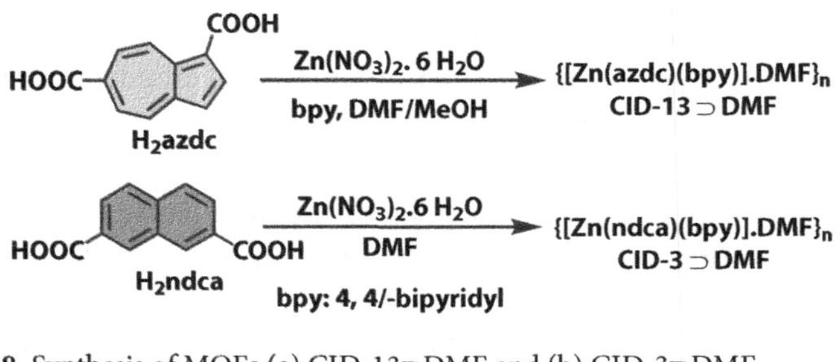

Figure 10.8 Synthesis of MOFs (a) CID-13⊃DMF and (b) CID-3⊃DMF.

10.5.2 Covalent Organic Frameworks

Covalent organic frameworks (COFs) are two or three-dimensional multi-functional porous materials formed by dynamic covalent bonds. The first COF was reported in 2005 [44]. Since then, several COFs have been developed due to their various interesting properties, such as high porosity, greater surface area, well-defined carbon-rich framework, diverse topology, and easily tunable building blocks. These unique properties of COFs enable them to be applicable in various fields such as heterogeneous catalysis [45], gas adsorption [46], gas separation [47], biomedical applications [48], and optoelectronic devices [49].

The 2D covalent organic frameworks with conductive and semiconductive properties have found wide applications in optoelectronic devices [50]. Their uniform structural arrangements and unique stacking patterns result in efficient charge transport within the in-plane and out-of-plane carrier mobile channels. This versatility makes them important in creating advanced optoelectronic devices such as solar cells, organic light-emitting diodes, and sensors. The band gaps show dependencies on the backbone of the molecular structure, the type of linkages, and the stacking pattern of 2D COFs. Imine-linked 2D COFs often exhibit surpassing band gaps than the practical narrow ones (1.0 to 1.5 eV) required for optoelectronic applications, in the range of 2.0-3.0 eV

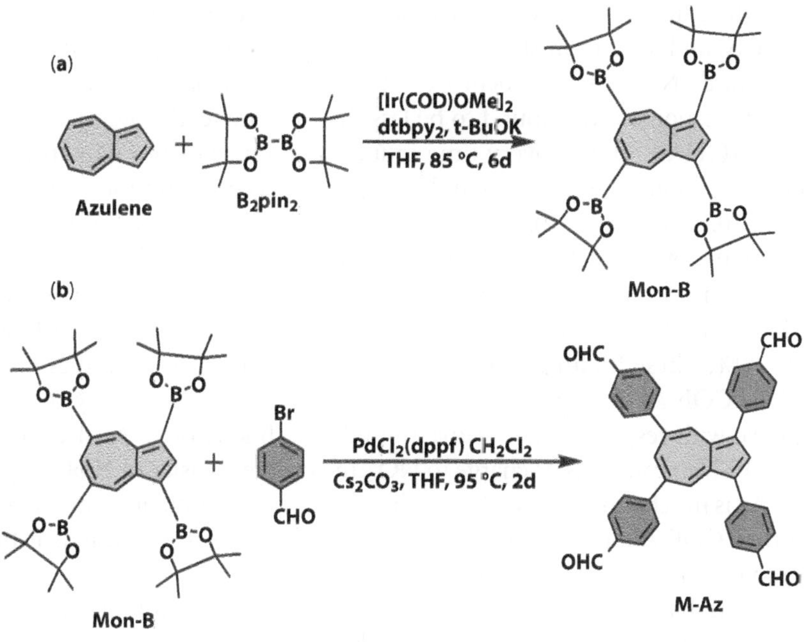

Figure 10.9 (a) Synthesis of Mon-B; (b) synthesis of M-Az.

[51a, b]. The trapping of electrons induced by the imine linkages is primarily to be held responsible for this undesirably larger band gap of 2D COFs [52]. This effect restrains the movement of electrons within the framework structure, eventually leading to a wider band gap. To mitigate the electron trapping effect, several studies emphasize reducing the band gap of 2D COFs. The approaches include the introduction of push-pull electronic structure [53], modulation of electronic structure [54], and doping technique [55]. In 2024, Feng *et al.* employed a fresh strategy by synthesizing the monomers through a simple scheme (Figure 10.9) that has disrupted building block symmetry, facilitating the controlled synthesis of 2D COFs having dual pores (Figure 10.10) [56]. By excluding D_{2h} vertices and transforming them into rectangles, they have generated new vertices with either C_{2v} or C_{2h} symmetry. These modifications led to the formation of isomeric COFs with Kagome (kgm) topology, based on central moieties, naphthalene and azulene, which exhibited reduced symmetry compared to D_{2h} vertices. The synthesis of the azulene precursor, **M-Az** used for the synthesis of COF is shown in Figure 10.9. The structural backbone of these COFs is shown in Figure 10.6. The band gap of the azulene-based COF (COF-Az) was 1.37 eV, significantly smaller than that of the isomeric COF-Nap (2.28 eV). The ordered channels of the isomeric framework are beneficial for gas adsorption and desorption, and they offer a platform for examining how structure influences the characteristics. In gas-sensing devices, both COFs were employed as active ingredients. For NO_2 detection over 10 weeks, the azulene-based COFs demonstrated tolerance, high recovery rate, specificity, and long-term stability. The COF-Az sensor was found to be specific towards NO_2 gas as evident from the 58.7 % response, while for other gases the responses were found to be less than it's one tenth value at 10 ppm (Figure 10.11 a). The recovery time found to be ranged from 44 s to 375 s depending upon the exposure to different ppm level concentration of NO_2 gas (Figure 10.11 b). Figure 10.11 (c) illustrated the effect of relative humidity on the response of the COF-Az sensor. It was found to be reduced from 58.6% to 44.0% on increasing the RH up to 40%, the response was increased to 49.3% as the RH was 60%, and attained stability stable at 49.3% level when the RH was set to 80%. These findings inferred the saturation of water molecules inside PI-IDEs/COF-Az sensor thus increasing conductivity and finally a stable response with respect to the relative humidity. While N of C=N in COF-Az and COF-Nap can transfer charge to NO_2, the charge transfer of N of C=N in COF-Az is much more efficient, according to the XPS examination of the NO_2 adsorbed COF. About 85.3 ppb was discovered to be the theoretical detection limit in near-field communication.

Azulene-based COFs are also investigated as a photothermal material due to the unique optoelectronic behavior of azulene. To improve the

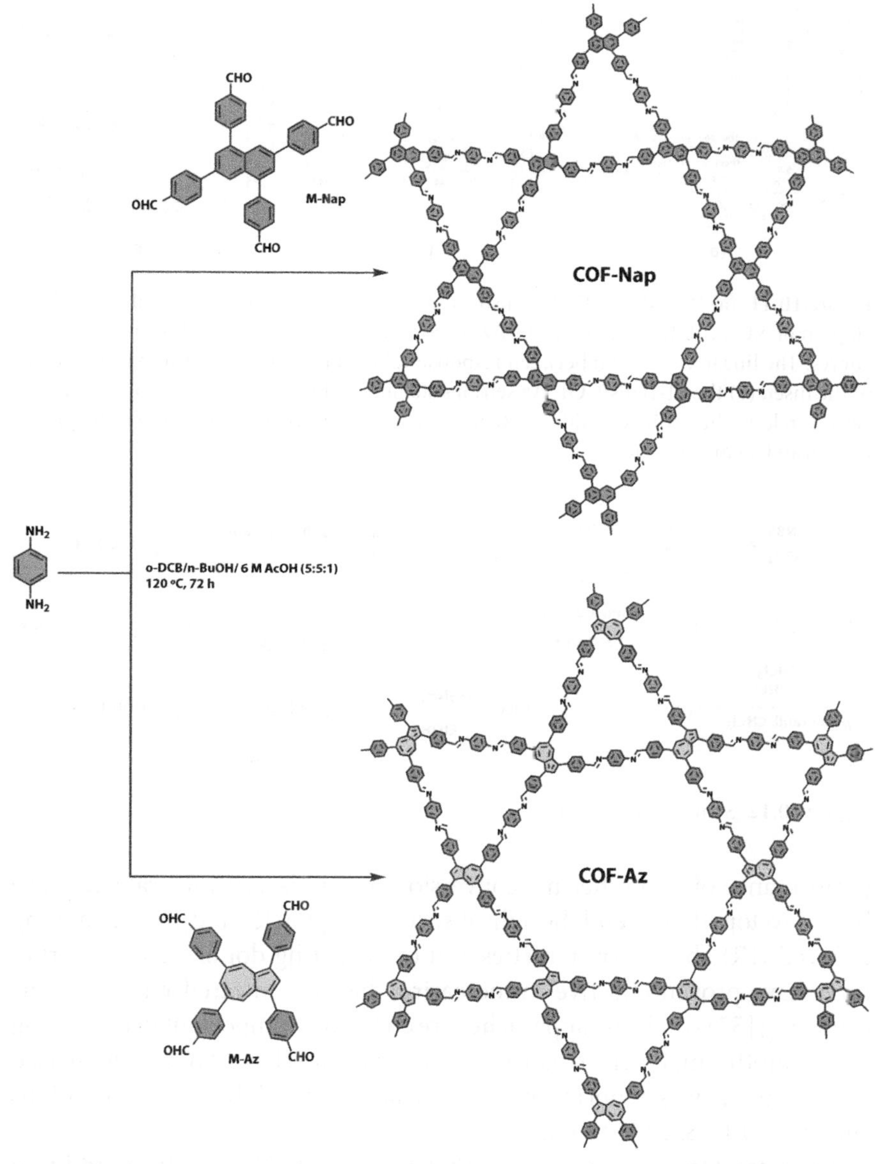

Figure 10.10 Synthesis of COF-Nap and COF-Az, isomeric dual-pore 2D COFs.

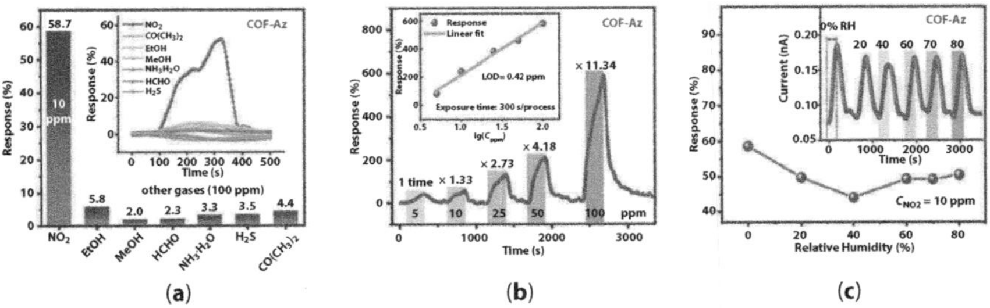

Figure 10.11 (a) PI-IDEs/COF-Az sensor responses to 100 ppm of other analytes and 10 ppm of NO_2; (b) the PI-IDEs/COF-Az sensor's transient response and recovery curves. The linear association between response values and NO_2 concentrations is shown in the inset; (c) the PI-IDEs/COF-Az sensor's stability test results in 10 ppm of NO_2 at various relative humidity conditions. Reproduced with permission [56]. Copyright 2023, American Chemical Society.

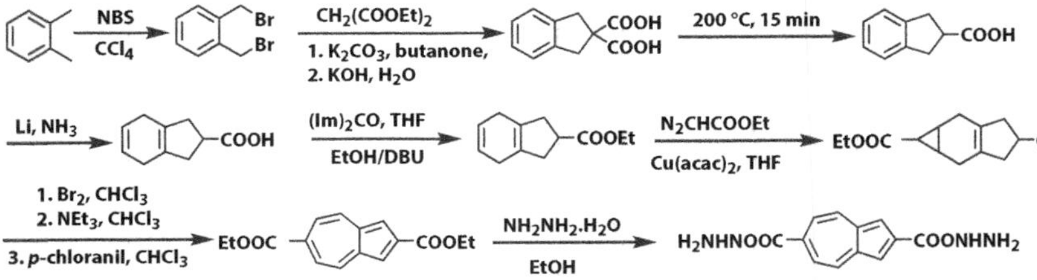

ADCH

Figure 10.12 Synthesis of ADCH.

performance of photothermal conversion materials, several strategies have been developed to extend the light absorption region [57a–d]. Incorporating infrared (IR) absorption moieties and constructing donor-acceptor structures have proven effective in increasing the near-infrared regions' light-capturing [57a–c]. These approaches are collectively important in enhancing the photothermal performance of various materials. An azulene-based COF-Azu-TP was reported by Tian *et al.* for photothermal applications (Figures 10.12 & 10.13) [58].

In contrast to the isomeric naphthalene-based COF-Na-TP, the addition of azulene to the framework moiety results in interlayer dipole interactions and broad-spectrum light absorption capacity, making COF-Azu-TP an effective photothermal conversion material.

Solid-state UV-Vis-IR spectroscopy has also been used to study the photophysical characteristics of COF-Azu-TP and COF-Na-TP. Strong and broad absorption that COF-Azu-TP showed between 200 and 1000 nm in contrast to COF-Na-TP's absorption range from 200 to 550 nm.

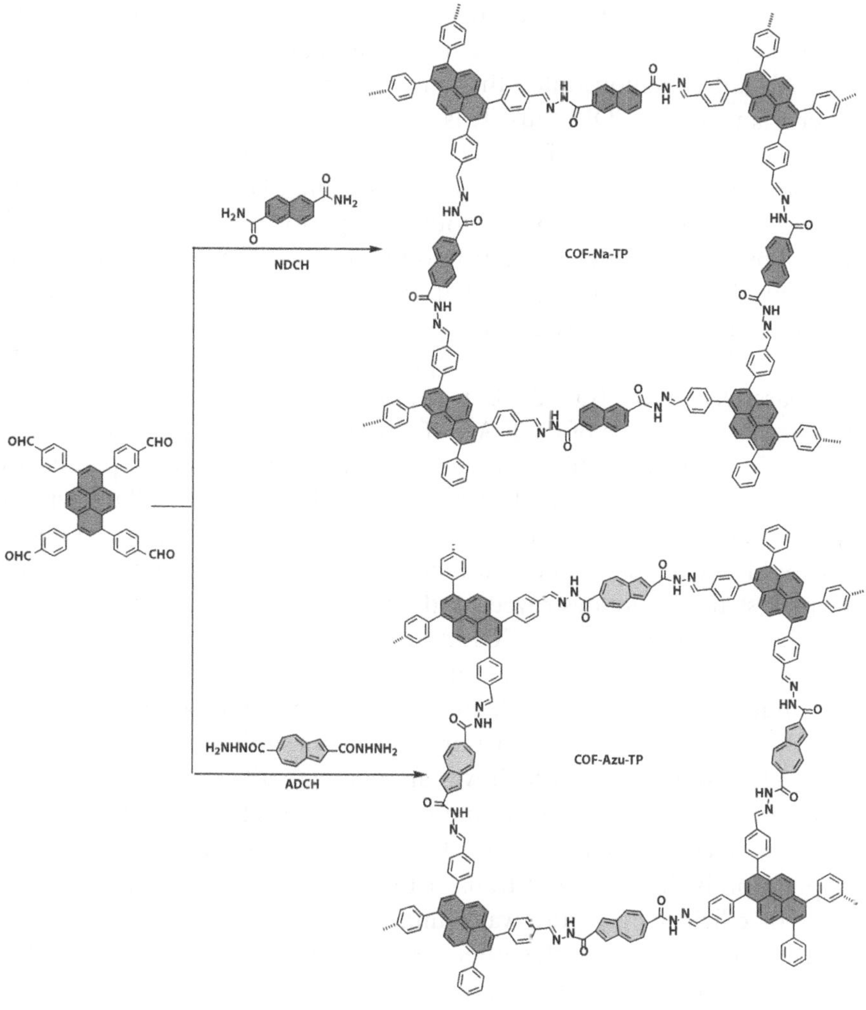

Figure 10.13 The COF-Azu-TP and COF-Na-TP synthesis strategies. It should be noted that the diagram's azulene moieties orientations could be dispersed randomly.

The optical band gap for COF-Azu-TP and COF-Na-TP was found to be 1.46 eV and 2.38 eV, respectively, based on their Tauc plots. COF-Azu-TP with its highly advantageous narrow band gap promotes non-radiative decay, thereby increasing the possibilities of heat generation, enhancing the light-harvesting capacity. To evaluate the potential of photothermal conversion, the photothermal behavior of COF-Azu-TP and COF-Na-TP was compared. Upon irradiation by 808 nm laser with 1 Wcm^{-2} energy power, the temperature of COF-Azu-TP powder quickly enhanced up to 200°C within 10 s then reached the saturation at approximately 230°C by

30 s. However, under identical conditions, COF-Na-TP only rose to 70°C clearly emphasizing azulene's crucial role in light to heat energy conversion. Also, it is demonstrated that an equilibrium temperature of 195°C, lower than that of COF-Azu-TP, is attained by POP-Azu-TP, the amorphous counterpart of COF-Azu-TP. This observation indicates a higher photothermal conversion efficiency of crystalline COF than its amorphous counterpart. The outstanding photothermal performance of COF-Azu-TP was followed by an investigation into this material's possible application in the production of solar steam. For these applications, the COF-Azu-TP was supported by a substrate like polyurethane (PU) porous foam and *in situ* COF/PU composites were prepared with different amounts of COF. Comparatively, COF-Na-TP/PU was also made under comparable conditions. As the dosage of COF-Azu-TP increased, the as-prepared foam's colour changed from white to brown to dark. Optical microscope images revealed the homogeneous dispersion of COF micro-crystallites within the PU pore and at higher concentrations of COF-Azu-TP, the composites could not maintain good elasticity because at higher concentrations the pore structure got destroyed. With the experimental findings in mind, they have also implemented COF-Azu-TP and COF-Azu-TP/PU as light absorber and heat generators for an interfacial water evaporation system. A cylindrical shaped foam of 32 mm size was floated on water and subsequently exposed to 1 sun irradiation. The foam's surface temperature progressively rose to 41.8°C which was approximate 9°C rise over the blank polyurethane foam. Because of the foam's high heat insulation, the heat energy is retained inside, which is beneficial for interfacial photothermal evaporation by limiting heat transfer to the surrounding area. The solar-to-vapor efficiency for pure water under one sun irradiation was determined to be 97.2%, while the water evaporation rate of COF-Azu-TP (10)/PU was observed to be 1.43 kg m^2 h^{-1}. The fact that this value is more than the blank PU foam (0.59 kg m^2 h^{-1}) and COF-Na-TP-10/PU (0.61 kg m^2 h^{-1}) makes it noteworthy. The COF-Az-TP (10 mg)/PU composite was also evaluated for potential desalinating seawater. Natural saltwater taken from the Yellow Sea evaporates at a rate of 1.22 kg m^{-2} h^{-1}, with an 83.3% solar-to-vapor efficiency. After sun desalination, ion concentrations of Na^+, K^+, Mg^{2+}, and Ca^{2+} decreased by three to four orders of magnitude compared to the baseline ion concentrations. The World Health Organization and the United States Environmental Protection Agency have defined drinking water guidelines, which these levels fully meet [59a, 59b]. As a successful technique for desalinating saltwater, this result highlights the promise held by COF-Azu-TP/PU foam in removing pure water from saltwater. Programmable, stimuli-responsive shape memory

polymers (SMPs), materials employed in industries including biomedical and aerospace applications, are transformed by using COF-Azu-TP's high photothermal conversion capacity. With stimulation from heat, light, or electricity, a conventional SMP can be programmed into a temporary shape that it can subsequently switch back to permanently. The ability to remotely control light, focus it precisely, and switch it quickly makes it an especially appealing external trigger. Blends of polylactic acid (PLA) and polycaprolactone (PCL) were mixed with COF-Azu-TP as a filler to investigate its potential use in SMPs. Common semicrystalline SMPs like PLA can have their glass transition temperature lowered by the addition of PCL, which in turn can change the SMP matrix's deformation temperature. Through solvent volatilization, a COF-Azu-TP/PLA/PCL composite was created. This composite was compared to another PLA/PCL composite made with the same technique but without COF-Azu-TP.

The inherent push-pull nature of azulene makes it an excellent candidate for application in molecular electronics. This unique structure offers low aromatic delocalization energy, favoring charge transport behavior, as well as allowing charge transport from electron-donating to electron-accepting parts in a controlled manner [60a–c]. Zhao *et al.* reported an azulene-based 2D conjugated covalent organic framework, COF-Azu, for organic memristors prepared by a liquid-liquid interface polymerization strategy [61]. Organic/polymer memristors are categorized into digital-type memristors and analog-type memristors [62]. While analog-type memristors can mimic basic synaptic functions, digital-type memristors behave as resistive RAM, offering non-volatility, high density, fast speed, low power consumption, large ON/OFF, high endurance, and long retention. Zhao *et al.* examined the electronic performance of the COF-Azu film coated onto the ITO substrate (Al/COF-Azu/ITO), which has an active layer thickness of 160 nm and an average roughness of 14.3 nm. Using a Keithley 4200 semiconductor parameter analyzer in ambient conditions without encapsulating the device, the memristor exhibited a switching memory effect, which is nonvolatile and resistive in nature [63]. With an ON/OFF current ratio of 50, the switch-on and switch-off voltages were, respectively, -0.50 V and +1.95 V. Even at the 300th switching cycle, the ON/OFF ratio remained remarkably consistent throughout the procedure, demonstrating strong device dependability [64]. Azulene, with its seven-membered, electron-deficient ring and five-membered, electron-rich ring, provides electronic features that control transport *via* its electron-accepting and -donating subsystems [65]. A partial transfer of charge between these rings induces redox behavior and a high dipole moment [66]. In the OFF state, no voltage is applied. An enhancement in the electric field causes a greater

ICT impact and the generation of many holes and electrons as the negative voltage rises across the COF-Azu film. These carriers progressively create conductive filaments, which lower the active layer's resistance. Due to the creation of conductive filaments stabilized by azulene's enormous dipole moments, the device transitions from a high-resistive state (OFF) to a low-resistive state (ON) at high enough voltages, guaranteeing non-volatility. Applying a sufficient positive bias reverses this, recombining the electrons and holes, and returns the device to its initial high-resistive state.

10.5.3 Conjugated Microporous Polymer

Conjugated microporous polymers (CMPs) are a kind of POPs connected by a conjugated π skeleton, having embedded nanopores. They are highly investigated for their various interesting properties, such as organic composition, customizable pore structures, adjustable hetero atom doping, and tunable optical and electronic properties. These features of CMPs hold great potential for their different usefulness, such as gas adsorption [67], heterogeneous catalysis [68], optoelectronic property [69], and storage for electrical energy [70].

Ghasimi *et al.* reported the synthesis of an azulene-containing conjugated porous polymer and further used it as a catalyst in visible light-promoted Stille-type coupling reaction [71]. The compound was synthesized using the Suzuki cross-coupling reaction, as shown in Figure 10.14(a), between

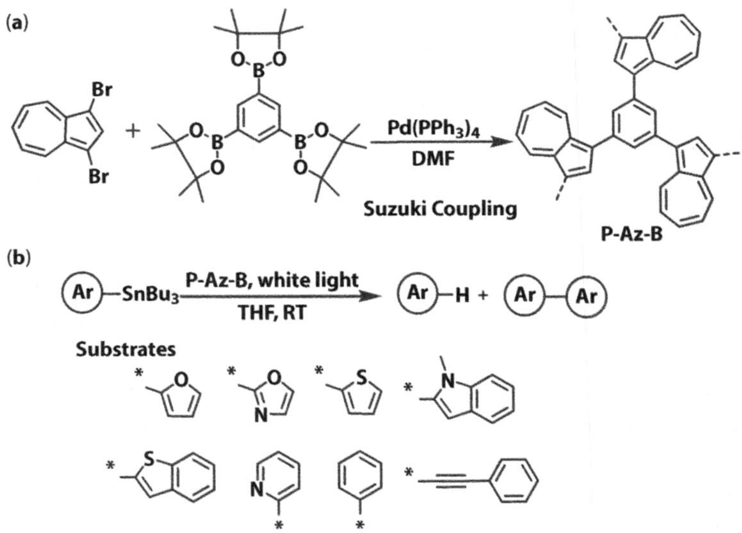

Figure 10.14 (a) Synthesis of P-Az-B (b) destannylation reaction of various aryl stannanes using P-Az-B as photocatalyst.

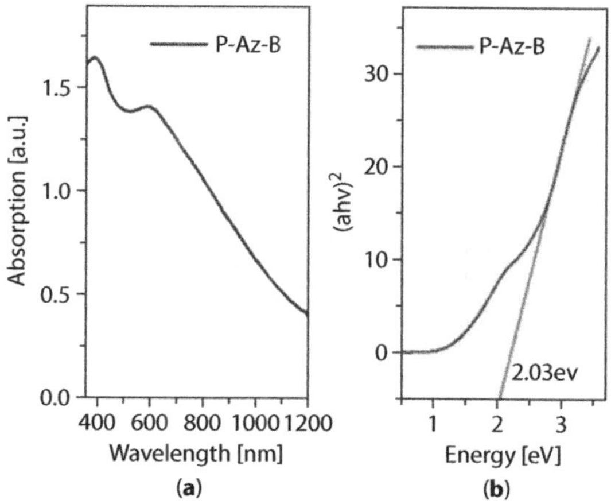

Figure 10.15 (a) UV-Vis diffuse reflectance spectrum plot and (b) Tauc plot P-Az-B. Reproduced with permission [71]. Copyright 2017, Wiley-VCH.

1, 3-dibromoaazulene and 1, 3, 5-phenyltriboronic acid tris(pinacol) ester. P-Az-B was used in photocatalytic destannylation of various aryl stannanes as shown in Figure 10.14 (b). The UV-Vis diffuse reflectance spectrum of P-Az–B is shown in Figure 10.15(a), reflects a broad absorption profile spanning from visible to the near-infrared range (<1200 nm). The optical bandgap of P-Az-B calculated from the Tauc plot is 2.03 eV, Figure 10.15(b). The HOMO and LUMO energies of P-Az-B were calculated from cyclic voltammetry and found to be +1.14 V and −1.10 V, respectively, with reference to a saturated calomel electrode. These numbers represent the photo-generated redox potentials of P-Az-B in photocatalytic devices. Noticeably, P-Az-B is observed to have comparable redox potentials with those of transition metal composite photocatalysts.

Kumar *et al.* reported a series of conjugated porous polymers (CPPs) based on 2,4,6-tris(5-ethynylthiophen-2-yl)-1,3,5-triazine core moiety [72], which was coupled with various conjugated building blocks, including naphthodithiphene (**CPP1**), benzodithiophene (**CPP2**), bithiophene (**CPP3**), tetraphenyl porphyrin (**CPP4**), azulene (**CPP5**) (Figure 10.16), and perylene diimide (**CPP6**), by using the Sonogashira-coupling reaction to produce the targeted CPPs. The synthesis of azulene containing conjugated porous polymer **(CPP5)** is shown in Figure 10.16. The co-monomer variation strategy allows the facile band gap engineering of CPPs. Porous molecules containing inherent donor-acceptor characteristics, like azulene, and highly fused and π-extended conjugated systems, like perylene diimide and porphyrin, impart excellent thermal

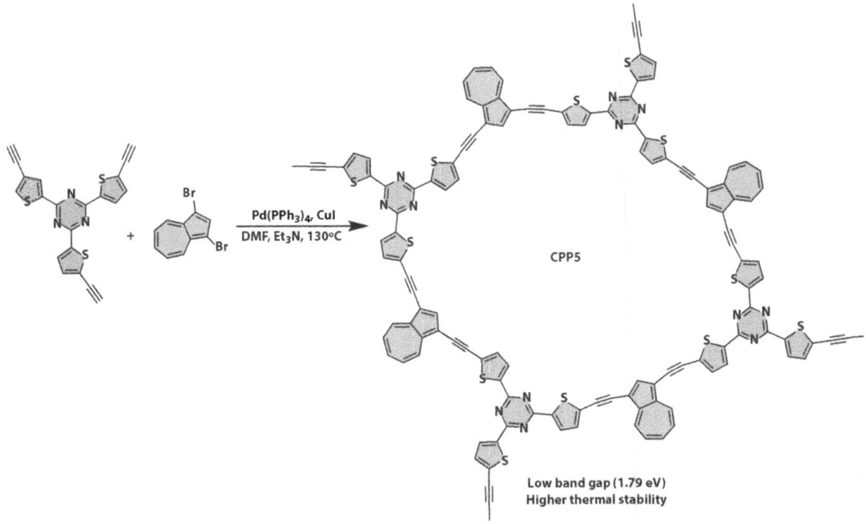

Figure 10.16 Synthesis of CPP5.

and chemical stability to the materials. The solid-state UV-visible spectra of the CPPs exhibited a broad absorption range from 350 nm to 900 nm, indicating the presence of extended π-conjugation in the core of the polymers. The Kubelka–Munk plot shows that band gaps of the polymers range from 1.73 eV to 2.19 eV. The anticipated trend for the bang gaps was CPP5<CPP7<CPP6<CPP8<CPP2<CPP1<CPP3<CPP4. The intrinsic donor-acceptor nature of the azulene moiety influences the electronic structure of the polymer CPP5, exhibiting a lower band gap relative to the other polymers, like CPP6/7 and CPP8, which have larger conjugated comonomers. The thermal gravimetric analysis proved higher thermal stability for CPP5 and CPP8, reaching as high as 500°C while retaining 50% of their initial mass. The impressive higher thermal stability is attributed to the highly rigid, conjugated, and crosslinked moiety. The scanning electron microscopy analysis showed agglomerated cloud-like morphology, and BET analysis at 77 K produced a surface area of 317 m^2/g with no hysteresis loop in N_2 adsorption-desorption isotherm, indicating the highly rigid porous network in the CPP5 polymer. These properties, like narrow band gap, higher thermal stability of azulene-containing polymers, make them promising materials for applications in organic electronics, energy storage devices.

10.5.4 Hydrogen-Bonded Organic Frameworks

Hydrogen-bonded organic frameworks (HOFs) are a class of 2D or 3D crystalline porous materials that are entirely made up of organic building blocks connected *via* hydrogen bonding interactions. Hydrogen bonding, as reversible, is furnished with easy regenerative and solution processing abilities [73a, 73b]. Supramolecular synthons, the basic building blocks of HOFs, are classified into two categories according to the way they generate hydrogen bonds: homologous synthons, which are created by self-assembly of identical units, and heteromorphic synthons, which are made up of various H-bond acceptor-donor pairings [74]. Since it is easy to grow high-quality single crystals of HOFs, single-crystal X-ray diffraction can be used to determine the superstructure with clarity. By carefully crafting molecular building pieces that contain rigidity, auxiliary interactions, and interpenetrations inside their frameworks while retaining the stability of the extended networks, HOFs with persistent porosities can be created.

The applications of HOFs are broad and impactful, encompassing catalysis [75] and gas adsorption [76], where their porous structure can be leveraged to enhance the performance of batteries [77] and supercapacitors

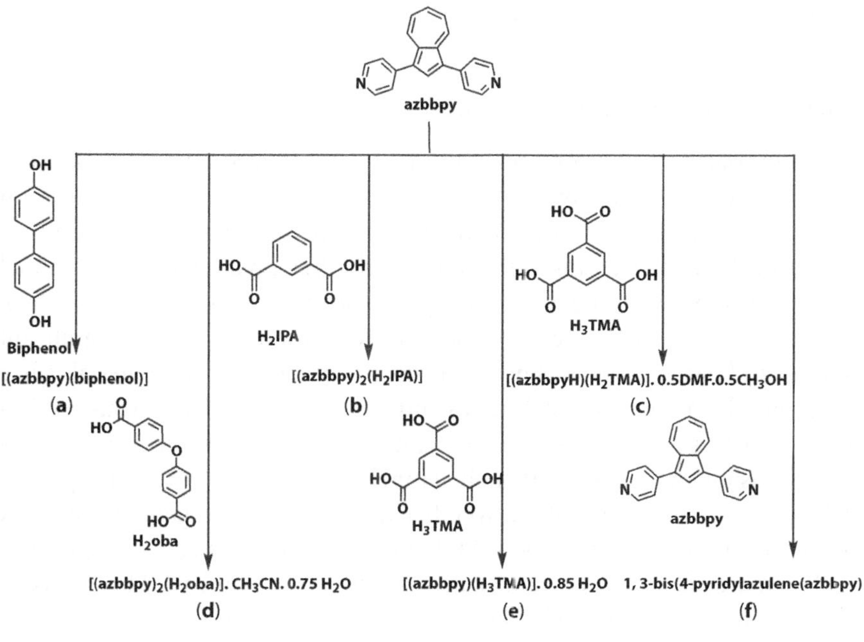

Figure 10.17 Synthesis of azulene-based supramolecular cocrystals that contain aromatic di- and tri-carboxylic acids and 4,4′-biphenol.

[78]. Due to its biocompatibility and customizable properties, HOFs have great potential in bio-medical applications such as drug delivery [79]. Additionally, due to the fine pore structure, HOFs are also applicable in storing and separating fine chemicals [80], which is crucial for industrial processes and environmental remediation.

A unique binary series of cocrystal and salt was synthesized by Ion *et al.* *via* the supramolecular interaction between 1,3-bis(4-pyridyl)azulene(azbbpy) H-bond acceptor, and several aromatic hydrogen bond donors (Figure 10.17) [81]. The azbbpy forms hydrogen bonding with the coformers, producing different supramolecular architectures during 3D packing and assembly. In their work, a 1:1 combination of 4,4′-biphenol (biphenol) and the binary cocrystal of azbbpy was used to create a 2D supramolecular network. Furthermore, a 2:1 mixture of azbbpy and 1,3-isophthalic acid (H_{2ipa}) and a 1:1 mixture of azbbpy and 4,4′-oxo-bis-benzoic acid (H_2oba) were used to create zigzag chains. Two separate systems were generated when azbbpy was co-crystallized in varying ratios with 1,3,5-benzene tricarboxylic acid (H_3TMA): cocrystals and salts. A vast and intricate hydrogen bond network was present in every crystal structure, incorporating both sturdy O–H···N and O–H···O hydrogen bonds based on homosynthons and heterosynthons. However, the weak nature of hydrogen bonds imparts low stability to the HOFs, which eventually restricts their practical applications to a certain extent.

10.5.5 Covalent Triazine Framework

Covalent triazine frameworks are important types of porous framework materials that are constructed by aromatic triazine rings with planar π-conjugation properties [82]. Owing to their defined structure, higher stability, and ultrahigh porosity with N atoms in the frameworks, these materials have potential applications in diverse fields such as gas adsorption, gas separation, sensing, and catalysis [83, 84]. The covalent triazine framework attracts significant interest and has emerged as a novel category of organic porous materials since the groundbreaking work of Thomas and coworkers on microporous polytriazine networks in 2008 [85]. The preparation of covalent triazine frameworks (CTFs) using molten salts at high temperatures (400–600°C) has sparked debate due to their true nature, whether they exist as an organic framework or transform into porous carbon (PC) during synthesis. Jiang *et al.* reported a CTF prepared from 1, 3-dicyanoazulene as monomer, namely CTF-Azs, using molten $ZnCl_2$ at 400-600°C (Figure 10.18) [86]. Chemical structural analysis reveals that CTFs prepared at lower temperatures (around 400°C) exhibit polymeric characteristics, which indicates their organic framework nature. In

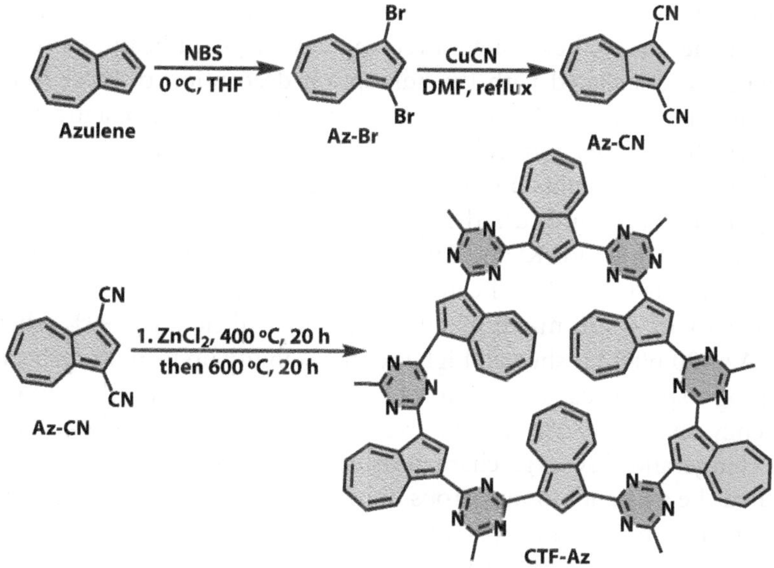

Figure 10.18 Synthesis of CTF-Az.

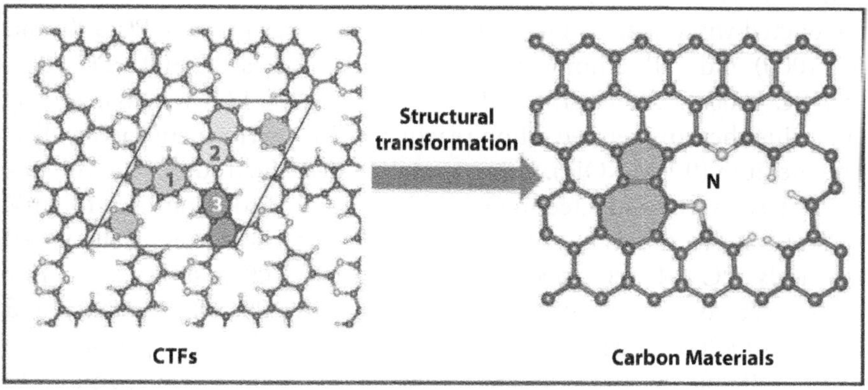

Figure 10.19 Schematic diagram of the structural conversion from CTFs to carbon materials. Reproduced with permission Ref. [86]. Copyright 2022, Wiley-VCH.

contrast, those synthesized at higher temperatures (around 600°C) display features typical of carbon materials, which shows the significant structural transformation at elevated temperatures (Figure 10.19).

At higher temperatures, CTF-Azs retains its high porosity, but the polymeric characteristics disappear. Notably, these high-temperature-treated CTF-Azs exhibit exceptional performance as cathode materials for lithium-ion (Li^+) and potassium-ion (K^+) batteries. The cyclic voltammogram of the material exhibited a rectangular shape, indicating a (de)Li^+/K^+ behavior associated with the adsorption-desorption of these ions, as illustrated in Figure 10.20 (a) and (b). Furthermore, galvanostatic charge-discharge tests were conducted at current densities ranging from 0.2 to 10 A/g^{-1}, providing insights into the electrochemical performance, as shown in Figure 10.20 (c) and (d). The specific capacity was observed to be 174 and 118.5 mA h g^{-1} for K^+ and Li^+ ions at a discharge rate of 0.2 Ag^{-1}. The CTF-Az-400/600|K half-cell can endure 100 cycles with reversible stability capacity of ~ 120 mA h g^{-1} and coulombic efficiency > 99% whereas CTF-Az-400/600| Li^+ showed higher cyclic stability up to 5000 cycles with retaining a specific capacity up to 112.2 mA h g^{-1}, as well as the coulombic efficiency level of 99.5% (Figure 10.20 (e), (f)). CTF-Azs exhibits capacitive charge and discharge characteristics. The EPR spectra of CTF- Azs revealed the existence of numerous unpaired electrons, indicating that the mechanism is not purely capacitive either.

Azulene rearrangement into the naphthalene occurs below 400°C, but space-confining effects allow the azulene unit to exist in the framework even at 1000°C. Jiang *et al.* developed PCs with controlled topological and heteroatom defects by annealing at 800, 900, and 1000°C. The synthesis was conducted without ammonia activation (denoted as PC-800, PC-900, and PC-1000) and with ammonia activation (denoted as PC-800a, PC-900a, and PC-1000a). The electrocatalytic activity of the PCs was assessed by analyzing their oxygen reduction reaction (ORR) performance in an alkaline medium (0.1 M KOH). Using a rotating disk electrode, cyclic voltammetry tests were conducted in O_2 and N_2-saturated solutions at a 100 mV s^{-1} scan rate with a catalyst loading of 0.364 mg cm^{-2}. Among the tested materials, PC-900, PC-1000, PC-900a, and PC-1000a showed superior ORR peak potentials (~0.682 V), outperforming PC-800, PC-800a, and CTF-Az-400/600. This higher potential indicates better ORR performance in PC-900 and PC-1000 variants. However, PC-800, PC-800a, and CTF-Azs, despite their slightly lower peak potentials, exhibit notable suitability for ORR due to their structural and electronic properties. Electrochemical impedance spectroscopy was performed to evaluate ORR kinetics. Both PC-800 and PC-800a displayed larger semicircles, indicating higher resistance and slower charge transfer, which hindered their ORR performance. In contrast, PC-900a exhibited a smaller semicircle, reflecting faster charge transfer. Additionally, PC-900a demonstrated excellent durability, retaining 81% of its ORR current after 40,000 s of operation. Notably, PC-900a outperformed Pt/C in methanol tolerance, showing only slight current degradation. They have also assessed oxygen evolution reaction (OER)

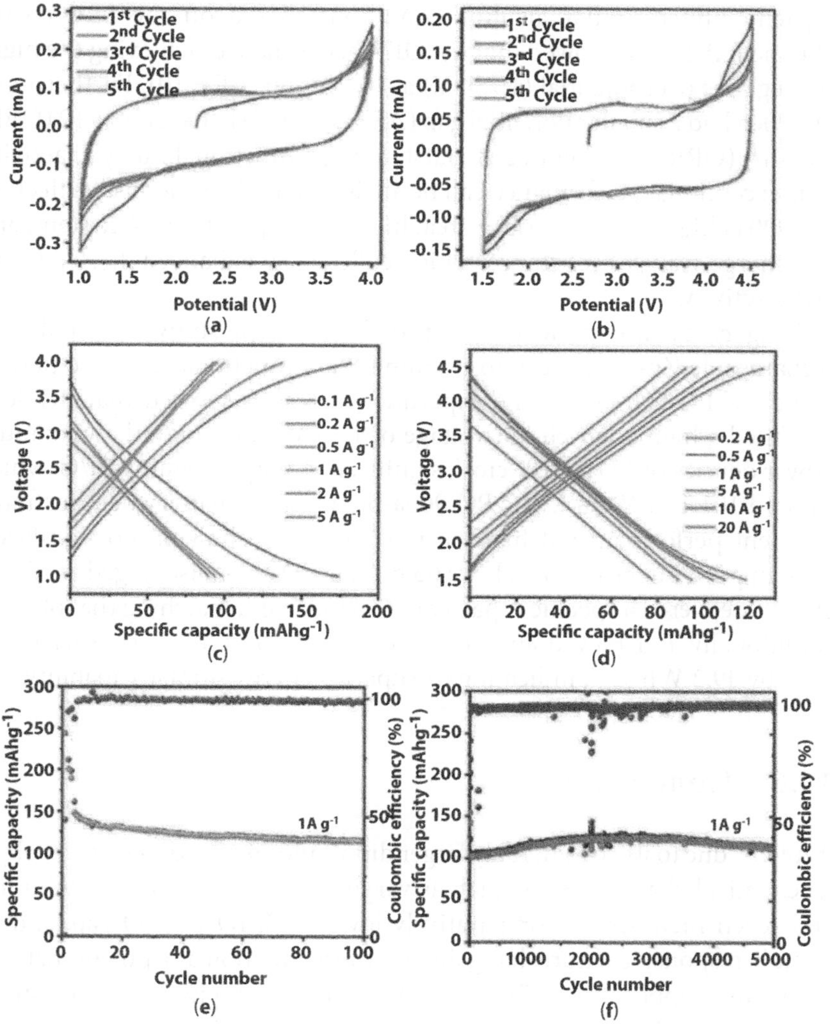

Figure 10.20 (a) Electrochemical performance of CTF-Az-400/600 | K^+ (voltage range 1-4 vs K/K^+) -First Five cyclic voltammogram at a scan rate of $1 mVs^{-1}$; (b) electrochemical performance of CTF-Az-400/600 | Li half-cell-cyclic voltammogram of CTF-Az-400/600 at a scan rate of $1 mVs^{-1}$; (c) galvanostatic charge-discharge profiles at varying current densities of CTF-Az-400/600 | K^+; (d) galvanostatic charge-discharge curves at different current densities of CTF-Az-400/600 | Li^+; (e) cycling stability test of CTF-Az-400/600 | K^+ at 1 A g^{-1} up to 100 cycles; of (f) CTF-Az-400/600 | Li^+ at 1 A g^{-1} up to 5000 cycles. Reproduced with permission Ref. [86]. Copyright 2022, Wiley-VCH.

capacity under similar alkaline conditions. Based on the LSV curves obtained, PC-900a showed the best OER performance, achieving the highest negative overpotential (1.754 V) at 10 mA cm^{-2} (E_j= 10). Furthermore, PC-900a had a smaller potential gap (ΔE = 973 mV) between E_j = 10 (OER) and $E_1/_2$ (ORR), highlighting its outstanding bifunctional catalytic activity. This exceptional performance can be attributed to the synergistic effects of PC-900a's high specific surface area, high N and pyridinic-N content, and low charge-transfer resistance, which collectively enhanced its electrocatalytic activity.

Jiang *et al.* also performed a proof-of-concept study by building a rechargeable Zn-air battery to examine the bifunctional catalytic performance of PC-900a [86]. The experimental results showed that PC-900a exhibited a high open-circuit voltage of 1.45 V and achieved a maximum power density of 141.8 mW cm^{-2}, significantly outperforming Pt/C, which showed 101.1 mW cm^{-2}. The PC-900a-based Zn-air battery demonstrated excellent performance at high current densities, showing no significant drop in power density even above 65 mA cm^{-2}. When discharged at 75 mA cm^{-2}, it delivered a specific capacity of 650 mA h g^{-1}, matching that of Pt/C. Additionally, it achieved an energy density of 666.4 Wh kg^{-1}, surpassing Pt/C by 49.2 Wh kg^{-1}, indicating its superior energy storage capability.

10.6 Conclusion

Azulene, due to its considerably larger dipole moment and its distinct physicochemical characteristics, such as varying reactivities on rings with five and seven members, a comparatively lower HOMO-LUMO gap, and its stimuli-responsive nature, has gained utmost attention as a potent material for various applications. Photochemical and photophysical applications like organic FETs, near-IR absorption, solar cells containing organic semiconductors, catalysis, sensing, and electrochemical applications like battery devices. Despite getting the necessary importance, the difficulties associated with synthesizing azulene-based derivatives have limited the number of azulene-based materials. Still, with the possibility of different substitution positions and the development of substituent synthesis strategies, various azulene derivatives have been synthesized in recent years. Azulene moieties in conjugated oligomers and polymers have different orientations, which have changed the orbital density distribution of the molecular front and changed the properties and applications of these materials. The POP exhibiting excellent gas adsorption property presents itself as a material of choice for heterogeneous catalysis, gas storage, and ion separation.

Using the azulene moiety to synthesize POPs can lead to blending properties to make them more effective in their applications. Nonetheless, preparing azulene-based polymers and frameworks still presents many difficulties that demand precise and thorough study to overcome them. Thus, synthesizing azulene derivatives remains a salient subject for the research community, understanding and standardization of which can open new developmental routes for azulene and its related materials.

AI disclosure statement: The author acknowledges the use of AI-based tools for minor language editing or grammatical correction. However, all intellectual contributions, analyses, and conclusions presented in this chapter are entirely the author's own.

References

1. Farmahini, A.H., Krishnamurthy, S., Friedrich, D., Brandani, S., Sarkisov, L., Performance-based screening of porous materials for carbon capture. *Chem. Rev.*, 121, 10666, 2021.
2. Liu, Z., Yuan, X., Zhang, S., Wang, J., Huang, Q., Yu, N., Zhu, Y., Fu, L., Wang, F., Chen, Y., Wu, Y., Three-dimensional ordered porous electrode materials for electrochemical energy storage. *NPG Asia Mater.*, 11, 1, 12, 2019.
3. Zhou, C., Ma, L., Feng, Y., Kuo, C.-Y., Ku, Y.-C., Liu, C.-E., Cheng, X., Li, J., Si, Y., Huang, H., Huang, Y., Zhao, H., Chang, C.-F., Das, S., Liu, S., Chen, Z., Enhanced Polarization Switching Characteristics of HfO_2 Ultrathin Films via Acceptor–Donor Co-Doping. *Nat. Commun.*, 15, 1, 2893, 2024.
4. Banerjee, S., Anayah, R.I., Gerke, C.S., Thoi, V.S., From molecules to porous materials: Integrating discrete electrolytic active sites into extended frameworks. *ACS Cent. Sci.*, 6, 10, 1671, 2020.
5. Zhao, D., Thallapally, P.K., Petit, C., Gascon, J., Advanced porous materials: design, synthesis and applications in sustainability. *ACS Sustain. Chem. Eng.*, 7, 9, 7997, 2019.
6. Zdravkov, B., Čermák, J., Šefara, M., Janků, J., Pore classification in the characterization of porous materials: A perspective. *Open Chem.*, 5, 2, 385, 2007.
7. Lan, K. and Zhao, D., Functional ordered mesoporous materials: present and future. *Nano Lett.*, 22, 8, 3177, 2022.
8. Dutta, S., Ghosh, S.K., Fajal, S., Porous organic polymers (POPs) for environmental remediation. *Mater. Horiz.*, 10, 4083, 2023.
9. Rabbani, M.G., Sasse, R.K., Behera, S., Jena, P., Liu, J., Thallapally, P.K., Islamoglu, T., Shehab, M.K., Kaid, M.M., Farha, O.K., El-Kaderi, H.M., High performance porous organic polymers (POPs) for environmental remediation of toxic gases. *Langmuir*, 40, 15, 8024, 2024.

10. Chang, Z., Zhang, D.S., Chen, Q., Bu, X.H., Microporous organic polymers for gas storage and separation applications. *Phys. Chem. Chem. Phys.*, 15, 5430, 2013.
11. Bandyopadhyay, S., Singh, C., Jash, P.M., Hussain, M.W., Paul, A., Patra, A., Redox-active, pyrene-based pristine porous organic polymers for efficient energy storage with exceptional cyclic stability. *Chem. Commun.*, 54, 6796, 2013.
12. Machado, T.F., Valente, A.J., Serra, M.E.S., Murtinho, D., Triazene-linked porous organic polymers for heterogeneous catalysis and pollutant adsorption. *ACS Appl. Polym. Mater.*, 6, 7, 4171, 2024.
13. Gomez-Suarez, M., Chen, Y., Zhang, J., Porous organic polymers as a promising platform for efficient capture of heavy metal pollutants in wastewater. *Polym. Chem.*, 14, 4000, 2023.
14. Gayle, J., Roy, S., Gupta, S., Hassan, S., Rao, A., Demingos, P.G., Miller, K., Guo, G., Wang, X., Garg, A., Singh, C.V., Imine-linked 2D conjugated porous organic polymer films for tunable acid vapor sensing. *ACS Appl. Mater. Interfaces*, 16, 2, 2726, 2024.
15. Miller, K., Gayle, J.M., Roy, S., Abdellah, M.H., Hardian, R., Cseri, S., Demingos, P.G., Nadella, H.R., Lee, F., Tripathi, M., Gupta, S., Guo, G., Bhattacharyya, S., Wang, X., Dalton, A.B., Garg, A., Singh, C.V., Vajtai, R., Szekely, G., Ajayan, P., Tunable 2D conjugated porous organic polymer films for precise molecular nanofiltration and optoelectronics. *Small*, 20, 2401269, 2024.
16. Tian, J., Lin, F., Yu, S.B., Yu, J., Tang, Q., Li, Z.T., Water dispersible and soluble porous organic polymers for biomedical applications. *Aggregate*, 3, 5, 187, 2022.
17. Liu, X., Liu, C.F., Xu, S., Cheng, T., Wang, S., Lai, W.T., Huang, W., Porous organic polymers for high-performance supercapacitor. *Chem. Soc. Rev.*, 51, 3181, 2022.
18. Yang, C., Wang, K., Lyu, W., Liu, H., Li, J., Wang, Y., Jiang, R., Yuan, J., Liao, Y., Nanofibrous porous organic polymers and their derivatives: From synthesis to applications. *Adv. Sci.*, 11, 19, 2400626, 2024.
19. Zhang, Y.Q., Yang, C.C., Ma, J.Y., Tian, W.Q., The enhancement of nonlinear optical properties of azulene based nanographene by N atoms: a finishing touch. *Chem. Sci.*, 15, 2100, 2024.
20. Pino-Rios, R., Báez-Grez, R., Szczepanik, D.W., Solá, M., Designing singlet fission materials with an anti-Kasha behaviour. *Phys. Chem. Chem. Phys.*, 26, 15386, 2024.
21. Shevyakov, S.V., Li, H., Muthyala, R., Asato, A.E., Croney, J.C., Jameson, D.M., Liu, R.S., Orbital control of the color and excited state properties of formylated and fluorinated derivatives of azulene. *J. Phys. Chem. A*, 107, 18, 3295, 2003.

22. Plattner, P.A. and St. Pfau, A., Zur kenntnis der flüchtigen pflanzenstoffe V. Über die darstellung des grundkörpers der Azulen-Reihe. *Helv. Chim. Acta*, 20, 224, 1937.
23. Ziegler, K. and Hafner, K., Ziegler-Hafner azulene synthesis. *Angew. Chem.*, 67, 301, 1955.
24. Shoji, T., Okujima, T., Ito, S., Development of heterocycle-substituted and fused azulenes in the last decade (2010-20202). *Int. J. Mol. Sci.*, 21, 19, 7087, 2020.
25. Razus, A.C., Azulene, reactivity and scientific interest inversely proportional to ring size; part 2: The seven-membered ring. *Symmetry*, 15, 7, 1391, 2023.
26. Barybin, M.V., Nonbenzenoid aromatic isocyanides: new coordination building blocks for organometallic and surface chemistry. *Coord. Chem. Rev.*, 254, 1240, 2010.
27. Cowper, P., Jin, Y., Turton, M.D., Kociok-Köhn, G., Lewis, S.E., Azulenesulfonium salts: accessible, stable and versatile reagents for cross-coupling. *Angew. Chem. Int. Ed.*, 55, 7, 2564, 2016.
28. McDonald, R.N., Petty, H.E., Wolfe, N.L., Paukstelis, J.V., Nonbenzenoid aromatic systems. X. formation, nuclear magnetic resonance spectral identification, and reactions of both Meisenheimer type and methyleneazulenate anions. *J. Org. Chem.*, 39, 1877, 1974.
29. Zheng, W.-T., Huang, K., Dai, S., Solvothermal and Template-Free Synthesis of N-Functionalized Mesoporous Polymer for Amine Impregnation and CO_2 Adsorption. *Microporous Mesoporous Mater.*, 290, 109653–109659, 2019.
30. Alloush, A. M., Abdulghani, H., Amasha, H. A., Saleh, T. A., Al Hamouz, O. C., Microwave Assisted Synthesis of Novel Porous Organic Polymers for Effective Selective Capture of CO_2. *J. Ind. Eng. Chem.*, 113, 215–225, 2022.
31. Zhang, P. and Dai, S., Mechanochemical Synthesis of Porous Organic Materials. *J. Mater. Chem. A*, 5, 31, 16118–16127, 2017.
32. Liu, L., Yu, R., Yin, L., Zhang, N., Zhu, G., Porous organic framework membranes based on interface-induced polymerization: Design, synthesis and applications. *Chem. Sci.*, 15, 6, 1924, 2024.
33. Li, H., Wang, K., Sun, Y., Lollar, C.T., Li, J., Zhou, H.C., Recent advances in gas storage and separation using metal organic-frameworks. *Mater. Today*, 21, 2, 108, 2018.
34. Cheng, Y., Datta, S.J., Zhou, S., Jia, J., Shekhah, O., Eddaoudi, M., Advances in metal-organic framework-based membranes. *Chem. Soc. Rev.*, 51, 19, 8300, 2022.
35. Ellis, J.E., Crawford, S.E., Kim, K.J., Metal-organic framework thin films as versatile chemical sensing materials. *Mater. Adv.*, 2, 19, 6169, 2021.
36. Chen, R., Chen, X., Wang, Y., Wang, B., Biomimetic metal-organic frameworks for biomedical applications. *Trends Chem.*, 5, 6, 460, 2023.
37. Pascanu, V., González Miera, G., Inge, A.K., Martín-Matute, B., Metal-organic frameworks as catalysts for organic synthesis: A critical perspective. *J. Am. Chem. Soc.*, 141, 18, 7223, 2019.

38. Barman, S., Furukawa, H., Blacque, O., Venkatesan, K., Yaghi, O.M., Berke, H., Azulene based metal-organic frameworks for strong adsorption of H_2. *Chem. Commun.*, 46, 7981, 2010.
39. Hasegawa, J.-Y., Higuchi, M., Hijikata, Y., Kitagawa, S., Charge-polarized coordination space foe H_2 adsorption. *Chem. Mater.*, 21, 1829, 2009.
40. (a) Hübner, O., Glöss, A., Fichtner, M., Klopper, W., On the interaction of dihydrogen with aromatic systems. *J. Phys. Chem. A*, 108, 3019, 2004; (b) Wong, M., Kuiken, B.E.V., Buda, C., Dunietz, B.D., Multi adsorption and coadsorption of hydrogen on model conjugated systems. *J. Phys. Chem. C*, 113, 12571, 2009.
41. Mathias, L.J. and Overberger, C.G., Simple syntheses of 1,3- bis (perfluoroacyl) azulenes and 1,3- azulenedicarboxylic acid. *J. Org. Chem.*, 45, 1701, 1980.
42. (a) Dincă, M. and Long, J.R., Strong H_2 binding and selective gas adsorption within the three-dimensional coordination solid $Mg_3(O_2C\text{-}C_{10}H_6\text{-}CO_2)_3$. *J. Am. Chem. Soc.*, 127, 9376, 2005. (b) Dincă, M., Dailly, A., Liu, Y., Brown, C.M., Neumann, D.A., Long, J.R., Hydrogen storage in a metal-organic framework with exposed Mn^{2+} coordination sites. *J. Am. Chem. Soc.*, 128, 16878, 2006. (c) Lee, Y.-G., Moon, H.R., Suh, P., A comparison of the H_2 sorption capacities of isostructural metal-organic frameworks with and without accessible metal sites: $[\{Zn_2(abtc)(dmf)_2\}_3]$ and $[\{Cu(abtc)(dmf)_2\}_3]$ versus $[Cu_2(abtc)_3]$. *Angew. Chem. Int. Ed.*, 47, 7741, 2008.
43. Kishida, K., Horike, S., Nakagawa, K., Kitagawa, S., Synthesis and adsorption properties of azulene-containing porous interdigitated framework. *Chem. Lett.*, 41, 4, 425, 2012.
44. Côté, A.P., Benin, A.I., Ockwig, N.W., O'Keeffe, M., Matzger, A.J., Yaghi, O.M., Porous, crystalline, covalent organic framework. *Science*, 310, 1166, 2005.
45. Guo, J. and Jiang, D., Covalent organic frameworks for heterogeneous catalysis: principle, current status, and challenges. *ACS Cent. Sci.*, 6, 6, 869, 2020.
46. Aksu, G.O. and Keskin, S., Rapid and accurate screening of the COFs space for natural gas purification: COF informatics. *ACS Appl. Mater. Interfaces*, 16, 15, 19806, 2024.
47. Ying, Y., Peh, S.B., Yang, H., Yang, Z., Zhao, D., Ultrathin covalent organic framework membranes *via* a multi-interfacial engineering strategy for gas adsorption. *Adv. Mater.*, 34, 25, 2104946, 2022.
48. Shi, Y., Yang, J., Gao, F., Zhang, Q., Covalent organic frameworks: Recent progress in biomedical applications. *ACS Nano*, 17, 3, 1879, 2023.
49. Keller, N. and Bein, T., Optoelectronic processes in covalent organic frameworks. *Chem. Soc. Rev.*, 50, 1813, 2021.
50. Wang, S., Xu, X., Yue, Y., Yu, K., Shui, Q., Huang, N., Chen, H., Semiconductive covalent organic frameworks: Structural design, synthesis and applications. *Small Struct.*, 1, 2, 2000021, 2020.

51. (a) Luo, R., Lv, H., Liao, Q., Wang, N., Yang, J., Li, Y., Xi, K., Wu, X., Ju, H., Lei, J., Intrareticular charge transfer regulated electrochemiluminescence of donor-acceptor covalent organic frameworks. *Nat. Commun.*, 12, 1, 6808, 2021. (b) Zhou, Z.-B., Tian, P.-J., Yao, J., Lu, Y., Qi, Q.-Y., Zhao, X., Toward azo-linked covalent organic frameworks by developing linkage chemistry *via* linker exchange. *Nat. Commun.*, 13, 1, 2180, 2022.
52. Abbaszadeh, D., Kunz, A., Kotadiya, N.B., Mondal, A., Andrienko, D., Michels, J.J., Wetzelaer, G.-J. A. H., Blom, P.W.M., Electron trapping in conjugated polymers. *Chem. Mater.*, 31, 17, 6380, 2019.
53. Jin, S., Ding, X., Feng, X., Supur, M., Furukawa, K., Takahashi, S., Addicoat, M., El-Khouly, M.E., Nakamura, T., Irle, S., Charge dynamics in a donor-acceptor covalent organic framework with periodically ordered bicontinuous heterojunctions. *Angew. Chem. Int. Ed.*, 52, 7, 2017, 2013.
54. Chen, X., Addicoat, M., Irle, S., Nagai, A., Jiang, D., Control of crystallinity and porosity of covalent organic frameworks by managing interlayer interactions based on self-complementary π-electronic force. *J. Am. Chem. Soc.*, 135, 2, 546, 2013.
55. Wang, M., Wang, M., Lin, H.-H., Ballabio, M., Zhong, H., Bonn, M., Zhou, S., Heine, T., Canovas, E., Dong, R., Feng, X., High-mobility semiconducting two-dimensional conjugated covalent organic frameworks with p-type doping. *J. Am. Chem. Soc.*, 142, 52, 21622, 2020.
56. Feng, B., Chen, X., Yan, P., Huang, S., Lu, C., Ji, H., Zhu, J., Yang, Z., Cao, K., Zhuang, C., Isomeric dual-pore two-dimensional covalent organic frameworks. *J. Am. Chem. Soc.*, 145, 49, 26871, 2023.
57. (a) Xia, R., Li, C., Yuan, X., Wu, Q., Jiang, B., Xie, Z., Facile preparation of thienoisoindigo-based nanoscale covalent-organic framework with robust photothermal activity for cancer therapy. *ACS Appl. Mater. Interfaces*, 14, 19129, 2022; (b) Ding, N., Zhou, T., Weng, W., Lin, Z., Liu, S., Maitarad, P., Wang, C., Guo, J., Multivariate synthetic strategy for improving crystallinity of zwitterionic squaraine-linked covalent organic frameworks with enhanced photothermal performance. *J. Small*, 18, 2201275, 2022; (c) Feng, J., Ren, W.-X., Gao, J.-L., Li, F., Kong, F., Yao, B.-J., Dong, Y.-B., Core-shell structured covalent-organic framework as a nano-agent for single-laser-induced phototherapy. *ACS Appl. Mater. Interfaces*, 13, 17243, 2021; (d) Sun, T., Xia, R., Zhou, J., Zheng, X., Liu, S., Xie, Z., Protein-assisted synthesis of nanoscale covalent organic frameworks for phototherapy of cancer. *Mater. Chem. Front.*, 4, 2346, 2020.
58. Tian, P.-J., Han, X.-H., Qi, Q.-Y., Zhao, X., An Azulene-Based Crystalline Porous Covalent Organic Framework for Efficient Photothermal Conversion. *Small*, 20, 21, 2307635, 2023.
59. (a) World Health Organization, Drinking-water quality guidelines, https://www.who.int/teams/environment-climate-change-and-health/water-sanitation-and-health/water-safety-and-quality/

drinking-water-quality-guidelines, 2024. (b) U.S. Environmental Protection Agency, Drinking water. Report on the Environment, https://www.epa.gov/report-environment/drinking-water, 2024.

60. (a) Schwarz, F., Koch, M., Kastlunger, G., Berke, H., Stadler, R., Venkatesan, K., Lörtscher, E., Charge transport and conductance switching of redox-active azulene derivatives. *Angew. Chem. Int. Ed.*, 55, 11781, 2016; (b) Xin, H., Hou, B., Gao, X., Azulene based π-functional materials: Design, synthesis and applications. *Acc. Chem. Res.*, 54, 1737, 2021. (c) Huang, J., Huang, S., Zhao, Y., Feng, B., Jiang, K., Sun, S., Ke, C., Kymakis, E., Zhuang, X., Azulene-based molecules, polymers and frameworks for optoelectronic and energy applications. *Small Methods*, 4, 2000628, 2020.
61. Zhao, Z., El-Khouly, M.E., Che, Q., Sun, F., Zhang, B., He, H., Chen, Y., Redox-active azulene-based 2D conjugated covalent organic framework for organic memristors. *Angew. Chem. Int. Ed.*, 62, 7, 202217249, 2023.
62. Chen, Y., Liu, G., Wang, C., Zhang, W., Li, R.-W., Wang, L., Polymer memristors for information storage and neuromorphic applications. *Mater. Horiz.*, 1, 489, 2014.
63. Chen, Y., Zhang, B., Liu, G., Zhuang, X., Kang, E.T., Graphene and derivatives: Switch on and off. *Chem. Soc. Rev.*, 41, 4688, 2012.
64. Li, G., Chen, Y., Gao, S., Zhang, B., Li, R.-W., Zhuang, X., Recent advances in resistive switching materials and devices: from memories to memristors. *Eng. Sci.*, 4, 4, 2018.
65. Schwarz, F., Koch, M., Kastlunger, G., Berke, H., Stadler, R., Venkatesan, K., Lörtscher, E., Charge transfer and conductance switching of redox-active azulene derivatives. *Angew. Chem. Int. Ed.*, 55, 11781, 2016.
66. Tobler, H.J., Bauer, H., Günthard, H., The microwave spectrum and dipole moment of azulene. *J. Mol. Spectrosc.*, 18, 239, 1965.
67. Wang, X., Zhao, Y., Wei, L., Zhang, C., Jiang, J.K., Nitrogen-rich conjugated microporous polymers: impact of building blocks on porosity and gas adsorption. *J. Mater. Chem. A*, 3, 21185, 2015.
68. Gan, S., Zeng, Y., Liu, J., Nie, J., Lu, C., Ma, C., Wang, F., Yang, G., Click-based conjugated microporous polymers as efficient heterogeneous photocatalyst for organic transformations. *Catal. Sci. Technol.*, 12, 1202, 2022.
69. Zhang, Y., Sun, Q., Li, Z., Zhi, Y., Li, H., Li, Z., Xia, H., Liu, X., Light emitting conjugated microporous polymers based on an excited state intramolecular proton transfer strategy and selective switch-off sensing of anions. *Mater. Chem. Front.*, 4, 10, 3040, 2020.
70. Saber, A.F., Ahmed, A., Kuo, S.W., EL-Mahdy, A.F., Tetraphenylcyclopentadiene-based conjugated microporous polymers for high-performance energy storage carbons. *Polym. Chem.*, 14, 35, 4079, 2023.
71. Ghasimi, S., Bretschneider, S.A., Huang, W., Landfester, K., Zhang, K.A.I., A conjugated microporous polymer for palladium-free, visible-light promoted photocatalytic Stille-type coupling reactions. *Adv. Sci.*, 4, 6, 1700101, 2017.

72. Kumar, N.R., Agrawal, A.R., Karmakar, H.S., Choudhury, A., Zade, S.S., Tailoring band gaps in thienyltriazine based conjugated porous polymers through comonomer variation. *Chem. Select,* 9, 2–12, 2024.
73. (a) Wang, H., Li, B., Hu, T.-L., Yao, Z., Zhou, W., Xiang, S., Chen, B., A flexible microporous hydrogen-bonded organic framework for gas sorption and separation. *J. Am. Chem. Soc.*, 137, 9963, 2015. (b) Karmakar, A., Illathvalappil, R., Anothumakkool, B., Sen, A., Samanta, P., Desai, A.V., Kurungot, S., Ghosh, S.K., Hydrogen-bonded organic frameworks (HOFs): A new class of porous crystalline proton-conducting materials. *Angew. Chem. Int. Ed.*, 55, 10667, 2016.
74. Desiraju, G.R., Supramolecular synthons in crystal engineering - A new organic synthesis. *Angew. Chem. Int. Ed.*, 34, 2311, 1995.
75. Huang, C., Zhao, C., Deng, Q., Zhang, H., Yu, D., Ren., J., Qu., X., Hydrogen-bonded organic framework-based bioorthogonal catalysis prevents drug metabolic inactivation. *Nat. Catal.*, 6, 8, 729, 2023.
76. Abhari, P.S., Gholizadeh, S., Rouhani, F., Li, Y.L., Morsali, A., Liu, T.F., Recent progress in gas separation platforms based on hydrogen-bonded organic frameworks (HOFs). *Inorg. Chem. Front.*, 10, 6134, 2023.
77. Liu, X., Yang, X., Wang, H., Hisaki, I., Wang, K., Jiang, J., A robust redox-active hydrogen-bonded organic framework for rechargeable batteries. *J. Mater. Chem. A*, 10, 1808, 2022.
78. Halder, A., Ghosh, M., M, A.K., Bera, S., Addicoat, M., Sasmal, H.S., Karak, S., Kurungot, S., Banerjee, R., Interlayer hydrogen-bonded covalent organic frameworks as high-performance supercapacitors. *J. Am. Chem. Soc.*, 140, 35, 10941, 2018.
79. Chen, X., Zheng, Q., Cai, W., Sheng, J., Wang, M., Biodegradable hydrogen-bonded organic framework for cytosolic protein delivery. *ACS Appl. Mater. Interfaces*, 15, 47, 54346, 2023.
80. Lin, R.B. and Chen, B., Hydrogen-bonded organic frameworks: Chemistry and functions. *Chem.*, 8, 8, 2114, 2022.
81. Ion, A.E., Dogaru, A., Shova, S., Madalan, A.M., Akintola, O., Ionescu, S., Voicescu, M., Nica, S., Buchholz, A., Plass, W., Andruh, M., Organic co-crystals of 1,3-bis(4-pyridyl)azulene with a series of hydrogen bond donors. *CrystEngComm*, 20, 4463, 2018.
82. Taylor, D., Dalgarno, S.J., Xu, Z., Vilela, F., Conjugated porous polymers: incredibly versatile materials with far-reaching applications. *Chem. Soc. Rev.*, 49, 3981, 2020.
83. Liao, L., Li, M., Yin, Y., Chen, J., Zhong, Q., Du, R., Liu, S., He, Y., Fu, W., Zeng, F., Advances in the synthesis of covalent triazine frameworks. *ACS Omega*, 8, 4527, 2023.
84. Kulkarni, R., Noda, Y., Barange, D.K., Kochergin, Y.S., Lyu, P., Balcarova, B., Nachtigall, P., Bojdys, M.J., Real-time optical and electronic sensing with a

β-amino enone linked, triazine-containing 2D covalent organic framework. *Nat. Commun.*, 10, 3228, 2019.

85. Kuhn, P., Antonietti, M., Thomas, A., Porous, Covalent triazine-based frameworks prepared by ionothermal synthesis. *Angew. Chem. Int. Ed.*, 47, 3450, 2008.
86. Jiang, K., Peng, P., Tranca, D., Tong, G., Ke, C., Lu, C., Hu, J., Liang, H., Li, J., Zhou, S., Kymakis, E., Zhuang, X., Covalent triazine frameworks and porous carbons: Perspective from an azulene-based case. *Macromol. Rapid Commun.*, 43, 1022, 2022.

11

Calixazulenes and Azuliporphyrins

Nitika Grover

Department of Chemistry, Birla Institute of Technology, Jhunjhunu, Rajasthan, India

Abstract

Azuliporphyrins are porphyrin analogs in which one of the pyrrole rings is replaced by an azulene unit. Similar to pyrrole's α-positions, azulene's 1,3-positions undergo electrophilic substitution that facilitates forming azulene-containing porphyrinoid systems. First, Azuliporphyrin was accessed *via* a "3+1" MacDonald condensation reaction between tripyrranes and 1,3-azulenedicarbaldehyde. Later, Lindsey's porphyrin condensation method was also applied to synthesizing Azuliporphyrins. These porphyrinoids exhibit rich organometallic chemistry that leads to the access of a plethora of metal complexes. Azuliporphyrins' periphery can easily be modified *via* oxidative ring contractions that permit the conversion of azuliporphyrins into carba-, oxacarba-, thiacarba-, and selenacarbaporphyrins. Calix[4]azulene is a calixarene derivative that consists of four azulene chromophores linked by methylene groups. The first azulene analog of the calixarenes was reported by Asao *et al.* in 1998 and was named "azulenophane". In 2002, Lash and Colby developed a straightforward and convenient way for synthesizing the calix[4]azulene. The unique chemical and electronic properties make azulene-containing porphyrinoids attractive candidates for various applications.

***Keywords*:** Azuliporphyrins, calix[4]azulene, Lindsey condensation, core-modified porphyrinoids

Email: nitika.grover@pilani.bits-pilani.ac.in

Neha Rani Kumar and Abhijeet R. Agrawal (eds.) Azulene-Based Materials: Design, Synthesis, and Applications, (311–332) © 2026 Scrivener Publishing LLC

11.1 Introduction

Porphyrins are tetrapyrrole macrocycles consisting of four pyrrole subunits interconnected by methine bridges (Figure 11.1a). This class of compounds is characterized by a large, aromatic macrocyclic structure, resulting in unique electronic and optical properties. Therefore, it exhibits diverse applications in chemistry, biology, and materials science. Several research groups across the globe have dedicated their extensive efforts to modifying the porphyrin core [1]. Carbaporphyrinoids are one of the classes of core-modified porphyrins, where carbon atoms replace one or more pyrrolic nitrogen atoms. These systems are created by the replacement of a pyrrolic unit in the porphyrin framework with various other structures like cyclopentadiene, indene, azulene, cycloheptatriene, or benzene, which results in the formation of a new class of porphyrin-like macrocycles [2]. Carbaporphyrinoids form stable complexes with various metals, including silver, copper, and palladium. These complexes display distinct structural and spectroscopic properties. The study of carbaporphyrinoids provides fundamental insights into aromaticity, tautomerism, and macrocyclic reactivity, which may have implications for applications in areas such as catalysis and materials science. Over the past 20 years, significant progress has been made in studying carbaporphyrinoids. These systems include various analogs like N-confused porphyrins, true carbaporphyrins, benziporphyrins, tropiporphyrins, and azuliporphyrins have been explored in the literature.

Azuliporphyrins and calixazulenes are the two exemplary classes of core-modified porphyrins. Azuliporphyrins are a family of porphyrin analogs in which an azulene moiety replaces one of the pyrrolic subunits, while calixazulenes are modified calixarenes with incorporated azulene moieties (Figures 11.1b and 11.1c) [3–5]. Although both azuliporphyrins and

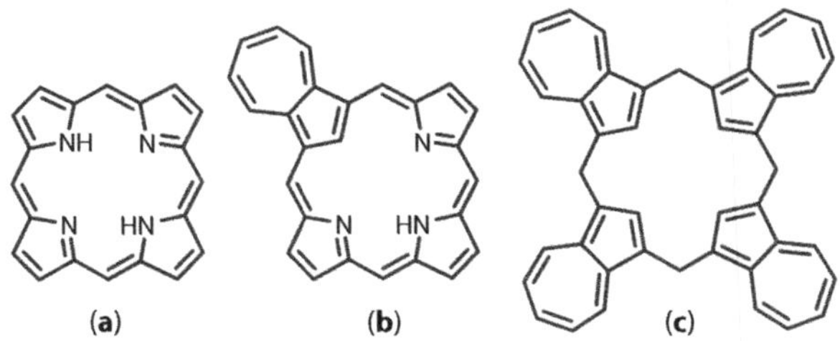

Figure 11.1 Illustrative examples of (a) porphyrin, (b) azuliporphyrin, and (c) calixazulene skeleton.

calixazulenes involve azulene-based structures, they differ significantly in their molecular frameworks, properties, and potential applications. This chapter details the synthesis and properties of azuliporphyrins and calixazulenes and also sheds light on the effect of these modifications on their reactivity and aromatic characteristics.

11.2 Azuliporphyrin

Azuliporphyrins, first reported in 1997, are a family of porphyrin analogs in which an azulene moiety replaces one or more of the pyrrolic subunits [3]. Electrophilic substitution takes place at 1,3-positions of azulene, which is structurally similar to the 2,5-positions of pyrrole, and this characteristic facilitates the incorporation of azulene into the porphyrinoid system (Figure 11.2). Azuliporphyrins exhibit unique reactivity, such that these porphyrin analogs can form organometallic complexes with various metals under mild reaction conditions and undergo selective oxidations at the core C-atom to obtain 21-oxyazuliporphyrins. First, azuliporphyrin was accessed *via* a “3+1” MacDonald condensation reaction between tripyrranes and 1,3-azulenedicarbaldehyde. Later, “2+2” condensation and Lindsey’s porphyrin condensation method were also applied to synthesizing azuliporphyrins [4]. These porphyrinoids exhibit rich organometallic chemistry that leads to the access of a plethora of metal complexes. Azuliporphyrins’ periphery can easily be modified *via* oxidative ring contractions that convert azuliporphyrins into carba-, oxacarba-, thiacarba-, and selenacarbaporphyrins. The unique reactivity, structural, and photophysical chemistry of azuliporphyrins has put them in the spotlight for the past two decades [5]. Some of the significant synthetic efforts to access azuliporphyrins are discussed in detail.

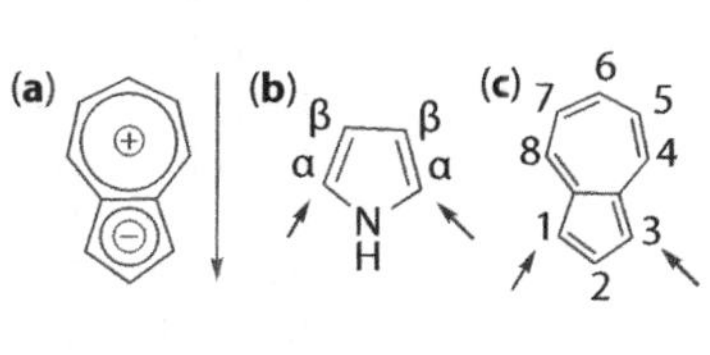

Figure 11.2 (a) Dipolar character of azulene; (b) and (c) preferred reactive sites for electrophilic substitution in pyrrole and azulene.

11.2.1 Synthetic Developments of Azuliporphyrins

Lash and Chaney reported a 3+1 MacDonald-type condensation reaction between 1,3-azulenedicarbaldehyde (**11.1**) and tripyrrane (**11.2**) in the presence of Trifluoroacetic acid (TFA), followed by the addition of triethylamine and oxidation with 2, 3-Dichloro-5,6-dicyano-1,4-benzoquinone (DDQ), which led to the formation of azuliporphyrin (**11.3**) (Scheme 11.1) in 28% yield [6]. Later, a modified protocol was reported that gave azuliporphyrins (**11.4–11.11**) with yields of up to 77% [7].

Graham *et al.* employed an alternative method to access azuliporphyrins *via* azulitripyrrane. A reaction of acetoxymethylpyrroles (**11.12**) with azulene (**11.13**) in the presence of CH_3COOH yielded azulitripyrrane (**14**) (Scheme 11.2). TFA was used to deprotect the tert-butylester groups in **11.14** and the product was subjected to a condensation reaction followed by oxidation using DDQ to give azuliporphyrins **11.3**, **11.9**, and **11.10** in moderate yields [8].

This approach has been further extended to access core-modified azuliporphyrin, such as benzoazuliporphyrins (**11.17**) [9]. This method provided a more convenient synthesis of benzoazuliporphyrins, as the previous report included the fusion of azuliporphyrins and bicyclooctadiene units at 200 °C, followed by retro-Diels–Alder elimination of ethylene to obtain benzoazuliporphyrins. Next, a HCl-catalyzed "2 + 2" condensation reaction between dipyrrylmethane **11.18** and azulenylmethylpyrrole dialdehyde **11.19** yielded azuliporphyrin **11.3** (Scheme 11.3).

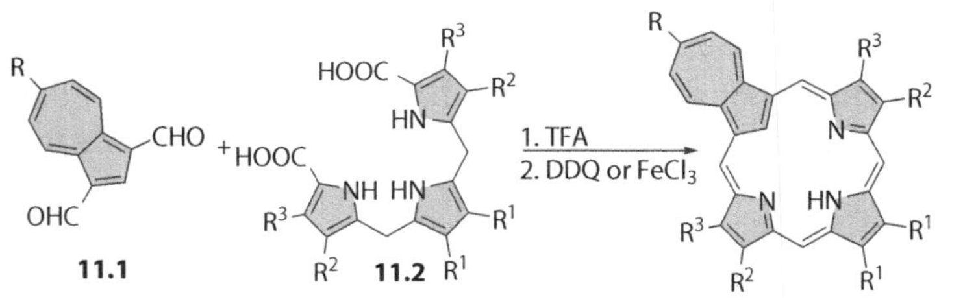

R = H, $R^1 = R^2$ = Et, R^3 = Me; **11.3**
R = H, R^1 = Ph, R^2 = Et, R^3 = Me; **11.4**
R = H, R^1 = Ph, R^2 = ^{t}Bu, R^3 = Me; **11.5**
R = H, R^1 = Et, $R^2 = R^3$ = Ph; **11.6**
R = H, $R^1 = (CH_2)_4$, R^2 = Et, R^3 = Me; **11.7**
R = ^{t}Bu, R^1 = Ph, R^2 = Et, R^3 = Me; **11.8**
R = ^{t}Bu, $R^1 = R^2$ = Et, R^3 = Me; **11.9**
R = Ph, $R^1 = R^2$ = Et, R^3 = Me; **11.10**
R = Ph, R^1 = Ph, R^2 = Et, R^3 = Me; **11.11**

Scheme 11.1 MacDonald "3 + 1" strategy to access azuliporphyrins.

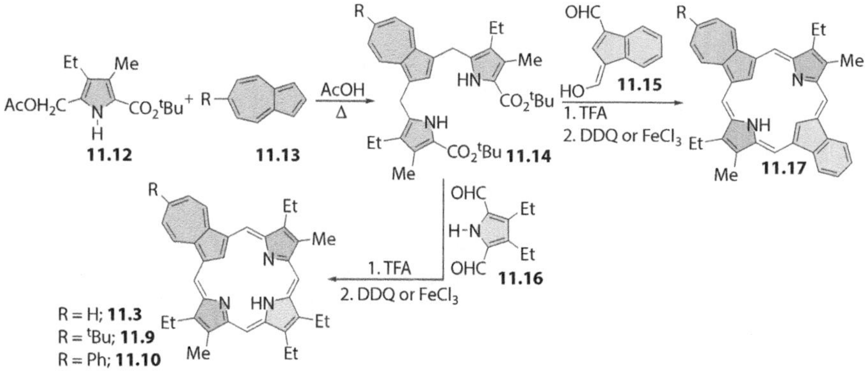

Scheme 11.2 Back-To-Front "3 + 1" method for synthesis of azuliporphyrins and benzoazuliporphyrins.

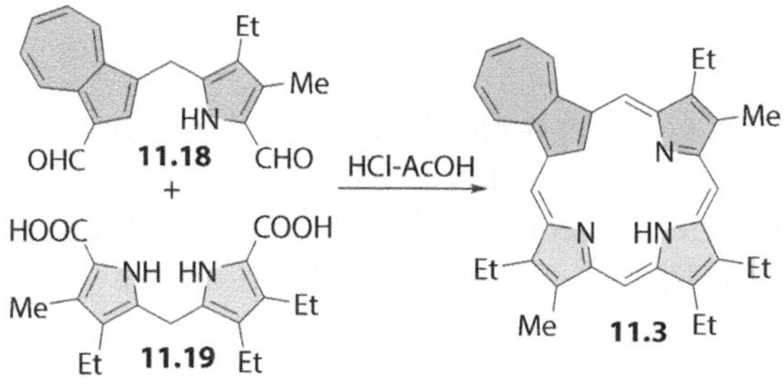

Scheme 11.3 "2 + 2" Condensation for preparing azuliporphyrins.

The Rothemund method has been adapted for the synthesis of tetraarylazuliporphyrins [10]. Later yields for tetraarylazuliporphyrins were improved using Lindsey condensation conditions [11]. A reaction of azulene (1 equiv), benzaldehyde (4 equiv), and pyrrole (3 equiv) in the presence of $BF_3{\cdot}Et_2O$, followed by DDQ oxidation, yielded 13% of tetraphenylazuliporphyrin. It was observed that the reaction did not work in DCM, but $CHCl_3$ appeared as a suitable solvent to conduct the reactions, potentially due to the presence of C_2H_5OH stabilizer in $CHCl_3$. The reaction took 16h, indicating that the rate of formation of azuliporphyrinogen was relatively slow. Different aldehydes were condensed under these reaction conditions to obtain azuliporphyrins **11.20–11.29** in 18–22% yield (Scheme 11.4). Among all 4-halobenzaldehydes gave superior yields.

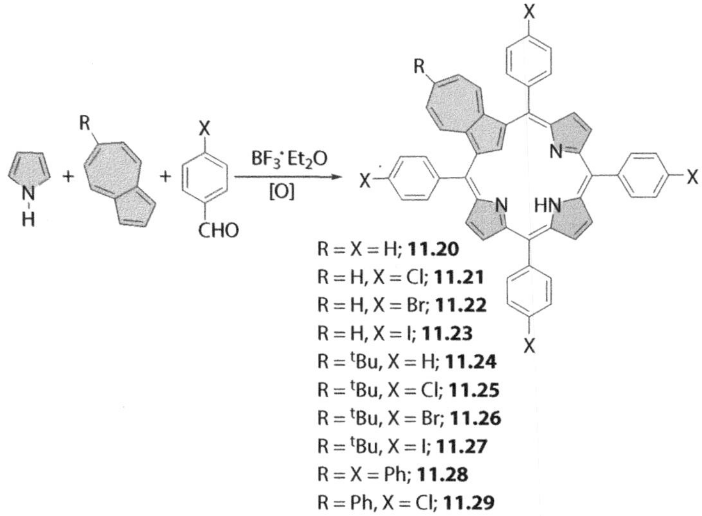

Scheme 11.4 Lindsey condensation conditions to access meso-substituted azuliporphyrins.

Chandrashekar *et al.* reported the synthesis of core-modified azuliporphyrins using [3 + 1] acid-catalyzed MacDonald-type condensation reaction [12]. A reaction of core-modified tripyrromethanes (**11.30–11.31**) with azulene-1,3-dicarboxaldehyde in the presence of TFA and DDQ yielded porphyrin (**11.32, 11.33**) (Scheme 11.5).

Lash *et al.* adopted a different strategy to access core-modified azuliporphyrins 11.34 and 11.35 in 45–55% yield using corresponding azulitripyrranes [13]. These heteroazuliporphyrins were unstable in basic conditions and readily converted into corresponding benzoheterocarbaporphyrins in the presence of KOH and tert-butyl hydroperoxide (Scheme 11.6).

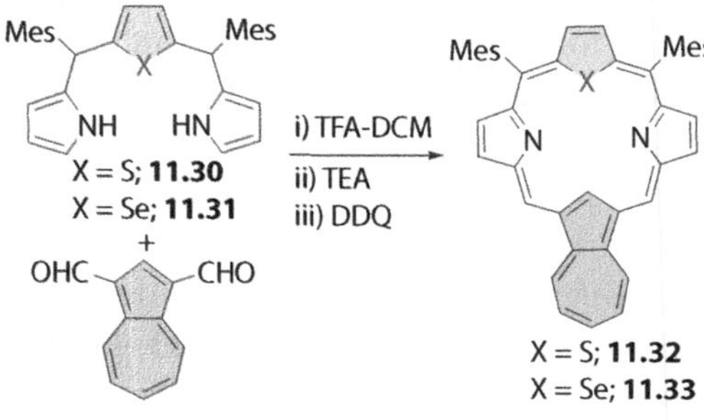

Scheme 11.5 Synthesis of core-modified azuliporphyrins.

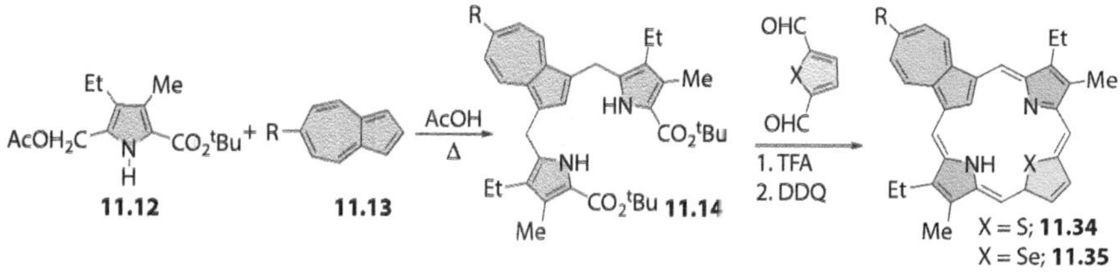

Scheme 11.6 Synthesis of core-modified azuliporphyrins using azulitripyrranes.

As shown in Scheme 11.7, 2+2 condensation was employed to access the diazuliporphyrins [14]. A HCl or HBr-catalyzed reaction of diazulenylmethane dialdehyde (**11.36**) with dipyrrylmethanes (**11.37**, **11.38**), followed by $FeCl_3$ oxidation, gave aromatic adj-diazuliporphyrins (**11.39**, **11.40**). The free base structures were expected to exist as mesoionic species; however, the synthesized compounds were separated as monoprotonated macrocycle salts (**11.41**, **11.42**).

Conventional porphyrin condensation has been applied to access heterodiazuliporphyrins. In one of the examples, a mixture of azulene and thiophene dicarbinols was treated with a catalytic amount of boron trifluoride etherate and DDQ led to the formation of macrocycle **11.44** [15]. Similar results were obtained when azulene reacted with furane dicarbinols (Scheme 11.8) [16]. Heterodiazuliporphyrins can be further oxidized to give compounds **11.43** and **11.44**, which can be reverted to **11.45** and **11.46** using tin (II) chloride.

Examples of C-alkylated azuliporphyrins have been synthesized using "3+1" Macdonald type condensation reaction between 2-alkylazulene and acetoxymethylpyrrole [17]. Similarly, a reaction between azulene

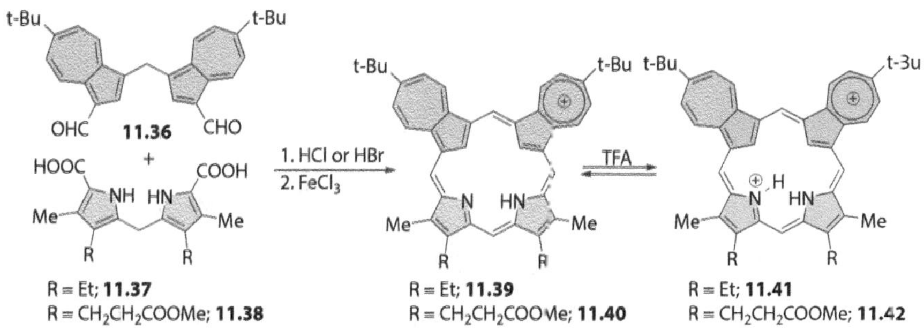

Scheme 11.7 Synthesis of adj-diazuliporphyrins.

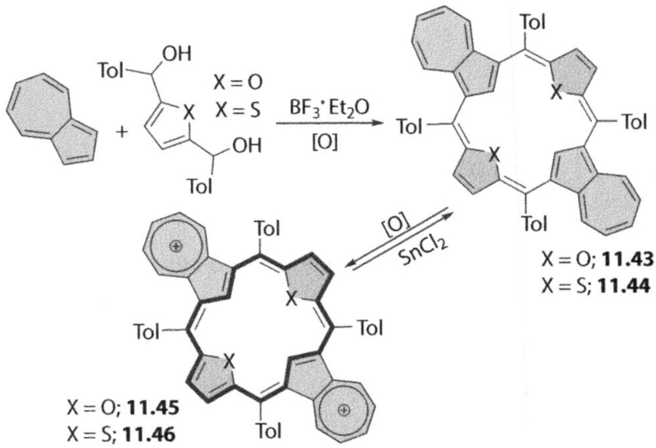

X = O; **11.43**
X = S; **11.44**

X = O; **11.45**
X = S; **11.46**

Scheme 11.8 Synthesis of heterodiazuliporphyrins.

dialdehyde and N-methyltripyrrane gave 23-methylazuliporphyrin. The free base alkylated azuliporphyrins were unstable and were isolated in dicationic form.

11.2.2 Reactivity of Azuliporphyrins

Azuliporphyrins can be converted into carbaporphyrinoid in the presence of pyrrolidine. Adding pyrrolidine to compound **11.3** resulted in the adduct **11.47**, which exhibits a carbaporphyrin-type conjugation with an 18π electron delocalization pathway (Scheme 11.9) [18]. The azuliporphyrin-pyrrolidine adduct cannot be isolated in the solid form as the removal of excess pyrroline brings it back to azuliporphyrin. A similar reactivity was seen when tetraphenylazuliporphyrin **11.20** was treated with pyrrolidine, thiophenol, benzylamine, and hydrazine [19].

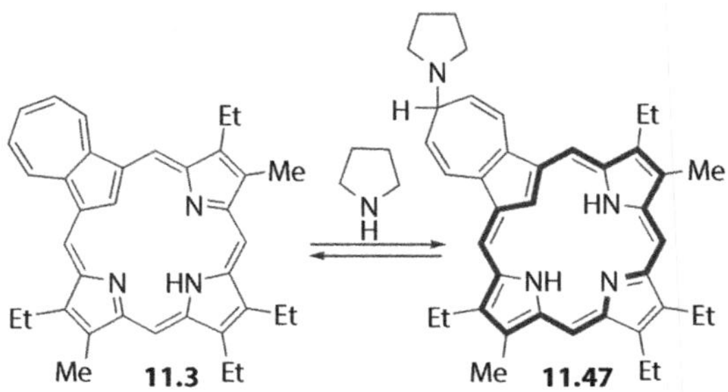

11.3 **11.47**

Scheme 11.9 Formation of azuliporphyrin-pyrrolidine adduct.

Next, a reaction of porphyrin **11.3** with tert-butyl hydroperoxide allowed benzocarbaporphyrins **11.50** to form. The mechanism given in Scheme 11.10 explains the unprecedented formation of **11.50**. After being attacked by the tertbutyl peroxide anion, the adduct **11.48** can undergo a Cope rearrangement, yielding the cyclopropane **11.49**. Eliminating tert-butyl alcohol yields the aldehyde **11.50**. TLC analysis shows a slight presence of a third carbaporphyrin. This could be due to an initial nucleophilic assault at a different place on the azulene ring. Similar results were obtained for the tetraarylazuliporphyrins. Notably, this route serves as the only known method for synthesizing tetraarylcarbaporphyrins.

Azuliporphyrins serve as dianionic ligands and can form stable organometallic derivatives. The reaction of **11.3** with $Ni(OAc)_2$ or $Pd(OAc)_2$, respectively, resulted in the synthesis of the corresponding metallated azuliporphyrins **11.51** and **11.52**, in high yields (Scheme 11.11) [20]. Platinum (II) complex **11.53** was synthesized using the same strategy; however, the yields remain low. The ^1H NMR of the palladium complex showed that meso-protons exhibit improved diatropic character. The X-ray analysis for palladium-(II) complex **11.52** revealed that the porphyrin skeleton exhibits a planar configuration with a twisted azulene unit [21]. In contrast to the Pd-complex **11.52**, the Ni-complex **11.51** showed a ruffled conformation.

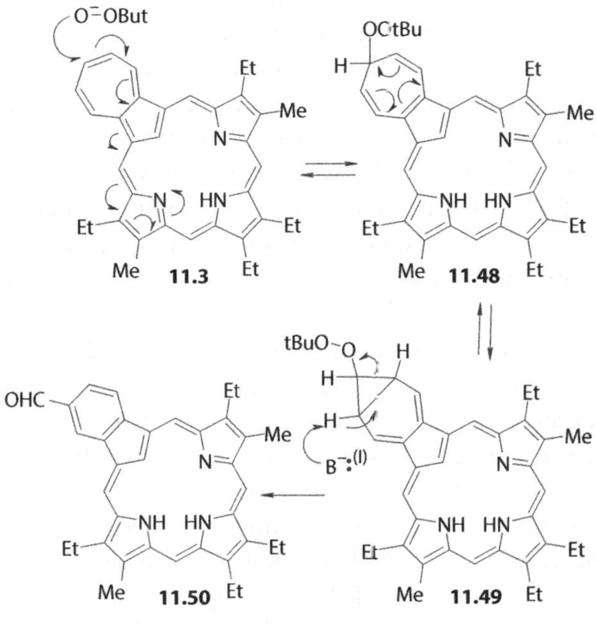

Scheme 11.10 Synthesis of benzocarbaporphyrins.

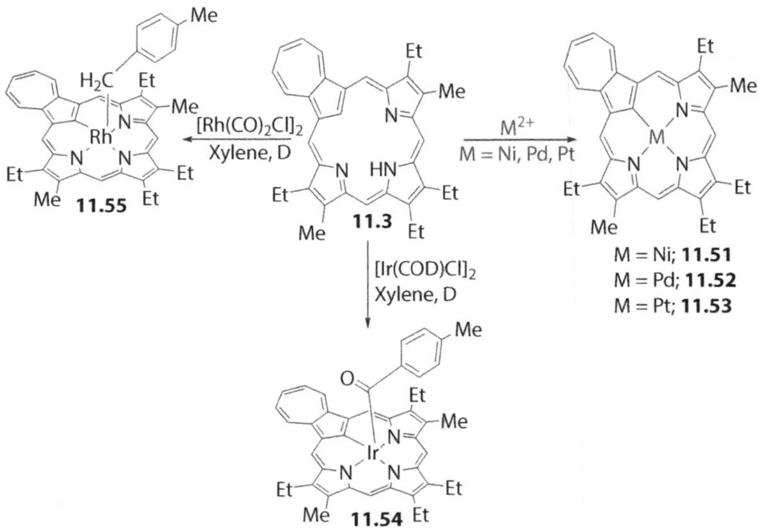

Scheme 11.11 Metalation of azuliporphyrins.

Iridium (III) azuliporphyrin **11.54** was synthesized *via* reaction of azuliporphyrin **11.3** with bis(1,5-cyclooctadiene)diiridium(I) dichloride in refluxing xylene [22]. Despite the low yield, the iridium complex exhibited interesting features such as two metal-carbon bonds, i.e, azulene carbon-iridium bond and axial benzoyl ligand were seen in one molecule. It was anticipated that the first methylbenzyl group from the xylene unit attached to the metal atom, which is oxidized by environmental oxygen results in the formation of the axial benzoyl unit.

A reaction of $[Rh(CO)_2Cl]_2$ with azuliporphyrin gave rhodium(III) complex **11.55** [23]. The rhodium complex exhibited a methylbenzyl unit as an axial ligand. Single-crystal analysis showed that axial ligands in **11.54** and **11.55** adopt an orthogonal arrangement with respect to the porphyrin plane.

A reaction of meso-tetraphenylazuliporphyrin **11.20** with 1 equiv. of $Ru_2(CO)_{12}$ gave the corresponding ruthenium(II) complex; however, the addition of excess $Ru_2(CO)_{12}$ led to the formation of $Ru_4(CO)_9$ cluster complex in which the azulene moiety acts as a π-coordination site to accommodate the $Ru_4(CO)_9$ cluster. Furthermore, the reaction of $Ru_3(CO)_{12}$ with nickel(II), palladium(II), and platinum(II) azuliporphyrins yielded the respective bimetallic complexes [24]. As shown in Scheme 11.12, a reaction of tetraphenylazuliporphyrin **11.20** with copper(II)acetate led to the formation of the corresponding copper(II) complex **11.56**. Further reaction

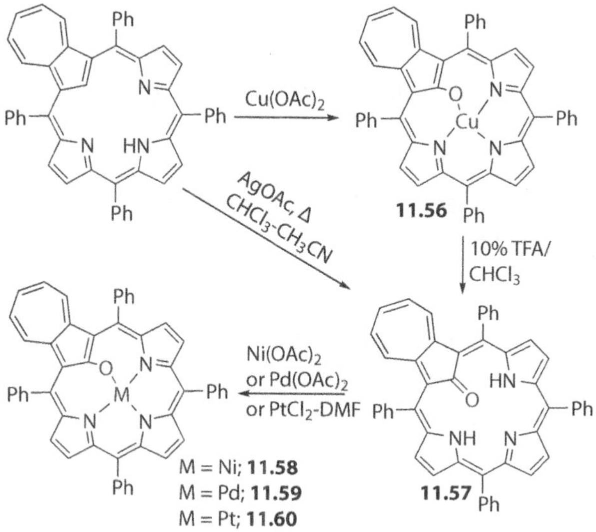

Scheme 11.12 Oxidative metalation of tetraphenylazuliporphyrins.

of Cu-complex with 10% TFA in $CHCl_3$ gave the respective 21-oxyazuliporphyrins **11.57**. Compound **11.57** also acts as a dianionic ligand and gives metal complexes (**11.58–11.60**) when treated with nickel (II) acetate, palladium (II) acetate, or platinum (II) chloride [25]. Furthermore, it was also observed that direct access to compound **11.57** could be obtained *via* the reaction of tetraarylazuliporphyrins **11.20** with silver(I) acetate in chloroform-acetonitrile [26].

11.2.3 Expanded Azuliporphyrins

A few examples of azulene-containing expanded porphyrinoids are also known in the literature. These macrocycles open exciting avenues for future porphyrin research. Azulisapphyrin dication **11.62** was synthesized *via* a TFA-catalyzed reaction between azulene dialdehyde **11.1** and tetrapyrrole dicarboxylic acid **11.61**. In this reaction, the intermediate was *in situ* oxidized using $FeCl_{3,}$ followed by hydrochloric acid protonation. (Scheme 11.13) [27].

In a similar vein, nonaromatic diazulihexaphyrin **11.64** was synthesized *via* TFA-catalyzed condensation between azulene diacrylaldehyde **11.63** with dipyrrylmethan **11.37** (Scheme 11.14) [28]. Overall, expanded azuliporphyrins remain unexplored and overlooked in the literature, and more efforts are required in this direction.

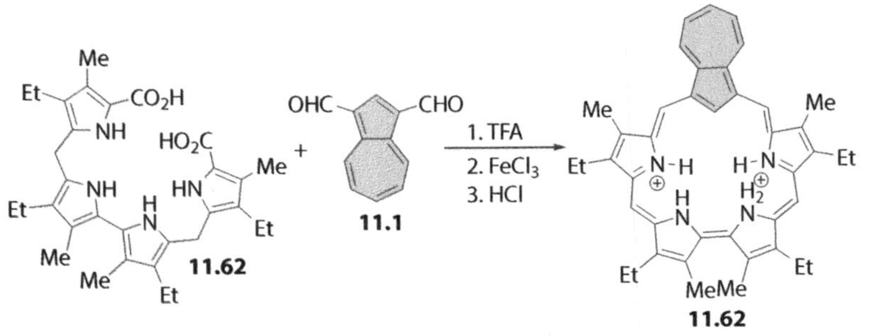

Scheme 11.13 Synthesis of azulisapphyrin dication.

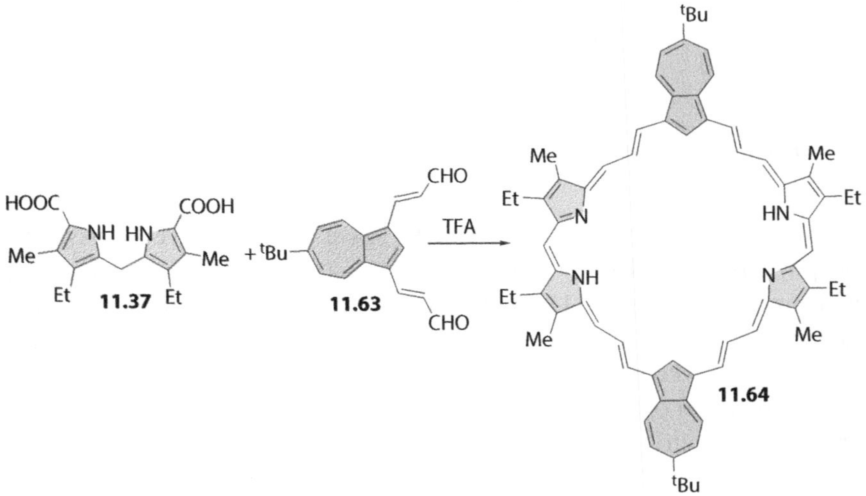

Scheme 11.14 Synthesis of diazulihexaphyrin.

11.3 Calixazulenes

Calix azulenes are macrocyclic, composed of azulene chromophores linked by methylene groups. These compounds demonstrate distinctive photochemical and photophysical properties as compared to their calixarene analogs [29]. In contrast to azulene units, calix[4]azulene shows extensive photostability. These molecules remain stable when exposed to the visible and near-UV range. Furthermore, these molecules do not show evident "anti-Kasha" fluorescence, which is commonly seen in azulene. Calix[4]azulene exhibited a bathochromic shift as compared to single azulene, indicating the interaction between azulene units in one molecule.

These compounds include solely carbon and hydrogen atoms and no intraannular hetero-atom functional groups that could help organize their deep, π-electron-rich voids. In this regard, they differ greatly from the better-known and researched calixarenes. As a result, these molecules provide an opportunity to assess their ability to form supramolecular complexes with various guest molecules such as fullerene C_{60} or tetraalkylammonium salts, as well as whether host calixazulene pre-organization is required [30].

11.3.1 Synthesis and Reactivity of Calixazulenes

Given its structural similarity with the quatyrin skeleton, several attempts have been made to obtain calixazulene system. The first synthesis of calixazulene was reported by Asao *et al.* in 1988. The 2-Methoxyazulene was subjected to a multistep synthetic strategy to access [1.1.1.1] (1,3)-2-methoxyazulenophane, which is an azulene derivative of calix [4] arene [31]. Later, it was reported that in the presence of Florisil, azulene reacts with paraformaldehyde to give calix[4]azulene **11.65** in a 74% yield (Scheme 11.15) [32]. A variety of acids have been used as catalysts to obtain calix[4]azulene; however, low yield and undesired formation of calix[5] azulene remain the main obstacles. To overcome these issues, a multistep synthetic strategy was employed to access **11.65**. In the first step, $NaBH_4$ was used for the reduction of azulene dialdehyde, which unexpectedly led to a self-condensation reaction to give diazulenylmethane **11.70**. It was proposed that diazulene product **11.70** was formed *via* ipso substitution of carbocation intermediate with azulene dicarbinol, following formaldehyde elimination. This process led to the formation of calixazulene, which was degraded on the silica (Scheme 11.14). However, the use of Florisil gave calix[4]azulene. This observation led to developing a one-pot synthesis of calix[4]azulene using Florisil as a catalyst [33].

A similar reactivity was observed for the 6-tert-Butylazulene, which gave compound **11.66** in 68% yield. However, yield remained low for 6-methylazulene, potentially due to the acidic nature of methyl substituents. Similarly, poor yields were observed for the 6-phenylazulene.

As stated above, structurally calixazulenes are a meso-reduced quatyrin, and the oxidation of calix[4]azulene unit should provide access to quatyrin skeleton. However, the oxidation of calix[4]azulene with DDQ resulted in the decomposition of the starting material. In another effort, **11.66** was treated with triphenylcarbenium tetrafluoroborate and triphenylcarbenium, which led to the formation of mono-, di-, and tricationic products **11.74–11.76** (Scheme 11.16). However, the isolation of one species remained a problem in this case. On the contrary, pure **11.75** was obtained

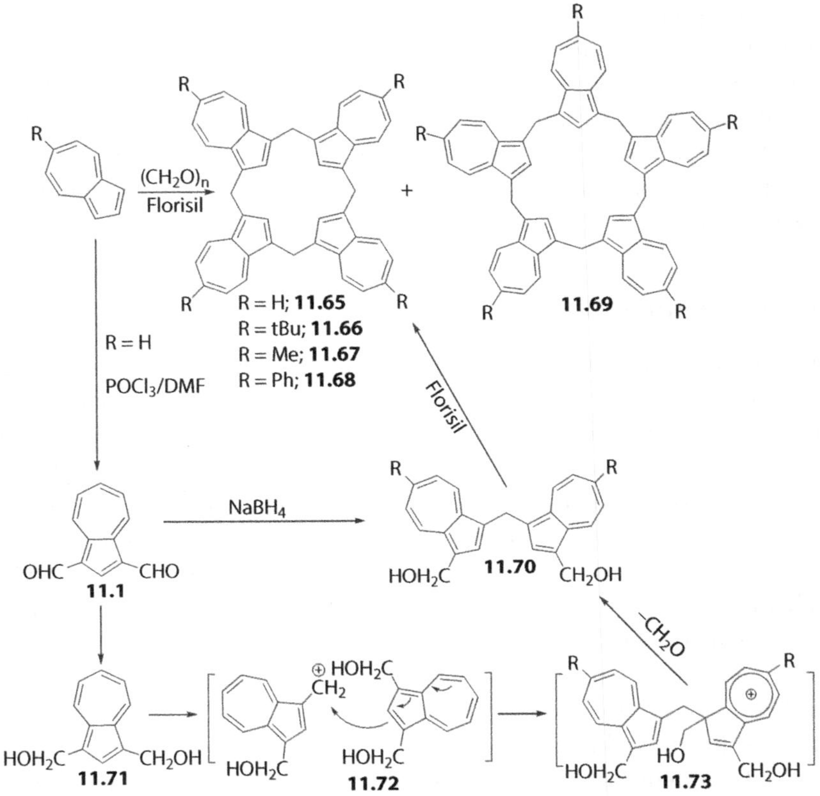

Scheme 11.15 Synthesis and reaction mechanism to access calix[4]azulenes.

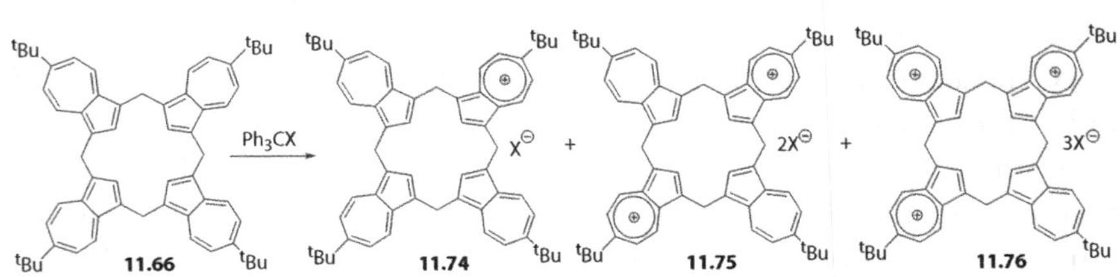

Scheme 11.16 Generation of conjugated cations from a calix[4]azulene.

when 5 equiv of Ph_3CPF_6 was reacted with **11.66**. Due to the structural similarity, compound **11.75** can be considered a dihydroquatyrin system and shows a strong absorption band at 616 nm.

A reaction of ferrocenecarbaldehyde with 6-tert-Butylazulene in the presence of acetic acid gave a corresponding calix[4]azulene **11.78** in 49% yield (Scheme 11.17) [34]. In this reaction, bis(1-azulenyl)

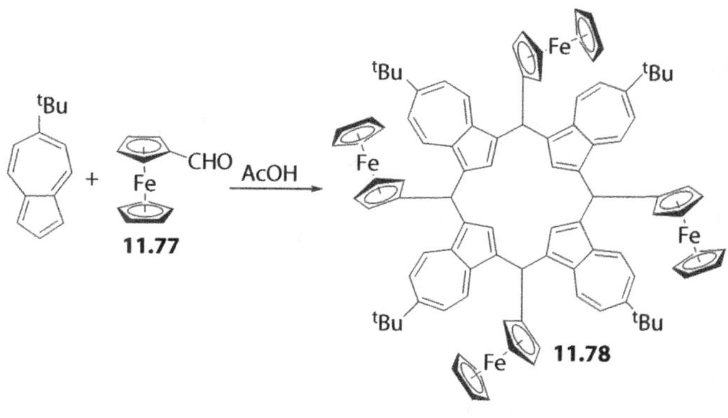

Scheme 11.17 Synthesis of ferrocene-substituted calixazulene.

ferrocenylmethane derivative was also obtained as a side product. Adding DDQ to compound **11.78** yielded bis(1-azulenyl)ferrocenylmethylium ion (5+) *via* hydride abstraction. The cyclic voltammetry of compound **11.78** showed multi-electron oxidation.

Latos–Grażyński *et al.* have reported that a reaction of azulene with aromatic aldehydes in reacted with aromatic aldehydes presence of boron trifluoride etherate yielded tetraarylcalixazulenes in **11.79–11.81** in 40–97%. The oxidation DDQ, followed by adding $HBF_4{\cdot}Et_2O$ to **11.79–11.81** access to the tetraazuliporphyrin tetracations **11.82–11.84**. Compounds **11.82–11.84** are non-aromatic due to the 16π-electron delocalization pathway (Scheme 11.18). Density Functional Theory (DFT) calculation showed that these compounds adopt a saddle-shaped conformation. ^{1}H NMR for compounds **11.82–11.84** showed inner core protons are downfield shifted, potentially due to the four positive charges associated with the macrocycles. In summary, NMR and DFT calculations supported that these cation macrocyles are nonaromatic and do not have effective conjugation pathways [35].

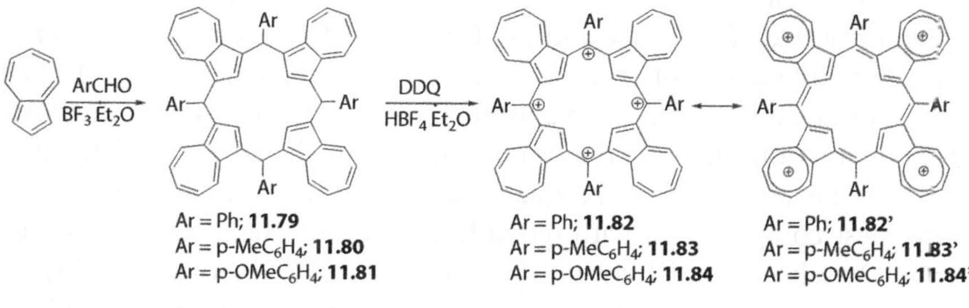

Scheme 11.18 Synthesis of tetraazuliporphyrin tetracations.

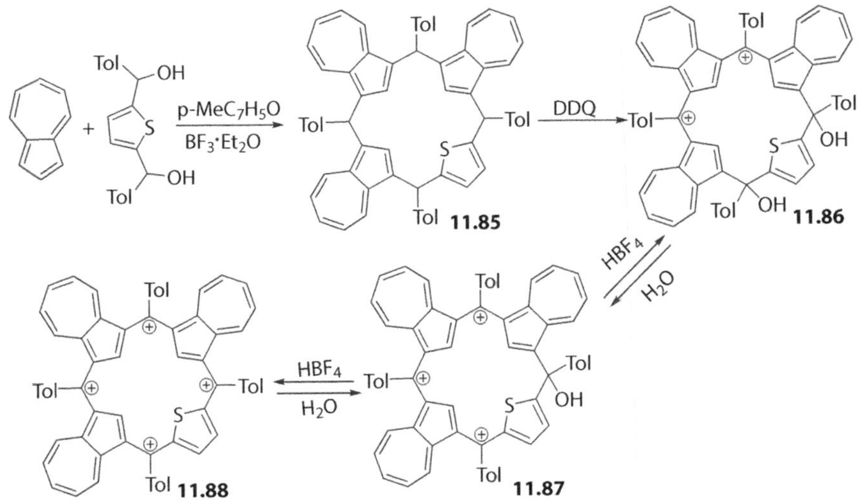

Scheme 11.19 Synthesis of a triazulene porphyrinoid tetracation.

Similarly, reactions between azulene, thiopene dicarbinol, and p-methylbenzaldehyde in the presence of boron trifluoroide etherate yielded compound **11.85**. Further oxidation of compound **11.85** with 8 equiv of DDQ resulted in compound **11.86**. Next is the subsequent addition of $HBF_4{\cdot}Et_2O$ to access trication **11.87** and tetracation **11.88** (Scheme 11.19). The inner core protons in these compounds were highly downfield shifted. The DFT calculation showed these macrocycles were nonplanar with alternate conformation of the azulene and thiophene rings [36]. The theoretical calculation of electron density showed a positive charge on the meso-carbon, indicating the non-aromatic nature of the compound.

11.3.2 Supramolecular Host-Guest Interaction

The similarity between calix[4]azulenes and calix[4]arenes makes them an obvious candidate for molecular recognition investigations. Due to the presence of deep cavity size and available π-electron cloud, calixazulenes offer more interactive sites as compared to their calixarene analogs. To this end, several studies are available in the literature that deals with the host-guest interaction studies of the complexation of Lash's calix[4]azulene [30].

It was observed that calix[4]azulene **11.65** formed a supramolecular host-guest complexation with tetraalkylammonium halides and tetrafluoroborate salts. ^1H NMR titration performed in $CDCl_3$ showed the formation of a 1:1 complex between tetramethylammonium chloride (TMACl) and **11.65**. UV-Vis titration was used to calculate the association constant between TMAX (X = Cl^-, Br^-, I^-) and **11.65**. The observed values followed the expected trend for the association constant for TMAX, Cl^-> Br^-> I^-. Furthermore, the authors performed the DFT calculation to support the experimental results [29].

In another study, it was observed that tetra-(5,7-diphenyl)calix[4]azulene and C_{60} form a 1:1 supramolecular complex (Scheme 11.20). A conventional mechanochemical hand-grinding technique was used for complex formation. Due to low solubility, solid-state analysis was performed for this complex. The inclusion of C_{60} in the calixarene cavity in columnar arrays was evidenced by solid-state NMR. It was proposed that additional 'edge-to-face' type interactions between the 2′,6′-protons of the phenyl group substituents of the azulenes and C_{60} played a crucial role in stabilizing supramolecular assembly. DFT calculation suggested that cone conformers of calizazulene and C_{60} form an energetically preferred 1:1 complex [37].

Next, the same research group reported the non-covalent interaction between compound **11.66** and C_{60}. The authors used conventional hand grinding in a mortar and pestle to access the supramolecular complex. It was observed that C_{60} and **11.66** interacted to form a supramolecular complex, having stoichiometry 1:1. Similar to earlier work, the authors used solid-state NMR to analyze the complex formation (Scheme 11.21). The DFT calculations showed that **11.66** has a greater tendency to interact with C_{60} as compared to simple azulene. It was also reported that the tert-butyl group played a crucial role in the stability of the complex, as these groups facilitate the additional π-methyl interactions [38].

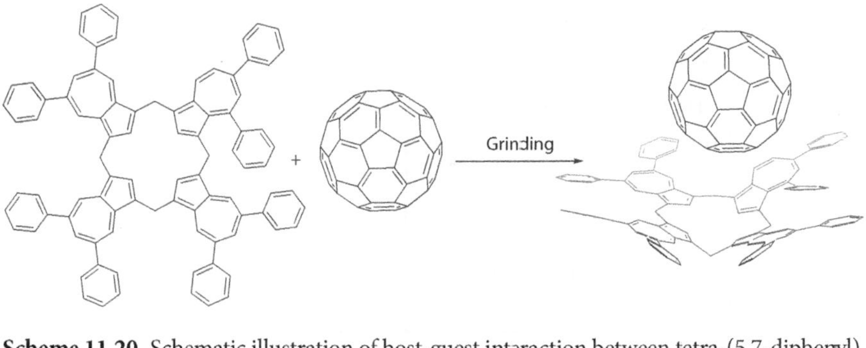

Scheme 11.20 Schematic illustration of host-guest interaction between tetra-(5,7-diphenyl)calix[4]azulene and C_{60}.

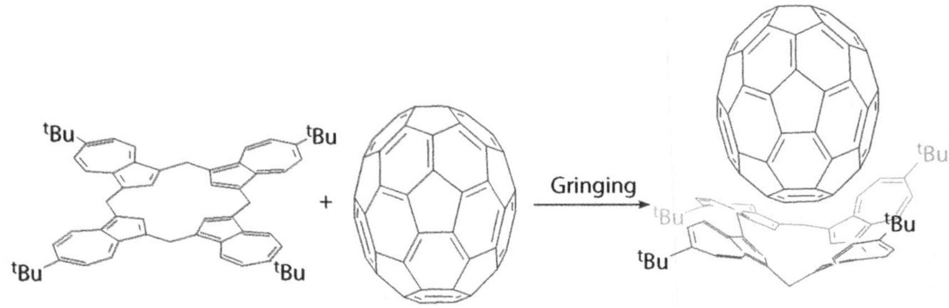

Scheme 11.21 Schematic illustration of host-guest interaction between **11.66** and C_{60}.

Kudo *et al.* reported that calix[5]azulene **11.69** can be synthesized *via* a H_2SO_4-catalyzed condensation reaction between bis(guaiazulene-3-yl) methane and azulene. The authors investigated the non-covalent interaction between **11.69** and various pyridinium salts. The UV-Vis studies were performed to analyze the stoichiometry and binding constants. As shown in Table 11.1, calix[5]azulene exhibits higher association constants as compared to calix[4]azulene. It was also observed that bromide salts showed higher affinity as compared to corresponding chloride salts [39].

Table 11.1 Complexation study on calixazulenes and pyridinium salts.

			Kassoc * 104 (M-1)	
Entry	**Guest molecule**	**pKa[40]**	**Calix[5]azulene**	**Calix[4]azulene**
1.	Pyridinium bromide	5.2	4.67	3.07
2.	Pyridinium chloride	5.2	2.93	2.55
3.	2-Acetylpyridinium bromide	3.7	4.79	3.59
4.	2-Methylpyridinium bromide	5.9	3.07	3.07
5.	2-Cyanopyridinium bromide	-0.3	3.89	3.08
6.	4-Cyanopyridinium bromide	1.9	3.59	3.09

11.4 Conclusion and Outlook

Azulene is an effective alternative to pyrrole in carbaporphyrinoid systems. The 1,3-positions of azulene have reactivity similar to α-positions of pyrrole, allowing for the formation of useful building blocks like azulitripyrranes, and can facilitate macrocycle formation, as evident in the synthesis of tetraarylazuliporphyrins and calix[4]azulenes. The diatropic characteristics of azuliporphyrins can be effectively enhanced upon protonation. Azuliporphyrins' reactivity has allowed the synthesis of various novel organometallic derivatives with a plethora of transition metal ions such as Ni(II), Pd(II), Pt(II), Ir(III), Rh(III), and Ru(II). The unique reactivity of azuliporphyrins led to oxidative ring contractions, which resulted in benzocarbaporphyrins. This method opens an alternate pathway to access aromatic carbaporphyrinoid systems. Selective oxidation of the interior azulene carbon produces 21-oxyazuliporphyrins, which can also serve as dianionic ligands for transition metal ions. Azulene has been integrated into various dicarbaporphyrinoid complexes, exhibiting intriguing reactivity and substantial aromatic properties. In addition, azuliporphyrins showed unique reactivity in the combined presence of base and peroxide, leading to the formation of tetraarylbenzocarbaporphyrins. The reaction of tetraarylbenzocarbaporphyrins with various metal ions provides access to corresponding organometallic derivatives. More than one azulene unit has also been incorporated into the macrocyclic system. To this end, diazulenylmethane dialdehyde was used as a building block to synthesize diazuliporphyrins. A HCl or HBr-catalyzed reaction of diazulenylmethane dialdehyde with dipyrrylmethanes led to the formation of monoprotonated diazuliporphyrins.

Pioneer work from Lash *et al.* provides easy access to calixazulenes. In recent years, calixazulene has been exploited in supramolecular chemistry. It has been demonstrated that the calixazulene cavity can accommodate a variety of guest molecules, such as quaternary ammonium salts and C_{60}. The azulene unit in calixazulene provides a p-interaction site that facilitates the additional stability of the supramolecular complex. A few examples of expanded azuliporphyrinoids are also known in the literature. These azulene-containing porphyrinoids have a distinct and complicated reactivity that compares favorably to better-studied porphyrin analog systems, such as N-confused porphyrins. Researchers have made progress in the synthesis of extended azuliporphyrins. Research on azulene-containing porphyrin analogs has yielded encouraging findings, but there is still more to learn. Future discoveries are expected to be equally significant.

AI disclosure statement: The author confirms that no artificial intelligence (AI)-generated content or AI-assisted writing tools were used in the preparation of this chapter (including summaries and the abstract). All text, analysis, and conclusions are the result of the author's original work.

References

1. Senge, M.O., Sergeeva, N.N., Hale, K.J., Classic highlights in porphyrin and porphyrinoid total synthesis and biosynthesis. *Chem. Soc. Rev.*, *50*, 4730, 2021.
2. Lash, T.D., Carbaporphyrinoid Systems. *Chem. Rev.*, 117, 2313, 2017.
3. Lash, T.D. and Hayes, M.J., Carbaporphyrins. *Angew. Chem. Int. Ed.*, *36*, 840, 1997.
4. Lash, T.D., Carbaporphyrins and related systems: Synthesis, characterization, reactivity and insights into the nature of porphyrinoid aromaticity, in: *Handbook of Porphyrin Science –With Applications to Chemistry, Physics, Material Science, Engineering, Biology and Medicine*, vol. 16, K.M. Kadish, K.M. Smith, R. Guilard (Eds.), pp. 1–329, World Scientific Publishing, Singapore, 2012.
5. Lash, T.D., Out of the blue! Azuliporphyrins and related carbaporphyrinoid systems. *Acc. Chem. Res.*, 49, 471, 2016.
6. Lash, T.D. and Chaney, S.T., Azuliporphyrin, A case of borderline porphyrinoid aromaticity. *Angew. Chem. Int. Ed.*, 36, 839, 1997.
7. Lash, T.D. and Bergman, K.M., Further observations on conformational and substituent effects in acid catalyzed "3 + 1" cyclizations of tripyrranes with aromatic dialdehydes. *J. Org. Chem.*, 77, 9774, 2012.
8. Graham, S.R., Colby, D.A., Lash, T.D., Azulene analogues of tripyrranes and carbaporphyrinoids therefrom. *Angew. Chem. Int. Ed.*, 41, 1371, 2002.
9. Okujima, T., Kikkawa, T., Nakano, H., Kubota, H., Fukugami, N., Ono, N., Yamada, H., Uno, H., Retro-Diels–Alder approach to the synthesis of π-expanded azuliporphyrins and their porphyrinoid aromaticity. *Chem. Eur. J.*, 18, 12854, 2012.
10. Colby, D.A. and Lash, T.D., Adaptation of the Rothemund reaction for carbaporphyrin synthesis: Preparation of meso-tetraphenylazuliporphyrin and related benzocarbaporphyrins. *Chem. Eur. J.*, 8, 5397, 2002.
11. Lash, T.D., Colby, D.A., Ferrence, G.M., Further studies on the synthesis of meso-tetraarylazuliporphyrins under Lindsey- Rothemund reaction conditions and their conversion into benzocarbaporphyrins. *Eur. J. Org. Chem.*, 2003, 4533, 2003.
12. Venkatraman, S., Anand, V.G., PrabhuRaja, V., Rath, H., Sankar, J., Chandrasekar, T.K., Teng, W., Senge, K.R., First structural characterization of core modified 10,15-meso aryl azuliporphyrins: Observation of C–H···

π interaction between pyrrole β-CH and mesityl ring. *Chem. Commun.*, 38, 1660, 2002.
13. Lash, T.D., El-Beck, J.A., Ferrence, G.M., Synthesis, structural characterization and reactivity of heteroazuliporphyrins. *Org. Biomol. Chem.*, 12, 316, 2014.
14. Zhang, Z., Ferrence, G.M., Lash, T.D., adj-Diazuliporphyrins, a new family of dicarbaporphyrinoids with unprecedented mesoionic characteristics. *Org. Lett.*, 11, 101, 2009.
15. Sprutta, N., Swiderska, M., Latos-Grazynski, L., Dithiadiazuliporphyrin: Facile generation of carbaporphyrinoid cation radical and dication. *J. Am. Chem. Soc.*, 127, 13108, 2005.
16. Sprutta, N., Siczek, M., Latos-Grazynski, L., Pawlicki, M., Szterenberg, L., Lis, T., Dioxadiazuliporphyrin: A near-IR redox switchable chromophore. *J. Org. Chem.*, 72, 9501, 2007.
17. Moriones, J.S.D., Latham, A.N., Lash, T.D., Synthesis of internally alkylated azuliporphyrins. *J. Porphyrins Phthalocyanines, 23*, 1, 2019.
18. Lash, T.D., The azuliporphyrin-carbaporphyrin connection. *Chem. Commun.*, 34, 1683, 1998.
19. Lash, T.D., Origin of aromatic character in porphyrinoid systems. *J. Porphyrins Phthalocyanines*, 15, 1093, 2011.
20. Lash, T.D., Colby, D.A., Graham, S.R., Ferrence, G.M., Szczepura, L.F., Organometallic chemistry of azuliporphyrins: Synthesis, spectroscopy, electrochemistry and structural characterization of nickel(II), palladium(II) and platinum(II) complexes of azuliporphyrins. *Inorg. Chem.*, 42, 7326, 2003.
21. Ferrence, G.M. and Lash, T.D., (23-tert-Butyl-5,10,15,20-tetraphenylazuliporphyrinato)-palladium(II) dichloromethane solvate. *Acta Crystallogr., Sect. E: Struct. Rep. Online*, 63, m1351, 2007.
22. Lash, T.D., Pokharel, K., Zeller, M., Ferrence, G.M., Iridium(III) azuliporphyrins. *Chem. Commun.*, 48, 11793, 2012.
23. Stateman, L.M., Ferrence, G.M., Lash, T.D., Rhodium(III) azuliporphyrins. *Organometallics*, 34, 3842, 2015.
24. Bialek, M. and Latos-Grazynski, L., Merging of inner and outer ruthenium organometallic coordination motifs within an azuliporphyrin framework. *Chem. Commun.*, 50, 9270, 2014.
25. Colby, D.A., Ferrence, G.M., Lash, T.D., Oxidative metalation of azuliporphyrins with copper (II) salts: Formation of a porphyrin analogue system with a unique fully conjugated nonaromatic azulene subunit. *Angew. Chem. Int. Ed.*, 43, 1346, 2004.
26. Lash, T.D., Colby, D.A., El-Beck, J.A., AbuSalim, D.I., Ferrence, G.M., Preparation, structural characterization, assessment of potential antiaromaticity and metalation of 21-oxyazuliporphyrins. *Inorg. Chem.*, 54, 9174, 2015.
27. Richter, D.T. and Lash, T.D., Synthesis of sapphyrins, heterosapphyrins and carbasapphyrins by a "4 + 1" approach. *J. Org. Chem.*, 69, 8842, 2004.

28. Zhang, Z., Ferrence, G.M., Lash, T.D., MacDonald-type reactions using bis-acrylaldehydes: Synthesis of an expanded sapphyrin and vinylogous hexaphyrins. *Org. Lett.*, 11, 1249, 2009.
29. Stevens, A.L., Yeow, C., White, J.M., Bradley, S.J., Ghiggino, K.P., Steer, R.P., Electronic spectroscopy and photophysics of calix[4]azulene. *J. Photochem. Photobiol., A*, 405, 112922, 2021.
30. Georghiou, P.E., Rahman, S., Alodhay, A., Nishimura, H., Lee, J., Wakamiya, A., Scott, L.T., Calixazulenes: azulene-based calixarene analogues – an overview and recent supramolecular complexation studies. *Beilstein J. Org. Chem.*, 14, 2488, 2018.
31. Asao, T., Ito, S., Morita, N., Synthesis of [1.1.1.1](1,3)-2-methoxyazulenophane: Azulene analogue of calixarenes. *Tetrahedron Lett.*, 29, 2839, 1988.
32. Lash, T.D., El-Beck, J.A., Colby, D.A., Synthesis of a tetraazulene porphodimethene analogue. *J. Org. Chem.*, 74, 8830, 2009.
33. Colby, D.A. and Lash, T.D., Calix[4]azulene. *J. Org. Chem.*, 67, 1031, 2002.
34. Shoji, T., Higashi, J., Ito, S., Toyota, K., Iwamoto, T., Morita, N., Synthesis and properties of ferrocenylmethylene-bridged calix[4]azulene and a new example of bis(1-azulenyl)ferrocenylmethylium ion. *Eur. J. Org. Chem.*, 2009, 5948, 2009.
35. Sprutta, N., Mackowiak, S., Kocik, M., Szterenberg, L., Lis, T., Latos-Grażyński, L., Tetraazuliporphyrin tetracation. *Angew. Chem. Int. Ed.*, 48, 3337, 2009.
36. Sprutta, N., Welnic, M., Bialek, M.J., Lis, T., Szterenberg, L., Latos-Grażyński, L., Carbocations confined in a thiatriazuliporphyrin frame. *Chem. Eur. J.*, 22, 6974, 2016.
37. Schneider, C., Nishimura, H., Lee, J., Scott, L.T., Wakamiya, A., Forbes, R., Georghiou, P.E., Mechanochemically-generated solid state complex of C60-fullerene with tetra-(5,7- diphenyl)calix[4]azulene, NMR, XRD and DFT studies. *Supramol. Chem.*, 30, 575, 2018.
38. Georghiou, P.E., Schneider, C., Shamov, G., Lash, T.D., Rahman, S., Giddings, S., Mechanochemical formation of a 1:1 C_{60}:Tert-butylcalix[4]azulene supramolecular complex: Solid-state NMR and DFT computational studies. *Supramol. Chem., 28*, 396, 2016.
39. Jin, Y., Akagawa, K., Mutai, T., Yoshikawa, I., Kudo, K., Synthesis, characterization of calix[5]azulene and its complexation with pyridinium salts. *Tetrahedron*, 88, 132146, 2021.
40. Sadler, S.A., Tajuddin, H., Mkhalid, I.A., Batsanov, A.S., Albesa-Jove, D., Cheung, M.S., Maxwell, A., Shukla, L., Roberts, B., Blackemore, D.C., Lin, Z., Marder, T.B., Steet, P.G., Iridium-catalyzed C–H borylation of pyridines. *Org. Biomol. Chem., 12*, 7318, 2014.

12

Functionalization and Utilization of Azulene in Organometallic and Coordination Chemistry

Sulendar K. Mahato* **and Rakesh Kumar Rai**

Department of Chemistry, IIT Hyderabad, Hyderabad, Telengana, Kandi, India

Abstract
Azulenes are an isomer of naphthalene, characterized by fused five- and seven-membered ring systems. They are known for their deep blue color and are found in various products and synthetic materials. Its optical, electrical, and biological properties are different from those of the isoelectronic isomer naphthalene due to its special structural arrangement. Thus far, azulenes and their equivalents have found widespread usage in fields like medicines, organic electronic materials, and fluorescence switching. Azulenes, with their unique aromatic properties, have found applications in organometallic chemistry. They serve as ligands for transition metal complexes, contributing to various catalytic reactions and material synthesis. Their π-electron system allows for interesting coordination chemistry. They can coordinate transition metal centers, forming stable complexes with various geometries and coordination modes. Their unique electronic properties and tunable π-system allow for tailored interactions with metal ions, offering opportunities for designing novel coordination complexes with specific properties and applications.

Keywords: Azulenes, fluorescence, metal complex, coordination complexes

12.1 Introduction

Azulene, an isomer of naphthalene, stands out due to its distinct photophysical properties, such as the blue color resulting from a narrow energy

**Corresponding author*: sulendarkr1991@gmail.com

Neha Rani Kumar and Abhijeet R. Agrawal (eds.) Azulene-Based Materials: Design, Synthesis, and Applications, (333–350) © 2026 Scrivener Publishing LLC

gap between the highest occupied molecular orbital (HOMO) and the lowest unoccupied molecular orbital (LUMO), and its anti-Kasha fluorescence. Azulene's unique electronic structure features an electron-rich five-membered ring and an electron-deficient seven-membered ring, contributing to its dipole moment of 1.08 D (Figure 12.1). This molecule exhibits a significant difference in reactivity at odd and even positions and shows stimulus-responsive behavior. Because of these characteristics, azulene-containing polymers are interesting candidates for use in organic electronic applications such as photovoltaic (PV) cells and organic field-effect transistors (OFET). Because azulene possesses a fused five- and seven-membered ring structure, it has been successfully incorporated into the polymer backbone *via* one or both rings [1, 2].

The significant dipole moment of azulenes, stemming from the polarization of their π-electron cloud, contributes to their high reactivity. This reactivity pattern manifests in distinct nucleophilic and electrophilic behavior across the molecule. In the five-membered ring, C1- and C3-positions are nucleophilic, while the C4-, C6-, and C8-positions of the seven-membered ring tend to be electrophilic. Consequently, electrophilic substitutions are generally observed at C1- and C3-positions, whereas nucleophilic additions are observed at the C4-, C6-, and C8-positions.

The unique "donor-acceptor" structure of azulene sets it apart from typical fused benzenoid compounds and makes it suitable for creating advanced optoelectronic materials. In contrast, azulene features a non-mirror-symmetrical frontier molecular orbital (FMO) geometry. The LUMO density is localized at the even positions and the HOMO density is localized at the even positions. This distinct FMO geometry results in azulene having a small energy gap between its ground state (S_0) and the first excited state (S_1), causing the $S_0 \rightarrow S_1$ transition absorption to occur in the visible range and giving azulene its characteristic blue color. (Figure 12.2). The narrow HOMO-LUMO gaps in azulene and its derivatives have made it an important moiety to be included in materials with applications in molecular electronics, functional components in conducting polymers, optoelectronic materials, and organic photovoltaics.

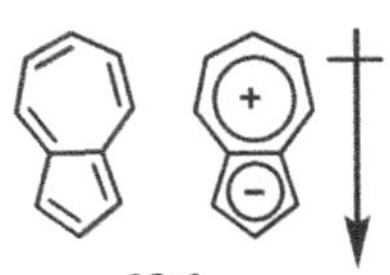

12.1

Figure 12.1 Dipolar character of azulene.

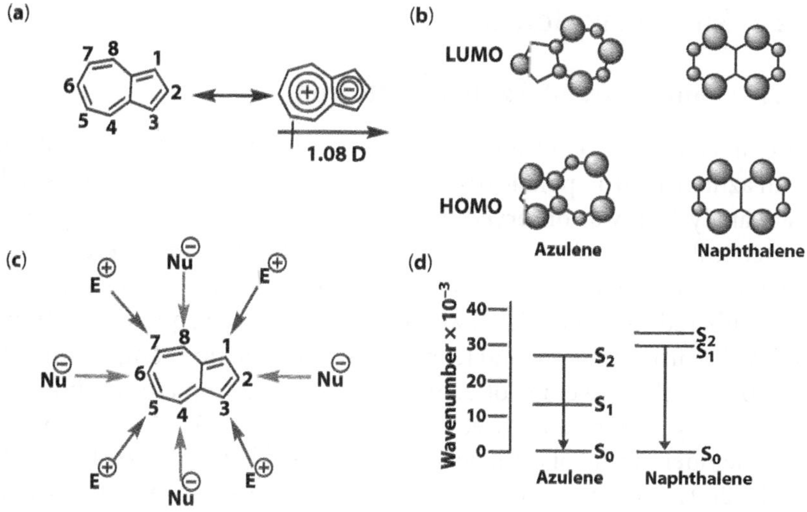

Figure 12.2 (a) Numbering sequence in azulene and its polarized resonance structure; (b) electron density in HOMO and LUMO of azulene (left) and naphthalene (right); (c) electrophilic and nucleophilic centers in azulene; (d) fluorescence was observed for naphthalene and azulene. Reproduced with permission [2]. Copyright 2021, American Chemical Society.

Azulene exhibits a strong transition from the ground state (S_0) to the second excited singlet state (S_2), leading to a significant energy gap between the first excited singlet state (S_1) and S_2 ($\Delta E(S_2{-}S_1) > 10{,}000\ cm^{-1}$). This large energy gap diminishes the rate of the non-radiative decay process from S_2 to S_1. Consequently, fluorescence from S_2 to the ground state (S_0) dominates over the radiationless $S_2 \rightarrow S_1$ transition.

In recent years, there has been a growing interest in azulene due to its distinctive electronic and optical characteristics, leading to an active exploration of azulene-based π-functional materials [2]. These derivatives find applications in diverse fields such as anion sensors, electrochromic materials, charge-transfer complexes, nonlinear optical materials, organic conductors, and near-infrared resonance materials. Additionally, they have been utilized in OFETs [3] and solar cells [4]. However, the progress in developing azulene-based organic materials is hindered by challenges in molecular design and synthesis.

12.2 Metallaazulenes

12.2.1 Complexes of Azulene with Iron

Churchill *et al.* have reported the nature of bonding present in metal-azulene in azulenedi-iron pentacarbonyl $C_{10}H_8Fe_2(CO)_5$ **12.2** [5]. From single-crystal analysis, it was observed that the iron atom of an Fe(CO) group is symmetrically bonded to all carbon atoms in the five-membered ring of the azulene system. One iron atom is positioned 1.798 Å below the planar five-membered ring of azulene, which has a root mean square deviation of 0.003 Å. The iron-carbon bond lengths in this interaction range from 2.088 to 2.100 Å, with an average of 2.093 Å. The second iron atom is bound to three carbonyl groups and three carbon atoms (C4, C5, and C6) from the seven-membered ring of azulene, with Fe-C bond lengths of 2.142 Å, 2.050 Å, and 2.200 Å, respectively Figure 12.3. The distance between the two iron atoms is 2.782 Å, indicating a metal-metal bond. This bonding allows each iron atom to achieve an 18-electron configuration, consistent with the diamagnetic nature of the complex.

12.2.2 Complexes of Azulene with Rhodium and Iridium

Cano *et al.* have reported the synthesis of cationic rhodium and iridium complexes containing azulene ligands [6]. Two approaches were followed while carrying out the synthesis:

1) Treatment of rhodium or iridium precursors like [MCl (diolefin)]$_2$ with silver salts ($AgPF_6$ or $AgClO_4$) and azulene.
2) Directly reacting preformed cationic complexes with $[M(diolefin)(Me_2CO)_2]^+$ with azulene.

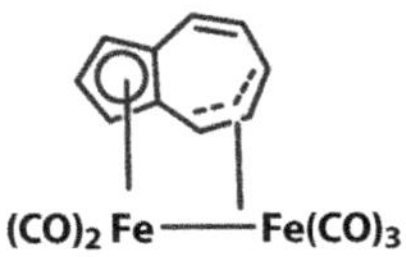

12.2

Figure 12.3 Structure of iron-azulene complex.

The addition of stoichiometric amounts of azulene to acetone solutions of $[Rh(diolefin)_2]^+$ (diolefin = tetrafluorobenzobarrelene, 1,5-cyclooctadiene, or 2,5-norbornadiene) gives deep red solutions from which red solids of **12.5** can be isolated. The synthetic scheme is as shown below (Scheme 12.1).

1/2 $[RhCl(diolefin)]_2$ **12.3** $\xrightarrow[Me_2Co]{AgA}$ $[Rh(diolefin)(Me_2CO)_x]A$ **12.4** $\xrightarrow{az}$ [Rh(diolefin)(az)]A **12.5**

(A = PF_6^-, diolefin = TFB (I), COD (II), NBD (III), DQ (duroquinone) (IV))

(A = ClO_4^-, diolefin = Me_3TFB (trimethyltetrafluorobenzobarrelene) (V))

az = azulene

Scheme 12.1 Schematics for the synthesis of complexes of rhodium with azulene.

The similar iridium derivative $[Ir(COD)az]ClO_4$ (**12.6**) was prepared by reaction of $[Ir(COD)az]ClO_4$ with $AgClO_4$ and azulene.

12.2.3 Synthesis of Metalla-Dual-Azulenes

The synthesis of metalla-dual-azulenes was reported by Xia *et al.* consisting of a metallaazulene and a metal-free organic azulene intertwined by sharing the tropylium motif. The synthesis was accomplished by subjection the diyne moiety to Friedel–Crafts cyclization followed by oxidative dehydrogenation of the resulting compound under mild conditions. The reaction of the diyne **12.7** with $[Ir(CH_3CN)(CO)(PPh_3)_2]BF_4$ in CH_2Cl_2 under N_2 at r.t. for 6 h gave compound **12.4** in approximately 81% yield. Further treatment of dialkyne with phenylacetylene and $AgBF_4$ in the presence of $HBF_4{\cdot}Et_2O$ at r.t. in CH_2Cl_2 for 15 min yielded the final compound (Figure 12.4) [7].

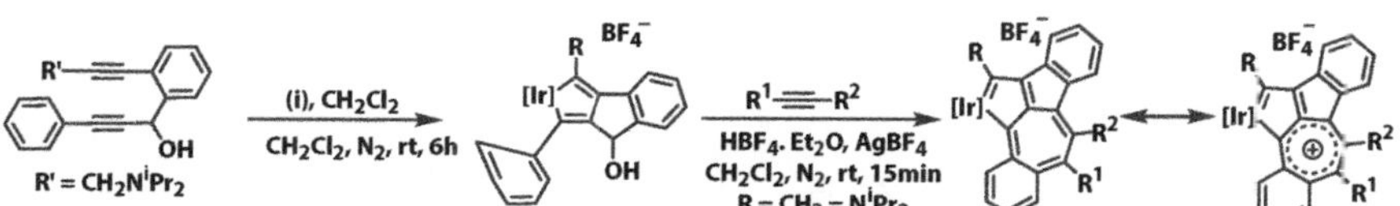

12.7 (i) $[Ir(CH_3CN)(CO)(PPh_3)_2]BF_4$ 12.8 12.9 12.10

Figure 12.4 Synthesis of metallazulenes.

The resulting fused-polycyclic metallacycles generally exhibit strong absorption bands in the UV-Vis spectra. These bands have distinct energy absorption characteristics that may be precisely adjusted by changing the substituents at C5, C6, and C7 on the seven-membered ring. Additionally, the complexes can selectively sense fluoride ions with a limit of detection of 9.73×10^{-7} M.

12.2.4 Organometallic Complexes of Azulene

Shetti *et al.* have reported the coordination polymers of poly{1,3-bis[2-(3-alkylthienyl)] azulene} (**12.11**). The organometallic complexes of $Ru_3(CO)_{12}$ in refluxing xylene yielded organometallic complexes. During the process, the ratio of reactants can be changed to alter the amount of ruthenium in these complexes. The degree of ruthenium coordination to the azulene units in the polymer backbone was evaluated using the ^{1}H NMR chemical shift values of the azulene unit, as coordinated azulene displays upfield-shifted resonance signals compared to unbound azulenes. The polymer's ruthenium content might be changed to modify the characteristics of these complexes **12.11–12.14**, according to absorption and electrochemical investigations. Greater cathodic shift and greater blue shift of the absorption bands were caused by a higher ruthenium concentration. As compared to the free azulenes, the ^{1}H-NMR of the ruthenium complexes showed an upfield shift of resonance signals as compared to the free azulenes (Figure 12.5) [8].

12.2.5 Azuliporphyrins

Azulene-containing porphyrinoids, which are known as azuliporphyrins, have garnered significant interest from researchers owing to their intriguing properties. This includes their capacity to form mildly-regulated organometallic derivatives and aromatic characteristics that fall between nonaromatic macrocycles and porphyrins. Azuliporphyrins consist of

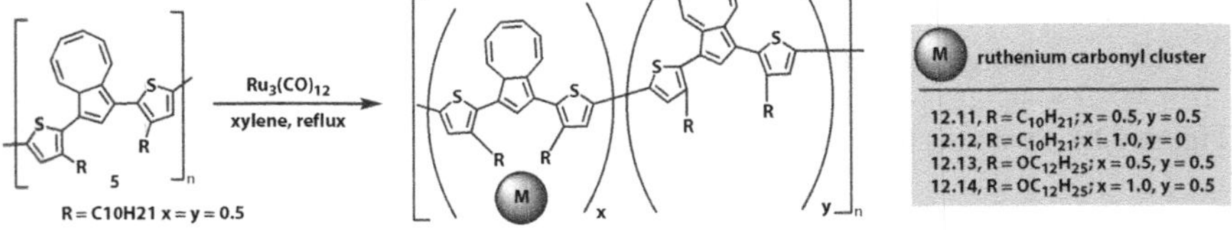

Figure 12.5 Synthesis of polymer ruthenium complexes.

three pyrrole units and one azulene unit. This modification results in a cross-conjugated system, meaning that the conjugation pathways are interrupted compared to the fully conjugated systems of regular porphyrins. Due to the incorporation of the azulene unit, the electronic properties of azuliporphyrins differ significantly from those of porphyrins. The azulene unit contributes to the cross-conjugation, leading to unique electronic and optical properties. Unlike porphyrins, which are known for their aromatic characteristics (delocalized π-electrons over the macrocycle), azuliporphyrins in their free base form do not exhibit the typical aromatic behavior. This is because the azulene unit disrupts the continuous π-electron delocalization that is essential for aromaticity. Azuliporphyrins are of interest in research due to their unique properties. They can be studied for potential applications in materials science, catalysis, and as models for understanding the impact of structural modifications on aromatic systems.

Szczepura *et al.* have reported the synthesis of nickel (II), palladium (II), and platinum (II) azuliporphyrins (Figure 12.6). The UV-Vis spectra of the palladium chelates were quite different from the nickel complexes and showed three bands in the range of 360-450 nm, together with broad absorptions from 500 to 800 nm. The nickel (II) azuliporphyrin showed a slight red-shifted absorption as observed by UV-Vis spectroscopy [9].

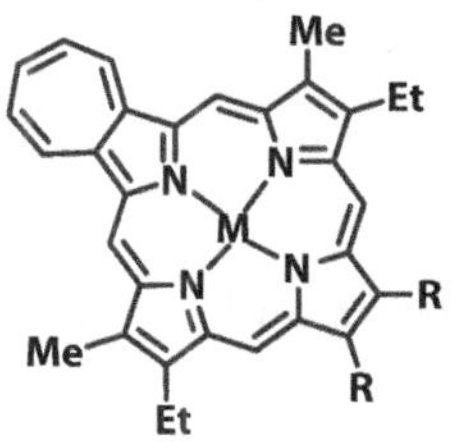

M = Ni(II), Pd(II), Pt(II)
R = Et (12.15), Ph (12.16)

Figure 12.6 Metal complexes of N-confused porphyrins.

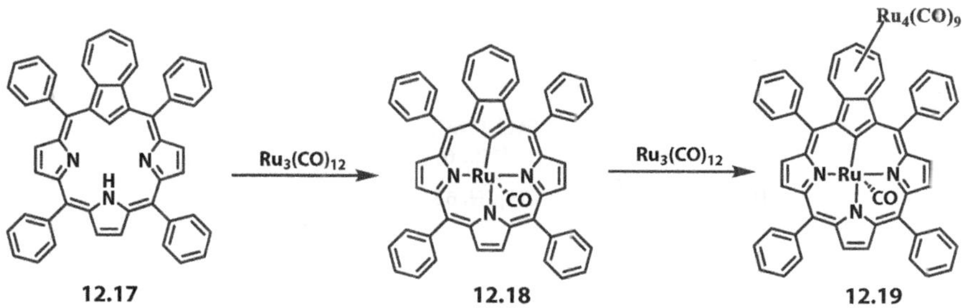

12.17 12.18 12.19

Figure 12.7 Reaction of ruthenium dodecacarbonyl with an azuliporphyrin.

Grazynski *et al.* have synthesized ruthenium-based azuliporphyrin wherein ruthenium (II) metal is inserted into an azuliporphyrin (TPAP) to give carbonyl ruthenium (II) azuliporphyrin [Ru (TPAP)(CO)]. The azulene moiety in the above complex serves as a π-coordination platform to accommodate the $Ru_4(CO)_9$ cluster. The synthetic route for the synthesis of azulene is shown in Figure 12.7 [10].

There are very few reports on metal-azulene σ-complexes wherein metal atoms are directly bonded. One such report is from Kanbara *et al.* wherein the metal ions are bonded to the carbon atom at the C7 position. The palladium(II) and platinum(II) complexes with an SCS-pincer ligand incorporating an azulene unit are shown in Figure 12.8 [11].

The synthesis of pincer complexes is shown below:

The pincer ligand containing thioamide groups (**12.24**) was prepared by treating the corresponding amide molecule with Lawesson's reagent. Further treatment of ligand **12.24** with $PdCl_2$ and LiCl in methanol under refluxing conditions produced the palladium-κ^{3}SCS pincer complex (12.25) with a 77% yield (as shown in Figure 12.9). Likewise, the platinum

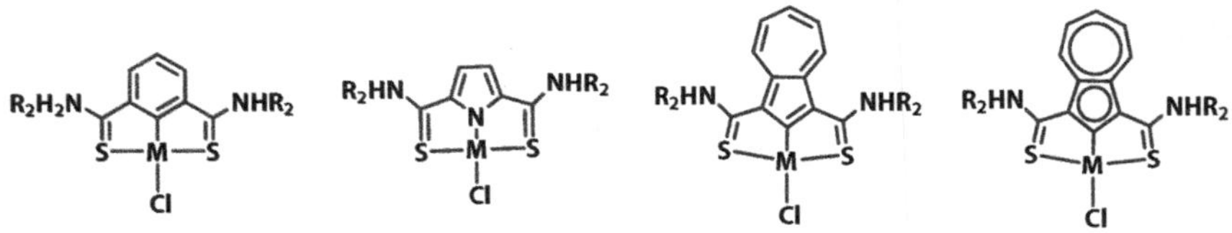

12.20 12.21 12.22 12.23

Figure 12.8 Molecular structures of the M-(κ^3 SCS) and M-(κ^3 SNS) complexes.

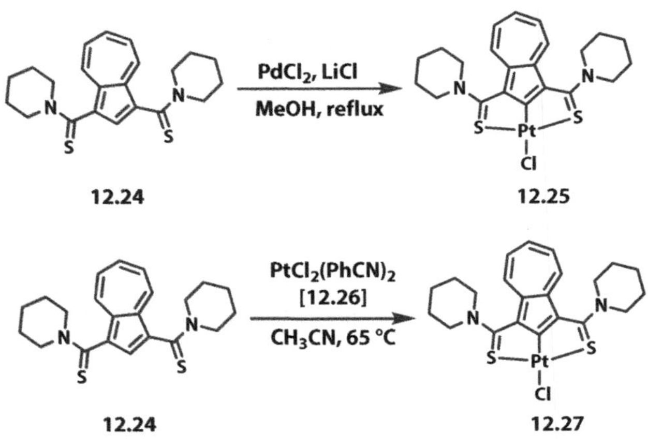

12.24 12.25

12.24 12.27

Figure 12.9 Synthesis of the pincer Pd and Pt complexes with an azulene unit.

complex (12.27) was obtained with an 85% yield by reacting ligand 12.24 with $PtCl_2(PhCN)_2$ in acetonitrile at 65°C.

The absorption spectra of ligand **12.24** exhibited a strong absorption band at 295 nm (ε = 43,800 L mol^{-1} cm^{-1}), arising from the π–π* transition of the azulene unit. This π–π* transition in complexes **12.25** and **12.27** was significantly red-shifted to around 350 nm and 353 nm, respectively. This bathochromic shift could be caused by the effective conjugation between the azulene and thioamide groups through metallation. The Pd complex also had a weak absorption at 480 nm and another broad absorption extending to 600 nm. Similarly, the Pt complex exhibited a broad absorption extending up to 650 nm.

In another report by Yip *et al.* eight Pt-azulene complexes were synthesized to understand the effect of metalation on the electronic structures, luminescence of organic chromophores. The authors have used two strategies to synthesize platinum (II)-azulene σ-complexes. First, $Pt^0(PEt_3)_4$ can be oxidatively added to 1-bromoazulene and 1,3-dibromoazulene to directly bind Pt (II) ions to the rim of the organic molecule at the 1- and 1,3-positions. A different synthesis method entails the coupling of trans-Pt (II)$(PEt_3)_2$(I) with 1,3-bis(triisopropylsilyl)azulene, or 1-bromo-3-triisopropylsilylazulene, catalyzed by copper(I) [12].

12.2.6 Synthesis of Selenium and Tellurium Core-Modified Azuliporphyrinogens

Chauhan *et al.* have reported the synthesis of diseleno/ditellura azuliporphyinogens. The synthesis of the targeted selenium and tellurium core-modified carbaporphyrinogens was carried out by the treatment of alkyl/aryl substituted selenophene/tellurophene diols with azulene in the presence of acid. The oxidation of **12.28** resulted in the formation of the corresponding carbaporphyrins **12.29** (Figure 12.10). Insights into the

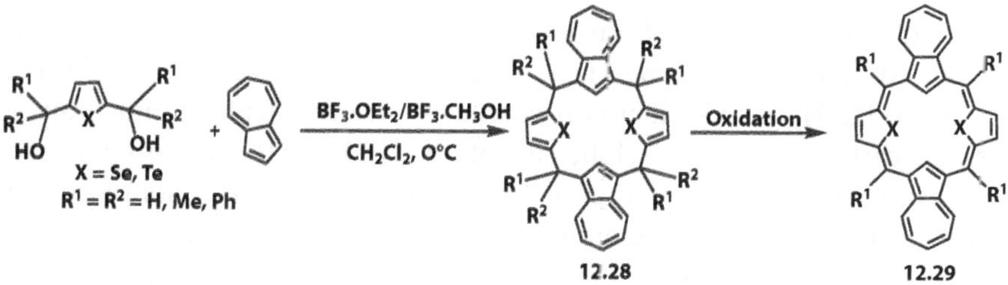

Figure 12.10 Synthetic scheme for the synthesis of Se/Te cored azuliporphyrins.

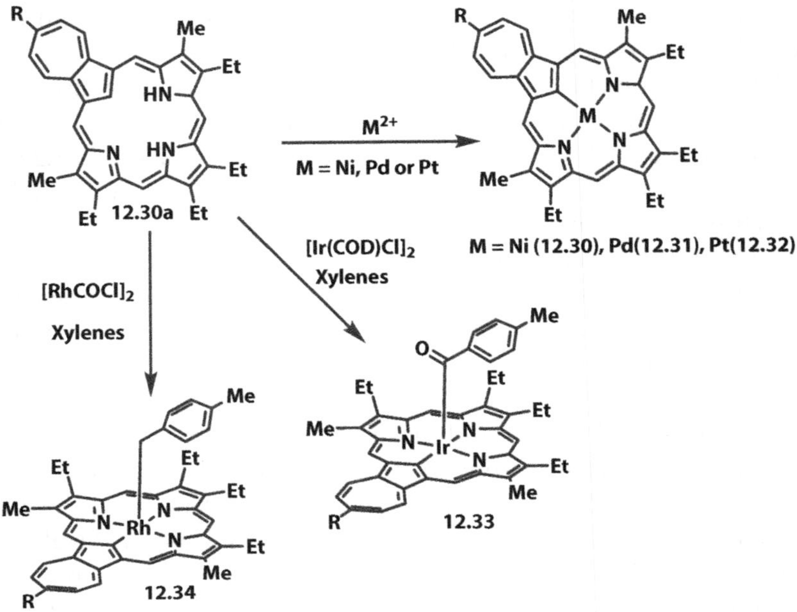

Figure 12.11 Synthesis of organometallic complexes of azuliporphyrin.

redox properties of these compounds revealed their ability to undergo successive oxidations [13].

Timothy *et al.* have also reported several organometallic complexes of azuliporphyrins (**12.30–12.32**) obtained by the reaction of **12.30a** with nickel (II) acetate, palladium (II) acetate, or platinum (II). Additionally, these porphyrinoids react with $[Ir(COD)Cl]_2$, or $[Rh(CO)_2Cl]$ complexes in refluxing xylene to yield compounds **12.33** and **12.34**, respectively, as shown in Figure 12.11 [14].

12.3 Metal Carbonyl Complexes of Azulene

In 1958, Burton and Wilkinson developed the first known transition metal complex of azulene, i.e., azulenedimolybdenum hexacarbonyl, $C_{10}H_8Mo_2(CO)_6$ (12.35). It was prepared by heating azulene and molybdenum carbonyl in ethylcyclohexane at 100–120 °C. This compound had interesting properties and was observed to be dark red in solution and black in the solid state. The molecular structure of this complex was predicted by King and Bisnette. The structure consists of one molybdenum-molybdenum bond of length 3.267 ± 0.0068, and the two Mo(CO)~ groups

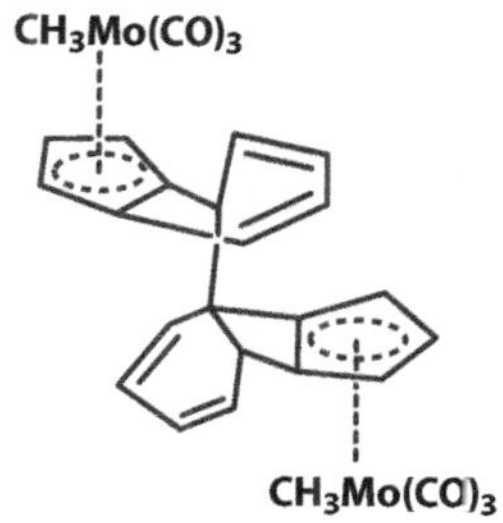

12.35

Figure 12.12 The bonding in $[C_{10}H_8Mo(CO)_3CH_3]$.

are present on the same side of the molecule. The π-electrons of the azulene ligand are factored into two sets of five. It was also revealed that one Mo(CO) group was bonded to the carbon atoms of the five-membered ring *via* a η-cyclopentadienyl- molybdenum bond. The remaining five carbon atoms of the azulene skeleton are bonded to the second $Mo(CO)_3$ unit *via* a noncyclic π-dienyl-metal linkage (Figure 12.12).

12.4 Coordination Complexes of Azulene

Coordination chemistry of azulenes has been mostly explored for their carbon-metal bonding, commonly classified as organometallic compounds. A ton of organometallic compounds of azulene have been reported in the literature, and the new angle in their coordination chemistry has been mainly brought out by substituting it with various functional groups. Zhang and Petoud have reported an interesting chemistry for the efficient sensitization of four near-infrared luminescent lanthanide cations (Nd^{3+}, Er^{3+}, Tm^{3+}, Yb^{3+}) *via* substituted azulene derivatives to form complex **12.36–12.38**, shown in Figure 12.13 [15]. The bidentate azulene-based ligand 2-hydroxyazulene-1,3-dicarboxylate (HAz) was shown to adopt a tetra-coordinated complex $M(Az)_4$ (M–Nd^{3+}, Er^{3+}, Tm^{3+}, Yb^{3+}). Further, they reported that the π→π* transition band of the free ligand at 428 nm is red-shifted to 452 nm after coordinating to the metal with an increased molar absorptivity.

Birzan *et al.* have produced a detailed review on the 1-vinylazulene-based metal sensors, including both heavy metals and lanthanides in their report [16]. In their survey, they have taken into account the azulene derivatives based on the condensation of azulen-1-carbaldehyde and heterocycles. Among the reported derivatives, the coordination compounds

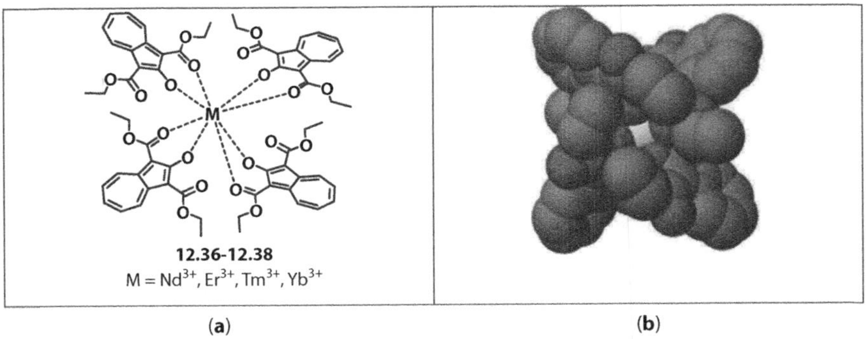

Figure 12.13 (a) Chemical structure of the complex $M(Az)_4$ (M – Nd^{3+}, Er^{3+}, Tm^{3+}, Yb^{3+}); (b) structural view of the complex $[Yb(Az)_4]^-$ created with molecular modeling (MM3 parameters). Reproduced with permission [15]. Copyright 2008, John Wiley and Sons.

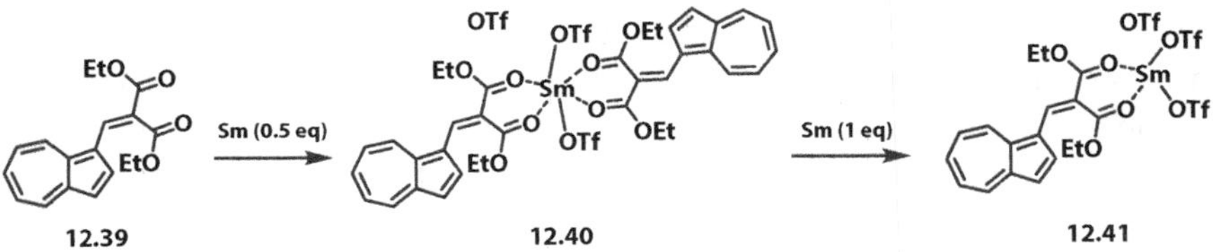

Figure 12.14 Scheme depicting the complexation reaction of malonic ester and Sm, complex **34**.

of malonic esters exemplify a pronounced sensor for the lanthanides (Sm, Eu, Tb) and heavy metals (Pb, Cu, Cd, Hg). Complex **12.41**, formed by the reaction of Sm and malonic ester, is shown in Figure 12.14.

The coordination nature of the azulene derivatives with the transition metals has also been reported in the literature. For instance, Haraguchi *et al.* worked with the not-so-common diradicals linked by non-alternant hydrocarbons, azulene-1,3-diyl linked bis(nitronyl nitroxide) (**NN_2Az**) and bis(iminonitroxide) (**IN_2Az**) diradicals, and investigated their complexation reaction and magnetic properties with Copper(II) hexafluoroacetylacetonate, $Cu(hfac)_2$. **NN_2Az** and **IN_2Az** were accounted to form a 1:2 $[(Cu(hfac)_2)_2(\mathbf{NN_2Az})]$ complex (**12.18**) and a 1:1 $[Cu(hfac)_2(\mathbf{IN_2Az})]\bullet C_6H_{12}$ complex (36), respectively, on reaction with $Cu(hfac)_2$, shown in Figure 12.15 [17].

Recently, Eppel *et al.* presented new gold(III) complexes (**12.44–12.46**), prepared by the reaction of $[^{tBu}(C\wedge N\wedge C)AuCl]$ and various substituted azulene moieties, shown in Figure 12.15 [18]. It involved a reaction between the trimethylsilyl (TMS)-protected alkyne and gold (III) complex, $[^{tBu}(C\wedge N\wedge C)AuCl]$.

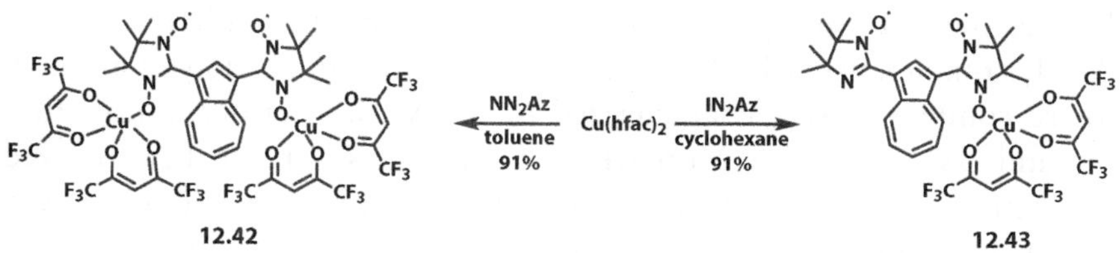

Figure 12.15 Complexation of $Cu(hfac)_2$ with diradical **NN_2Az** and **IN_2Az** to form complex **12.42** and **12.43**, respectively.

The TMS-protected alkyne was deprotected using silver(I) fluoride and is followed by a Sonogashira-type reaction with the *in situ* formed terminal alkyne to produce the corresponding [^{tBu}(C^N^C)-Au^{III}(azulenyl)] complexes (**12.44** and **12.46**). However, complex **12.45** was synthesized

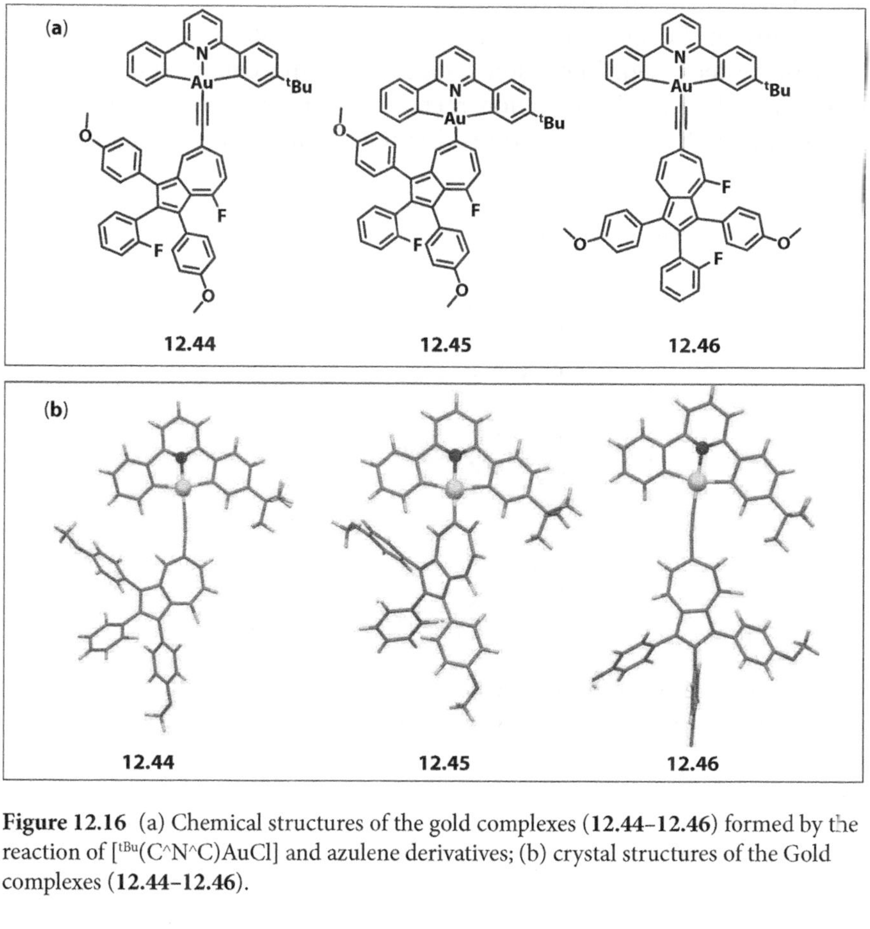

Figure 12.16 (a) Chemical structures of the gold complexes (**12.44–12.46**) formed by the reaction of [^{tBu}(C^N^C)AuCl] and azulene derivatives; (b) crystal structures of the Gold complexes (**12.44–12.46**).

by treating the Bpin-protected precursor compound with [^{tBu}(C^N^C) AuCl] complex using $Pd(OAc)_2$ and HP($t$$Bu_3$)$BF_4$ in a catalytic amount. These complexes were characterized by SC-XRD, electrochemical studies, and UV-Vis spectra. Moreover, the complexes were further analyzed for their photophysical properties. The crystal structure of the complexes, shown in Figure 12.16b, reveals that the Au-centers occupy a distorted square planar geometry. The C–Au–C bond angles are around 162-163°, and a nearly linear N–Au–C (ligand) alignment of 177-179°. In complex **12.45**, the azulene ligand is in a non-coplanar mode with respect to the C^N^C ligand plane, whereas the alkyne-substituted complexes **12.44** and **12.46** are coplanar. Moreover, complex **12.45** was observed to possess the smallest HOMO-LUMO gap of 1.92 eV among the three complexes. This was attributed to the missing alkyne moiety, which resulted in a smaller π-conjugated system.

Guzman-Espinoza *et al.* have performed a detailed density functional theory and quantum theory of atoms in molecules study to understand the nature of the spectator ligands in the coordination preference of the azulene system [19]. Their study focused on the complexes of azulene with Zr and Ti (early transition metals), Cr and Mo (middle transition metals), and Ni and Pd (late transition metals) as shown in Figure 12.17.

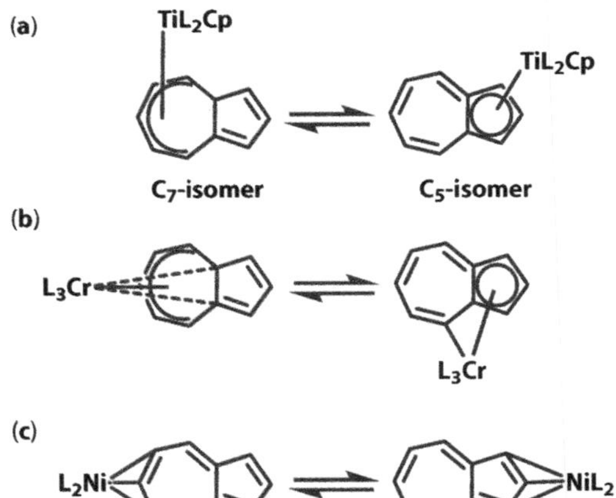

Figure 12.17 Haptotropic equilibrium for the hapto-isomers of (a) $[L_2Ti(Az)]^+$; (b) $L_3Cr(Az)$; and (c) $L_2Ni(Az)$ complexes, where L (CO, PH_3, and NH_3) and are the spectator ligands.

Each metal-azulene system was designed with the closed-shell carbonyl, phosphine, and amine complexes. In case of Ti and Zr, the complexes were cationic; therefore, cyclopentadienyl rings were incorporated to fulfil the 18-electron rule. The spectator ligands (CO, PH_3, and NH_3) modulate the backdonation capacity of the metal center depending on their π-accepting nature (CO > PH_3 > NH_3). They have reported the Gibbs free energy of the haptotropic equilibrium (ΔG) for the metal–azulene complexes (M–Az) to understand the coordination preference of the metal center to the azulene ring, i.e., whether the metal center will bind to the 7-member or 5-member ring. The ΔG value in Table 12.1 represents the difference between the C5 isomer (as the final state) and the C7 isomer (as the initial state). Interestingly, the ΔG values of all the complexes were negative in the case of carbonyl as spectator ligands, indicating that the strongly π–accepting nature of the carbonyl directed the coordination mode of the metal center to the 5-membered ring of the azulene system. On the contrary, ΔG values of all the complexes (except a few cases) were reported to be positive, defining the preferred binding site of the metal center on the 7-membered ring of the azulene.

Table 12.1 ΔG (kcal/mol) of the haptotropic equilibrium (Figure 12.17) of the azulene complexes— a) $M(Az)L_n$, where L = CO, PH_3, NH_3 and M = Cr, Mo, Ni, Pd; b) $M(Az)L_nCp$, where L = CO, PH_3, NH_3 and M = Ti, Zr. In the reported complexes, n = 3 for Cr, Mo, and n = 2 for Ni, Pd, Ti, Zr. Reproduced with permission [19]. Copyright 2008, John Wiley and Sons.

Metal	**CO**	**PH_3**	**NH_3**
Ti	-9.4	3.4	1.5
Zr	-2.2	17.9	21.7
Cr	-5.8	-1.0	11.7
Mo	-1.9	1.2	11.0
Ni	-3.8	-1.2	2.7
Pd	-1.2	-0.3	4.2

Calculations were reported with M06/Def2TZVP level of theory with the software Gaussian 09.

12.5 Conclusions

In this chapter, the diverse and dynamic chemistry of azulenes covers several key topics. Starting with an introduction to azulene's unique properties, we delved into metallaazulenes, highlighting their synthesis, structure, and potential applications. The functionalization of azulenes was discussed, showcasing various methods to modify azulene for enhanced reactivity and utility. Finally, the integration of azulene into polymers was examined, emphasizing its role in developing advanced materials with novel electronic and optical properties. This comprehensive overview underscores azulene's versatility and significant potential in modern chemistry and materials science.

AI disclosure statement: The author confirms that no artificial intelligence (AI)-generated content or AI-assisted writing tools were used in the preparation of this chapter (including summaries and the abstract). All text, analysis, and conclusions are the result of the author's original work.

References

1. Korichi, H., Zouchoune, F., Zendaoui, S.-M., Zouchoune, B., Saillard, J.-Y., The coordination chemistry of azulene: A comprehensive DFT Investigation. *Organometallics*, 29, 7, 1693, 2010.
2. Xin, H., Hou, B., Gao, X., Azulene-based π-functional materials: Design, synthesis, and applications. *Acc. Chem. Res.*, 54, 7, 1737, 2021.
3. Xin, H. and Gao, X., Application of azulene in constructing organic optoelectronic materials: New tricks for an old dog. *ChemPlusChem*, 82, 7, 945, 2017.
4. Puodziukynaite, E., Wang, H.-W., Lawrence, J., Wise, A.J., Russell, T.P., Barnes, M.D., Emrick, T., Azulene methacrylate polymers: Synthesis, electronic properties, and solar cell fabrication. *J. Am. Chem. Soc.*, 136, 31, 11043, 2014.
5. Churchill, M.R., The molecular structure of an azulene complex of iron; $C_{10}H_8Fe_2(CO)_5$. *Chem. Commun.*, 14, 450, 1966.
6. Oro, L.A., Valderrama, M., Cifuentes, P., Foces-Foces, C., Cano, F.H., Azulene as a ligand in cationic rhodium and iridium complexes. Crystal structure of $[Rh(TFB)(az)]PF_6$. *J. Organomet. Chem.*, 276, 1, 67, 1984.
7. Liu, H.-C., Ruan, K., Ma, K., Fei, J., Lin, Y.-M., Xia, H., Synthesis of metalla-dual-azulenes with fluoride ion recognition properties. *Nat. Commun.*, 14, 1, 5583, 2023.

8. Shetti, V.S., Chemical syntheses and salient features of azulene-containing homo- and copolymers. *Beilstein J. Org. Chem.*, 17, 2164, 2021.
9. Lash, T.D., Colby, D.A., Graham, S.R., Ferrence, G.M., Szczepura, L.F., Organometallic chemistry of azuliporphyrins: Synthesis, spectroscopy, electrochemistry, and structural characterization of nickel(II), palladium(II), and platinum(II) complexes of azuliporphyrins. *Inorg. Chem.*, 42, 22, 7326, 2003.
10. Białek, M.J., Białońska, A., Latos-Grażyński, L., Oxidation and oxygenation of carbonyl ruthenium(II) azuliporphyrin. *Inorg. Chem.*, 54, 13, 6184, 2015.
11. Kuwabara, J., Munezawa, G., Okamoto, K., Kanbara, T., Palladium(II) and platinum(II) complexes bearing a κ^3SCS pincer ligand with an azulene unit. *Dalton Trans.*, 39, 27, 6255, 2010.
12. Nguyen, V.H. and Yip, J.H.K., Platinated azulenes: Structures and electronic spectroscopy. *J. Organomet. Chem.*, 1004, 122921, 2024.
13. Ahmad, S., Singhal, A., Nisa, K., Chauhan, S.M.S., Synthesis of selenium and tellurium core-modified azuliporphyrinogens and benziporphyrinogens and corresponding carbaporphyrinoids. *Inorg. Chem.*, 57, 18, 11333, 2018.
14. Lash, T.D., Heteroporphyrins and carbaporphyrins, in: *Fundamentals of Porphyrin Chemistry*, P.J. Brothers and M.O. Senge (Eds.), pp. 385–451, Wiley, NJ, USA, 2022.
15. Zhang, J. and Petoud, S., Azulene-moiety-based ligand for the efficient sensitization of four near-infrared luminescent lanthanide cations: Nd^{3+}, Er^{3+}, Tm^{3+}, Yb^{3+}. *Chem. Eur. J.*, 14, 4, 1264, 2008.
16. Jordan, B.J., Pollier, M.A., Ofir, Y., Joubanian, S., Mehtala, J.G., Sinkel, C., Caldwell, S.T., Kennedy, A., Rabani, G., Cooke, G., Rotello, V.M., Lock and key" control of optical properties in a push-pull system. *Chem. Commun.*, 1653, 2008.
17. Haraguchi, M., Tretyakov, E., Gritsan, N., Romanenko, G., Gorbunov, D., Bogomyakov, A., Maryunina, K., Suzuki, S., Kozaki, M., Shiomi, D., Sato, K., Takui, T., Nishihara, S., Inoue, K., Okada, K., (Azulene-1,3-diyl)-bis(nitronyl nitroxide) and (Azulene-1,3-diyl)-bis(iminonitroxide) and their copper complexes. *Chem. Asian J.*, 12, 22, 2929, 2017.
18. Eppel, D., Oberhof, N., Dietl, M.C., Cieslik, P., Rudolph, M., Eberle, L., Krämer, P., Stuck, F., Rominger, F., Dreuw, A., Hashmi, A.S.K., Gold(III) meets azulene: A Class of [(^{tBu}C^N^C)Au^{III}(azulenyl)] pincer complexes. *Organometallics*, 40, 23, 3865, 2021.
19. Guzmán-Espinoza, C.V., Ledesma-Olvera, L.G., Barquera-Lozada, J.E., Effect of spectator ligands on haptotropic rearrangements of metal-azulene complexes. *Eur. J. Inorg. Chem.*, 44, 4579, 2021.

Part 4

THEORETICAL AND FUTURE PERSPECTIVES

13

Theoretical Insights into Azulene Chemistry

Manash Protim Borpuzari[1*] **and Abhijit Boruah**[2]

[1]*Department of Chemistry, Dhemaji College, Dhemaji, Assam, India*
[2]*Department of Chemistry, Jorhat Engineering College, Jorhat, Assam, India*

Abstract
Azulene, an isomer of naphthalene, with its non-classical 10 π-electrons, is of prime interest for theoretical chemists, due to its narrow HOMO-LUMO gap and anti-Kasha fluorescence from S_2 to S_0. This chapter provides a comprehensive review of azulene chemistry through the lens of the density functional theory (DFT) and wave function-based methods. Employing both density-based and wave function-based methods, one can understand both ground-state as well as excited-state molecular phenomena, shedding light on their aromaticity, resonance, and reaction mechanisms. Theoretical insights from these two approaches offer valuable understanding of azulene's behavior, providing a guideline for targeted applications in drug design and materials science.

Keywords: DFT, anti-Kasha fluorescence, B3LYP, GW, CI

13.1 Introduction

Among non-benzenoid hydrocarbons, azulene holds immense importance due to its remarkable structure. This unique structure nurtures its versatile applications across various industries, highlighting its relevance in scientific research and commercial endeavors. Azulene's medicinal properties and versatile chemical characteristics also make it highly valuable. As a well-studied isomer of naphthalene with 10 π-electrons, azulene features ring-fused unsaturated seven-membered and five-membered rings.

**Corresponding author*: mb45380@gmail.com

Neha Rani Kumar and Abhijeet R. Agrawal (eds.) Azulene-Based Materials: Design, Synthesis, and Applications, (353–376) © 2026 Scrivener Publishing LLC

Due to the delocalization of π-electrons, the cyclopentadienyl ring becomes electron-rich while the seven-membered counterpart (tropylium cation) becomes electron-deficient, injecting a notable donor-acceptor property (Figure 13.1) [1]. Furthermore, its distinctive electronic structure and small energy gap between ground state and first excited state impart a deep blue color to azulene due to a weak absorption in the visible range [2, 3]. The polarizability of its building blocks makes azulene much less symmetrical and gives a relatively high dipole moment of 1.08 D [4]. It directly affects its interaction patterns with other polar molecules and also leads to various unusual photophysical and chemical properties.

Azulene exhibits distinctive reactivity due to its distinctive electronic structure and unique frontier molecular orbitals. Unlike typical aromatic compounds, its resonance-stabilized bicyclic structure creates an unusual distribution of electron density across its carbon atoms. The odd positions, particularly the 1- and 3-positions, are electron-rich and readily undergo electrophilic reactions. These 1- and 3-positions also impart stimulus-responsive behavior to azulene in acidic environments. On the other hand, the 2-, 4-, 6-, and 8-positions, as shown in Figure 13.2, are electron-deficient and more favorable towards nucleophilic substitutions [5]. *Ab initio*

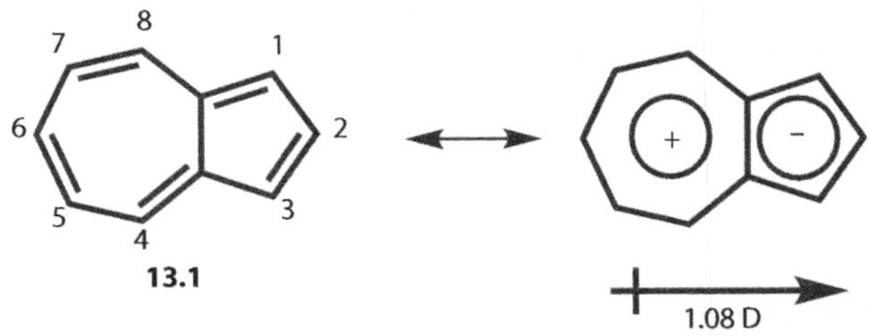

Figure 13.1 Numbering of azulene and its polarized resonance structure.

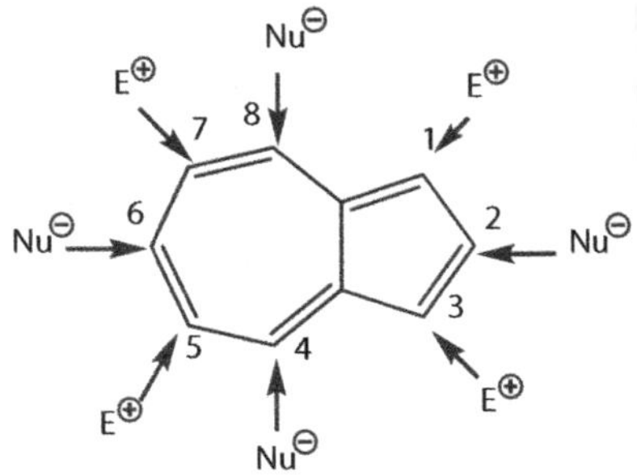

Figure 13.2 Reactivity of each position of azulene.

calculations, therefore, provide a remarkable insight on its reaction kinetics from both structural and energetic points of view.

In recent years, azulene has garnered significant attention due to its distinctive electronic and optical properties. Its unique donor-acceptor structure sets azulene apart from conventional fused benzenoids, making it a valuable component in developing advanced optoelectronic materials [6]. As depicted in Figure 13.3, naphthalene exhibits almost identical geometries for its highest occupied molecular orbital (HOMO) and lowest unoccupied molecular orbital (LUMO). In contrast, azulene displays a non-symmetrical frontier molecular orbital geometry, with HOMO density primarily localized at odd positions and LUMO density at even positions [7]. Furthermore, azulene features a narrow energy gap between its HOMO and LUMO, alongside an unusual fluorescence from the second excited state to the ground state, which has intensified research interest [8].

The synthesis of azulene-based π-functional materials has also become an active field of study [5, 9, 10]. Azulene derivatives have found extensive applications in various domains, including electrochromic materials [9], anion sensors [11], charge-transfer complexes [12], organic conductors [13], nonlinear optical materials [14], and near-infrared resonance materials [15]. Additionally, azulene derivatives are utilized in organic field-effect transistors (OFETs) [16] and solar cells [17]. Despite these advancements, the development of azulene-based organic materials faces challenges due to the complexities of molecular design and synthesis. In this chapter, we summarize the theoretical efforts made during the last decade to study the azulene chemistry.

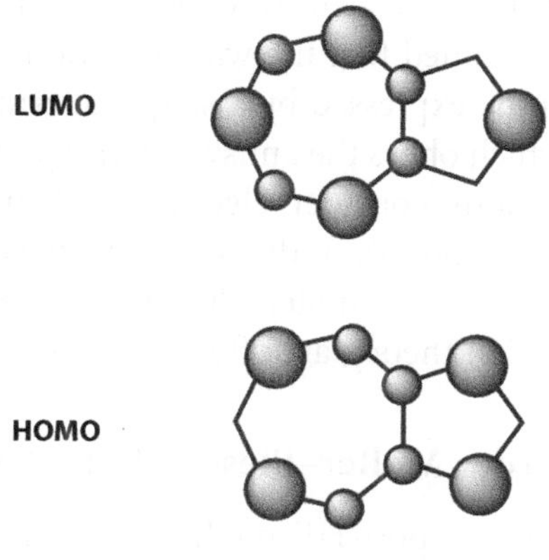

Figure 13.3 Probabilities of locating an electron in the HOMO and LUMO of azulene.

13.2 Computational Methods to Study Azulene Chemistry

Advancement of computational methods plays a crucial role in molecular modeling, rational drug design, and materials science. This advancement in the journey of computational methods has also been fueled by the innovations of highly configured computers and parallel processing. These methods provide detailed insights into reaction mechanisms, intermolecular interactions, and conformational changes. The computational approach has made a strong emphasis on the fundamental understanding of biological processes at the molecular level, drug discovery, and the design of new materials for future technologies. Techniques such as Hartree–Fock (HF), the second-order Møller–Plesset perturbation theory (MP2), complete active space-based method (CAS methods), density functional theory (DFT), and equation of motion (EOM) allow researchers to explore the structure, dynamics, and interactions of molecules at an atomic level [6, 12–15]. A comparison of different computational techniques is provided in Table 13.1

13.2.1 Hartree–Fock Theory (HF)

The HF [18] method is a foundational approach in quantum chemistry used to approximate the wave function and energy of a many-electron system. The minimum energy of the system is calculated by solving sets of non-linear equations by using a self-consistent field (SCF) approach, where the orbitals occupied by electrons are calculated iteratively under the approximation that the repulsive potential field between the electrons is assumed to be the average repulsions generated by all other electrons. Further, to solve these equations, it is assumed that the wave function corresponding to the minimum energy can be expressed by a single determinant, known as the Slater determinant, which obeys the antisymmetric principle of Pauli.

Although HF methods consider electron exchange, they completely neglect electron correlation. To address this, post-HF methods, such as configuration interaction (CI), Møller–Plesset perturbation theory (MP2), and coupled-cluster (CC) theory, are often employed.

13.2.2 Second-Order Møller–Plesset Perturbation Theory

The MP2 is an important post-HF method used in quantum chemistry to include static electron correlation effects that are neglected in the HF approach [19] MP2 is based on Rayleigh-Schrödinger perturbation theory. In this theory, the Hamiltonian of the system is divided into two parts, the first

Table 13.1 Advantages and disadvantages of the computational method.

Method	Advantages	Disadvantages
HF	• Conceptually simple. • Provides a qualitative description of electronic structures. • Computationally less time-consuming.	• Completely ignore the electron correlation. • Leads to inaccurate dispersion interactions.
MP2	• Includes electron correlation at the second-order level. • Provides improved accuracy over HF for dispersion interactions and moderate correlation effects.	• Computationally more time-consuming than HF and DFT. • Not reliable for systems with strong static correlation.
CAS	• Explicitly treats static correlation by considering a selected set of orbitals in full configuration interaction. • Well-suited for optimizing the transition states.	• Computationally expensive as the active space size increases.
DFT	• Computationally less time-consuming compared to wavefunction-based methods. • Widely applicable for large variety of molecular systems. • A good choice for ground state properties.	• Results depend on the choice of exchange-correlation functional. • Struggles with non-covalent interactions, excited states, and barrier heights.
TDDFT	• Extends DFT to excited-state calculations. • Relatively efficient for calculating electronic absorption spectra and excited-state properties.	• Limited accuracy for charge-transfer and Rydberg excitations. • Depends on the exchange-correlation functional used.
EOM	• Designed for calculating excited states, ionization potentials, and electron affinities. • Can be coupled with methods like CCSD (EOM-CCSD) for high accuracy.	• Computational cost is high, especially for larger systems. • Limited to relatively small to medium-sized molecules.

one is a zeroth-order part, which is the HF Hamiltonian, and the second one is a perturbation, which accounts for electron correlation effects. The total energy in the MP2 method is the sum of the HF energy and the correlation energy. The MP2 energy correction is computed as a second-order term in the perturbation expansion by calculating the interaction between the HF wave function and excited configurations where two electrons are promoted from occupied to virtual orbitals.

The MP2 energy correction can be expressed as:

$$E_{MP2} = \sum_{ij}^{occ} \sum_{ab}^{virt} \frac{|\langle ij||ab\rangle|^2}{\varepsilon_i + \varepsilon_j - \varepsilon_a - \varepsilon_b}$$

13.2.3 Complete Active Space-Based Method

Complete active space (CAS) [20] methods are a class of quantum chemical techniques designed to accurately describe systems where electronic correlation effects, particularly static (non-dynamical) correlation, are significant. These methods are particularly useful for molecules with near-degenerate electronic states. Unlike HF, instead of relying on a single determinant, CAS method calculates SCF from multi-determinant wavefunction, which is generated by dividing the molecular orbitals into three subsets: inactive, active, and virtual. Inactive orbitals are fully occupied, virtual orbitals are unoccupied, and active orbitals are those where the number of electrons and their distribution are explicitly considered. In this case, the wave function is generated by linear combination of Slater determinants which are obtained by distributing the active electrons among the active orbitals in all possible ways. Complete active space self consistent field (CASSCF) methods fully incorporate static correlation, however, it does not fully incorporate dynamic correlation [20]. Therefore, post-CASSCF methods like the complete active space with second-order perturbation theory (CASPT2) are better choices than CASSCF to study dynamic correlation.

13.2.4 Density Functional Theory

The DFT techniques [21] are mostly affordable as they have been one of the most popular electronic structure methods due to their reasonable accuracy, blended with modest cost in many applications. It has made a remarkable impact on the development of modern science by providing a simple framework for the many-body effects that dominate the computational cost of wave function-based methods. The DFT is different from other methods because it uses density as a variable instead of the wavefunction, where less computational power is required. Here, moreover, Kohn–Sham (KS) equations are solved iteratively from a guessed density using the exchange-correlation (E_{xc}) potential [22]. However, the exact form of E_{xc} is unknown, and various approximations like local density approximation (LDA), generalized gradient approximation (GGA), and hybrid functionals are developed to incorporate electron exchange and electron correlation. High-performance computing and parallel processing have further expanded the capability to model large systems over longer timescales and at higher accuracies. It is reported that DFT methods are highly accurate in determining structural, electronic, thermodynamic, and kinetic properties.

13.2.5 Time-Dependent Density Functional Theory (TDDFT)

The Time-dependent density functional theory (TDDFT) is the extension of the DFT to the time domain, making it possible to apply the DFT to study the electronic excited states. Information about these electronically excited states is obtained through the linear response formalism [23], which allows one to calculate the response of a system under a small perturbation acting on it such that it does not greatly affect the ground state structure. The TDDFT offers a relatively accurate and computationally efficient method for studying excited-state properties. Therefore, it may be a well-established method to investigate the complex electronic phenomena and optical properties of systems [24]. However, challenges may arise as the accuracy of TDDFT is largely dependent on the choice of exchange-correlation functionals.

13.2.6 Equation of Motion

Equation-of-motion methods are a sophisticated class of quantum chemical techniques, which are used to study excited states, ionized states, and electron attachment processes in molecular systems [20, 25]. For these calculations, a high-quality reference wavefunction obtained from CC calculations is considered. A suitable operator is applied to the reference state, which promotes electrons to higher orbitals or removes electrons, or adds electrons to generate the required excited, ionized, or electron-attached states. After that, a set of linear equations obtained from the similarity-transform Hamiltonian is solved, and the eigenvalues of the equation correspond to the excitation energies. The main advantage is that EOM calculations give accurate excited states; however, the calculations are computationally very demanding.

13.3 Recent Advances

13.3.1 Structural Properties

Usually, it is assumed that azulene prefers a structure with equalized C–C bond lengths along the perimeter (C_{2v}) instead of the bond alternating structure with C_s symmetry. The theoretical studies on the structural properties of azulene date back to 1982. During that time, theoretical studies were unable to properly describe the ground state structure of azulene due to overestimation of the calculated value of the dipole moment by wave function-based methods without electron correlation (*ab-initio* methods at 6-31G) [26]. The MP2 method, which includes electron correlation, in a 6-31G basis set, gave the value of dipole moment of 1.23 D; however, it was still far from the experimental value of 0.79D. On deeper analysis of the calculations, it was found that ab-inito methods without electron correlation calculated the dipole moment only from π system, which yields the dipole moment of 1.73 D. On the other hand, electron correlation from MP2 method transferred the charge density from HOMO to LUMO which reduced the dipole moment from 1.73D to 1.23 D. Similarly, the modified neglect of diatomic overlap [27] or slater-type orbital with three Gaussian function levels gave the lowest energy structure of azulene as C_s, and C_{2v} structure is at a slightly higher value (Energy Difference 3 kcal/mol). As expected, including electron correlation in the calculation gave a lower energy structure as C_{2v} [26].

Grimme [28] extensively studied the ground state structure of azulene using HF and MP2 methods with different basis sets and found that, as expected, HF methods wrongly predict the ground state structure, as well as the dipole moment of azulene, and the inclusion of electron correlation slightly improves the expected value. Further, he predicted that the main reason behind the error may be due to single determinant approximation, and therefore, he employed multi-reference configuration interaction (MRCI) calculations, which included electron correlation, as well as multiple determinant calculations. The MRCI calculations predicted the dipole moment of azulene to be 0.93 D, which is in far better agreement with the experimental value than MP2. Also, MRCI calculations predict C_{2v} geometry as the only minimum in the potential energy surface.

Inspired by this, CAS calculations were used using complete active space self-consistent field (CASSCF) and 2nd order complete active space perturbation theory (CASPT2). The ground and four lowest singlet excited states for azulene were optimized by using the CASSCF and CASPT2 methods, and found that the C_{2v} form of azulene is more stable than Cs. Also, the

CASPT2 calculation indicates that the inclusion of all the π correlation as well as the σ-π correlation is necessary for the determination of physical and chemical properties of azulene [29].

However, wave-function-based methods with electron correlations, as well as multireference methods, are computationally demanding. To solve this problem, Kozlowski *et al.* [30] compared the ground state geometry and the harmonic force field by UNO–CAS/6-31G* with a computationally less demanding density functional based B3LYP/6-31G* method. They have found that the structure obtained by B3LYP/6-31G* level shows C_{2v} symmetry. On the contrary, the structure obtained by UNO–CAS/6-31G* level is a bond alternating structure with Cs symmetry. Further, vibrational analysis predicted that UNO–CAS/6-31G* calculation has one imaginary frequency, whereas all frequencies are positive at B3LYP/6-31G* level of theory. This indicates that computationally less demanding density functional-based methods are quite accurate to study the structural and electronic properties of azulene.

The theoretical calculations conducted by Sousa *et al.* using G3 and G4 composite methods aimed to determine the standard gas-phase enthalpy of formation for azulene. The results indicated values of approximately 254.6 ± 7.9 kJ/mol and 258.9 ± 7.9 kJ/mol from the G3 and G4 calculations, respectively. These values were found to be in significant disagreement with some previously reported experimental data, suggesting potential errors in the experimental measurements for related compounds. The calculations also provided alternative estimates for the enthalpy of formation through different reaction schemes, yielding values around 444.8 kJ/mol and 438.3 kJ/mol, further supporting the need for careful consideration of the experimental data used in thermochemical assessments of azulene [31].

13.3.2 Noncovalent Interaction Between Azulene Dimer

Aromatic systems have been of great significance in biological systems such as DNA or proteins, attributed mainly to their noncovalent interactions. Due to the unique resonance properties of azulene, the study of the potential energy surface (PES) of azulene dimer may give valuable insight into the non-covalent interaction between azulene dimers, which in turn may give valuable insights for designing new π functional materials. As DFT accurately describes the structural properties of azulene, Piacenza *et al.* [32] have done full geometry optimizations at the dispersion corrected DFT-BLYP/TZV2P level of theory of 11 dimers of azulene and then calculated the interaction energies at the same level of theory. They have obtained a flattened PES, which may explain the observed crystal

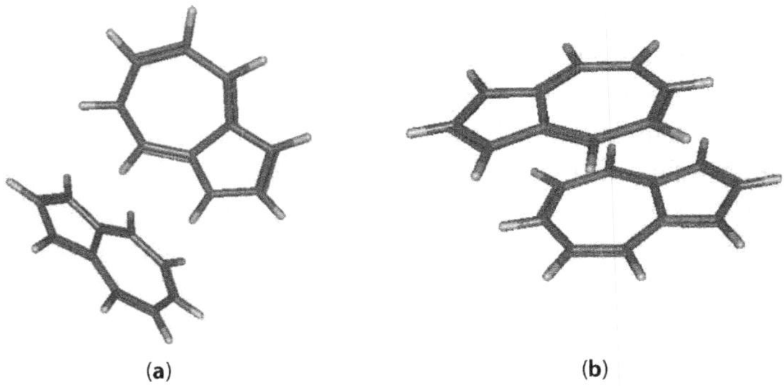

Figure 13.4 Structures of (a) T-shaped arrangement and (b) anti-parallel alignment of azulene dimer.

disordered structure of azulene [33, 34]. Around nine stacked complexes were investigated, out of which eight have binding energies higher than 7.4 kcal/mol, which exceeds the value of 7.1 kcal/mol for the best naphthalene dimer structure. The less stable structure was found to be the structure with T-shaped arrangements with CH...π contacts (Figure 13.4(a)). The structures with the lowest energies are found to be the structures in which the dipole moments of the two azulenes are in anti-parallel alignments (Figure 13.4(b)).

However, the structure with the lowest energy among the dimmers with a parallel alignment of dipole moments was found to be only about 0.5 kcal/mol less strongly bound. Interestingly, on energy decomposition analysis, it was found that the dipole-dipole interaction is of minor importance in the binding energies of the azulene dimer. Surprisingly, it was found that the energetically lowest structure corresponds to a configuration where the seven-membered rings are almost aligned on top of one another. On further inspection, they concluded that this lowest energy structure can be explained by the electrostatic potential of azulene. Among the two rings of azulene, the carbon atoms of the seven-membered ring are in the region with the least negative potential. Hence, these can come closer together than the five-membered rings. They also concluded that energetically favored complexes of π stacked structures (i) have displaced or rotated arrangements that minimize exchange-repulsion while still allowing good dispersion interactions, (ii) optimize electrostatic interactions by avoiding too-close contacts of regions with very negative electrostatic potential, and (iii) have significant orbital and charge-transfer contributions to binding. These findings were based on the results of the energy decomposition analysis and the potential energy curves.

They have also compared these results with high-level SCS-MP2 methods and found that due to molecule-independent C6 coefficients of dispersion correction parameter of BLYP-D, the BLYP-D method slightly underestimates the interaction energies as compared to the SCS-MP2 methods.

Further, in another study, DFT-D3 optimization of the dimmers and single-point CCSD(T)/CBS energy calculation scheme was employed to construct highly accurate potential energy surfaces of the azulene dimer. However, they have obtained only seven stationary points, instead of eleven, in the PES of azulene dimer [35].

13.3.3 Excited State Properties

The important excited state properties of azulene are the anti-Kasha behavior. Dunlop *et al.* [36] studied the ground and excited states of azulene in increasing order of complexity. According to Hückel's and Baird's rule, small conjugated cyclic hydrocarbons usually exhibit contrasting aromaticity/antiaromaticity in their ground singlet state (S0) and first-excited triplet state (T1) [37]. Also, the aromaticity of the first excited quintet state (Qu1) of azulene was earlier reported [38, 39]. Therefore, they computationally investigated the ground singlet state (S_0), first-excited triplet state (T_1), and first excited quintet state (Qu_1) to model the (anti)aromatic character of azulene in the lowest states of each multiplicity. They also made attempts to establish the relationship between azulene's excited-state (anti)aromaticity and its anti-Kasha behavior. They used the harmonic oscillator model of aromaticity, multicenter indices, fluctuation indices, I_{ring}, and the electron-density-of-delocalized-bonds scheme to analyze the aromatic indices of S0, T1, and Qu1. As expected, the ground state of azulene is aromatic with Hückel 10π delocalized electron, and on comparing the aromaticity indices with all possible delocalization circuits, it was found that the electrons are delocalized in the parameter of the azulene geometry. Moreover, poor π-electron delocalization was observed along the circuit involving the transannular bond (cyclopentadienyl and cycloheptatrienyl). On the other hand, aromaticity indices predicted that the first triplet excited state of azulene follows Baird's rules [40] and therefore, it is antiaromatic. Moreover, it was found that this antiaromaticity is partly reduced by transannular bond contraction, due to a decrease in the π delocalization along the parameter and also the increase in the π delocalization in the cyclopentadienyl and cycloheptatrienyl circuits of azulene. Moreover, the electron-density-of-delocalized-bonds calculation predicted that these circuits adopt the electronic structures of their corresponding cyclic radicals. Due to this, the geometry of T1 may become more relaxed, which can be predicted from the almost zero values of isomerization stabilization energies (ISE). However, T_1 azulene is still moderately antiaromatic. On studying the aromaticity of Qu1, they have found that it is aromatic with ring current being delocalized along its parameter. Further, the studies of magnetically induced current density (MICD) at the CASSCF(10,10) for the first singlet excited state (S_1) and the second singlet excited state (S_2) indicated that S_2 is aromatic with ring current along its perimeter, whereas S_1 is antiaromatic, similar to T_1. Moreover, the CASSCF (10,10) calculation indicated that S_2 of azulene has a significant multireference character with near-degenerate occupancy by unpaired electrons to HOMO − 1 and LUMO and to HOMO and LUMO + 1. This indicated that the multireference character of S_2 is almost similar to the Qu_1 state by adopting a similar normalized π-orbital occupancy. Moreover, due to antiaromaticity of S_1, azulene with two unpaired electrons in π and π* orbitals, and the two unpaired electrons exhibit low interelectron repulsion, which in turn lowers its S_1 energy [8]. Conversely, the unpaired electrons in S2 exhibit higher interelectron repulsion. The high difference in interelectron repulsion explains the large S_2–S_1 energy separation and, consequently, leads to a low rate of S_2–S_1 internal conversion (IC). To further study, the S_1–S_0 conical intersection geometry has been optimized, and the CASSCF aromaticity indices. In the optimized conical intersection geometry, S1 azulene adopts the electronic structure of two acyclic radicals separated by a double bond. This increases the total energy, and the increase is countered by a subsequent nonradiative transition to the aromatic ground state. The calculations also indicated that azulene exhibits low HOMO–LUMO separation, which favors the pairing of its two unpaired electrons (in S_1), and provides an efficient nonradiative pathway to S_0. Therefore, the calculation concluded that the conical intersection enabled S_1 antiaromaticity relief.

Higher level of CR-EOMCC(2,3) calculations using the 6-31G(d) and cc-pVDZ basis sets were performed to further study the excited state of azulene. It was found that δ-CR-EOMCC(2,3) results for the low-lying valence excitations in azulene were in very good agreement with the available CASPT2(10,10) [29]. The EOMCCSD calculation indicates S_9 is a doubly excited $^{1A}{}_1$ state dominated by the $(HOMO)^2 \rightarrow (LUMO + 1)^2$ $[(2a_2)^2 \rightarrow (3a_2)^2]$ and $(HOMO\text{-}1)^2 \rightarrow (LUMO)^2$ $[(3b_2)^2 \rightarrow (4b_2)^2]$ transitions. The oscillator strength of the $S_2 \rightarrow S_9$ probe transition is comparable to that characterizing the $S_0 \rightarrow S_2$ pump transition which indicates that the calculated doubly excited state is accessible from S_2 [41].

Pariser-Parr theoretical approach was employed to study the spectrum and structure of azulene by incorporating configuration interaction among singly excited configurations. The carbon skeleton of azulene were modeled with specific bond lengths, and the electronic states were analyzed using molecular orbitals (MOs). They also compared the results from Huckel molecular orbitals and perimeter molecular orbitals to assess their effectiveness in predicting electronic transitions and oscillator strengths. The results indicated that the calculated energies and oscillator strengths for azulene's electronic states show fair agreement with experimental data, particularly for the lowest excited states. Moreover, they also found that the intensity of the lowest state is underestimated, but it gave better results by using vibrational perturbations. The comparison of calculated and experimental values revealed that the Huckel MOs provide a better representation of azulene's electronic structure than perimeter MOs [42].

13.3.4 Rydberg States

The TDDFT calculations with the CAM-B3LYP method, and the MRD-CI method were performed at the 6-311G(d, p) basis set to study the Rydberg state determinations. Two leading configurations with π-π^* were observed for singlet valence states S_1 to S_4 with involvement of excitation from the two HOMOs to the two lowest LUMOs. The $^1B_1$ states ($\pi\sigma^*$, $\sigma\pi^*$) have been found to have weak non-zero oscillator strengths. The high-level coupled-cluster calculations predicted that the mid-energy region is dominated by Rydberg states [43]. The vibrational coupling within several of the singlet states were studied using TDDFT and found that many fundamental modes are excited, but with a very wide range of intensities. Numerous binary combinations occur, in which modes 5 ($17a_1$), 10 ($16a_1$), and 13 ($15a_1$) are prominent. Most of its intensity for the S_3 state is found on the rise of the strongest S_4 band at higher energy and much of the residual structure above S_3 after removal of the slope associated with S_4, is found to be similar to the IE_1 ionic state. They have proposed that two $2a_2$3p Rydberg states of the 3p set are present. The experimental energy region above S_4 is found to be more complex than the theoretical region.

Due to the anti-Kasha behavior of azulene, it is a prominent candidate for metal-free triplet photosensitizers. Computational studies have been performed on Az-derived photosensitizers, namely, Az-CHO and Az-CHS, implementing polarization consistent time-dependent optimally tuned range-separated hybrid (Figure 13.5). Calculations reveal the energetic reordering of low-lying $\pi\pi^*$ and $n\pi^*$ singlet states between Az-CHO and Az-CHS, indicating that Az-CHS is a nonfluorescent one, which may be due to the substantially smaller energy gap, causing enhanced S1 population *via* fast internal conversion. Moreover, a small singlet–triplet gap and large spin-orbit coupling for Az-CHX with X = O and S may cause high intersystem crossing rates [44].

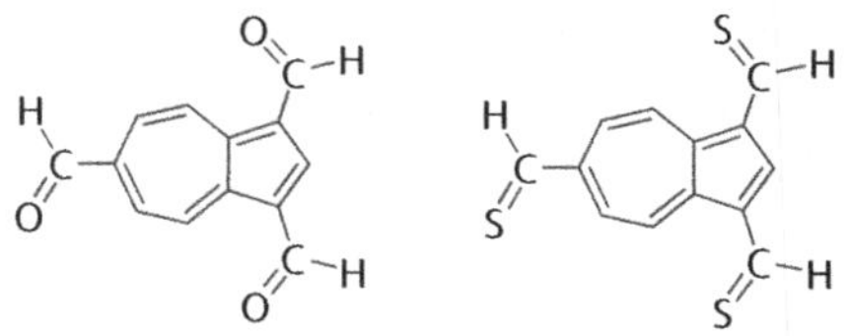

Figure 13.5 Structures of Az-CHO and Az-CHS.

13.3.5 Electron Transport

The electron transport characteristics of azulene have been studied by optimizing the structure of azulene and azulene dithiols at B3LYP/6-31G** level of theory. They made use of thiol anchors without hydrogen atoms to wire the optimized structure between two Au(1,1,1) electrode surfaces. The transmission spectra were measured in the region -2 to +2 eV with respect to the Fermi energy with varying current (I) through a two-probe system in both directions. Molecular projected self-consistent-Hamiltonian analysis, non-equilibrium Green's function, combined with exchange correlation DFT functional, were employed to compute the current. The Landauer–Buttiker formula has been used to calculate the current through a molecular junction.

$$I(V) = \frac{2e}{h} \int_{\mu L}^{\mu R} T(E,V)[f(E-\mu_L) - f(E-\mu_R)]dE$$

where

T(E,V) = transmission probability, *f*is = Fermi function, $\mu_{L/R}$ are the electrochemical potential of the left/right electrodes and the difference between them is given by eV with the applied bias voltage V, i.e., $\mu_L = \mu(0)$ -eV/2 and $\mu R = \mu(0)$ + eV/2. The $\mu R/L(0)$=E_F is the Fermi energy. This μL - μR energy region is called bias window.

The equations employed to calculate the transmission coefficients are

$$T[E,V] = T_r[\Gamma_L(E,V)G^R(E,V)\Gamma_R(E,V)G^A(E,V)]$$

where $\Gamma_{L/R}$ are the coupling functions associated with the L/R electrodes and $G^{R/A}$ are the retarded and advanced Green's functions of the system. High current is obtained from four connections (1,3-, 1,5-, 2,6-, and 4,7-), and it is found that 2,7-azulene dithiolate shows a diode-like behavior (RR = 4 at 2 V). Low to moderate current is observed for the remaining connections [45].

The nonlinear optical parameters and chemical reactivity descriptors were employed to study the electrical conductivity of a number of azulene derivatives by changing the donor-acceptor nature of the substituents [46]. In case of the dipole moment, as expected, the value of the dipole moment changes with a change in π-electron density of the molecule as a result of a donor-acceptor substitution position, e.g., a higher value of dipole moment is observed for azulene with electron withdrawing group NO_2 at the five-membered ring, whereas a lower value is observed with electron-donating NMe_2 at the seven-membered ring. The value of the dipole moment was maximum (12.92 D) when NO_2 is present as a substituent in the seven-membered rings and NMe_2 in the five-membered ring. In such cases, the entire molecule behaves as a single dipole, thus imparting the molecule with a high value of dipole moments. A comparison of the tensor component of polarizability and the average polarizability of each of the compounds was performed, and it was found that the percentage contributions of the component polarizability (α_{xx}, α_{yy}, and α_{zz}) to the average polarizability of each of the pairs are much the same. Moreover, the plot of average polarizability (α_{av}) vs. the energy of the molecule indicated a non-linear relation. This strictly violates the minimum polarizability principle. On further inspection, they have assumed that the violation of the minimum polarizability principle is due to the fact that at different substitutions, a change in the position of the donor-acceptor group causes a difference in the effective electric field. To validate that point, they have further calculated the electronic spatial extent ($\langle R^2 \rangle$) of the molecules. $\langle R^2 \rangle$ measures the spread of electron density on the molecule and is commensurate with the electronic volume. A plot of α_{av} vs. $\langle R^2 \rangle$ illustrates that an approximately fitted straight line, where dimethyl-(8-nitro-azulen-1-yl)-amine and dimethyl-(6-nitro-azulen-2-yl)-amine have the lowest and highest value of both $\langle R^2 \rangle$ and α_{av} value.

The study of 2,6′:2′,6″-terazulene is significant due to its potential applications in OFETs as an n-type semiconductor. Its unique electronic properties, derived from the intramolecular dipole moment of the azulene unit, make it a promising candidate for high-performance electronic devices. The B3LYP/6-31G(d) were used to analyze the electronic states of 2,6′:2′,6″-terazulene and calculated the dipole moment, HOMO and LUMO distributions, the intermolecular electronic coupling constants, and reorganization energies. The calculations indicated that 2,6′:2′,6″-terazulene exhibited a large dipole moment of 5.30 D, significantly affecting its electronic properties. The HOMO was localized primarily in the five-membered ring, while the LUMO was delocalized across the molecule, facilitating electron transport. Moreover, the n-type operation in OFET devices was found to be attributed to the favorable overlap of the LUMO in adjacent molecules, while the HOMO localization hindered hole transport. The optimal device performance was achieved with a carrier mobility of 0.29 $cm^2/V{\cdot}s$ and an on/off ratio of 2.5×10^7 at a substrate temperature of 100°C, demonstrating the importance of film morphology and molecular orientation in enhancing charge transport properties [16].

Density functional theory was used to optimize and analyze the structures of the terazulene isomers. From the calculations, they found that the LUMO of all terazulene isomers was fully delocalized over the entire molecule, indicating strong π-conjugation, which favors n-type transport. However, the HOMOs of the isomers displayed varying degrees of localization. For instance, TAz1 and TAz4's HOMO was localized at one end of the molecule, while TAz2 and TAz3 had HOMOs that were more spread out over the 2,2'-biazulene units. This implied that the HOMOs of TAz1 and TAz4, being localized, hindered hole transport through the crystal, favoring n-type transport whereas TAz2 and TAz3, with delocalized HOMOs and LUMOs, were predicted to exhibit ambipolar characteristics, allowing for both electron and hole transport in OFET devices [47].

The theoretical calculations in the study of 5,5'-di(2-azulenyl)-2,2'-bithiophene and 2,5-di(2-azulenyl)-thieno[3,2-b]thiophene were performed using DFT. In this case, the results also indicated that 5,5'-di(2-azulenyl)-2,2'-bithiophene and 2,5-di(2-azulenyl)-thieno[3,2-b]thiophene exhibit effective π-conjugation systems, which contribute to their small transition energies and high carrier mobilities in OFETs [48].

Azulene-BODIPY triads with acetylene spacer were experimentally prepared by varying the linking position of azulene. Theoretical calculations at B3LYP/6-31G(d) level of theory indicated that for both 2AzBDP2Az and 6AzBDP6Az, LUMO is delocalized over the entire molecule, whereas for 1AzBDP1Az and BDPAzBDP, LUMO is delocalized entirely on the BODIPY core (Figure 13.6). Moreover, for 1AzBDP1Az, 2AzBDP2Az and BDPAzBDP, HOMO is delocalized over the entire core but for 6AzBDP6Az, HOMO is delocalized over the azulelene core. From analyzing the overlapping of HOMO and LUMO, the authors concluded that 2AzBDP2Az may show ambipolar characteristics. Similarly, 1AzBDP1Az and BDPAzBDP may show hole transport characteristics, whereas 6AzBDP6Az may show electron transport characteristics. Finally, they concluded that by changing the connecting position between the azulene unit and BODIPY, the

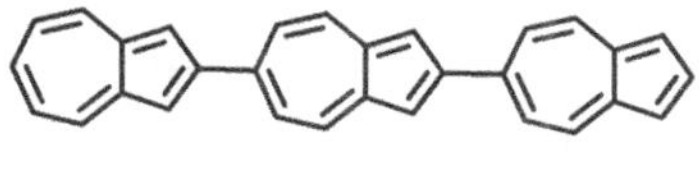

TAz1/13.4

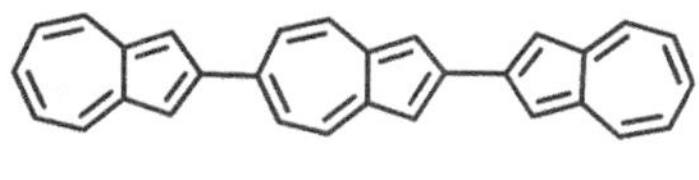

TAz2/13.5

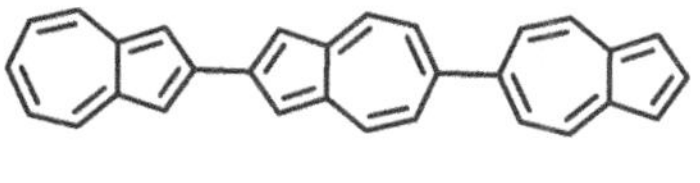

TAz3/13.6

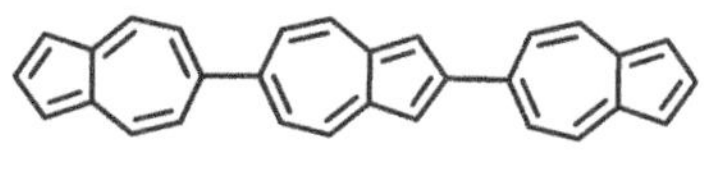

TAz4/13.7

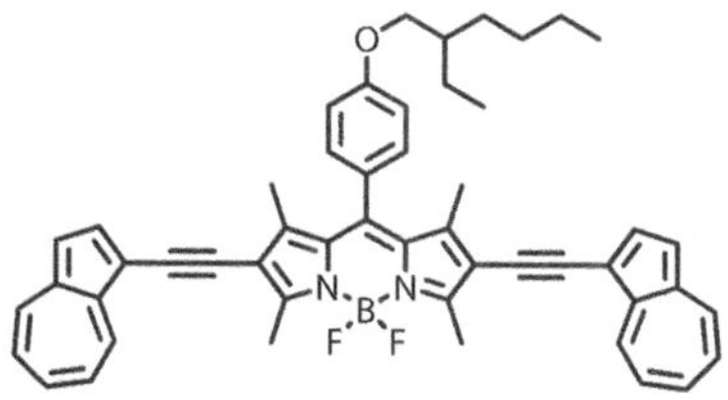

1AzBDP1Az/13.8

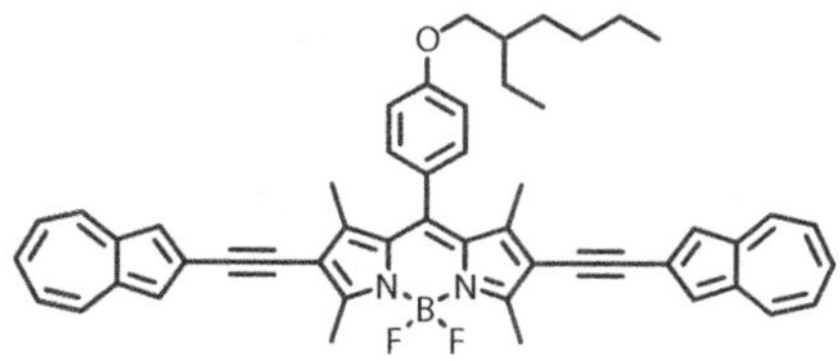

2AzBDP2Az/13.9

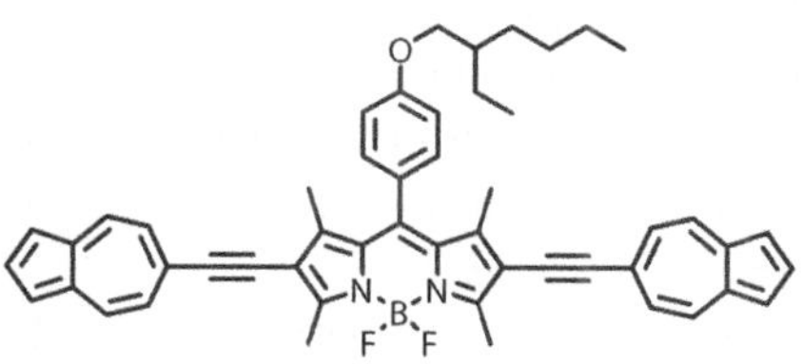

6AzBDP6Az/13.10

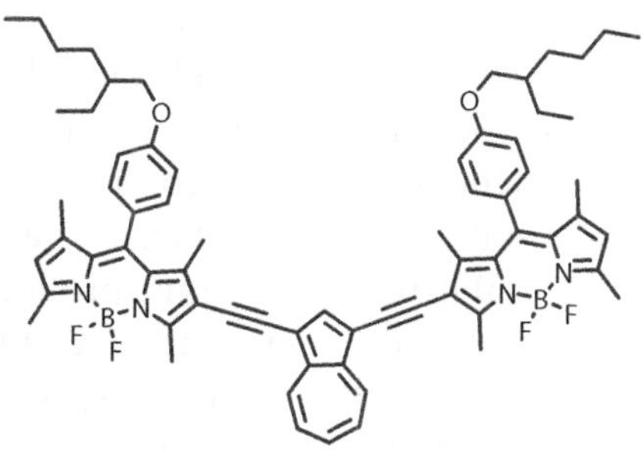

BDPAzBDP/13.11

Figure 13.6 Structures of TAz1, TAz2, TAz3, and TAz4 and azulene-BODIPY triads.

molecular orbital distributions can be changed, which in turn changes the polarity of OFETs based on the isomeric triads [49].

13.3.6 Reaction Mechanism

Due to the unique structure of azulene, electrophilic aromatic substitution reactions of azulene are of prime importance to study by theoretical chemists. Wave-function-based HF and MP2, and DFT-based B3LYP are used to study the electrophilic aromatic substitution reaction of benzene. From their calculations, they concluded that substituents at positions 4- or 6- made intermediates the least stable. Moreover, HF does not account for electron correlation; therefore, it indicated position 6 as the least stable,

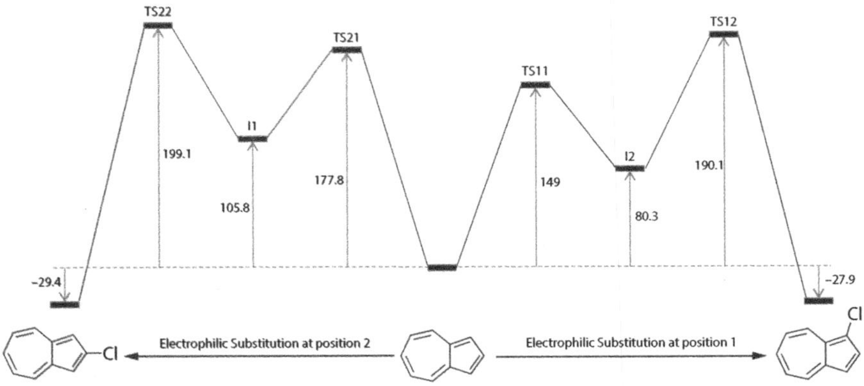

Figure 13.7 Energy diagram (in kcal/mol) for the chlorination of azulene at positions 1 and 2.

whereas both DFT and MP2 methods indicated position 4 is the most unstable one. For most of the electrophiles, substitution at position 2 was found to be the most stable one. However, on inspecting the barrier height of the reaction, they have concluded that the regioselectivity of the reaction of azulene is kinetically controlled and substitution at position 1 gives the most stable kinetic products, whereas substitution at position 2 gives the most stable thermodynamic products [50] (Figure 13.7).

I1 and I2 represent intermediate for substitution at position 1- and 2-, respectively, TS11 and TS12 represent the first transition state and the second transition state for substitution at position 1, respectively, and TS21 and TS22 represent the first transition state and the second transition state for substitution at position 2, respectively.

The coordination chemistry of azulene was studied using B3LYP of the molecular structure of LnMAz, MAz2, LnM2Az, and M2Az (Figure 13.8). Due to the asymmetric nature of azulene, the complex of azulene shows a very different coordinating behavior. They found that most of the time, M–M bonding is preferred over M-azulene bonding. Moreover, the study revealed the existence of multiple isomers of similar energy, with the potential for interconversion in solution through haptotropic shifts. On the other hand, coordination modes vary widely, with some configurations being energetically favorable depending on the specific metal and ligand environment, highlighting the complexity and dynamism of azulene-based complexes in organometallic chemistry [51].

The study of azulene rearrangement to naphthalene is significant in understanding the mechanisms of organic reactions, particularly those involving radical species and thermal processes [52]. The DFT calculations

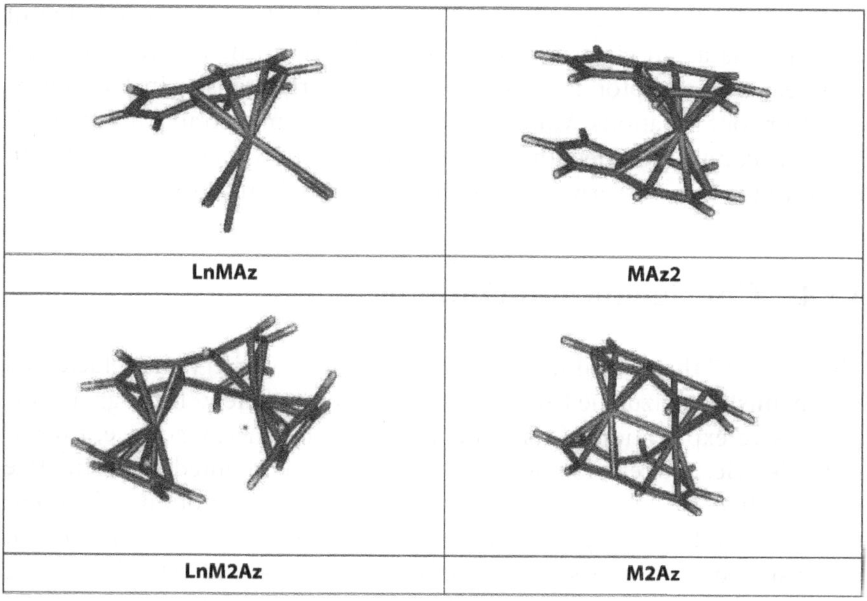

Figure 13.8 Structures of (1) LnMAz, (2) MAz2, (3) LnM2Az, and (4) M2Az.

were performed using the B3LYP functional at the 6-31G(d) basis set for geometry optimization and frequency calculations. The study involved analyzing various mechanisms, including radical-promoted pathways (methylene walk and spiran pathways) and intramolecular processes, to assess their activation energies and competitive nature under different thermal conditions. The author suggests that azulene rearrangement to naphthalene mainly occurs through radical-promoted pathways, particularly the methylene walk and spiran mechanisms, at elevated temperatures (400–500°C). Moreover, they observed that the calculated activation energies for intramolecular mechanisms were found to be significantly high ($\approx$ 350 kJ mol^{-1}), indicating that the arrangement through this pathway will be possible only if the temperatures are above 1100°C, which is impractical for most experimental conditions [53].

13.3.7 Hyperpolarizability

Austin Model 1 with a finite field approach and Zerner's intermediate neglect of differential overlap with configuration interaction were used to study the hyperpolarizability of azulene derivatives. Due to their nonalternant structure, they are promising candidates for use in nonlinear optics and optoelectronics. The calculations indicated that azulene derivatives possess large first hyperpolarizability values. The five-membered ring of azulene acts as an electron donor, while the seven-membered ring acts as an electron acceptor, facilitating efficient charge transfer in 1-donor-5-acceptor configurations. Moreover, their results demonstrated that extending the conjugation length and optimizing substituent positions can enhance the hyperpolarizability of these chromophores [54].

13.4 Conclusion and Outlook

The chapter "Theoretical Insights into Azulene Chemistry" has elucidated the chemistry of azulene from a theoretical point of view. Through a comprehensive exploration of its structure, potential energy surface, excited state, electric, and reaction mechanism, we have highlighted the distinctive features that set azulene apart from other aromatic compounds. The theoretical methods, ranging from HF to multi-reference wave-function-based to density-functional-based, have provided a deeper understanding of the azulene's chemical behavior, as well as the accuracy of the computational methods. For example, it is observed that azulene's chemistry can be better studied by using either multireference wavefunction methods with electron correlation or density functional-based methods. Moreover, based on the electronic distribution obtained from the density-based method, it can predict the behavior of azulene derivatives on p-type or n-type transport. These insights not only enhance our fundamental knowledge of azulene but also give a great insight into its potential applications in materials science, organic electronics, and photochemistry. Future research, guided by these theoretical methods, further improves understanding for targeted applications in drug design and materials science.

References

1. Garratt, P.J., *Aromaticity*, Wiley, New York, 1986.
2. Graham, S.R., Colby, D.A., Lash, T.D., Azulene analogues of tripyrranes and carbaporphyrinoids therefrom. *Angew. Chem. Int. Ed.*, 41, 1371, 2002.
3. Colby, D.A. and Lash, T.D., Calix[4]azulene. *J. Org. Chem.*, 67, 1031, 2002.
4. Piacenza, M. and Grimme, S., Van der Waals complexes of polar aromatic molecules: Unexpected structures for dimers of azulene. *J. Am. Chem. Soc.*, 27, 14841, 2005.
5. Lash, T.D., Out of the blue! Azuliporphyrins and related carbaporphyrinoid systems. *Acc. Chem. Res.*, 49, 471, 2016.

6. Xin, H., Hou, B., Gao, X., Azulene-based πfunctional materials: Design, synthesis, and applications. *Acc. Chem. Res.*, 54, 1737, 2021.
7. Shevyakov, S.V., Li, H., Muthyala, R., Asato, A.E., Croney, J.C., Jameson, D.M., Liu, R.S.H., Orbital control of the color and excited state properties of formylated and fluorinated derivatives of azulene. *J. Phys. Chem. A*, 107, 3295, 2003.
8. Michl, J. and Thulstrup, E.W., Why is azulene blue and anthracene white? A simple MO picture. *Tetrahedron*, 32, 205, 1976.
9. Ito, S. and Morita, N., Creation of stabilized electrochromic materials by taking advantage of azulene skeletons. *Eur. J. Org. Chem.*, 27, 4567, 2009.
10. Xin, H. and Gao, X.J.C., Application of azulene in constructing organic optoelectronic materials: New tricks for an old dog. *ChemPlusChem*, 82, 945, 2017.
11. Salman, H., Abraham, Y., Tal, S., Meltzman, S., Kapon, M., Tessler, N., Speiser, S., Eichen, Y., 1,3-Di(2-pyrrolyl)azulene: An efficient luminescent probe for fluoride. *Eur. J. Org. Chem.*, 11, 2207, 2005.
12. Schmitt, S., Baumgarten, M., Simon, J., Hafner, K., 2,4,6,8-Tetracyanoazulene: A new building block for "Organic Metals". *Angew. Chem. Int. Ed.*, 37, 1077, 1998.
13. Wang, F., Lai, Y.-H., Han, M.-Y., Stimuli-responsive conjugated copolymers having electro-active azulene and bithiophene units in the polymer skeleton: Effect of protonation and p-doping on conducting Properties. *Macromolecules*, 37, 3222, 2004.
14. Cristian, L., Sasaki, I., Lacroix, P.G., Donnadieu, B., Asselberghs, I., Clays, K., Razus, A.C., Donating strength of azulene in various azulen-1-yl substituted aationic dyes: Application in nonlinear optics. *Chem. Mater.*, 16, 3543, 2004.
15. Muranaka, A., Yonehara, M., Uchiyama, M., Azulenocyanine: A new family of phthalocyanines with intense near-IR absorption. *J. Am. Chem. Soc.*, 132, 7844, 2010.
16. Yamaguchi, Y., Ogawa, K., Nakayama, K., Ohba, Y., Katagiri, H., Terazulene: A high-performance n-type organic field-effect transistor based on molecular orbital distribution control. *J. Am. Chem. Soc.*, 135, 19095, 2013.
17. Nishimura, H., Ishida, N., Shimazaki, A., Wakamiya, A., Saeki, A., Scott, L.T., Murata, Y., Hole-transporting materials with a two dimensionally expanded π-system around an azulene core for efficient perovskite solar cells. *J. Am. Chem. Soc.*, 137, 15656, 2015.
18. Lewars, E., *Computational chemistry: Introduction to the theory and applications of molecular and quantum mechanics*, Springer International Publishing, Switzerland, 2016.
19. Møller, C. and Plesset, M.S., Note on an approximation treatment for many-electron systems. *Phys. Rev.*, 46, 618, 1934.
20. Cramer, C.J., *Essentials of computational chemistry: Theories and models*, 2nd Edition, John Wiley & Sons. Ltd, England, 2004.

21. Hohenberg, P. and Kohn, W., Density functional theory (DFT). *Phys. Rev. B*, 864, 136, 1964.
22. Kohn, W. and Sham, L.J., Self-consistent equations including exchange and correlation effects. *Phys. Rev. A*, 140, 1133, 1965.
23. Jørgensen, P. and Simons, J., *Second quantization-based methods in quantum chemistry*, Academic, New York, 1981.
24. Ghosh, S.K., Time-dependent density functional theory of many-electron systems, in: *Chemical reactivity theory: A density functional view*, CRC Press, Taylor & Francis Group, Boca Raton, 2009.
25. Jensen, F., *Introduction to computational chemistry*, second edition, John Wiley & Sons. Ltd, England, 2007.
26. Haddon, R.C. and Raghavachari, K., Planar cis-[10] annulene and azulene revisited. *J. Am. Chem. Soc.*, 104, 3516, 1982.
27. Glidewell, C. and Lloyd, D., MNDO study of bond orders in some conjugated bi-and tri-cyclic hydrocarbons. *Tetrahedron*, 40, 4455, 1984.
28. Grimme, S., *Ab initio* study of the structure and dipole moment of azulene. *Chem. Phys. Lett.*, 201, 67, 1993.
29. Murakami, A., Kobayashi, T., Goldberg, A., Nakamura, S., CASSCF and CASPT2 studies on the structures, transition energies, and dipole moments of ground and excited states for azulene. *J. Chem. Phys.*, 120, 1245, 2004.
30. Kozlowski, P.M., Rauhut, G., Pulay, P., Potential symmetry breaking, structure and definite vibrational assignment for azulene: Multiconfigurational and density functional results. *J. Chem. Phys.*, 103, 5650, 1995.
31. Clara, C.S., Sousaa, M., Agostinha, R.M., Victor, M.F., Morais, energetics and stability of azulene: From experimental thermochemistry to high-level quantum chemical calculations. *J. Chem. Thermodyn.*, 73, 101, 2014.
32. Piacenza, M. and Grimme, S., Van der Waals complexes of polar aromatic molecules: Unexpected structures for dimers of azulene. *J. Am. Chem. Soc.*, 127, 14841, 2005.
33. Robertson, J.M., Shearer, H.M.M., Sim, G.A., Watson, D.G., The crystal and molecular structure of azulene. *Acta Crystallogr.*, 15, 1, 1962.
34. Bräuniger, T., Pouoko, R., Luz, Z., Gutsche, P., Meinel, Zimmermann, H., Haeberlen, U., The dynamic disorder of azulene: A single crystal deuterium nuclear magnetic resonance study. *J. Chem. Phys.*, 112, 10858, 2000.
35. Chen, Y., Wang, Y.B., Zhang, Y., Wang, W., Accurate calculations of the noncovalent systems with flat potential energy surfaces: Naphthalene dimer and azulene dimer. *Comput. Theor. Chem.*, 1112, 52, 2017.
36. Dunlop, D., Ludvikova, L., Banerjee, A., Ottoson, H., Slanina, T., Excited-state (anti) aromaticity explains why azulene disobeys Kasha's rule. *J. Am. Chem. Soc.*, 145, 21569, 2023.

37. Rosenberg, M., Dahlstrand, C., Kilså, K., Ottosson, H., Excited state aromaticity and antiaromaticity: Opportunities for photophysical and photochemical rationalizations. *Chem. Rev.*, 114, 5379, 2014.
38. Möllerstedt, H., Piqueras, M.C., Crespo, R., Ottosson, H., Fulvenes, fulvalenes, and azulene: Are they aromatic chameleons? *J. Am. Chem. Soc.*, 126, 13938, 2004.
39. Soncini, A. and Fowler, P.W., Ring-current aromaticity in open-shell systems. *Chem. Phys. Lett.*, 450, 431, 2008.
40. Baird, N.C., Quantum organic photochemistry. II. Resonance and aromaticity in the lowest $3\pi\pi^*$ state of cyclic hydrocarbons. *J. Am. Chem. Soc.*, 94, 4941, 1972.
41. Piecuch, P., Hansen, J.A., Staedter, D., Faure, S., Blanchet, V., Existence of the doubly excited state that mediates the photoionization of azulene. *J. Chem. Phys.*, 138, 201102, 2013.
42. Pariser, R., Electronic spectrum and structure of azulene. *J. Chem. Phys.*, 25, 1112, 1956.
43. Palmer, M.H., Jones, N.C., Hoffmann, S.V., Aitken, R.A., Coreno, M., Simone, M., Grazioli, C., Patterson, L.J., The excited states of azulene: A study of the vibrational energy levels for the lower $\pi\pi^*$-valence states by configuration interaction and density functional calculations, and theoretical studies of the Rydberg states. *J. Chem. Phys.*, 157, 154307, 2022.
44. Debnath, P., Ahmed, R., Manna, A.K., Molecular-scale design of azulene-based triplet photosensitizers: Insights from time-dependent optimally tuned range-separated hybrid. *J. Phys. Chem. B*, 127, 10016, 2023.
45. Ahmed, M.N., Staykov, A., Yoshizawa, K., First-principles calculations of electron transport through azulene. *J. Phys. Chem. C*, 120, 9043, 2016.
46. Maity, R., Mandal, D., Misra, A., Effect of donor acceptor substitution position on the electrical responsive properties of azulene system: A computational study. *Mol. Phys.*, 117, 14, 1781, 2019.
47. Yamaguchi, Y., Takubo, M., Ogawa, K., Nakayama, K.I., Koganezawa, T., Katagiri, H., Terazulene isomers: Polarity change of OFETs through molecular orbital distribution contrast. *J. Am. Chem. Soc.*, 138, 35, 11335, 2016.
48. Yamaguchi, Y., Maruya, Y., Katagiri, H., Nakayama, K.I., Ohba, Y., Synthesis, properties, and OFET characteristics of 5, 5′-Di (2-azulenyl)-2, 2′-bithiophene (DAzBT) and 2, 5-Di (2-azulenyl)-thieno [3, 2-b] thiophene (DAzTT). *Org. Lett.*, 14, 9, 2316, 2012.
49. Kumar, N.R., Agrawal, A.R., Debnath, S., Choudhurya, A., Zade, S.S., Effect of connectivity variation in azulene-BODIPY triads and their optoelectronic properties. *New J. Chem.*, 47, 2456, 2023.
50. Shiraz, N.Z., Sharifzadeh, E.S., Koosha, N., Theoretical study on the regioselectivity of electrophilic aromatic substitution reactions of azulene. *Acta Chim. Slov.*, 60, 166, 2013.

51. Korichi, H., Zouchoune, F., Zendaoui, S.M., Zouchoune, B., Saillard, J.Y., The coordination chemistry of azulene: A comprehensive DFT investigation. *Organometallics*, 29, 7, 1693, 2010.
52. Heilbronner, E., Plattner, P.A., Wieland, K., Umlagerung von azulen in naphthalin. *Experientia*, 3, 2, 70, 1947.
53. Alder, R.W., East, S.P., Harvey, J.N., Oakley, M.T., The azulene-to-naphthalene rearrangement revisited: A DFT study of intramolecular and radical-promoted mechanisms. *J. Am. Chem. Soc.*, 125, 18, 5375, 2003.
54. Wang, P., Zhu, P., Ye, C., Asato, A.E., Liu, R.S., Theoretical investigation and molecular design of some azulene derivatives with large hyperpolarizabilities. *J. Phys. Chem. A*, 103, 35, 7076, 1999.

14

Future Perspectives of Azulene Chemistry

Kalyan Jyoti Kalita[1] and Neha Rani Kumar[2]*

[1]Department of Chemistry, Indian Institute of Technology, Hyderabad, Telangana, India
[2]Department of Chemistry, Dhemaji College, Dhemaji, Assam, India

Abstract
This chapter discusses the rich and multidimensional field of azulene chemistry, with a special focus on future prospects in organic synthesis, optoelectronics, energy storage, and medicinal applications. Sustainable synthesis approaches, new advances in functionalization techniques, and unique electronic and photophysical properties of azulene derivatives are incorporated into the discussion. The impact of artificial intelligence on the development and research acceleration of azulene chemistry is also considered.

Keywords: Azulene chemistry, aromatic hydrocarbons, azulene derivatives, sustainable synthesis, optoelectronic applications, organic materials science, functionalization techniques, non-alternant aromaticity

14.1 Introduction

Azulene, a bicyclic aromatic hydrocarbon, is distinguished by its unique deep blue color, high dipole moment (1.08 Debye), and non-alternant aromaticity. It is, thus, distinct from the more generally studied benzene-based systems (Figure 14.1). The unique molecular architecture, a fusion of an electron rich five-membered ring and an electron deficient seven-membered ring, gives rise to fascinating electronic and optical properties, making azulene a subject of enduring interest in organic chemistry [1–7].

Historically, azulene has received special attention based on its interesting structure and reactivity. Early investigations are based on the origin of

**Corresponding author*: nehakumar0926@gmail.com

Neha Rani Kumar and Abhijeet R. Agrawal (eds.) Azulene-Based Materials: Design, Synthesis, and Applications, (377–394) © 2026 Scrivener Publishing LLC

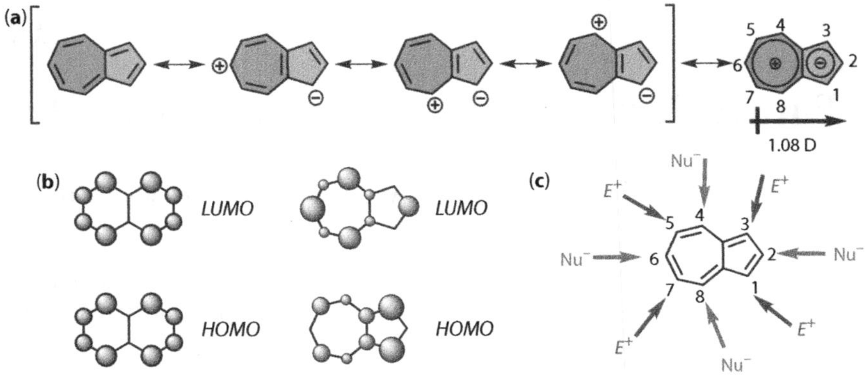

Figure 14.1 (a) Resonance structures of azulene; (b) probabilities of locating an electron in the highest occupied molecular orbital (HOMO) and the lowest occupied molecular orbital (LUMO) of naphthalene (left) azulene (right); and (c) reactivity of each position of azulene (here, E+ indicates electrophile, and Nu- indicates nucleophile).

non-alternant aromaticity and resonance effects. In recent decades, azulene derivatives have transitioned from purely academic curiosities to key components in materials science, photophysics, and medicinal chemistry. Their applications involve organic electronics like photovoltaic systems, organic light-emitting diodes (OLEDs), and other emerging application areas like bioimaging and photodynamic therapy (PDT) [1–7].

Interest in azulene synthesis has been further driven by a growing focus on sustainable and green chemistry, such as developing eco-friendly methods, including renewable starting materials, and energy-efficient processes, such as microwave-assisted synthesis. These advancements reflect a broader trend toward integrating sustainability into chemical manufacturing, aligning with global efforts to reduce environmental impact while fostering innovation [1–7]. More recent analyses of year-wise trends in publications present a steady rise in articles on azulene, establishing its increasing salience in areas of science outside chemistry (Figure 14.2).

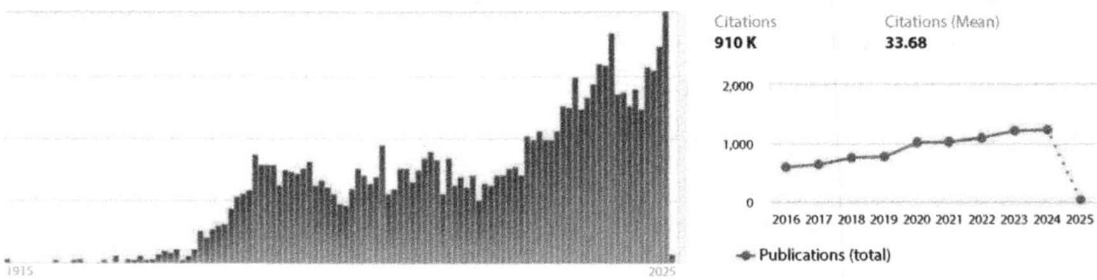

Figure 14.2 Number of publications per year on azulene (Data until 8th January 2025).

14.2 Advances in Synthesis

14.2.1 Green Synthetic Routes for Azulene and Its Derivatives

The search for green and environment-friendly synthetic methods has become a cornerstone of modern organic chemistry. In the case of azulene, much progress has been made in recent years toward developing green synthetic routes that are environmentally friendly while being maximally efficient [8, 9].

Transition-metal catalysis has played a significant role in advancing the green synthesis of azulene and its derivatives. The palladium- and copper-catalyzed reactions especially presented outstanding efficiency and selectivity. In summary, cross-coupling reactions catalyzed by palladium, such as the Suzuki–Miyaura and Sonogashira couplings, were used for the assembly of the azulene framework, starting from readily available halogenated intermediates and organoboranes or alkynes. These methods provide highly mild reaction conditions that tolerate broad functional groups, which is necessary for preparing complex derivatives of azulene related to optoelectronics and materials science [10].

Organocatalysis now emerges as an attractive alternative for meeting this challenge of avoiding expensive and toxic transition metals. Organocatalytic methods use small organic molecules as catalysts in bond-forming reactions, avoiding the problems of metal residues in pharmaceutical and electronic applications. For example, aminocatalysis has been applied to activate enals and enones toward cycloaddition reactions that form the azulene nucleus. These metal-free processes favor green chemistry principles but extend the possibilities of azulene derivatives through access to asymmetric synthesis and chiral centers.

Energy-efficient techniques have also revolutionized the synthesis of azulene compounds. Microwave-assisted synthesis is very popular, as it significantly reduces the time for such reactions from hours to minutes. Thus, fast and uniform heating becomes possible; rates of reaction increase, and often, the yields and purities increase, too. Microwave irradiation effectively involved cycloadditions and rearrangements, forming the azulene ring system [11].

Similarly, sonochemistry or ultrasonic irradiation has stimulated reactions under even milder conditions. Cavitation—nucleation, growth, and explosive collapse of cavities in a liquid—due to ultrasonic waves, produces local hotspots of extreme temperatures and pressures, thereby potentially accelerating the rate of reactions and yield of the desired products. Applications of ultrasound to synthesizing azulene derivatives, such as electrophilic aromatic substitution and cyclizations, are being explored with considerable promise [12].

The other frontier for green chemistry is solvent-free synthesis, which avoids using volatile organic compounds, a significant source of environmental pollution. Mechanochemistry involves grinding reactants in a mortar and pestle or employing a ball mill to make reactions possible by providing mechanical force. It has recently been applied to synthesize substituted azulenes providing an atom-economical synthesis with minimal waste formation. By conducting reactions without solvents, chemists can significantly reduce E-factors (the ratio of waste to desired product), contributing to more sustainable industrial processes [8].

14.2.2 Challenges in Regioselective Synthesis and Recent Solutions

Azulene has an unusual electronic structure, containing fused five-membered and seven-membered rings. It is thus, considerably challenging to achieve the regioselective functionalization. It does not resemble benzene because ring systems are not equivalent, and so it shows polarized electronic distribution along with a higher dipole moment of 1.08 Debye. Polarization, thus, significantly impacts the site selectivity of electrophilic and nucleophilic substitutions. Thus, it becomes challenging to control the functionalization [13–20].

One of the key challenges is achieving regioselectivity at the 2- and 6-positions of the azulene ring—positions crucial for modifying its electronic and optical properties. Due to higher electron density in these areas, traditional electrophilic substitution tends to occur preferentially at the 1 and 3 positions (Figure 14.1c). To solve this problem, chemists have developed strategies based on directed ortho-metalation. In this approach, a metal— usually lithium or magnesium—is used to deprotonate the azulene at a specific position, with the process being guided by a directing group that coordinates with the metal. Subsequent reactions can then introduce functional groups precisely at the desired positions [13–20].

For example, protecting the azulene nucleus with a silyl or alkoxy group as a directing moiety enables lithiation at the 2-position. The organolithium intermediate formed may undergo electrophilic quenching with halogens, acyl groups, and other electrophiles. This approach has allowed for the synthesis of 2,6-disubstituted azulene derivatives in a highly regioselective manner—important for organic electronics applications, in which substitution patterns of the material dictate its properties [13–20].

Cross-coupling reactions have also proven to be a valuable tool in overcoming regioselectivity issues. Position-selective halogenation of azulene gives rise to essential intermediates for palladium-catalyzed cross-coupling reactions.

Thus, for example, selective bromination at the 2-position followed by Suzuki–Miyaura coupling with aryl boronic acids enables the introduction of aryl groups at this less reactive site. Improvement in catalyst design and discovery of ligand systems could facilitate the reactivity and selectivity of such transformations [13–20].

Electrophilic aromatic substitution reactions can be fine-tuned through protecting groups and reaction conditions that favor substitutions at less reactive positions. Lewis acid-based catalysts can activate azulene at the 4-, 5-, and 6-positions. Improving catalyst efficiency is another crucial area of research for expanding the versatility of azulene functionalized products. Similarly, introducing electron-rich or electron-deficient groups can modulate the electron density distribution within the azulene ring. This strategy can selectively make the azulene moiety undergo subsequent reactions at a specific site.

Recent advances in photo redox catalysis provide new strategies for regioselective functionalization. Using energy from light, photo redox catalysts can activate radical intermediates at mild reaction conditions. The application of this approach has been established for C-H activation and subsequent functionalization at positions, which had been challenging to access. This methodology increases regioselectivity and enhances functional group tolerance and reaction efficiency.

Connectivity patterns: 2,6-connected azulene derivatives have emerged as materials with versatile functions and tailored properties owing to their enhanced stability, tunable optoelectronic properties, and reactivity (Figure 14.3). Optimization of electronic interactions by this connectivity pattern makes them perfect candidates for many applications. They have shown promising results in organic photovoltaics and OLEDs, where broad absorption spectra and efficient charge transport enhance device performance. Their inherent redox activity and stability render them suitable for energy

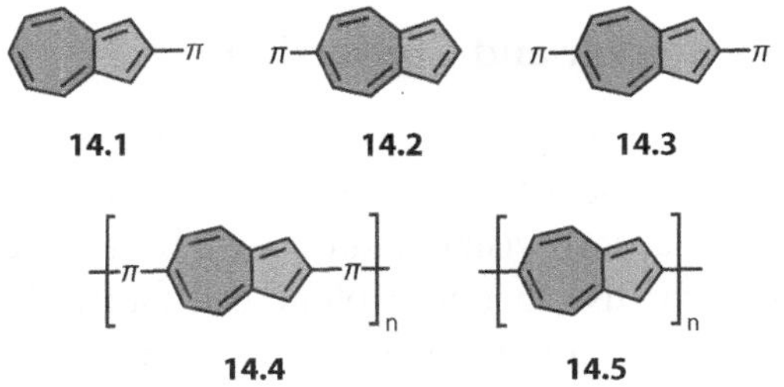

Figure 14.3 2,6 functionalized azulenes: (14.1) 2-pi substituted azulene; (14.2) 6-pi substituted azulene; (14.3) 2,6-pi substituted azulene; (14.4) polyazulenes with 2,6-pi substituted connectivity; (14.5) polyazulenes with 2,6-connectivity.

storage in batteries and supercapacitors. Specificity in molecular interactions supports applications in chemical sensors for pollutants, biomolecules, and environmental monitoring [6, 21, 22]. Thus, 2,6-connected azulenes offer endless opportunities, viz. conductivity, magnetism, luminescence behavior, memory devices essential for advanced computing, and quantum technologies. They are perfect for organic spintronic devices because of their unique electronic properties. Regioselective coupling and functionalization techniques, such as halogenation and cross-coupling, allow the introduction of precise modifications, which have enabled incorporation into larger systems like dendrimers and polymers for flexible electronics. Modifying and functionalizing azulene at specific positions can improve stability and functionality for advanced organic optoelectronics and energy storage applications. Realizing the potential of 2,6-connected azulenes requires interdisciplinary collaboration. A key focus is scaling production sustainably while tailoring properties for specific technologies. These derivatives represent a promising frontier in materials science, addressing global challenges in energy and technology with their unique properties and structural versatility.

14.2.3 Functionalization Techniques

Selective halogenation, cross-coupling, and electrophilic aromatic substitution have been developed as the most critical methods for azulene functionalization. These techniques enable the incorporation of azulene into larger conjugated systems, which improves their optoelectronic properties. Besides the small molecular systems, recent literature suggests that azulene-based dendrimers and polymers can also open new avenues for exploring their potential in flexible and wearable electronics.

14.3 Applications and Future Directions

14.3.1 Energy Storage and Conversion

Due to their stable redox properties, azulene derivatives emerge as promising candidates for organic batteries and supercapacitors. Because azulene-based materials can absorb visible light over a broad spectrum, they are being investigated for use in photovoltaic cell fabrication. This has the potential to increase the power conversion efficiency of organic solar cells towards the ongoing effort to reach closer to the Shockley–Queisser limit. Similarly, their use in OLEDs highlights their potential in next-generation lighting and display technologies.

The function of azulene derivatives in catalysis and sensing applications is also receiving more attention as research reveals their unique qualities; this could result in creative solutions for chemical detection and environmental monitoring. Energy technology breakthroughs are made possible by the continuous investigation of azulene derivatives. It creates new opportunities in materials science and medicine, where their unique molecular properties may stimulate the development of smart materials and innovative medicinal compositions. Because cooperation between chemists, material scientists, and engineers could hasten the development and commercialization of azulene-based technologies, this diverse potential emphasizes the value of multidisciplinary research [23, 24].

14.3.2 Organic Optoelectronics

Azulene-based semiconductors and conductive polymers are at the forefront of organic optoelectronics due to their distinct optoelectronic properties and structural flexibility. Azulene's non-alternant aromaticity gives rise to unique energy level alignments, allowing for effective charge transport and high dipole moments that can be leveraged in various optoelectronic devices. Because of these characteristics and their anti-Kasha emission, azulene and its derivatives are considered to be a unique class of organic materials that may emit highly excited states, making them a desirable candidate for use in cutting-edge lighting technologies [23–34].

The anti-Kasha's rule behavior of azulene has historical significance, as it was one of the first molecules studied to exhibit emission from the S2 state rather than the S1 state (Figure 14.4). This phenomenon is due to the unique electronic coupling between the molecular orbitals of azulene and its asymmetric structure, which lowers the internal conversion rate. This property opens pathways for designing efficient light-emitting materials and broadens the scope for fundamental studies on excited-state dynamics in organic systems [35–39].

The recent development of azulene-based materials for thermally activated delayed fluorescence (TADF) represents a significant breakthrough in optoelectronics. The TADF materials harvest triplet excitons, constituting about 75% of the excitons generated in electroluminescent devices, converting them into singlet excitons through reverse intersystem crossing. This can lead to high photoluminescence quantum yield by successfully harvesting both singlet and triplet excitons, which is critical for high-performance OLEDs. With their anti-Kasha emission and tunable energy levels, Azulene derivatives are uniquely suited for this purpose, offering the potential for brighter and more energy-efficient displays

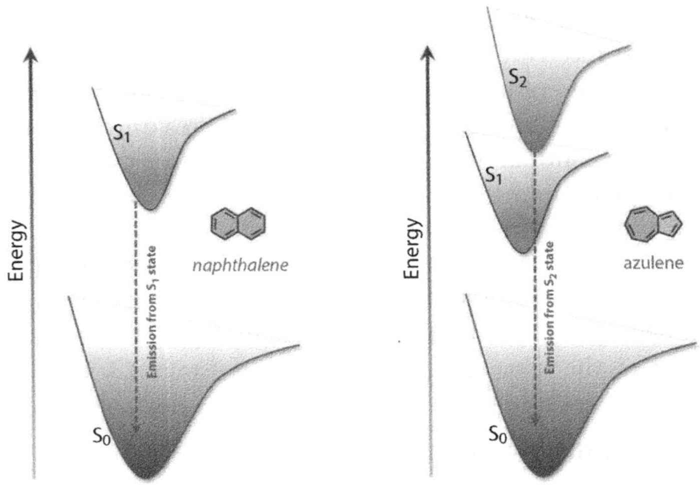

Figure 14.4 Emission observed from naphthalene (left) and azulene (right).

and lighting. In addition to TADF, azulene derivatives are being investigated for their phosphorescent properties. Heavy atoms or external spin-orbit coupling agents in azulene-based systems can promote intersystem crossing, enabling efficient phosphorescence. These materials are up-and-coming for low-energy, long-lasting light emission, making them suitable for safety indicators, bioimaging, and night-vision-compatible lighting applications [34–40].

14.3.3 Integration into Wearable and Flexible Devices

The compatibility of azulene-based materials with solution-processing techniques, such as spin-coating and inkjet printing, accelerates their adoption in wearable and flexible electronic devices. These materials' ability to form thin films with excellent mechanical flexibility and optical transparency makes them ideal for applications in flexible displays, bendable solar panels, and lightweight wearable sensors.

Azulene's high dipole moment and strong absorption in the visible region also make it suitable for organic photodetectors. These photodetectors are critical for applications ranging from optical communication systems to medical diagnostics, where flexibility and lightweight properties are increasingly important. Furthermore, azulene-based thin-film transistors (Figure 14.5) are being developed to integrate with flexible circuits, enabling the fabrication of foldable electronic devices. Lately, azulene-based single-molecule devices have demonstrated switchable

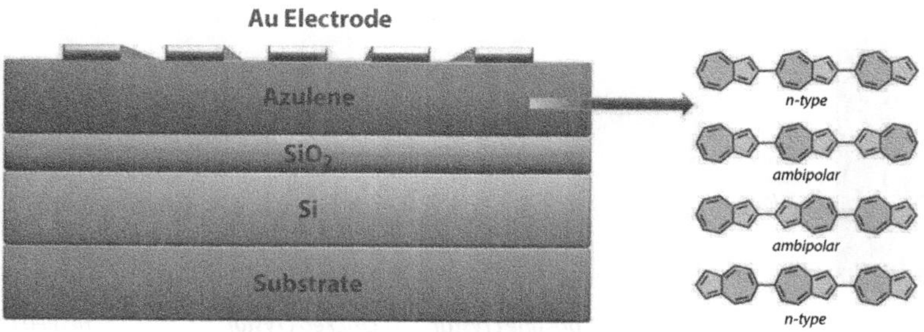

Figure 14.5 Emission observed from naphthalene (left) and azulene (right). Reproduced with permission [25]. Copyright 2013, American Chemical Society.

molecule–electrode coupling modes controlled by interfacial charge distribution and ionic environments, as highlighted in recent research. These findings underline azulene's role in advancing electrolyte-responsive optoelectronic systems and provide insights into tuning device functionalities for emerging applications [23–33].

14.3.4 Enhancing Efficiency in Energy Harvesting Devices

Azulene's broad absorption spectrum and efficient charge separation capabilities are being harnessed to improve the performance of organic photovoltaic cells (OPV). Specifically, azulene derivatives with optimized 2,6-connectivity patterns have demonstrated promise in donor-acceptor systems for OPVs, where, depending on the substitution pattern, they can function as either donor or acceptor components. Functionalized materials based on azulene are being investigated for their applications in photothermal and thermoelectric energy conversion and solar energy harvesting [23–33].

14.3.5 Intelligent Optoelectronic Systems and Sensors

Incorporating azulene derivatives into advanced sensors can revolutionize the healthcare, environmental monitoring, and telecommunications industries. For instance, optical sensors employ materials based on azulene to detect chemical and biological analytes with high sensitivity and selectivity. Real-time detection in challenging conditions is made possible by the precise interactions with target molecules by azulene's high dipole moment and adjustable optical characteristics. Intelligent systems using azulene-based optoelectronics are also being developed for dynamic energy management systems, responsive display panels, and adaptable lighting. The development of intelligent optoelectronic devices relies

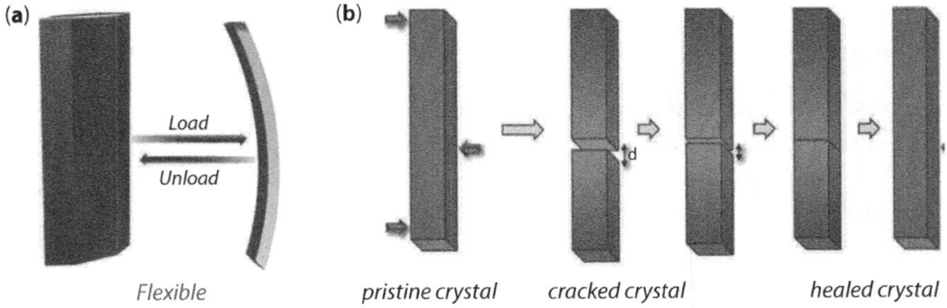

Figure 14.6 Schematic illustration of mechanical compliance on load application (mechanical force): (a) flexible, and (b) self-healing.

heavily on azulene's responsiveness to external stimuli, including light, mechanical force, and temperature [23–33]. In this regard, integrating mechanical compliance, such as flexibility and/or self-healing properties, into functional azulene derivatives is a promising area of research that can enhance their practical applications and durability in optoelectronics, sensors, and other technologies (Figure 14.6). Crystal engineering becomes handy in designing and realizing such an amalgamation of mechanical properties in functional materials. Some empirical ideas to achieve such a complex symphony include: a) incorporating flexible linkers (e.g., alkyl chains, ether linkages) chains into the molecular structure of the azulenes; b) utilizing hydrogen bonding and/or van der Waals interactions to create flexible supramolecular networks; c) introducing dynamic covalent bonds or reversible non-covalent interactions that can break and reform, allowing the material to heal itself. These interactions can help dissipate stress and contribute to flexibility and/or self-healing [41–46].

Furthermore, **azulene-based helical chiral materials** are emerging as a fascinating class in organic optoelectronics. Integrating the azulene moiety into helical structures enhances chiroptical properties due to the synergy of the helical architecture and azulene's intrinsic dipole moment. This chirality introduces optical activity to manipulate circularly polarized light, essential for applications in chiral photonics and circularly polarized organic light-emitting diodes. Devices that depend on chiral interactions perform better thanks to the helical framework's ability to fine-tune electrical and optical characteristics [47, 48].

The creation of these materials creates new avenues for catalysts and enantioselective sensors. In the pharmaceutical industry, where enantiomer discrimination is crucial, the combination of helical chirality and the unique electrical characteristics of azulene can improve the selectivity and sensitivity in detecting chiral compounds. Furthermore, helical architectures may efficiently transport charges down the chiral axis in organic field-effect transistors and other devices that need directed conductivity. Researchers are developing materials with previously unheard-of functions using azulene's geometric chirality and electrical quirks. These developments broaden the use of chiral optoelectronic materials in next-generation electrical and photonic devices while advancing our fundamental knowledge. An essential step towards incorporating molecular chirality into practical optoelectronic applications is investigating azulene-based helical chiral materials [47, 48].

Using its distinct conjugated structure with an electron-rich five-membered ring and an electron-deficient seven-membered ring, covalent organic frameworks (COFs) offer a flexible foundation for azulene-based COFs. Azulene-based COF-Azu, synthesized *via* liquid-liquid interface polymerization, exhibits non-volatile resistive switching in memristors due to electric field-induced charge transfer, enabling applications in neuromorphic computing with up to 80% image recognition accuracy. Similarly, COF-Azu-TP demonstrates exceptional photothermal conversion efficiency (97.2% under one-sun irradiation) and rapid shape recovery when incorporated into innovative materials. When compared to COF based on naphthalene, azulene-based COF's broad-spectrum light absorption and strong dipole moment significantly offer new avenues, such as the potential for cutting-edge uses in optoelectronics, photothermal systems, and sustainable technologies [49, 50].

14.3.6 Biomedical Applications

Of late, azulene derivatives have gained significant interest in the biomedical field due to their versatile bioactive properties and unique photophysical characteristics. One of the most prominent applications of azulene compounds is their role as anti-inflammatory agents. Naturally occurring azulenes, such as guaiazulene and chamazulene, found in chamomile and yarrow oils, have traditionally been used to soothe irritated skin and mucous membranes. They help create therapies for ailments including dermatitis, arthritis, and other inflammatory illnesses because of their capacity to regulate inflammatory reactions [51–57].

The anticancer effects of azulene derivatives are being investigated in oncology. Because of its distinct electrical structure, azulene may interact with biological macromolecules and may even stop the growth of cancer cells [51–57].

The distinctive photophysical properties of azulene, particularly its deep blue fluorescence, make it highly suitable for PDT and bioimaging applications. In PDT, particular light wavelengths trigger photosensitizers to release reactive oxygen species, which can kill sick cells specifically. Azulene derivatives can function as excellent photosensitizers because of their significant absorption in the visible spectrum and effective intersystem crossing to triplet states. This minimizes harm to nearby healthy tissues while enabling focused tumors or microbial diseases [51–57].

The luminescence of azulene makes it a valuable tool in bioimaging for very accurate biological process visualization. Researchers have created fluorescent probes and biological markers to identify specific cellular components or track medication delivery channels by conjugating azulene molecules with targeted ligands or proteins. This is especially helpful in domains where precise cellular and molecular visualization is essential, such as molecular biology and medical diagnostics [56–58].

Beyond imaging and therapy, azulene-based compounds are being investigated for their potential in developing responsive materials for biomedical sensors. The high dipole moment and tunable electronic properties of azulene allow for the creation of sensors that can detect changes in the environment, such as pH shifts, the presence of metal ions, or interaction with specific biomolecules. These sensors can be utilized in real-time monitoring of physiological conditions or detecting disease-associated biomarkers.

Additionally, azulene derivatives have potential antioxidant properties that could shield cells from oxidative stress by scavenging free radicals. Its antioxidant effect adds another therapeutic potential, particularly in preventing or reducing damage associated with chronic disorders, including neurological and cardiovascular ailments. Because of its anti-inflammatory and antioxidant qualities, azulene compounds help improve health and avoid disease. The antibacterial and anti-inflammatory qualities of azulene help to speed up tissue recovery and lower the risk of infection in wound healing applications.

Another exciting field of study is the integration of azulene derivatives into drug delivery. In conclusion, azulene derivatives have various biological uses because of their distinct structural, electrical, and photophysical characteristics. The continued exploration of azulene in the biomedical field offers the potential for innovative solutions to some of the most pressing health challenges.

14.4 Role of Computational Techniques, Including Artificial Intelligence, in Azulene Chemistry

The research and development in this decade, particularly azulene chemistry, is undergoing a revolution thanks to artificial intelligence (AI) and machine learning. These technologies provide hitherto unheard-of capacity for forecasting synthetic routes, reactivity, and molecular characteristics. Predictive algorithms trained on sizable datasets are being created to speed up the identification of novel derivatives to estimate azulene's stability, electronic configurations, and photophysical behavior. To rationalize azulene-based materials for applications ranging from pharmaceuticals to advanced materials, computational techniques, viz., density functional theory and molecular dynamics simulations enhanced by machine learning, are essential. These AI-enabled methods can accelerate research, like predicting structure-property correlations. AI-guided prediction of a desired property beforehand, with specialized functions for particular uses, including biomedical devices or organic electronics, can significantly reduce the timeframe of the innovation. Automation powered by AI is revolutionizing the synthesis of azulene. Machine learning techniques can enable robotic systems to create and optimize synthetic routes independently on a large scale. For example, high-throughput screening, guided by AI, is being used to identify optimal reaction conditions and catalyst systems for finding efficient synthetic routes of many potential drug molecules. In materials science, AI allows azulene to be integrated into larger frameworks by predicting its compatibility with other molecular systems. This is particularly valuable in designing azulene-containing polymers, dendrimers, and conjugated systems for optoelectronic and energy storage applications. The integration of AI in azulene chemistry is not without challenges. Predictors' quality strongly depends on access to high-quality data, which implies the necessity of collaborative efforts to build large datasets. Also, keeping AI models interpretable and user-friendly remains one of the key concerns, given their possible adoption by a broader scientific community [59–62].

14.5 Conclusion

Today, azulene chemistry offers a lot of challenges and opportunities at the crossroads between traditional synthesis and modern innovation, with its unique properties offering versatile applications across diverse fields. Challenges persist, mainly in regioselective synthesis, scalability, and commercial applications, but opportunities for future research on azulene are promising. Collaborations from computational chemistry, synthetic chemists, spectroscopists, material scientists, device engineers, and biomedical scientists will be imperative in harnessing azulene's full potential for breakthroughs in organic electronics, renewable energy, and medicinal chemistry. With the growing concern about sustainability in chemical research, azulene is at the forefront of developing eco-friendly technologies and will remain one of the backbones of 21st-century chemistry.

AI disclosure statement: The author acknowledges the use of AI-based tools for minor language editing or grammatical correction. However, all intellectual contributions, analyses, and conclusions presented in this chapter are entirely the author's own.

References

1. Nozoe, T. and Itô, S., Recent advances in the chemistry of azulenes and natural hydroazulenes, in: *Fortschritte der Chemie Organischer Naturstoffe/Progress in the Chemistry of Organic Natural Products/Progrès dans la Chimie des Substances Organiques Naturelles*, L. Zechmeister (Ed.), pp. 32–119, Springer Vienna, Vienna, 1961.
2. Korichi, H., Zouchoune, F., Zendaoui, S.-M., Zouchoune, B., Saillard, J.-Y., The coordination chemistry of azulene: A comprehensive DFT investigation. *Organometallics*, 29, 7, 1693, 2010.
3. Lemal, D.M. and Goldman, G.D., Synthesis of azulene, a blue hydrocarbon. *J. Chem. Educ.*, 65, 10, 923, 1988.
4. Liu, R.S.H., Colorful azulene and its equally colorful derivatives. *J. Chem. Educ.*, 79, 2, 183, 2002.
5. Michl, J. and Thulstrup, E.W., Why is azulene blue and anthracene white? A simple MO picture. *Tetrahedron*, 32, 2, 205, 1976.
6. Xin, H., Hou, B., Gao, X., Azulene-based π-functional materials: Design, synthesis, and applications. *Acc. Chem. Res.*, 54, 7, 1737, 2021.
7. Ong, A., Tao, T., Jiang, Q., Han, Y., Ou, Y., Huang, K., Chi, C., Azulene-fused acenes. *Angew. Chem. Int. Ed.*, 61, 44, e202209286, 2022.
8. Kar, S., Sanderson, H., Roy, K., Benfenati, E., Leszczynski, J., Green chemistry in the synthesis of pharmaceuticals. *Chem. Rev.*, 122, 3, 3637, 2022.
9. Sheldon, R.A., Green and sustainable manufacture of chemicals from biomass: State of the Art. *Green Chem.*, 16, 3, 950, 2014.
10. Elwahy, A.H.M., Abdelhamid, I.A., Shaaban, M.R., Recent advances in the functionalization of azulene through Pd-catalyzed cross-coupling reactions. *ChemistrySelect*, 6, 47, 13664, 2021.

11. Kappe, C.O., Dallinger, D., Murphree, S.S., *Practical Microwave Synthesis for Organic Chemists: Strategies, Instruments, and Protocols*, 1st ed., Wiley, Weinheim, 2008.
12. Luche, J.-L., *Synthetic Organic Sonochemistry*, Springer US, Boston, MA, 1998.
13. Ayogu, J.I. and Onoabedje, E.A., Recent advances in transition metal-catalysed cross-coupling of (hetero)aryl halides and analogues under ligand-free conditions. *Catal. Sci. Technol.*, 9, 19, 5233, 2019.
14. Dunås, P., Murfin, L.C., Nilsson, O.J., Jame, N., Lewis, S.E., Kann, N., Azulene functionalization by iron-mediated addition to a cyclohexadiene scaffold. *J. Org. Chem.*, 85, 21, 13453, 2020.
15. Székely, A., Péter, A., Aradi, K., Tolnai, G.L., Novák, Z., Gold-catalyzed direct alkynylation of azulenes. *Org. Lett.*, 19, 4, 954, 2017.
16. Yu, I.F., Wilson, J.W., Hartwig, J.F., Transition-metal-catalyzed silylation and borylation of C–H bonds for the synthesis and functionalization of complex molecules. *Chem. Rev.*, 123, 19, 11619, 2023.
17. Kumar, N.R., Agrawal, A.R., Choudhury, A., Zade, S.S., The effect of base and nucleophile on the nucleophilic substitution of methoxytropone derivatives: An easy access to 4- and 5-substituted multifunctional azulenes. *J. Org. Chem.*, *85*, 14, 9029, 2020.
18. Kumar, N.R., Agrawal, A.R., Zade, S.S., Abnormal nucleophilic substitution on methoxytropone derivatives: Steric strategy to synthesize 5-substituted azulenes. *Chem. Eur. J.*, 25, 62, 14064, 2019.
19. Ito, S., Shoji, T., Morita, N., Recent advances in the development of methods for the preparation of functionalized azulenes for electrochromic applications. *Synlett*, 2011, 16, 2279, 2011.
20. Shetti, V.S., Chemical syntheses and salient features of azulene-containing homo- and copolymers. *Beilstein J. Org. Chem.*, 17, 2164, 2021.
21. Koch, M., Blacque, O., Venkatesan, K., Impact of 2,6-connectivity in azulene: Optical properties and stimuli responsive behavior. *J. Mater. Chem. C*, 1, 44, 7400, 2013.
22. Sun, Q., Hou, I.C.-Y., Eimre, K., Pignedoli, C.A., Ruffieux, P., Narita, A., Fasel, R., On-surface synthesis of polyazulene with 2,6-connectivity. *Chem. Commun.*, 55, 89, 13466, 2019.
23. Xin, H. and Gao, X., Application of azulene in constructing organic optoelectronic materials: New tricks for an old dog. *ChemPlusChem*, 82, 7, 945, 2017.
24. Huang, J., Huang, S., Zhao, Y., Feng, B., Jiang, K., Sun, S., Ke, C., Kymakis, E., Zhuang, X., Azulene-based molecules, polymers, and frameworks for optoelectronic and energy applications. *Small Methods*, 4, 10, 2000628, 2020.
25. Yamaguchi, Y., Takubo, M., Ogawa, K., Nakayama, K., Koganezawa, T., Katagiri, H., Terazulene isomers: Polarity change of OFETs through molecular orbital distribution contrast. *J. Am. Chem. Soc.*, 138, 35, 11335, 2016.

26. Katagiri, H., Azulene-based materials for organic field-effect transistors, in: *Advances in Organic Crystal Chemistry*, M. Sakamoto and H. Uekusa (Eds.), pp. 341–358, Springer Singapore, Singapore, 2020.
27. Dong, J.-X. and Zhang, H.-L., Azulene-based organic functional molecules for optoelectronics. *Chin. Chem. Lett.*, 27, 8, 1097, 2016.
28. Kumar, N.R., Agrawal, A.R., Debnath, S., Choudhury, A., Zade, S.S., Effect of connectivity variation in azulene-BODIPY triads and their optoelectronic properties. *New J. Chem.*, 47, 5, 2456, 2023.
29. Hou, B., Zhou, Z., Yu, C., Xue, X.-S., Zhang, J., Yang, X., Li, J., Ge, C., Wang, J., Gao, X., 2,6-azulene-based homopolymers: Design, synthesis, and application in proton exchange membrane fuel cells. *ACS Macro Lett.*, 11, 5, 680, 2022.
30. Matsuura, Y., Single-molecule tunnel magnetoresistance of azulene. *Physica E*, 105, 219, 2019.
31. Xia, J., Capozzi, B., Wei, S., Strange, M., Batra, A., Moreno, J.R., Amir, R.J., Amir, E., Solomon, G.C., Venkataraman, L., Campos, L.M., Breakdown of interference rules in azulene, a nonalternant hydrocarbon. *Nano Lett.*, 14, 5, 2941, 2014.
32. Remacle, F., Speiser, S., Levine, R.D., Intermolecular and intramolecular logic gates. *J. Phys. Chem. B*, 105, 24, 5589, 2001.
33. Zhang, C., Kong, Y., Xiang, J., Chen, S., Kornyshev, A.A., Ulstrup, J., Gao, X., Zhang, G., Li, Y., Li, J., Switchable modes of azulene-based single molecule–electrode coupling controlled by interfacial charge distribution. *Chem. Sci.*, 16, 1353, 2025.
34. Valiev, R.R., Nasibullin, R.T., Cherepanov, V.N., Kurtsevich, A., Sundholm, D., Kurtén, T., Fast estimation of the internal conversion rate constant in photophysical applications. *Phys. Chem. Chem. Phys.*, 23, 11, 6344, 2021.
35. Beer, M. and Longuet-Higgins, H.C., Anomalous light emission of azulene. *J. Chem. Phys.*, 23, 8, 1390, 1955.
36. Viswanath, G. and Kasha, M., Confirmation of the anomalous fluorescence of azulene. *J. Chem. Phys.*, 24, 3, 574, 1956.
37. Dunlop, D., Ludvíková, L., Banerjee, A., Ottosson, H., Slanina, T., Excited-state (anti)aromaticity explains why azulene disobeys Kasha's rule. *J. Am. Chem. Soc.*, 145, 39, 21569, 2023.
38. Pino-Rios, R., Báez-Grez, R., Szczepanik, D.W., Solá, M., Designing potentially singlet fission materials with an anti-Kasha behaviour. *Phys. Chem. Chem. Phys.*, 26, 21, 15386, 2024.
39. Veys, K. and Escudero, D., Computational protocol to predict anti-Kasha emissions: The case of azulene derivatives. *J. Phys. Chem. A*, 124, 36, 7228, 2020.
40. Nickel, B. and Klemp, D., The lowest triplet state of azulene-H8 and azulene-D8 in liquid solution. I. survey, kinetic considerations, experimental technique, and temperature dependence of triplet decay. *Chem. Phys.*, 174, 2, 297, 1993.

41. Mondal, S., Tanari, P., Roy, S., Bhunia, S., Chowdhury, R., Pal, A.K., Datta, A., Pal, B., Reddy, C.M., Autonomous self-healing organic crystals for nonlinear optics. *Nat. Commun.*, 14, 1, 6589, 2023.
42. Kalita, K.J., Mondal, S., Reddy, C.M., Vijayaraghavan, R.K., Temperature-regulated dual phosphorescence and mechanical strain-induced luminescence modulation in a plastically bendable and twistable organic Crystal. *Chem. Mater.*, 35, 2, 709, 2023.
43. Kalita, K.J., Mondal, S., Reddy, C.M., Vijayaraghavan, R.K., Thermally activated delayed fluorescence in a mechanically soft charge-transfer complex: Role of the locally excited state. *Chem. Sci.*, 14, 47, 13870, 2023.
44. Kalita, K.J., Arikkottira, M.R., Vijayaraghavan, R.K., Organic charge transfer complex towards functional optical materials. *CrystEngComm*, 26, 35, 4751, 2024.
45. Kalita, K.J., Giri, I., Vijayaraghavan, R.K., Influence of non-covalent interactions in dictating the polarity and mobility of charge carriers in a series of crystalline NDIs: A computational case study. *RSC Adv.*, 11, 53, 33703, 2021.
46. Deka, P., Das, S., Sarma, P., Kalita, K.J., Vijayaraghavan, R.K., Reddy, C.M., Thakuria, R., Exceptional thermo-mechano-fluorochromism and nanomechanical analysis of mechanically responsive J and H type polymorphic systems. *Cryst. Growth Des.*, 24, 6, 2322, 2024.
47. Duan, C., Zhang, J., Xiang, J., Yang, X., Gao, X., Azulene-embedded [*n*] helicenes (*n* =5, 6 and 7). *Angew. Chem. Int. Ed.*, 61, 18, e202201494, 2022.
48. Ma, J., Fu, Y., Dmitrieva, E., Liu, F., Komber, H., Hennersdorf, F., Popov, A.A., Weigand, J.J., Liu, J., Feng, X., Helical nanographenes containing an azulene unit: Synthesis, crystal structures, and properties. *Angew. Chem. Int. Ed.*, 59, 14, 5637, 2020.
49. Zhao, Z., El-Khouly, M.E., Che, Q., Sun, F., Zhang, B., He, H., Chen, Y., Redox-active azulene-based 2D conjugated covalent organic framework for organic memristors. *Angew. Chem. Int. Ed.*, 62, 7, e202217249, 2023.
50. Tian, P., Han, X., Qi, Q., Zhao, X., An azulene-based crystalline porous covalent organic framework for efficient photothermal conversion. *Small*, 20, 21, 2307635, 2024.
51. Leino, T.O., Sieger, P., Yli-Kauhaluoma, J., Wallén, E.A.A., Kley, J.T., The azulene scaffold from a medicinal chemist's perspective: Physicochemical and *in vitro* parameters relevant for drug discovery. *Eur. J. Med. Chem.*, 237, 114374, 2022.
52. Bakun, P., Czarczynska-Goslinska, B., Goslinski, T., Lijewski, S., *In vitro* and *in vivo* biological activities of azulene derivatives with potential applications in medicine. *Med. Chem. Res.*, 30, 4, 834, 2021.
53. Meher, S. and Sharma, N.K., Azulene tethered *N*-aryl nucleobases: Synthesis, morphology and biochemical evaluations. *New J. Chem.*, 47, 12, 5593, 2023.
54. Bala, D., Jinga, L.-I., Popa, M., Hanganu, A., Voicescu, M., Bleotu, C., Tarko, L., Nica, S., Design, synthesis, and biological evaluation of new azulene-containing chalcones. *Materials*, 15, 5, 1629, 2022.

55. Slon, E., Slon, B., Kowalczuk, D., Azulene and its derivatives as potential compounds in the therapy of dermatological and anticancer diseases: New perspectives against the backdrop of current research. *Molecules*, 29, 9, 2020, 2024.
56. Murfin, L.C., Weber, M., Park, S.J., Kim, W.T., Lopez-Alled, C.M., McMullin, C.L., Pradaux-Caggiano, F., Lyall, C.L., Kociok-Köhn, G., Wenk, J., Bull, S.D., Yoon, J., Kim, H.M., James, T.D., Lewis, S.E., Azulene-derived fluorescent probe for bioimaging: Detection of reactive oxygen and nitrogen species by two-photon microscopy. *J. Am. Chem. Soc.*, 141, 49, 19389, 2019.
57. Cui, Z., Wang, Y., Wang, G., Feng, B., Lewis, S.E., Wang, K., Jiang, K., James, T.D., Zhang, H., Amphiphilic azulene-based fluorescent probe for simultaneous monitoring of fluctuations in carboxylesterase activity in diverse biological samples from a single organism. *Anal. Chem.*, 96, 49, 19732, 2024.
58. Murfin, L.C. and Lewis, S.E., Azulene—A bright core for sensing and imaging. *Molecules*, 26, 2, 353, 2021.
59. Nigam, A., Pollice, R., Friederich, P., Aspuru-Guzik, A., Artificial design of organic emitters *via* a genetic algorithm enhanced by a deep neural network. *Chem. Sci.*, 15, 7, 2618, 2024.
60. Zheng, P., Zubatyuk, R., Wu, W., Isayev, O., Dral, P.O., Artificial intelligence-enhanced quantum chemical method with broad applicability. *Nat. Commun.*, 12, 1, 7022, 2021.
61. Baum, Z.J., Yu, X., Ayala, P.Y., Zhao, Y., Watkins, S.P., Zhou, Q., Artificial intelligence in chemistry: Current trends and future directions. *J. Chem. Inf. Model.*, 61, 7, 3197, 2021.
62. Staszak, M., Staszak, K., Wieszczycka, K., Bajek, A., Roszkowski, K., Tylkowski, B., Machine learning in drug design: Use of artificial intelligence to explore the chemical structure–biological activity relationship. *WIREs Comput. Mol. Sci.*, 12, 2, e1568, 2022.

Index

1,3-diazaazulenes, 85, 90
1-azaazulenes, 85, 90, 92, 93
2-acetoacetamidotropone, 86
2-Aminotropones, 86
2-Chlorotropones, 85, 90
2-troponylhydrazides, 88
3-acetyl-2-chloro-1-azaazulene, 86
8-azaheptafulvenes, 93

Acenes, 209
Acetoacetate, 101
AFM, 236
$AgBF_4$, 214
$AgPF_6$, 226
$AlCl_3$, 220
Alder, 232
Aldol condensation, 61
Amino-Heck reaction, 92
Anion sensing, 186
Anisotropy, 206
Annulation, 107
Anthryl, 209
Anticancer, 253, 254
Anti-Kasha, 120, 353, 363, 365
Anti-Kasha emission, 383
Anti-Kasha fluorescence, 277, 279
Antioxidant, 261
Antiviral, 110
Apoptosis, 254, 255
Aromatic, 202
Aromatic chameleon, 53, 55
Aromaticity, 4, 5, 10, 11, 15
Aromaticity/antiaromaticity, 313, 318
Aromatization, 64, 75
Artemisia arborescens, 248
Artificial intelligence, 377, 389
Asymmetric, 229
Azaazulenes, 82, 85, 100, 109
AzAC, 225, 226
Azacoronene, 225
Aza–Wittig reaction, 108, 109
Aziridine, 59, 60
Azulene, 19, 20, 21, 44, 199, 200, 201, 203, 205, 207, 209, 211, 213, 215, 217, 219, 221, 223, 225, 227, 228, 229, 231, 233, 235, 237, 245, 277, 279, 280, 281, 282, 297, 298, 299, 302, 303, 304, 305, 306, 307, 308, 309, 333
Azulene-based, 199, 200, 202, 204, 206, 208, 210, 212, 214, 216, 218, 220, 222, 224, 226, 228, 230, 232, 234, 236
Azulene cation, 13
Azulene fused systems, 153
  azobenzene systems, 166
    correlation between the pH and electronic spectra, 170
    synthetic pathways, 172
  diarylethene systems, 161
    bidirectional visible light switching, 166
    donor-acceptor conjugation, 163
    proton assisted isomerization, 165
    synthetic pathways, 164
  stilbene systems, 154
    fluorescence switching, 158–160
    quantum chain process, 155

water soluble photoswitches, 156
Azulene sulfonium salts, 281
Azulenedi-iron pentacarbonyl, 336
Azulenedimolybdenum hexacarbonyl, 342
Azulene-embedded, 226
Azuli porphyinogens, 341
Azuliporphyrin, 313, 338–340

B3LYP, 353, 361, 365, 366, 367, 368, 369, 370, 371
Baird's rules, 363
Band gap, 277, 279, 288, 291, 295, 296, 302
Battery applications, 300, 301, 302, 306
BDA, 207, 208
BDA1, 209, 210
BDA2, 209, 210
Bebryazulene, 25
Benzo, 208, 210
Benzotriazulene, 61
BF3, 208, 209, 217, 236
BF4, 211, 214, 223
Biazulene diimides, 71
Bioavailability, 249, 252, 266
Bioimaging, 192
Bioimaging and biosensors, 374, 384, 388
Biological activities, 20, 24, 44–45
Bipin, 199
Bischler–Napieralski reaction, 96, 99, 100
Bistriflate, 221
BODIPY, 199
Borylazulene, 213
BPin, 222
Bpin, 202, 203, 207, 212, 213, 223, 226, 230
Bpin-azulene, 225
Bromoazulene, 213
BTA, 223, 224
Buchner ring expansion, 96
BuLi, 236
BuOK, 205, 212
BuOLi, 213
Burgess reagent, 208
Butterfly-shaped, 205
Butylphenyl, 220

C60, 204
Calixazulene, 322
Carbaporphyrinoids, 312
Carbaporphyrins, 341
Cardiovascular, 262
CAS, 356, 357, 358, 360, 361
CASPT2, 358, 360, 361, 364
CASSCF, 358, 360, 364
CCSD(T), 363
CDH, 228, 229
Chamazulene, 21, 26–27, 38–43, 45, 246, 250, 253
Chamomile (*Matricaria recutita L.*), 26–28
$CHCl_3$, 202, 218
Chiral, 202, 203, 379, 386, 387
Chromatography, 63
Clar, 227, 228
Clar sextets, 228
Clar's sextet rule, 10
COD, 223
Col_{hd}, 225
Condensation reaction, 314–317
Conical intersection, 8
Conjugated microporous polymers (CMPs), 277, 279, 294, 295, 296, 308
Coordination polymers, 338
Cope, 216
Coupling, 229
Covalent organic frameworks (COFs), 277, 278, 279, 287, 288, 289, 290, 291, 292, 293, 294, 306, 307, 387
Covalent triazine frameworks (CTFs), 277, 299, 300, 301, 302, 306, 307
Cross-coupling, 102
Cross-coupling reactions, 379, 380
CuCl, 235
CuI, 225, 231, 236

Curcuma kwangsiensis, 27
Curcuma species
Curcuma xanthorrhiza, 27
Cyclic voltammetry, 295, 300
Cycloaddition, 85, 86, 93, 100, 107, 108
Cyclodehydrog, 229
Cycloheptatrienone, 88, 89, 90
Cyclooxygenase, 248, 256
Cyclopentadienyl anion, 4, 10
Cytotoxic, 253, 254

DBA, 209, 210
DCM, 206, 209, 211, 212, 218, 219, 221, 222, 226, 231, 235, 236
DCV, 219, 220
DDQ, 207, 208, 209, 216, 217, 218, 219, 228, 230, 231, 233, 234, 235, 314
Dehydrotryptophan ester, 98
Dermatitis, 262
Desalination, 292
DFT, 138, 139, 205, 220, 222, 230, 353, 356, 357, 358, 359, 361, 363, 366, 368, 369, 370
Diazaazulenes, 83, 84, 85, 90, 101
Diels–Alder reaction, 232
Diketene, 86
Dipolar, 4, 5, 12
Dipole moment, 10, 11, 15, 277, 279, 293, 354, 360, 362, 367
DME, 203, 221, 222
DMF, 213, 223, 224, 226, 231
DMPA, 236, 237
Donor-acceptor, 132
Donor-acceptor structure, 279, 288, 293, 296, 307
Double [5]helicene, 205
DPPF, 204

Echinofuran, 25
Electrocatalysis, 300, 301, 302, 306
Electrochemical, 214
Electron localization function, 11
Electrophilic substitution, 312
Enamines, 90, 91
Energy harvesting devices, 385
EOM, 356, 357, 359, 364
Epigenetic, 111
Essential oils, 26–28, 39, 45
Ethynyltrimethylsilane, 236
EtOH, 217, 220, 236
Extraction methods

$FeCl_3$, 202, 218, 219, 225, 226
Fermi energy, 366
Fermi function, 366
Flash vacuum pyrolysis, 96
Flexible and wearable electronics, 382
Fluorescence, 54
Fluoride sensor, 187
FMOs, 220
Friedel–Craft's acylation, 127
Frontier molecular orbital (FMO), 12, 120, 279, 281
Fullerene (C_{60}), 326–327
Functionalization/substitution, 281, 282, 302, 303, 304, 305
Functionalization techniques, 377, 382

Gas adsorption, 277, 278, 284, 285, 286, 287, 288, 294, 297, 298, 299, 302, 304, 305, 306, 307, 308
Gibbs, 232
Glaser self-coupling, 233
Glidewell–Lloyd rule, 10, 11, 15
Gorgiabisazulene, 25
Gorgiallylazulene, 25
Gorgonians (marine organisms), 21–23
Green's function, 366
Grubbs' catalyst, 207
Guaianolide, 22, 27
Guaiazulene, 4, 6, 20–24, 28–34, 67
Guaiazulene derivatives, 21–24
Guiaiazulene (GA), 246

Hartwig–Buchwald reaction, 106
HBF_4, 211, 214

HCl, 203
Heavy metal sensing, 181
Heisenberg models, 201
Helenalin, 30
Heteroarylazulenes, 70
Heterocumulenes, 87, 88
Heterogeneous catalysis, 278, 279, 287, 294, 302
Hexafluoroacetylacetonate, 68
HF, 356, 357, 358, 360, 369, 372
HOMO, 125, 130, 133, 203, 211, 353, 355, 360, 364, 365, 367, 368
HOMO-LUMO, 279, 302
HOMO–LUMO gap, 12, 13
Host–guest interaction, 326
HPLC, 203, 231
HRMS, 223
Huckel, 365
Hydrogen storage (H_2), 284, 285, 286
Hydrogen-bonded organic frameworks (HOFs), 277, 297, 298, 309
Hyperpolarizability, 122, 124, 371, 372

ICl, 233, 235
IEPCM, 232
Isolation of chamazulene, 26–27
Isolation of guaiazulene, 21–23
Isosteric heat of adsorption (Q_{st}), 286

Kasha's rule, 15
Ketolactone, 23, 27
Knoevenagel condensation, 223
KOAc, 217
KOH, 221, 222
KOtBu, 221

Landauer-Buttiker formula, 366
LEDs, 230, 231
$LiAlH_4$, 221
Ligands, 204
Linderazulene, 22–23, 25
Linderene, 38
Lindsey condensation, 315
Liquid-liquid interface polymerization, 283, 284, 293
Lithium-ion battery, 300, 301
Local aromaticity, 10, 11
LUMO, 125, 126, 130, 133, 353, 355, 360, 364, 365, 367, 368

MacDonald condensation, 314
Magnetically induced current density, 11
Marine sources (*Euplexaura, Paramuricea, Alcyonium,*etc.), 21–23
Mechanochemical synthesis, 283, 284
Mechanochemistry, 380
Medicinal applications (anti-inflammatory, anticancer, antioxidant, antiulcer), 20, 24, 44–45
Medicinal chemistry, 5, 13, 15
Memristors, 293, 294, 308
Metalation, 318–321
Metallaazulene, 68, 336
Metalla-dual-azulenes, 337
Metal-organic frameworks (MOFs), 277, 278, 284, 285, 286, 287, 288, 306, 307
MICD, 364
Microporous/mesoporous/macroporous, 278, 285
Microwave-assisted synthesis, 282, 283, 378, 379
MOF-assisted extraction, 28
Monoazaazulenes, 82
MP2, 356, 357, 358, 360, 363, 369, 370
MRCI, 360
MSA, 235
MsCl, 221
Multireference, 361, 364, 372
Muriceidines A–C, 26
Muriceidone A, 26

NaH, 226

Nanographenes, 199, 201, 203, 205, 207, 209, 211, 213, 215, 217, 219, 221, 223, 225, 226, 227, 229, 231, 233, 235, 237
NaOMe, 209, 210, 223
Naphthalene, 200
NBS, 203, 204
NDA, 209, 210
Negishi coupling, 102, 104, 105
NGs, 234, 236
NHTs, 213
NICS, 204, 206, 226
NIR, 201, 206, 221, 226
NIS, 211
Nitride sensor, 191
NMR, 204, 205, 223, 231
NO_2 sensing, 288, 289, 290
Non-alternant aromaticity, 377, 378, 383
Non-alternant hydrocarbons, 344
Non-benzenoid, 4
Noncovalent interaction, 361
Non-volatile memory, 293, 294
Nozoe, 4, 6, 12
NSAIDs, 256
N-silylenamines, 90
Nucleus-independent chemical shift, 11

OAc, 203, 211, 213, 217, 236
OEt_2, 208, 209, 217, 236
OFET, 355, 367, 368, 369
OHC, 223, 236
OMe, 205, 207, 212, 213, 215, 222, 223, 225
Organic electronics, 13, 14, 15
Organic field effect transistors, 128, 129, 130, 131, 132, 334
Organic nonlinear optical, 121, 123, 124
Organic optoelectronics, 382, 383, 386
Organic photovoltaics, 381
Organocatalysis, 379
OTf, 212, 219, 220, 230, 231
Oxidation (ring contraction), 318–319
Oxygen evolution reaction (OER), 300, 301, 302
Oxygen reduction reaction (ORR), 300, 301, 302

PAH, 207, 214, 215, 217, 218, 219, 220, 221, 226, 227, 231
PAHs, 199, 200, 201, 202, 203, 205, 206, 207, 209, 210, 211, 212, 213, 215, 217, 218, 219, 220, 221, 222, 223, 224, 225, 227, 229, 231, 233, 235, 237
Paramuricea chamaeleon, 22
Pariser–Parr–Pople, 201
PBr_3, 224, 225
PCy_3, 207
Pd-catalyst, 204, 225
$PdCl_2$, 207
PDE_2, 109
PES, 361, 363
Pfau and Plattner, 4, 6
Pharmacokinetic, 249, 266
Pharmacological effects, 20, 24, 44–45
Phenethyl amine, 59
Phenyl, 231
Phenylacetylene, 68
Phosphate sensor, 186
Phosphodiesterase 2, 109
Phosphorescent properties, 384
Photocatalysis, 294, 295
Photochromic molecules, 152
Photodissociation, 13
Photodynamic therapy, 248
Photodynamic therapy (PDT), 378
Photophysical properties, 13, 15, 54, 57, 73, 323
Photosensitizers, 365
Photothermal conversion, 288, 289, 290, 291, 292, 307, 387
Photovoltaic cells, 334
Phthaldialdehyde, 61

Pictet–Spengler condensation, 98, 99
Pincer ligand, 340
Platt's perimeter rule, 10
$POCl_3$, 213
Polarizability, 354, 367
Polyaromatic hydrocarbons (PAHs), 53
Polyazaazulene, 100, 101
Polycyclic, 202
Polycyclic hydrocarons (PHs), 203, 206, 208, 216, 217
Porosity, 278, 279, 285, 287, 296, 299
Porous organic polymers (POPs), 277, 278, 279, 292, 302, 303, 304, 305, 306, 307, 308
Porphyrin, 312
Porphyrin units, 124, 140
Positive-negative, 125, 130, 134
Positive-negative-positive, 125, 134
Potassium-ion battery, 300, 301
Power conversion efficiency, 136, 137
PPh_3, 202, 204, 207, 209, 210, 211, 212, 213, 217, 220, 221, 223, 226, 230, 231, 235, 236
PrCN, 230
$PtCl_2$, 236
Pyridinium ylide, 85, 91, 92
Pyrrolidine adduct, 318

Raman spectroscopy, 234
Reactivity descriptor, 367
Regioselective synthesis, 380, 389
Regioselectivity, 89, 370
Resonance structures, 11
Rydberg states, 357, 365

Sapphyrin, 322
$SbCl_3$, 226
SbF_6, 209, 218
Scholl reaction, 225, 227, 231, 232
Secondary building units (SBUs), 285
Second-order NLO, 124
Self-healing, 386
Sensing/chemical sensing, 277, 288, 289, 290, 299, 302, 304, 305, 306
Sesquiterpenes/guaianolides, 22–23, 28, 40
Shape memory polymers (SMPs), 292, 293
Singlet state, 363
Solar steam generation, 292
Solvothermal synthesis, 282, 283, 285
Sonochemistry, 379
Sonogashira coupling, 232, 234
Sonogashira–Hagihara reaction, 105
Soxhlet, 26–27
Spectator ligands, 347
SPhos, 203, 204, 207, 217, 236
Steam distillation, 26–27
Stille coupling, 281, 294
Stokes shift, 224
Stones–Wales rearrangement, 231
Supercritical fluid extraction, 27
Supramolecular chemistry, 326–328
Supramolecular synthons, 297, 298
Surface area (BET), 278, 285, 296, 302
Sustainable synthesis, 377
Suzuki coupling, 62, 102, 103
Suzuki-Miyaura coupling, 281, 294
Suzuki–Miyaura cross-coupling reaction, 205, 206, 208, 216, 218, 222, 223, 233, 234
Synthesis, 199, 200, 202, 203, 204, 205, 207, 209, 211, 212, 213, 214, 215, 217, 218, 219, 220, 221, 222, 223, 225, 226, 230, 231, 235
Synthesis of chamazulene, 38–43
Synthesis of guaiazulene, 28–34

Taraxacum wallichii, 26
TBAF, 233, 235
TD-DFT, 359, 365
Terazulene, 204, 367, 368
Tetraarylazuliporphyrins, 315
TfO, 219, 220
TfOH, 218, 219, 230, 231, 234, 235

Thermal stability, 279, 296
Thermally activated delayed fluorescence (TADF), 383
THF, 202, 205, 207, 209, 212, 213, 217, 219, 221, 222, 223, 235, 236
Thin-film transistors, 384
TIPS, 235
TMS, 225, 236
Toluene, 228
Transition state, 370
Transition-metal catalysis, 379
Transmission coefficients, 366
Transmission probability, 366
Triazaazulenes, 83, 84
Triplet state, 363
Tripyrrane, 314
Tropolones, 56
Troponoids, 85
Tropylium cation, 4, 10
TsHN, 213
$TsNHNH_2$, 213

Ulcer, 259, 260
Ullman coupling, 229

Vetivazulene, 4, 6
Vilsmeier reaction, 62, 213
Virginolide, 32, 41

Wittig reaction, 211

Xantphos, 211
XPhos, 204

Ziegler–Hafner, 6
Zn-air battery, 302

Printed and bound by CPI Group (UK) Ltd, Croydon, CR0 4YY
17/09/2026
14964276-0005